Life Histories of

North American Gallinaceous Birds

Life Histories of North American Gallinaceous Birds

by

Arthur Cleveland Bent

Dover Publications, Inc.
New York

Published in the United Kingdom by Constable and Company, Limited, 10 Orange Street, London W. C. 2.

This Dover edition, first published in 1963, is an unabridged and unaltered republication of the work first published in 1932 by the United States Government Printing Office, as Smithsonian Institution United States National Museum *Bulletin 162.*

International Standard Book Number: 0-486-21028-6

Manufactured in the United States of America

Dover Publications, Inc.
180 Varick Street
New York 14, N. Y.

ADVERTISEMENT

The scientific publications of the National Museum include two series, known, respectively, as *Proceedings* and *Bulletin*.

The *Proceedings* series, begun in 1878, is intended primarily as a medium for the publication of original papers, based on the collections of the National Museum, that set forth newly acquired facts in biology, anthropology, and geology, with descriptions of new forms and revisions of limited groups. Copies of each paper, in pamphlet form, are distributed as published to libraries and scientific organizations and to specialists and others interested in the different subjects. The dates at which these separate papers are published are recorded in the table of contents of each of the volumes.

The series of *Bulletins*, the first of which was issued in 1875, contains separate publications comprising monographs of large zoological groups and other general systematic treatises (occasionally in several volumes), faunal works, reports of expeditions, catalogues of type specimens and special collections, and other material of similar nature. The majority of the volumes are octavo in size, but a quarto size has been adopted in a few instances in which large plates were regarded as indispensable. In the *Bulletin* series appear volumes under the heading *Contributions from the United States National Herbarium*, in octavo form, published by the National Museum since 1902, which contain papers relating to the botanical collections of the Museum.

The present work forms No. 162 of the *Bulletin* series.

ALEXANDER WETMORE,
Assistant Secretary, Smithsonian Institution.

WASHINGTON, D. C., *February 12, 1932.*

CONTENTS

Order Galliformes—Continued.

Family Perdicidae—Continued.

INTRODUCTION

This is the ninth in a series of bulletins of the United States National Museum on the life histories of North American birds.

Previous numbers have been issued as follows:

107. Life Histories of North American Diving Birds, August 1, 1919.

113. Life Histories of North American Gulls and Terns, August 27, 1921.

121. Life Histories of North American Petrels and Pelicans and their Allies, October 19, 1922.

126. Life Histories of North American Wild Fowl (part), May 25, 1923.

130. Life Histories of North American Wild Fowl (part), June 27, 1925.

135. Life Histories of North American Marsh Birds, March 11, 1927.

142. Life Histories of North American Shore Birds (pt. 1), December 31, 1927.

146. Life Histories of North American Shore Birds (pt. 2), March 24, 1929.

The same general plan has been followed, as explained in previous bulletins, and the same sources of information have been utilized. The nomenclature of the new Check List of the American Ornithologists' Union has been followed, but it has seemed best to continue in the same order of arrangement of families and species as given in the old (1910) check list.

This is the first group in which any considerable number of subspecies have had to be treated. An attempt has been made to give as full a life history as possible of the best-known subspecies and to avoid duplication by writing briefly of the others and giving only the characters of the subspecies, its range, and any habits peculiar to it. In many cases certain habits, probably common to the species as a whole, have been recorded for only one subspecies; such habits are mentioned under the subspecies on which the observations were made. The distribution gives the range of the species as a whole, with only rough outlines of the ranges of the subspecies, which can not be accurately defined in many cases.

The egg dates are the condensed results of a mass of records taken from the data in a large number of the best egg collections in the country, as well as from contributed field notes and from a few published sources. They indicate the dates on which eggs have been actually found in various parts of the country, showing the earliest and latest dates and the limits between which half the dates fall, the height of the season.

The plumages are described in only enough detail to enable the reader to trace the sequence of molts and plumages from birth to maturity and to recognize the birds in the different stages and at the different seasons. No attempt has been made fully to describe adult plumages; this has been already well done in the many manuals. The names of colors, when in quotation marks, are taken from Ridgway's Color Standards and Nomenclature (1912) and the terms used to describe the shapes of eggs are taken from his Nomenclature of Colors (1886 edition). The bold-faced type in the measurements of eggs indicates the four extremes of measurements.

Many of those who contributed material for previous bulletins have continued to cooperate. Receipt of material from more than 320 contributors has been acknowledged previously. In addition to these, our thanks are due to the following new contributors: Clinton G. Abbott, W. C. Adams, M. C. Badger, J. H. Baker, Paul Bartsch, Glenn Berner, E. J. Booth, O. M. Bryens, R. L. Coffin, L. V. Compton, J. V. Crone, P. D. Dalke, Ben East, F. L. Farley, J. A. Gillespie, F. J. Herman, R. B. Horsfall, F. N. Irving, Miss A. M. Keen, E. A. Kitchin, T. T. McCabe, Norman McClintock, E. A. McIlhenny, G. W. Monson, G. W. Morse, J. J. Murray, L. T. S. Norris-Elye, W. W. Perrett, G. B. Pickwell, Gower Rabbitts, W. M. Rosen, W. B. Savary, E. W. Schmidt, W. E. Sherwood, F. H. Shoemaker, W. A. Squires, J. W. Sugden, J. G. Suthard, L. M. Terrill, C. W. Tindall, I. R. Tomkins, Miss F. May Tuttle, J. H. Wales, N. A. Wood, and Miss M. W. Wythe.

Through the courtesy of the Bureau of Biological Survey, the services of Frederick C. Lincoln were again obtained to compile the distribution paragraphs. With the matchless reference files of the Biological Survey at his disposal, his many hours of careful and thorough work have produced results far more satisfactory than could have been attained by the author, who claims no credit and assumes no responsibility for this part of the work.

Dr. Charles W. Townsend and Dr. Winsor M. Tyler rendered valuable assistance in reading and indexing, for this group, the greater part of the literature on North American birds, which saved the author many hours of tedious work. Doctor Townsend also contributed the entire life histories of four species and Doctor Tyler also contributed one. Dr. Alfred O. Gross contributed two life histories and wrote up the diseases of the ruffed grouse. Dr. Arthur A. Allen contributed the courtship of the ruffed grouse. E. A. McIlhenny and the Rev. P. B. Peabody loaned the author valuable negatives and Maj. Allan Brooks loaned two beautiful drawings. Thanks are due to the late Frank C. Willard for many hours of careful work in collecting, arranging, and figuring a great mass of data on egg dates and measurements; the author would be glad to have some one volunteer to undertake this work in future. The author is much indebted to Dr. Charles W. Richmond, of the United States National Museum, for many hours of careful and sympathetic work in reading the proofs and correcting errors in this and all previous volumes; his expert knowledge has been of great value.

The manuscript for this volume was completed in January, 1931. Contributions received since then will be acknowledged later. Only information of great importance could be added. The reader is reminded again that this is a cooperative work; if he fails to find in these volumes anything that he knows about the birds, he can blame himself for not having sent the information to

THE AUTHOR.

Life Histories of

North American Gallinaceous Birds

LIFE HISTORIES OF NORTH AMERICAN GALLINACEOUS BIRDS (ORDERS GALLIFORMES AND COLUMBIFORMES)

By Arthur Cleveland Bent

Taunton, Massachusetts

Order GALLIFORMES

Family PERDICIDAE, Quails

PERDIX PERDIX PERDIX (Linnaeus)

EUROPEAN PARTRIDGE

HABITS

Attempts to introduce the European gray partridge into North America have met with marked success in certain favorable localities and with many dismal failures in other places less congenial to it. Dr. John C. Phillips (1928) has summarized the whole history of these attempts. Of the earlier unsuccessful importations he says:

The earliest attempt at introduction, which so far as known was made by Richard Bache, son-in-law of Benjamin Franklin, who stocked his plantation on the Delaware River near what is now the town of Beverly, N. J., with Hungarian partridges, dates back to the latter part of the eighteenth century. There were subsequent attempts in Virginia and New Jersey, most important of which was Pierre Lorillard's effort in 1879 at Jobstown, N. J. Later attempts commenced in a small way in 1899, but the real fever of importation along the Atlantic coast began about 1905 and has lasted up to the present, although the period 1907 and 1914 saw the height of the industry. In Eastern States importations of these hardy little birds have been put down all the way from Portland, Me., and northern New York to South Carolina, Georgia, Florida, and Mississippi. In Connecticut, Pennsylvania, and New Jersey, the work was done on a large scale and, at first, with encouraging results. In a few places the birds undoubtedly bred the first season, and in other places as in the Connecticut Valley, they persisted for 8 or 10 years in considerable numbers; eventually they vanished, however, between 1915 and 1920.

The results on the western plains and prairies have been quite successful, of which he writes:

The results in the far Western States and in western and central Canada may be briefly summarized. The most remarkable success followed immediately upon the first introductions into Alberta, near Calgary, in 1908–9. On April 20, November 16, and December 10, 1908, Calgary sportsmen liberated about 70 pairs over a small area mostly south and west of Calgary. More came on April 20, 21, and 22, 1909, and in all some 207 pairs seem to have formed the basis for this wonderful result. The first birds were placed some 15 miles south of Calgary, and after the first large plantings, 40 pairs in one place and 30 not far away (High River and west of that place), the rest were planted mostly in lots of 10 pairs. This stock came from Hungary. Some time later the Northern Alberta Game and Fish Protection League liberated a fresh importation of 230 birds in Alberta near Edmonton, but the stock from Calgary had in the meantime spread north to that city. The gain in territory from this nucleus has been little short of marvelous. The birds have now spread at least 60 miles northwest of Edmonton (Pembina River) and breed there. There has been an open season on them in Alberta for years, and they are now by far the commonest of imported game birds in western Canada. The spread from this initial plant has carried the Hungarian partridge into Saskatchewan and all over its western part as far north as township 60 and south to the international boundary. All this happened within only five years from the time the bird was first recorded in the Province.

A. G. Lawrence writes to me:

First liberated in Manitoba in April, 1924, when the Game Protective League released at Warren, Manitoba, 40 pairs imported direct from Czechoslovakia. A second shipment was received in January, 1925, 17 pairs being later released at Neepawa, Manitoba, and 26 pairs at Warren. These birds are apparently well adapted to the prairies and seem to be establishing themselves in the areas in which they were liberated.

The experience with this partridge in the State of Washington well illustrates the fact that it will flourish, increase, and spread in the type of open country that it prefers, but will barely hold its own or will die out entirely in less favorable regions. D. J. Leffingwell says in his notes:

We find the introduction of the partridge has been most successful in the dry nonforested areas with an elevation of 1,000 or more feet above sea level and where the game enemies are rare. The lack of vermin and the large open fields in which the birds may feed are probably the most important factors. The birds should not be introduced outside of the Temperate or Transition Zone.

In the comparatively humid regions of western Washington attempts to introduce Hungarian partridges have not been very successful. S. F. Rathbun says of this section:

Western Washington is a picturesque region of mountains, hills, valleys, and streams. Originally it was clothed with a dense and luxuriant forest mostly coniferous, but now a great change in this respect is apparent. As has been so often the case in the past in a new country, the development of the region began along the lines of least resistance—in this instance it being where land and water met—and now to a large extent the tall forests have been replaced by broad cultivated areas that steadily encroach upon the still undeveloped ones.

On the other hand the birds have prospered and spread in the eastern part of the State, of which he writes:

Eastern Washington, on the contrary, is a section quite devoid of forests except along many of its streams and some of the more rugged parts, and even then this growth lacks the luxuriance of that of the west side; in fact, being scanty by comparison. And many parts of eastern Washington are more or less elevated and open, wide-sweeping plateaus rolling in turn to the water-courses.

In Oregon the story is much the same. William L. Finley writes to me:

During the years 1913–14 we liberated 1,522 of these birds in various counties throughout the State. In the Willamette Valley and places in southern Oregon the climate is mild, and the country is varied with patches of timber, fields, and gardens, which from all reports is very similar to the European home of these birds. In the eastern part of Oregon where the partridges were liberated the altitude is a little higher; it is colder in winter; the hills are covered with broad grain fields with quite a lot of wild sagebrush country surrounding, also more or less trees and brush in the canyons. It came rather as a surprise to find that the partridges did not increase and thrive in the Williamette Valley and southern Oregon, but they multiplied quite rapidly all through the northeastern part of Oregon, and especially in the southeastern part of Washington, where quite a number of these birds were imported and released.

According to Charles J. Spiker (1929) Hungarian partridges have been introduced successfully in northern Iowa, where they have spread into six counties, as well as three counties in southern Minnesota. He says of this bird:

There is no more charming bird on the Iowa landscape than the Hungarian Partridge, nor one which better deserves protection at the hands of those who have brought it from its native haunts to become acclimated and adjusted to new environments. While it is not highly colored, like the Ring-necked Pheasant, yet it is a beautiful bird and merits a great deal of enthusiasm from an aesthetic point of view as well as the more mercenary point of view of the sportsman. In size it is somewhat larger than the Bob-white, and has some of the characteristics of this species. Seen as it flies directly away from the observer, especially as it first takes off from the ground or spreads its tail in alighting, it presents its very distinguishing field mark. This is the rich russet of the tail feathers, visible only in flight, and concealed by the upper coverts when at rest, but greatly resembling the sheen of that of the Red-tailed Hawk. If one be so fortunate as to behold the bird on a bank about on a level with his eyes or slightly above him, as it has upon two or three occasions occurred with me, he will note the black crescent just below the breast, practically in the middle of the belly, but so located that the bird must be in just the exact position for this mark to show itself.

Courtship.—The Rev. F. C. R. Jourdain has sent me the following quotation from F. Menteith Ogilvie:

In March courtship proper will have begun. In the great majority of cases, the birds will have definitely selected their partners. Here and there, where the males are in excess, constant fights will take place, often resulting in the elder male ousting the younger from the possession of the female, a most undesirable occurrence when it happens, looked at from the breeding point of view. The old males are not only more pugnacious and stronger birds, but they are also either infertile or much less fertile than the young male and the result of the union is likely to be a small laying, a still smaller hatching, and a large percentage of rotten eggs. Throughout March, while pairing is going on fighting is generally continuous and severe. These fights are very amusing to watch—the two males, bristling with fury, feathers raised and wattles showing, rush at each other striking and buffeting with their wings, generally jumping a few inches from the ground. The "round" may last 3 or 4 minutes; the lady, close by picking up a seed here and there and preening herself, is apparently unconscious of the furious rivalry she is exciting. The fighters now separate a little distance and recommence feeding and peace seems to be declared, till one or other approaches too near the female, when war is instantly declared again. So the battle continues with intervals over a considerable period, possibly a week or more, until one of the two is finally vanquished and the happy pair are left to their honeymoon. I have often watched fights of this kind, and I never could see that the Partridges inflicted any real damage on each other; their principal offensive weapon seemed to be their wings. Their bills they rarely used, and their feet they didn't appear to use at all. The studied inattention of the female is most amusing to watch, and I conclude she exercises no choice in the matter at all, beyond promising her hand to the better man.

Nesting.—The nest of this partridge is a very simple affair, a slight depression in the ground, lined with a few dead leaves, dry grass, or straw. It is usually placed among bushes, or in long grass, fields of clover, or in standing grain. Mr. Jourdain says in his notes:

It should be noted that during the time of laying (which may last for three weeks) the eggs are carefully covered up by the hen bird with grass or dead leaves. When she comes to the nest to lay she scratches away the covering, deposits an egg, and then replaces it again. Until the clutch is complete the eggs are laid anyhow. When the hen is about to incubate she arranges them with the greatest care and for a single day curiously enough, leaves the eggs uncovered and then begins to incubate. She is a good mother and sits very closely, especially after the first few days. The male bird takes no part in brooding but remains close at hand for defense if necessary.

Eggs.—Of the eggs Mr. Jourdain says that

normally the clutch ranges from 8 or 9 to about 20. I have known cases of as many as 21 and 22, which may have been the produce of one hen but the higher numbers which occasionally are met with, 26 to 40 (!), are undoubtedly due to two hens laying together in one nest. In color they are uniformly olive, sometimes darker, sometimes lighter, but occasionally clutches have been found with almost white eggs, while a bluish type has also been recorded. One hundred British eggs measured by myself averaged 36.8 by 27.4 millimeters. The eggs showing the greatest extremes measured 38.9 by 28.4 and 37.7 by 29.4, 33.8 by 26.3, and 37.5 by 25.7 millimeters.

Young.—Mr. Jourdain writes:

Incubation lasts not less than 24 full days, as a rule, though Hanroth gives 23½ as the period in Germany. In England most birds hatch out on the 25th day. When the young are hatched both parents take charge and are most active and courageous in defense of the young. On one occasion I heard a pair on the far side of a hedge, and looking over the top I was surprised to find that the bold little cock flew straight at my head with loud outcry while the hen busied herself in getting the young under cover as soon as possible.

Several of the early British writers have referred to an incident, related by Yarrell (1871) as follows:

A person engaged in a field, not far from my residence, had his attention arrested by some objects on the ground, which, upon approaching, he found to be two Partridges, a male and female, engaged in battle with a Carrion Crow; so successful and so absorbed were they in the issue of the contest, that they actually held the Crow till it was seized and taken from them by the spectator of the scene. Upon search, young birds, very lately hatched, were found concealed amongst the grass. It would appear, therefore, that the Crow, a mortal enemy to all kinds of young game, in attempting to carry off one of these, had been attacked by the parent birds, and with this singular result.

Plumages.—In Witherby's handbook (1920) the downy young is described as follows:

Crown chestnut with a few small black spots sometimes extending to lines; back of neck with a wide black line down centre, at sides pale buff marked black; rest of upper-parts pale buff with some rufous and black blotches or ill-defined lines, at base of wings a spot, and on rump a patch, of chestnut; forehead and sides of head pale yellow-buff (sometimes tinged rufous) with spots, small blotches, and lines of black; chin and throat uniform pale yellow-buff; rest of under-parts slightly yellower, bases of down sooty.

And the juvenal plumage, in which the sexes are alike, is thus described:

Crown black-brown finely streaked buff, each feather having buff shaft-streak; back of neck, mantle, back, rump and upper tail-coverts buff-brown, with whitish to pale buff shaft-streaks inconspicuously margined blackish; lores and sides of head dark brown streaked whitish; chin, throat and centre of belly whitish to pale buff; breast, sides and flanks and under tail-coverts brown-buff slightly paler than mantle and with whiter shaft-streaks, faintly margined brown on flanks; tail much like adult but feathers tipped buff and with subterminal dusky bar and spots and central ones speckled and barred dusky; primaries brown with pale buff tips and widely spaced bars on outer webs; secondaries with pale buff bars extending across both webs and vermiculated brown, shafts pale buff; scapulars, inner secondaries and wing-coverts brown-buff with wide brown-black bars and mottlings and pale shaft-streaks widening to white spots at tips of feathers.

A postjuvenal molt, which is complete except for the outer two primaries, produces a first winter plumage. The sexes are now differentiated and resemble the two adults, except for the more pointed tips of the outer primaries. This molt begins when the young bird is about half grown and is sometimes prolonged through December.

Adults have a partial prenuptial molt in May and June, sometimes in April, and a complete molt from July to November or December. Several observers have experienced some difficulty in distinguishing the sexes among adults, chiefly because many females have the dark chestnut patch on the belly more or less well developed. There seems to be some difference of opinion as to whether this character is more pronounced in old or in young females. But the sexes can always be distinguished by two characters; the light chestnut on the sides of the head is lighter and more restricted in the female; the median wing coverts of the female are dark brown or black, with widely spaced, pale-buff bars; whereas these coverts in the male have no transverse bars, but only a pale-buff shaft streak.

Food.—In Witherby's handbook (1920) the food of the partridge is summarized by Mr. Jourdain, as follows:

> Chiefly shoots and leaves of grass and clover as well as seeds of many species including *Polygonum, Trifolium, Alchemilla, Galium, Spergula, Persicaria, Poa,* etc. Turnip leaves, young shoots of heather, bramble and blaeberry, hawthorn berries, and corn also eaten. In spring and summer insects are also taken, including diptera (*Tipulidae* and larvae), coleoptera and hymenoptera (ants and their pupae being very favourite food). Also aphides. Once recorded as eating pears on tree!

Crops and stomachs of American birds also contained wheat, barley, and oats, mainly waste grain, seeds of wild buckwheat, pigweed, and other weeds, and grasshoppers. It is said that these birds do not pull up sprouting corn as the pheasants do. Their food habits seem to be wholly beneficial.

Behavior.—Macgillivray (1837) says, of the gray partridge, tnat it

> is fond of rambling into waste or pasture grounds, which are covered with long grass, furze, or broom; but it does not often enter woods, and never perches on trees. It runs with surprising speed, when alarmed or in pursuit of its companions, although in general, it squats under the apprehension of danger, or when nearly approached takes flight. Its mode of flying is similar to that of the Brown Ptarmigan; it rises obliquely to some height, and then flies off in a direct course, rapidly flapping its wings, which produce a whirring sound.

Yarrell (1871) writes:

> During the day a covey of Partridges, keeping together, are seldom seen on the wing unless disturbed; they frequent grass-fields, preferring the hedge-sides, some of them picking up insects, and occasionally the green leaves of plants; others dusting themselves in any dry spot where the soil is loose, and this would seem to be a constant practice with them in dry weather, if we may judge by the numerous dusting places, with the marks and feathers to be found about their haunts; and sportsmen find, in the early part of the shooting-season, that young and weak birds are frequently infested with numerous parasites. In the afternoon the covey repair to some neighbouring field of standing corn, or, if that be cut, to the stubble, for the second daily meal of grain; and, this completed, the call-note may be heard, accord-

ing to White, as soon as the beetles begin to buzz, and the whole move away together to some spot where they jug, as it is called—that is, squat and nestle close together for the night; and from the appearance of the mutings, or droppings, which are generally deposited in a circle of only a few inches in diameter, it would appear that the birds arrange themselves also in a circle, of which their tails form the centre, all the heads being outwards—a disposition which instinct has suggested as the best for observing the approach of any of their numerous enemies, whatever may be the direction, and thus increase their security by enabling them to avoid a surprise. In the morning early they again visit the stubble for a breakfast, and pass the rest of the days as before. Fields of clover or turnips are very favourite places of resort during the day. Mr. Harvie-Brown informs the editor that when the snow lay upon the ground he has known a covey to roost regularly on a limb of a large tree; and he has also seen Partridges "treed" by a dog.

Considerable discussion has appeared in print on the effect, on our native game birds, of introducing Hungarian partridges. Some claim that where the partridges are increasing the native grouse are disappearing. Most of our grouse are subject to periodic fluctuation in numbers from other causes; and it does not seem to have been definitely proved that the partridges are the cause of any local decrease in grouse. There are certainly plenty of suitable nesting sites for all these ground-nesting species; there is no proof that any shortage of food supply has led to any disastrous competition between them; and there is no evidence that the smaller partridges ever attack the larger grouse, which should be more than a match for them. Though there is always danger in introducing a foreign species, it would seem that the little gray partridge is more likely to prove a complementary than a competitive species.

Enemies.—Partridges, like all other ground-nesting species, are preyed upon by the whole long list of furred and feathered enemies, but they are such prolific breeders that their natural enemies are not likely seriously to reduce their numbers. Their habits of feeding in the open during the day and roosting in the open at night, make them especially exposed to the attacks of hawks and owls. The ring-necked pheasant may have to be reckoned with, as an enemy of the partridge. Mr. Spiker (1929) writes:

Northwestern Iowa has not until fairly recently been afflicted with this pernicious bird, but they are on the increase, and farmers have told me that with the coming in of the Ring-necked Pheasant, the partridges are departing. Perhaps a concrete example would be admissible here. Mr. Raymond Rowe, a farmer living a few miles northwest of Sibley, while plowing late last fall (1927), observed something of a commotion in a little swale a short distance from his plowing. Prompted by curiosity he walked over to the place and flushed half a dozen partridges and three Ring-necked Pheasants. On the ground before him lay the bleeding bodies of three partridges newly killed. It was just dusk, and doubtless the smaller species had crept into the long grass to spend the night and had been fallen upon by the pheasants who were already there. Stories are also told of the destruction of the nests of the Hungarian Partridge by pheasants.

Voice.—Mr. Spiker (1929) says that the voice is

not unmusical, and yet not conspicuous unless listened for, it is especially noticeable on a still spring evening, when there is little or no breeze, and the shadows of dusk follow the disappearance of the sun. There is a single two-syllabled chuckling note which may be represented somewhat by the syllables "kee-uck," the second syllable being rather raspy or throaty as compared to the first, which is high pitched and nasal. Upon being flushed, the bird takes off with the startling whirr of wings characteristic of this family, uttering the while a rapid cackling which diminishes to the above given notes repeated several times and with a gradually increasing interval between them. In the immediate vicinity of Ashton it is not unusual to hear from four to eight of these birds calling at the same time and from as many different directions.

Walter H. Rich (1909) writes of some birds in captivity:

In their coop they used a great variety of language; they clucked like a Grouse; they chattered like a Blackbird; they snapped their bills like an Owl; they "jawed" like a Parrot; they made a guttural note of alarm like the "br-r-r-r" of a startled Pigeon; they hissed like a Black Duck guarding her nest, or like a Thomas cat whose dignity is ruffled not quite enough for anger; and, in addition, they are said to "crow" at evening.

Game.—Provided that the Hungarian partridge does not seriously interfere with the welfare of our native species, it seems to be a wise and valuable addition to our list of game birds. I have never hunted it, but those who have speak very highly of it. It is a strong, swift flier, smart and sagacious, well fitted to test the skill of the best sportsmen. It is a fine bird for the table. Unfortunately it will survive and flourish only in certain favorable sections, mainly the northwestern grainfields and grassy plains. There it can probably survive much more intensive hunting than either the prairie chicken or the sharp-tailed grouse.

One of the men who helped to introduce the partridge in Washington wrote to Mr. Rathbun as follows:

From the standpoint of a game bird I believe them to be the gamest of them all. The law of the covey is very strong, and when they flush all of them go at the same time. There seems to be less than a fraction of a second between the time the first one and the last one makes his get-away. They will always be able to take care of themselves, since they become very wild when much shooting is done. When one is winged or slightly wounded so that he can not fly he will run a mile sometimes before a hunter's dog overtakes him. During the winter months they come right into the towns and eat at the back doors of the residences. They will help themselves to strawstacks, haystacks, and anything edible. At night they burrow in the snow, sometimes making little tunnels 4 or 5 feet long under 2 feet of snow. I have hunted upland game birds in the West covering a period of 37 years, but I believe the Hungarian partridge, considered from every standpoint as a game bird, is the premier one of the Pacific coast.

Winter.—Mr. Spiker (1929) says:

The species is gregarious during the winter, beginning to flock in October and continuing till the last of February. During this season they frequent the stalk fields left after the picking of the corn. When the gregarious spirit is upon them they are exceedingly wary and are up and away almost as they see the hunter enter the field. The startling noise with which they take flight and their extremely rapid coursing across the field make them a very difficult target, and, although many attempts are made by poachers, few birds fall as victims. By the latter part of February, however, there comes a change when they begin breaking up and pairing off, and at this time they appear to lose some of their wariness.

DISTRIBUTION

Introduced more or less unsuccessfully in the Eastern States, from Maine and New York southward to Florida and Mississippi, also in California. In the Central States, from Minnesota and Michigan to Kansas and Arkansas, most attempts at introduction have failed, except in extreme northwestern Kansas, in parts of Iowa (Osceola and Lyon Counties), and in southeastern Wisconsin (Waukesha County). Introduced birds have done well in southern British Columbia (Fraser Valley) and in eastern Washington and Oregon (east of the Cascade Mountains). The most remarkable success has been attained in Alberta, Saskatchewan, and Manitoba, where the birds have flourished and spread over a wide territory.

Egg dates.—Washington: 4 records, May 25 to June 10.

COLINUS VIRGINIANUS VIRGINIANUS (Linnaeus)

EASTERN BOBWHITE

HABITS

In the springtime and early in summer bobwhite deserves his name, which he loudly proclaims in no uncertain terms and in a decidedly cheering tone from some favorite perch on a fence post or the low branch of some small tree. But at other seasons I prefer to call him a quail, the name most familiar to northern sportsmen, or a partridge, as he is even more appropriately called in the South. But European sportsmen would say that neither of these names is strictly accurate, so we may as well call him bobwhite, which is at least distinctive. By whatever name we call him, he is one of our most popular and best beloved birds. From a wide distribution in the East, he has followed the plow westward with the clearing of the forests and the cultivation of the fertile lands of the Middle West; and more recently he has been successfully introduced into many far-western States.

Bobwhite is one of the farmer's best friends; his economic status is wholly beneficial; he is not known to be injurious to any of our crops, as what grain he eats is mostly waste grain, picked up in the stubble fields after the crops are harvested. It seems to me, however, that too much stress has been laid on his services as a destroyer of weed seeds. Nature has provided so lavishly in the distribution of weed seeds that only a very small fraction of them can find room to germinate, and the seeds picked up by birds, which never glean thoroughly, only leave room for others to grow. I doubt if even a square foot of ground has ever been kept clear of weeds by birds. The hoe and the cultivator will always have to be used. But bobwhite has a fine score to his credit as a destroyer of grasshoppers, locusts, potato beetles, plant lice, and other injurious insects.

It has been suggested by some bird protectionists that the bobwhite should be removed from the game-bird list and be rigidly protected at all seasons as a song bird and an insectivorous bird. But we must not lose sight of its economic value as a game bird and the pleasure and healthful exercise that it gives to thousands of sportsmen. There are hundreds of other birds that bring joy to the hearts of amateur bird admirers and many others that are nearly or quite as useful as insect destroyers, so why should we deprive the sportsmen of their most popular upland game bird, when they have not more than two or three species at best in any one section of the country? Edward H. Forbush (1927) has summed up the matter very well, as follows:

> As a popular game bird of the open country Bob-white has no rival. Probably about 500,000 sportsmen now go out annually from cities east of the Rocky Mountains to hunt this bird. This necessitates a great annual expenditure for hunters' clothing, guns, ammunition, dogs, and guides. It adds to the revenue of farmers and country hostelries. In some of the southern states Bob-white pays the taxes on many farms where the farmers sell their shooting rights to sportsmen. Perhaps there is no bird to which the American people are more deeply indebted for both aesthetic and material benefits. He is the most democratic and ubiquitous of all our game birds. He is not a bird of desert, wilderness, or mountain peak which one must go far to find. He seeks the home, farm, garden, and field; he is the friend and companion of mankind; a much needed helper on the farm; a destroyer of insect pests and weeds; a swift flying game bird, lying well to a dog; and, last as well as least, good food, a savory morsel, nutritious and digestible.

One does not have to go far afield to find the haunts of bobwhites, for they shun the deep forest areas, seldom resort to the woods except to escape from danger, and are rarely found on the wide open prairies. They seem to love the society of human beings and their cultivated fields. During spring and summer they are particularly domestic and sociable, when it is no uncommon occurrence to hear their loud, ringing calls almost under our windows, to see one perched on a fence post near the house or on the low branch of an apple tree in the orchard, or to find them running along the driveway or a garden path. They are very tame and confiding at that season and seem to know that they are safe. At other seasons they resort to more open country and seek more seclusion. In New England they prefer the vicinity of farms, where they find suitable feeding grounds in old weed patches and stubble fields where crops of buckwheat, millet, rye, wheat, oats, or other grains have been harvested. But near at hand they must have suitable cover, thick, swampy tangles or brier patches in which to roost at night, or dense thickets or woodlots in which to seek refuge when pursued. In the South, according to M. P. Skinner, they "like weedy corners of cornfields next to a tangle of blackberry briars, cane, cat briars, and brush, into which they can retreat at a moment's notice. They also like cotton fields, especially if a corner be grown up to broom sedge and low brush." The cultivated fields of the South are usually well overgrown with weeds in the fall, where the partridges find both food and shelter in the old fields of cowpeas, ground nuts, and other crops, overgrown with crabgrass, foxtail grass, Japan clover, plume, and wild grasses.

Courtship.—It is not until spring is well advanced that the coveys, which have kept together all winter, begin to break up and scatter. Then it is that the young cock, which has now acquired full maturity and vigor, begins to feel the urge of love and, separating from his companions, sets about the important business of securing a mate. Dressed in his best springtime attire, his bosom swelling with pride, he selects his perch, a fence post, the low branch of a tree, or some convenient stump, from which to send out his love call to his expected bride. *Bob-white! Ah, bob-white bob-bob-white!* It rings out, loud and clear, repeated at frequent intervals, while he listens for a response, perhaps for half an hour or more in vain. At length he may hear the coveted sound, the sweet, soft call of the demure little hen. With crest erected and eyes aglow, he flies to meet her and display his charms, fluttering and strutting about her and coaxing her with all the pomp and pride of a turkey gobbler. But she is shy and coy, and does not yield at first. Perhaps she runs away, and then ensues a lively game of chase. Aretas A. Saunders tells in his notes of such a chase that he saw under favorable circumstances. The hen kept about 5 feet ahead of the cock, running rapidly, faster than he had seen this species move at any other time, back and forth, in and out, around some clumps of grass. Though he watched for 15 minutes, the cock did not seem to gain an inch. Doubtless he did eventually.

But bobwhite's road to happiness is not always so smooth. As his clarion call of defiance rings out across the fields an answering call, *bob-bob-white*, reaches his jealous ears, the voice of an unknown rival. Back and forth the challenges are exchanged, as the brave little warrior advances to meet his foe. Louder, sharper, and angrier are their cries, as they dodge about, bursting with rage and eager for the fray, seeking a vantage point for the attack. At last they clinch in furious combat, like small game cocks, savagely biting and tearing with sharp little beaks, scratching with claws, and buffeting with strong little wings. The fighting is fast and furious for a time until one gives up exhausted and slinks away. Finally the brave little conqueror enjoys the spoils of victory, the acceptance of his suit by the modest little hen, who now knows that she has picked a winner.

Herbert L. Stoddard (1931), in his excellent and exhaustive monograph on the bobwhite, published by the committee on the Cooperative Quail Investigation, has added to our knowledge a vast fund of information on the habits of this valuable species, its enemies, diseases, and means for preserving and increasing it, based on a five years' study in cooperation with the Biological Survey. Anyone interested in this subject should study this voluminous report, as our space will permit only brief extracts from it.

As to the breaking up of the coveys in the spring, which "are usually composed of the remnants of several hatchings," he says that "many of the birds are not closely, if at all, related." At this season, "cocks, which had been peaceable companions previously, became pugnacious," and frequent fights occurred. In the inclosures the fights were harmless, as a rule, but in the wild "an occasional combat no doubt proves fatal, for two dead cock quail that had been picked up afield were brought to us with the flesh bitten to the bone at the junction of head and neck."

Referring to the courtship display, he writes:

> This display is a frontal one. The head is lowered and frequently turned sideways to show the snowy-white head markings to the best advantage, the wings are extended until the primary tips touch the ground, while the elbows are elevated over the back and thrown forward, forming a vertical feathered wall. The bird, otherwise puffed out to the utmost in addition to the spread, forward-thrust wings and lowered, side-turned head, now walks or advances in short rushes toward the hen, and follows her at good speed in full display in case she turns and runs.

Some evidence was obtained to indicate that some mated pairs remain mated during winter and for at least two breeding seasons. As to the devotion of mated pairs, he says:

> Two weeks to a month may elapse, depending on the weather, between the time of pairing and the beginning of nesting. During this period the pairs appear inseparable, the hen usually taking the lead in foraging expeditions, with the cock a devoted follower. He is very attentive at this time, as indeed he is all during the breeding season, unless he takes up incubation duties, when he appears to lose interest in the opposite sex. It is amusing to see him catch a grasshopper or other large insect after a lively chase. He puffs himself up and, holding the insect out in a stiff, wooden manner, starts a soft, rapidly repeated *cu-cu-cu-cu* to attract his mate, who rushes to him and eats the dismembered insect. This common habit may be frequently observed all during the breeding season, the hen usually being the one to get the insects caught by the cock, even when the pair are rearing a brood.

Nesting.—The bobwhite's nest is a very simple affair, but artfully concealed and seldom found, except by accident, as the bird is a very close sitter and usually does not leave the nest until almost trodden upon. The favorite nesting sites seem to be along old fence rows, where the grass grows long and thick or is mixed with tangles of vines or briers, in neglected brushy corners of old fields, under discarded piles of brush, or in the tangled underbrush that, mixed with grass, grows on the edges of woods, thickets, or swamps. The nest is often placed in open fields of tall grass, where the hay cutter sometimes destroys it, in cultivated fields of grain or alfalfa, or at the base of a tree in the farmer's orchard, if the grass is long enough to conceal it. A nest is often found in an unexpected place. Once, at my cottage on Cape Cod, I worked for two days weeding my garden within 3 feet of a boundary fence and was surprised the next day, on cutting the grass along the fence, to uncover a quail's nest, with 15 eggs, from which the bird had never stirred. I was told one day that there was a quail's nest under a brush pile at our golf club and went up to photograph it. I found a pile of pine boughs that had been cast aside just off the edge of an elevated putting green. I walked around it carefully several times trying to see the bird, but I never found it until I lifted the right bough and flushed her. I saw her several times afterwards and believe she raised her brood successfully.

George Finlay Simmons (1915) tells of a nest found by him in Texas "under the edge of a bale of hay in an old shed on the prairie," which he discovered by flushing the bird. Charles R. Stockard (1905) writes from Mississippi:

> In fields of sedge grass or oats many pairs will often nest very close together. June, 1895, I found in a thirty acre field of sedge grass sixteen nests of the Bob-white, all containing large sets, ranging from twelve to twenty-two eggs, and the total number of eggs in this field must have been about three hundred.

Out of 602 nests, studied by Stoddard (1931) and his associates, 97 were in woodland, 336 in broom-sedge fields, 88 in fallow fields, and "about 4 per cent in cultivated fields, mostly in the grassy growth around stumps in corn or cotton fields, but occasionally under trash cast aside by plows or cultivators." In the few cases where nest construction was under observation the work was done entirely by the male under the supervision of his mate.

The construction of a typical nest is very simple. Having selected a suitable spot, where the vegetation is thick enough to afford effective concealment, a hollow is scooped out and lined with dead grass or other convenient material; after that the dead and growing grass or other vegetation is woven into an arch over the nest, often completely concealing it, and leaving only a small opening on the side, just large enough for the bird to enter or leave the nest; while incubating, the bird looks out through this opening; if there are any vines or briers growing about the nest, these are also woven into the arch to make it firmer and more impenetrable. F. W. Rapp describes in his notes a more elaborate nest, resembling a marsh wren's nest in construction and shape and very firmly built; it was located in a fence row and was made of oak leaves and June grass, neatly woven together into a ball, flattened on the bottom, with a hole on one side. Often the simplest nest is made by entering a thick clump of grass and flattening down a hollow in the center, without disturbing the grass tops at all.

Major Bendire (1892) quotes Judge John N. Clark as having seen a male bobwhite building a nest, as follows:

> In May, 1887, while on a hill back of my house one morning, I heard a Quail whistle, but the note, which was continually repeated, had a smothered sound. Tracking the notes to their source, I found a male Bob White building a nest in a little patch of dewberry vines. He was busy carrying in the grasses and weaving a roof, as well as whistling at his work. The dome was very expertly fashioned, and fitted into its place without changing the surroundings, so that I believe I would never have observed it, had he kept quiet.

He also speaks of a nest, found in Louisiana, which "was entirely constructed of pine needles, arched over, and the entrance probably a foot or more from the nest proper."

Eggs.—The bobwhite ordinarily lays from 12 to 20 eggs, 14 to 16 being perhaps the commonest numbers; as few as 7 or 8 and as many as 30, 32, and even 37 eggs have been found in a nest; but these large numbers are probably the product of more than one female and are deposited in layers. The eggs are mainly subpyriform in shape, sometimes quite pointed or again more rounded. The shell is smooth, with very little gloss, and decidedly hard and tough. The color is dull white or creamy white, rarely "light buff" or "pale ochraceous-buff." They are never spotted, but are usually more or less nest stained. The measurements of 55 eggs in the United States National Museum average 30 by 24 millimeters; the eggs showing the four extremes measure **32.5** by 24, 31 by **26**, and **26** by **22.5** millimeters.

Bobwhites occasionally lay their eggs in other birds' nests. H. J. Giddings (1897) reports the finding of a quail's egg in a towhee's nest; and the editor in a footnote refers to one laying in a domestic turkey's nest. E. B. Payne (1897) adds that he "found in a meadowlark's nest five of the meadowlark's eggs and four of the quail's." Mr. Rapp mentions in his notes a quail's nest shown to him that contained 12 eggs of the quail and 2 of the domestic hen. Herbert L. Stoddard has a photograph of a bantam's egg in a quail's nest.

Young.—It is generally supposed that at least two broods of young are raised in a season, perhaps three in the southern part of the quail's range, as very early and very late broods are of common occurrence. But, as the quail has many enemies and many nests are broken up or deserted, it may be that the late broods are merely belated attempts to raise a family; in which case, perhaps one brood in the North and two in the South is more nearly the average.

Most authorities agree that the period of incubation is about 23 or 24 days. Both sexes share this duty. In the study of 276 nests by Mr. Stoddard (1931), in southern Georgia and northern Florida, so far as could be ascertained 73 were entirely in charge of the cock and 175 in charge of the hen. If any fatal accident befall the hen, as too often happens, then the cock assumes full charge of the eggs and afterwards takes care of the young. It is said, too, that after the young are two or three weeks old the mother hands the brood over to the care of their father and starts to lay a second set of eggs; but I doubt if this has been definitely proved.

Young quail leave the nest almost as soon as they are hatched, and the eggshells are generally left in the nest, although occasionally a chick is seen running away with part of the shell on its back. They are carefully tended by their devoted parents, who use every known artifice to distract an enemy. Dr. T. M. Brewer (Baird, Brewer, and Ridgway, 1905) relates the following to illustrate an extreme case of parental boldness:

> Once as I was rapidly descending a path on the side of a hill, among a low growth of scrub-oak I came suddenly upon a covey of young Quail, feeding on blueberries, and directly in the path. They did not see me until I was close upon them, when the old bird, a fine old male, flew directly towards me and tumbled at my feet as if in a dying condition, giving at the same time a shrill whistle, expressive of intense alarm. I stooped and put my hand upon his extended wings, and could easily have caught him. The young birds, at the cry of the parent, flew in all directions; and their devoted father soon followed them, and began calling to them in a low cluck, like the cry of the Brown Thresher. The young at this time were hardly more than a week old, and seemed to fly perfectly well to a short distance.

Their ability to fly at such an early age is due to the fact that their wings begin to sprout almost as soon as they are hatched; I have seen young chicks not more than 2 inches long with wings reaching to their tails; they are very active and vigorous and grow very rapidly. They are experts at hiding; a warning note from the watchful parent, who previously has kept the brood together by frequent gentle twitterings, sends them to cover instantly; instinctively they dart under some fallen leaf, beneath a tuft of grass, into some thick vegetation or little hollow, where they remain motionless until told by their parent that danger has passed. Edwyn Sandys (1904) has described this so well that I quote the following:

> If those who may stumble upon a brood of quail will take a sportsman-naturalist's advice, they will promptly back away for a few yards, sit down, and remain silently watchful. No search should be attempted, for the searcher

is more likely to trample the life out of the youngsters than to catch one. But if he hide in patience, he may see the old hen return, mark her cautiously stealing to the spot, and hear her low musical twitter which tells that the peril has passed. Then from the scant tuft here, from the drooping leaf yonder, apparently from the bare ground over which his eyes have roved a dozen times, will arise active balls of pretty down until the spot appears to swarm with them. And the devoted mother will whisper soft greetings to each, and in some mysterious manner will make the correct count, and then with nervous care shepherd them forward to where there is safer cover. And they will troop after her in perfect confidence, to resume their bug-hunting and botanical researches as though nothing important had transpired.

Young quail are busy foragers, and they grow rapidly. Within a few days after leaving the nest they are capable of a flight of several yards. A brood flushed by a dog will buzz up like so many overgrown grasshoppers, fly a short distance, then dive into cover in a comical imitation of the tactics of their seniors. As insect catchers they are unrivalled, their keen eyes and tireless little legs being a most efficient equipment even for a sustained chase. The parents scratch for them and call them to some dainty after the manner of bantam fowls, and the shrewd chicks speedily grasp the idea and set to work for themselves. A tiny quail scratching in a dusty spot is a most amusing sight. The wee legs twinkle through the various movements at a rate which the eye can scarcely follow, and the sturdy feet kick the dust for inches around. When a prey is uncovered it is pounced upon with amazing speed and accuracy, while a flying insect may call forth an electric leap and a clean catch a foot or more above the ground. As the season advances grain, seeds of various weeds, berries, wild grapes, and mast are added to the menu, in which insects still remain prominent. After the wheat has been cut the broad stubbles become favorite resorts, especially when they are crowded with ragweed. Patches of standing corn now furnish attractive shelter and the suitable dusting places so necessary to gallinaceous birds.

Plumages.—Only in the smallest chicks can the pure natal down be seen. In a typical chick the forehead and sides of the head are from "ochraceous-tawny" to "ochraceous-buff," with a stripe of brownish black from the eye to the nape; a broad band from the hind neck to the crown, terminating in a point above the forehead, is "chestnut," deepening to "bay" on the edges; there is a similar broad band of the same colors from the upper back to the rump; the rest of the upper parts is mottled with "chestnut," dusky, and buff; the chin and lower parts are pale buff or buffy white. In some specimens from the South the back and rump are almost wholly "chestnut," mixed with some black.

The juvenal plumage begins to appear on the wings and scapulars at a very early age, even before the chick has increased perceptibly in size; I have seen chicks 2 or 3 inches long that had wings extending beyond the tail and that would soon be able to fly. In this plumage the sexes are alike, except that, according to Dr. Jonathan Dwight (1900), "the males are apt to be richer colored than are females, with grayer tails, whiter chins, blacker throat bands, and often a slight dusky barring on the breast." The first feathers to appear on the back and scapulars are black on the inner web, broadly tipped with white, and mottled with brown ("russet" to "tawny") and dusky on the outer web, with white shaft stripes, broadening at the tip; as these feathers grow out longer the black appears only as a large subterminal spot. In full juvenal plumage the crown is centrally dusky, laterally gray ("hair brown" to "drab"), mottled or variegated with black; the throat is white in the male and buffy white in the female; the breast and flanks are "drab" to "light drab," with whitish shaft streaks; the belly is paler or white; the tail is gray, mottled with white; and the primaries are mottled with pale buff on the edges.

Even before the juvenal plumage is fully acquired the postjuvenal molt into the first winter plumage begins. This molt is complete except for the outer pair of primaries on each wing, which are retained all through the first year; and it takes place at any time from late in summer until November, depending on the time at which the young were hatched. The first winter plumage is scarcely distinguishable from that of the adult, and the sexes are widely differentiated; but the colors above are duller with paler edgings, and the underparts are more buffy and somewhat less barred. Young birds can be distinguished from adults all through the first winter and spring by the outer pair of primaries, the first and second, on each wing, which are still juvenal (pointed).

The first prenuptial molt, as well as all subsequent prenuptial molts, amounts to the renewal of only a few feathers about the head and throat. The first postnuptial molt, the following summer and fall, chiefly in September, is complete and produces the adult winter plumage. Adults then continue to have similar molts each year, a very limited head molt in spring and a complete postnuptial molt from August to October. The slight seasonal difference between spring and fall plumages is mainly due to wear and fading.

Among the thousands of quail shot and the large series preserved in collections, some odd types of plumage are to be found, such as males with black or buff throats, very dark or melanistic types, others in which the browns are replaced with buff or the buffs with white, producing a pallid type; partial albinos are occasionally seen and very rarely one that is wholly pure white. Erythrism is reported and illustrated by Stoddard (1931).

Food.—Quail are very regular in their feeding habits. Every sportsman knows this and takes advantage of it, for he knows when and where to look for them. They do not leave their roosting place very early in the morning, as they prefer to wait until the rising sun has, at least partially, dried the dew off the grass; in winter or late in fall, when every blade of grass, twig, or spray of vegetation is white with hoarfrost, and when the feeble rays of the sun are late in rising, they are slow to venture out. But usually by an hour after sunrise they are afoot toward some convenient weed patch, stubble field, berry patch, or cultivated field. Here they feed for an hour or two, filling their crops, and then retire to some sheltered spot for a midday siesta, digesting their food, dusting or preening their plumage, or merely basking in the sun or dozing. About two hours before sunset they return to their feeding grounds again for another feast before going to roost at dusk.

The food of the bobwhite has been exhaustively studied, and a mass of material has been published on it. Space will not permit any detailed account of it here; I can give only a general idea of it. The most complete account of it that I have seen is given by Sylvester D. Judd (1905) of the Biological Survey, to which the reader is referred. He says that the bobwhite is "one of our most nearly omnivorous species. In addition to seeds, fruit, leaves, buds, tubers, and insects, it has been known to eat spiders, myriapods, crustaceans, mollusks, and even batrachians." In analysis of 918 stomachs, collected during every month in the year, in 21 States and in Canada, the food for the year as a whole consisted of vegetable matter, 83.59 per cent, and animal matter, 16.41 per cent, mixed with some sand and gravel. Of the vegetable food, grain constituted 17.38 per cent, seed 52.83 per cent and fruit 9.57 per cent; the grain was probably mostly waste kernels, and the seeds were mainly weed seeds; not a single kernel of sprouting grain was found in any of the crops or stomachs; and there is no evidence that quail ever do any damage to standing crops. The fruits eaten were practically all wild fruits. The animal matter was distributed among beetles, 6.92 per cent; grasshoppers, 3.71 per cent; bugs, 2.77 per cent; caterpillars, 0.95 per cent; and other things, 2.06 per cent. From October to March the food is almost entirely vegetable matter, but late in spring and in summer it is made up largely of insects, August showing 44.1 per cent of insect food. The insects eaten are mostly injurious species, many of which are avoided by other insectivorous birds, such as "the potato beetle, twelve-spotted cucumber beetle, striped cucumber beetle, squash lady-bird beetle, various cutworms, the tobacco worm, army worm, cotton worm, cotton bollworm, the clover weevil, cotton boll weevil, imbricated snout beetle, May beetle, click beetle, the red-legged grasshopper, Rocky Mountain locust, and chinch bug."

Since the above was written, the author has seen Stoddard's (1931) much more elaborate account of the food and feeding habits of quail in the Southeastern States, contributed by C. O. Handley. Doctor Judd's report covered a wider territory, and the stomachs were obtained for each month of the year, but most of them were taken late in fall and in winter, and there were no stomachs of young birds examined. Mr. Handley's report is based on the examination of the food of 1,625 adult and 42 young bobwhites; it covers 53 pages and is far too voluminous and too elaborate for me even to attempt to quote from it. It should be carefully studied. A condensed table gives the monthly and yearly percentages of the various items in the food. The total yearly averages show 85.59 per cent of vegetable and 14.41 per cent of animal food. The principal items in the vegetable food are: Fruits, 19.41; legumes, 15.17; mast, 13.42; grass seeds, 10.65; and miscellaneous seeds, 10.24 per cent; and in the animal food: Orthoptera, 7.43; Coleoptera, 2.98; Hemiptera, 1.96; and other insects, 1.06 per cent. For all the interesting details the reader is referred to this exhaustive report.

E. L. Moseley (1928) gives a striking illustration of the value of bobwhites as destroyers of potato beetles in Ohio, where these birds have increased enormously under 10 years of rigid protection. He says:

For several years past potatoes have been raised successfully on many farms in Ohio without spraying for beetles, or taking any measures to combat the insects. In fact many patches have been practically free from the "bugs." Bob-whites have been observed to spend much of the time among the potato vines. They have been seen to follow a row, picking off the potato beetles. When the potato patch was located near woodland there was no trouble with the beetles; but when the patch was near the highway or buildings, even on the same farm, the insects were troublesome. On farms where the Bob White found nesting sites and protection, the potato vines, if not too near the buildings, were kept free from the insects. A patch of potatoes surrounded by open fields, without bushes, tall weeds, or crops that might shelter the Bob White, was likely to be infested with beetles. A farmer living eight miles south of Defiance raised about fifty Bob Whites on his place. During the two years that these birds were there he had no trouble with insects on either potatoes or cabbage. The following autumn a number of the birds were killed by hunters, while others were frightened away. The next summer the potato beetles were back in numbers. The farmer is again raising Bob Whites and protecting them from hunters.

Mrs. Margaret M. Nice (1910) found that a captive bobwhite ate 568 mosquitoes in two hours, another 5,000 plant lice in a day, and another 1,000 grasshoppers and 532 other insects in a day; also that it ate from 600 to 30,000 weed seeds each day, according to the size of the seeds and the bird's capacity. I can not give here a complete list of the food of the bobwhite, as given by Doctor Judd (1905), but a few of the most important seeds are those of various grasses, rushes, sedges, sorrel, smartweed, bindweed, chickweed, lupine, clover, vetches, spurges, maples, ashes, oaks, pines, violets, morning-glory, ragweed, sunflower, beggarticks, and foxtail and witch grass. Among the fruits are waxmyrtle, barberry, bayberry, mulberry, thimble berries, blackberries, wild strawberries, rose hips, wild apples, cherries, poison ivy, sumacs, holly, black alder, bittersweet, frost grapes, blueberries, huckleberries, elderberries, viburnums, honeysuckle, partridgeberry, and woodbine. Wherever the foregoing plants are cultivated or allowed to grow in profusion, bobwhites will find abundant food all through the year and will be encouraged to remain, with profit to the farmer and joy to the sportsman.

The more important items of insect food have been mentioned above. From 35 to 46 per cent of the summer food of adults consists of insects, but the young chicks eat a much larger proportion of this food. Small beetles of various kinds, weevils, small grasshoppers, caterpillars, ants, stink bugs, spiders, and thousand legs have been identified in the food of small chicks.

Behavior.—When a flock of quail suddenly bursts into the air from almost underfoot the effect is startling and gives the impression of great strength and speed. They have been referred to as feathered bombshells. Such sudden flights of a whole bevy in unison are due to the fact that they have crouched, trusting to their wonderful powers of concealment, until the very last moment, when they are forced to make a quick get-away. From their crouching attitude they are in position to make a strong spring into the air, giving them a good start, which their short but powerful wings continue in a burst of speed. Such bombshell flights are the rule when the birds are feeding in close formation, or when suddenly disturbed in their roosting circles. At other times their flight is much less startling but often quite as swift. I have often seen a single quail, or a pair or two, rise and fly away as softly and as silently as any other bird, when not alarmed. Their flight is not long protracted and generally ends by scaling down on set wings into the nearest cover. In settling, a flock usually scatters, to be joined together later by the gather call. Often single birds and sometimes a whole flock will alight in a tree, if alarmed. When leaving the tree their flight is silent and usually scaling downward. That they are not capable of long flights is shown by the fact that they become very much exhausted in flying across wide rivers and have even been known to drop into the water in attempting such flights.

Stoddard (1931) made a number of tests to determine the speed of bobwhites in flight. "These showed a speed for mature birds ranging from 28 to 38 miles an hour. It seems fair to estimate that the sportsman's hurtling mark sometimes exceeds 40 miles an hour, and birds just ahead of 'blue darters' are believed to go even faster for short distances."

Quail do much of their traveling on foot, and they are great travelers. They cover considerable ground in a day's routine, and a bevy may be found in any one of several feeding places. In some sections they are said to make seasonal migrations from one type of country to another, the journeys being made largely on foot. It is no uncommon occurrence to see a pair in spring, or a flock in fall, running along or across a country road. They make a very smart and trim appearance, with bodies held erect and heads held high, as they run swiftly along on their strong little legs. If too hard pressed they rise, flit gently over a fence or wall, and disappear. One can not help admiring their graceful carriage and their efficiency as runners. I believe they prefer to escape from their enemies by running, until too hard pressed; a bird dog will often trail a running bevy for a long distance.

Their characteristic method of roosting in a close circle, with bodies closely packed and heads facing out, is well known. For this they select some sheltered spot under an evergreen tree or thick bush, or in some dense tangle of briers or underbrush. Sometimes they select a small island in a river or a pond for a roosting place. If not disturbed they will occupy the same spot for many nights in succession, as evidenced by an increasing circle of droppings. Miss Althea R. Sherman told me that she had seen young quail, on the day they were hatched, assume the circular arrangement of a roosting covey, heads outward and tails in the center of the circle. An interesting account of how this circle is formed is given by Dr. Lynds Jones (1903) based on an observation by Robert J. Sim under especially favorable circumstances:

First one stepped around over the spot selected, then another joined him, the two standing pressed close together, forming the first arc of the circle. Another and another joined themselves to this nucleus, always with heads pointing out, tails touching, until the circle was complete. But two were left out! One stepped up to the group, made an opening, then crowded himself in, with much ruffling of feathers. One remained outside, with no room anywhere to get in. He, too, ran up to the circle of heads, then round and round, trying here and there in vain; it was a solid mass. Nothing daunted, he nimbly jumped upon the line of backs pressed into a nearly smooth surface, felt here and there for a yielding spot, began wedging himself between two brothers, slipped lower and lower, and finally became one of the bristling heads. In this defensive body against frost and living enemy we may leave them.

But quail do not always roost on the ground. Mr. Sandys (1904) says that

it is no uncommon thing to find them regularly roosting in such places as a mass of wild grape vines attached to a fence or a tree, in some thick, bushy tree, in an apple tree near the poultry, sometimes in the fowl-house, barn, or stable, on the lower rails of a weedy fence, on top of logs, and occasionally on the bare rails of a fence.

The ability of quail to hide and escape detection under the most scanty protection is truly remarkable. One is often surprised to see a bird or a whole covey arise from a spot that seems to offer no chance for concealment. Their ability to withhold their scent under such circumstances will be referred to later. Mr. Forbush (1927) relates some interesting observations on a bobwhite that spent a winter in his yard and became quite tame. He escaped the notice of a wandering dog by squatting on bare ground. A slow, quiet settling of his whole body was followed by the widening of the shoulders and an indrawing of the head, and, shaking out his feathers, he squatted on the snowy ground "as flat as a pancake." The white markings of the throat and head were cunningly concealed, the top of the head projecting barely enough beyond the general outline to allow him a comprehensive view of his surroundings. Once he effaced himself from sight in a little hollow at the foot of a tree, where he was invisible even through a glass at 40 feet away, until he "grew" out of the ground and walked away. Again he faded from view in a cleft in a stump less than 3 inches deep. Where there are dry leaves or grass concealment is easy.

Voice.—The most characteristic and best-known note of the bobwhite is the spring call, or challenge note, of the male, from which its name is derived. It is heard all through the breeding season in summer. It is subject to considerable individual variation and has been variously interpreted as *bob-white*, *more-wet*, *no-more-wet*, *peas-most-ripe*, *buck-wheat-ripe*, *wha-whoi*, *sow-more-wheat*, and others. This call is subject to considerable variation; the number of the preliminary *bobs* varies from one to two or rarely three; sometime these first syllables are entirely omitted and we hear only the loud *white*, which again may be shortened to *whit*. Aretas A. Saunders, who has made a study of the voice of the bobwhite, has sent me some elaborate notes on it. He says that the pitch in this call, counting all his records, varies from G″ to F‴, one tone less than an octave. One 3-note call covered this whole range, but the 2-note calls generally begin on A″, most commonly have the *white* note begin a tone higher, and slur up a single tone or a minor third. Sometimes the second note gives more accent and time to the first part of the slur, and sometimes the lower note of the slur is on the same pitch as the first note. The least range of pitch is shown in a 2-note call beginning on C, starting the slur on C♯, and ending on D. What he calls the slur comes, of course, in the last, or *white*, note.

The *bobwhite* note is almost invariably given while the bird is standing on some favorite perch, but R. Bruce Horsfall writes to me that while visiting in Virginia, on August 2, he saw a male bobwhite fly across an old orchard, with few remaining trees but much uncut grass, uttering this note in flight, fully a dozen calls in rapid succession, ceasing only with the termination of the flight.

Although many writers refer to the "bobwhite" note as the call of the cock bird to his sitting mate, Stoddard (1931) says:

We respectfully express our belief, based upon all the data we have been able to obtain personally, that the "bobwhite" call note is *largely* the call of the unmated cocks; ardent fellows eager to mate, but doomed to a summer of loneliness, from lack of physical prowess or an insufficient number of hens to go around.

The sweetest and loveliest call, entirely different from the foregoing or the following, is the 4-syllable whistle of the female, used to answer the male in spring and to call the young later in the season. My father, who was an expert whistler as well as a keen sportsman, could imitate this note to perfection. He often amused himself, when bobwhites were whistling in spring, by concealing himself in some thick brush and answering the *bobwhite* call of the male with this enticing note. It was amusing to see the effect on the cock bird, as he came nearer at each repetition of the answer to his call, looking in vain for his expected mate, and sometimes coming within 20 feet before detecting the deception. Once two cocks came to look for the anticipated hen; then a lively fight ensued, all on account of an imaginary bride. This call consists of four notes, the first and third short, soft, and on a low key, and the second and fourth longer, louder, richer, and on a much higher key. I have seen it written *je-hoi-a-chin*, or *whoooeee-che*, but to me it sounds more like *a-loie-a-hee*. It is a beautiful, soft, rich note, with a decided emphasis on the second syllable, of a liquid quality with no harsh sounds.

The third whistling note is the well-known gather call, so often heard during the fall when the flock has become scattered and the birds are trying to get together again, particularly toward night when they are gathering to go to roost. It has also been called the scatter call. It is a loud, emphatic whistle of two parts, slurred together, with an emphasis on the first. It has a human quality and to my mind is much like the whistle that I use to call my dog. It sounds to me like *quoi-hee*. To Mr. Sandys (1904)

it sounds very like *ka-loi-hee, ka-loi-hee*, especially when the old hen is doing the calling. There are many variations of it, too, *whoil-kee* representing a common one. It is an open question if the cock utters this call, although some accomplished sportsmen have claimed that he does. The writer has been a close observer of quail and would think nothing of calling young birds almost to his feet, yet he has never been able to trace this call to the old male; that is, as a rallying call to the brood. He is well aware that young males use it in replying to the mother, but he has yet to see a male of more than one season utter it.

Mr. Saunders has three records of this call, which he describes in his notes as a "repeated, slurred whistle, with usually an l-like sound between the notes, so it sounds like *coolee*." His records show ranges in pitch from A♯ to C''', or from B♭ to D♯.

In addition to these three very distinct and striking calls, there is often heard a subdued, conversational chatter while the birds are running and feeding. Doctor Judd (1905) heard, as a part of the courtship performance, "a series of queer responsive 'caterwalings,' more unbirdlike than those of the yellow-breasted chat, suggesting now the call of a cat to its kittens, now the scolding of a caged gray squirrel, now the alarm notes of a mother grouse, blended with the strident cry of the guinea hen. As a finale sometimes came a loud rasping noise, not unlike the effort of a broken-voiced whip-poor-will."

Sandys (1904) says:

A winged bird running, or an uninjured one running from under brush, preparatory to taking wing, frequently voices a musical *tick-tick-tick-a-voy*. A bird closely chased by a hawk emits a sharp cackling, expressive of extreme terror. Quite frequently a bevy just before taking wing passes round a low, purring note—presumably a warning to spring all together. When the hen is calling to scattered young, she sometimes varies the cry to an abrupt *Ko-lang*, after which she remains silent for some time. This the writer believes to be a hint to the young to cease calling—that the danger still threatens, and is prompted by her catching a glimpse of dog or man. A bevy travelling afoot keeps up what may be termed a twittering conversation, and there is a low alarm note, like a whispered imitation of the cry of a hen when a hawk appears.

Stoddard (1931) describes the above-mentioned notes more elaborately, with slightly different interpretations. He also describes several others. The crowing or caterwauling note, a rasping call that varies considerably, is uttered habitually by the cocks at all seasons. He mentions several variations of the scatter call, used to bring together scattered birds or as a morning awakening call, and says:

One of the most interesting features of the "scatter" call and its variations is that it evolves by imperceptible degrees from the shrill, piping "lost call" of the baby chicks. This starts out with the newly hatched chicks as an anxious piping *hu-hu-hu-hu-whe-whe-whe-whee-wheee* with rising inflection like *do-re-mi* of the musical scale.

Of the decoy ruse call, he writes:

One of the strangest calls of bobwhites, and a very important one from the standpoint of their preservation. is the fine cheeping *p-s-i-e-u, p-s-i-e-u, p-s-i-e-u* call, uttered by adults and their baby chicks in unison as the brood is stumbled upon by man or beast. This note, proceeding alike from both the frantic parents as they beat about in the dust trying to lure the enemy away, and by the fleeing chicks as they scatter and hide, proves most confusing to the senses, and is a real quail "sleight of hand" that is apt to leave the confused disturber in such a frame of mind that he questions whether he saw fleeing chicks, or whether it was all just a trick of the eye. Deciding it was the latter, most enemies pursue the seemingly wounded parents, which sail away on perfect wing after the enemy has been decoyed from the vicinity of the brood. Thousands of chicks must be saved yearly by this cleverly executed ruse, in which parents and chicks display perfect teamwork, even before the latter are a day old.

The alarm note is started "as soon as the chicks have scattered and hidden or the parents have failed to decoy an intruder away. It consists of a monotonous *t-o-i-l—ick*, *ick*, *ick*, *ick; t-o-i-l—ick*, *ick*, *ick*, *t-o-i-l-i-c*, *t-o-i-l-i-c*, *t-u-e-l-i-c-k;* or *t-o-i-l-i-c*, *ip*, *ip*, *ip*, *tic*, *tic*, *tic*, *t-u-e-l-i-c*, *t-u-e-l-i-c*, *ick*, *ip*, etc., uttered with machinelike regularity for a time, or as long as danger appears to be imminent."

He also mentions a distress call, a "piteous whistled *c-i-e-u*, *c-i-e-u*, uttered loudly and as rapidly as the mouth can open and close," given as old or young birds are captured; also a "cheeping" or cackling call of the developing chick, referred to as the "flicker call." Then there is the "battle cry," of the unmated cocks, a harsh, screaming note, uttered in flight; the food call, "a soft, clucking *cu*, *cu*, *cu*, *cu*, and a variety of soft conversational notes."

Fall.—When fall comes the bobwhite becomes a quail. Its habits change entirely, as it forsakes the haunts of man and becomes a wild bird. It is no longer a sociable and trusting friend of human beings, so it resorts to the fields and woods, where it can find shelter in the brushy tangles. It travels now in coveys made up of family parties or in larger flocks of more than one family.

Quail are not supposed to be migratory, in the usual sense of the word, and in many sections, New England, for instance, I believe that they are practically sedentary throughout the year. In some sections, however, they seem to perform short migrations to better feeding grounds, or perhaps to escape adverse winter conditions. Audubon (1840) writes:

This species performs occasional migrations from the north-west to the south-east, usually in the beginning of October, and somewhat in the manner of the Wild Turkey. For a few weeks at this season, the north-western shores of the Ohio are covered with flocks of Partridges. They ramble through the woods along the margin of the stream, and generally fly across towards evening. Like the Turkeys, many of the weaker Partridges often fall into the water, while thus attempting to cross, and generally perish; for although they swim surprisingly, they have not muscular power sufficient to keep up a protracted struggle, although, when they have fallen within a few yards of the shore, they easily escape being drowned. As soon as the Partridges have crossed the principal streams in their way, they disperse in flocks over the country, and return to their ordinary mode of life.

This habit is also mentioned by Amos W. Butler (1898), who says that in Indiana they desert the uplands in fall and congregate in large numbers in the Ohio River bottoms; many attempt to cross the river into Kentucky; some perish in the attempt and others reach the farther shore in an exhausted condition. H. D. Minot (1877) says:

> In Delaware and Maryland, however, coveys of Quail often appear, who are distinctively called by the sportsmen there "runners." On the western side of the Chesapeake, an old sportsman assured me that covey after covey passed through the country, where food and shelter were abundant, crossing the peninsula on foot, but often perishing by the wholesale in attempting to pass the wider inlets, and he added in proof of this that he had taken as many as forty at a time from the middle of the river near his house.

But everywhere quail become very restless in fall and are much given to erratic wandering from no apparent cause. They are less crazy in this respect than ruffed grouse; I have never known them to fly against buildings and be killed; but I have frequently seen them in my yard and garden in the center of the city. Mr. Butler (1898) says that "they are found in trees and among the shrubbery in gardens, in outbuildings, and among lumber piles. I have seen them in the cellar window-boxes and over the transoms of the front doors of the houses." These wanderings may be due to a latent migratory instinct.

Game.—Everything taken into consideration, the quail, partridge, or bobwhite is undoubtedly the most universally popular of all North American game birds, in spite of the fact that many sportsmen consider the ruffed grouse the prince of game birds. The sophisticated grouse may be the more difficult bird to bag, but the quail, with its southern subspecies, has a much wider distribution, nearer to the haunts of man, is generally more numerous and more prolific, lies better to the dog, flies swiftly enough to make good marksmanship necessary, and is an equally delicious morsel for the table.

One who has never tried it can hardly appreciate the joy and the thrills of a day in the field, with a congenial companion and a brace of well-trained bird dogs, in pursuit of this wonderful game bird. The keen, sparkling October air and the vigorous exercise stimulate both body and mind. The tired business man breathes more freely as he starts out from the old farmhouse across the fields for his holiday with the birds. On a frosty morning, when the grass and herbage are sparkling white with hoarfrost, it is well not to start too early, as quail are not early risers and do not like to get their feet and plumage wet. But when the sun is well up it is time to look for them, for they may be traveling along some brushy old fence toward their favorite buckwheat stubble, one of the best places to find them. When you reach the field where the birds are expected to be found, the most interesting part of the sport begins; the intelligent dogs have learned to quarter the ground thoroughly and hunt in every likely spot where their bird sense leads them; excitement becomes intense, as they show by their careful movements that they have scented game; and, finally, the sudden stop and the rigid pose, with nose pointed toward the birds, bring the climax, as the sportsmen step up and the covey bursts into the air with a whir of wings. A good shot may bag two or even three birds on the first rise; I have seen men that boasted of stopping as many as five with an automatic repeater, but I have never seen one do it; and I have seen many clean misses. The rest of the covey have flown straight to the nearest cover, perhaps scattered in several directions, some into a patch of scrub oaks on a hillside, some into the tangled underbrush in a swampy hollow, and others into the nearest woodlot. The men should mark them down, but had better leave them for a while until they begin to run about and leave a little scent; otherwise they will be very hard to find. Picking up these scattered singles is hard enough at best; it requires good work on the part of the dogs and gives the hunter many difficult shots in unexpected places. The man that can put two quail in his pocket for every four shells fired is a good shot.

Perhaps the birds have not been found in the buckwheat stubble. Each covey has several feeding places and it is necessary to cover considerable ground, hunting the wheat, rye, oat, and corn stubbles, especially if overgrown with ragweed or other weeds, as well as any other old neglected fields and weed patches where the birds can find food and shelter. Sometimes the dogs will show signs of game in a likely spot but fail to find the birds; quail often make short flights from one field to another, thus breaking the scent. Sometimes a flushed covey will be marked down very carefully in a fairly open field and be immediately followed up; but a careful search by experienced men and good dogs will fail to reveal the presence of a single bird. This has caused much controversy as to the power of quail to withhold their scent. The explanation probably is that the rapid passage through the air dissipates most of the scent from the plumage; the birds, being frightened, crouch low on the ground with feathers closely pressed against the body, shutting in body odors; and as they have not run any there is no foot scent. It has often happened that, in a later search over the same ground, after the birds have begun to run about, they have been readily found. There has been no willful or even conscious withholding of scent.

For about four hours during the middle of the day, quail retire from their feeding grounds for their noonday rest. The hunters may as well do likewise, until the birds come out to feed again about two hours before sunset. The hours of waning daylight often furnish some of the best and most interesting shooting; the scattered covey is anxious to get together before roosting time; and the hunters get the final thrills of the day as they hear the sweet, gentle gather call, *quoi-hee*, *quoi-hee*, from a distant patch of scrub oaks, an answering call from the brier patch in the swale, and another from the edge of the near-by woods. They are content to call it a day and leave the gentle birds to settle down for the night.

Enemies.—Quail have numerous enemies, furred, feathered, and scaled, but fortunately they are such persistent and prolific breeders that they can stand the strain from natural enemies if man will give them half a chance.

Stray cats, or domestic cats run wild, are doubtless the most destructive enemies of quail. They catch and devour enormous numbers of both young and old birds, as they hunt them day and night. Mr. Forbush (1927) gives some striking illustrations of this and speaks of one big cat that is said to have killed more than 200 bobwhites. Dogs that are allowed to run loose and hunt independently kill a great many old and young birds. Stray cats in the woods and fields should be shot on sight. Domestic cats and dogs should be restrained during the nesting season.

In Jamaica the mongoose is said to have virtually exterminated the introduced quail. Foxes, minks, and weasels kill some birds, but they probably find rabbits easier to catch and more to their liking. Raccoons, opossums, skunks, and rats destroy a great many eggs.

Among bird enemies the crow is one of the worst. Crows are very clever in hunting up nests and destroy a great many eggs; they have even been known to kill the adult birds in winter. Crows, in my opinion, should be shot whenever possible, for they can not be much reduced at best. Cooper's hawk is probably the worst of the hawks. The goshawk and the sharp-shinned hawk are almost as bad. Red-tailed hawks have been known to kill quail, but they are too slow to catch very many, and they are useful as rodent destroyers. Great horned and other owls must be reckoned with, but the former is very fond of skunks, and all the owls keep the destructive rodents in check. Quail have learned that brier patches and thick tangles offer good protection against their enemies in the air.

Any of the larger snakes will eat the eggs and probably destroy a great many, but here again we must give them credit for living largely on the rodent enemies of the bobwhite. Major Bendire (1892) speaks of a large rattlesnake, killed in Texas, that had swallowed five adult quail at one meal, and another that had taken four bobwhites and a scaled quail.

In his chapter on mortality, Stoddard (1931) states that of 602 nests studied about "36 per cent were more or less successful and about 64 per cent unsuccessful." The failures were due to nest desertion, destruction by natural enemies, destruction by the elements, rains, floods, or droughts, and disturbance by human beings, by farm work, or by poultry and cattle. Among the destroyers of eggs he mentions, in addition to the enemies named above, blue jays, turkeys, and red ants; the ants enter the egg as soon as the membrane is punctured by the emerging chick, which is literally eaten alive; out of 278 nests studied by Louis Campbell in 1928, 34 were taken over by ants.

After hatching, young quail are preyed upon by most of the more active enemies named above, to which must be added turkeys, guinea fowl, pheasants, and shrikes. The chief winged enemies of the older young and adults are Cooper's and sharp-shinned hawks, and in the North the goshawk. The Buteos are mainly, or wholly, beneficial. Mr. Stoddard (1931) exempts the sparrow hawk from blame and says: "In several instances individuals took up quarters temporarily on the fence posts of propagating enclosures and made forays against the large grasshoppers on the ground beneath, without harming the quail chicks in the least." In favor of the marsh hawk, he writes:

In view of the fact that not more than 4 quail were discovered in approximately 1,100 pellets, marsh hawks can hardly be accused of making any serious inroads on the number of quail in the region. On the other hand, one or more cotton rats were found in 925 of these pellets. Since cotton rats destroy the eggs of quail, the marsh hawk is probably the best benefactor the quail has in the area, for it is actively engaged in reducing the numbers of these rodents. Remains of at least 14 snakes, most of which were colubrines, were discovered. These also are probably eaters of quail eggs.

Diseases.—The chapters on parasites and diseases, in Stoddard's (1931) report, were contributed by Dr. Eloise B. Cram, Myrna F. Jones, and Ena A. Allen, of the Bureau of Animal Industry. They are well worth careful study, but are too long (110 pages) and too technical for any adequate presentation here. Suffice it to say that bobwhites are attacked by many of the same parasites and suffer from many of the same diseases as ruffed grouse. Among the Protozoa the most important are those which cause malaria, coccidiosis, and blackhead. Nematodes, or roundworms, were found "in a high percentage of the birds examined"; 16 species were identified, and their life histories explained. In the intestines five species of tapeworms were found and similarly described. As external parasites, lice, ticks, mites, and fleas are mentioned. Among the nonparasitic diseases the following are fully described: Foot disease, bird pox, dry gangrene, chicken pox, "nutritional roup," aspergillosis, "quail disease," and tularemia. This brief summary and other references to Stoddard's (1931) work give a very inadequate idea of the wealth of material that his exhaustive report contains; it must be read to be appreciated; some of the interesting chapters can not even be summarized here.

Winter.—In the southern portions of their range, where quail enjoy open winters, their habits and haunts are about the same as during fall; but in the northern regions of ice and snow they have a hard struggle for existence and many perish from hunger and cold in severe winters. Quail have been known to dive into soft snowdrifts for protection from severe cold; Sandys (1904) says he has caught them in such situations. More often, at the approach of a snowstorm, they huddle together in some sheltered spot and let the snow cover them. This gives them good protection from the wind and cold; but if the snow turns to rain, followed by a severe freeze the birds are imprisoned and often perish from hunger before they can escape. Birds seldom freeze to death, if they can get plenty of food, but cold combined with hunger they can not stand. Mr. Forbush (1927) tells an interesting story of a man who had been feeding a covey of quail; for 10 days after a heavy snowstorm, followed by a thaw and freeze, they failed to come to their usual feeding place; believing them to be imprisoned under the snow he went to the place where they were accustomed to sleep and broke the crust; the next day they came to feed and a search showed that they had found the place where he had broken the crust for them.

Quail often find more or less open situations where they get some shelter, under logs or fallen trees, under thick evergreens, in tangles of briers, in brush piles, or under banks with southern exposure; in such places they find bare ground and can pick up some food, as well as the gravel or grit that they need. They avoid open places and do not like to travel on snow, where they are so conspicuous; but they have to go out to forage for food, such as the seeds of weeds, projecting above the snow, rose hips, dried berries, seeds of sumac, bayberries, and other plants. When hard pressed they often visit the barnyard to feed with the poultry. Farmers, sportsmen, boy scouts, and many other persons make a practice of feeding quail regularly in winter. They should have a shelter, made of brush, evergreen boughs, or corn stalks, open at both ends so that the birds can escape at either end. The ground under this should be kept bare and well supplied with almost any kind of grain and plenty of grit. Quail will come regularly to such places and the lives of many will be saved.

DISTRIBUTION

Range.—Chiefly the Eastern United States, ranging west to eastern Texas, eastern Colorado, and the Dakotas. The great interest shown by sportsmen in the bobwhite has resulted in introductions over the entire country. Though many of these experiments have resulted in failure, others have been notably successful, and in some regions the introduced birds have spread out and met those that are indigenous, thus causing an actual extension of range. The problem is further complicated by the fact that in addition to true *virginianus*, introductions of so-called Mexican quail (*Colinus v. texanus*) have been made in several States. As these have readily crossed with the native birds, and as individuals show a great deal of variation, it is frequently difficult to determine the natural limits of the different races. The following summary, however, presents a reasonably accurate picture of the natural range of the species:

North to southeastern Wyoming (Horse Shoe Creek); South Dakota (Faulkton); North Dakota (Bartlett and Larimore); Wisconsin (Danbury and Menominee); Michigan (Douglas Lake and Alcona County); southern Ontario (Mount Forest, Listowel, Toronto, and Port Hope); Vermont (Londonderry); and Maine (West Gardiner). **East** to Maine (West Gardiner and West Fryeburg); rarely eastern New Hampshire (Hampton); Massachusetts (Gloucester, Boston, and Cape Cod); Rhode Island (Newport); New York (Shelter Island and Roslyn); New Jersey (Red Bank, Vineland, and Sea Isle City); Delaware (Lincoln); Virginia (Belle Haven, Eastville, Cape Charles, Norfolk, and Dismal Swamp); North Carolina (Currituck Sound and Raleigh); South Carolina (Waverly Mills, St. Helena Island, and Frogmore); Georgia (Savannah, Riceboro, Jekyl Island, and Okefenokee Swamp); and southern Florida (Miami). **South** to southern Florida (Coconut Grove, Indian Key, and Key West); Alabama (Bon Secour and Mobile); Mississippi (Biloxi and Bay St. Louis); Louisiana (New Orleans, Houma, Abbeville, Mermentau, and Iowa Station); southeastern Texas (Galveston, Corpus Christi, and Brownsville); Tamaulipas (Tampico, Altamira, and Victoria); and southern Nuevo Leon (Mier y Noriega). **West** to Nuevo Leon (Mier y Noriega and Montemorelos); Coahuila (Sabinas); western Texas (Langtry, Lozier, Fort Stockton, and Pecos); probably southeastern New Mexico (Carlsbad); eastern Colorado (Monon, Beloit, Yuma, and Crook); and southeastern Wyoming (Uva and Horse Shoe Creek).

Introductions.—Introductions have been made in Colorado (Upper Arkansas Valley, Wet Mountains, Pueblo, Denver, Estes Park, Loveland, Fort Collins, Greeley, Saguache, and Grand Junction); Utah (Salt Lake Valley); Montana (Sappington, Anaconda, Flathead Lake, and Kalispell); Idaho (Boise, Nampa, Coeur d'Alene, and Rathdrum); Oregon (Lake Alvord, Snake River Valley, Pendleton, Scio, Dayton, and Portland); Washington (Starbuck, Whidbey Island, Walla Walla, Cheney, Spokane, Osoyoos Lake, Olympia, Tacoma, Seattle, and Blaine); and California (San Felipe and Gilroy).

This species also has been transplanted to southwestern Canada (British Columbia and Manitoba); the West Indies (Cuba, Jamaica, New Providence, Haiti, Porto Rico, Barbados, St. Kitts, Antigua, St. Croix, and Guadeloupe); New Zealand; England (Norfolk); Sweden; Germany (Hanover); France; and China (Kashing). The introductions in the West Indies (in some instances the Cuban form, *C. v. cubanensis*) have been more or less successful, but so far as known all European and Asiatic attempts to acclimatize this bird have been total failures.

The foregoing distribution covers the range of the entire species, which is subdivided into three subspecies. *C. v. floridanus* occupies the whole of Florida from the vicinity of Gainesville, Palatka, and Tarpon Springs southward. *C. v. texanus* occupies central and southern Texas, from southeastern New Mexico to northeastern Mexico. Typical *virginianus* occupies the rest of the range in Eastern and Central United States.

Egg dates.—Massachusetts, Rhode Island, and Connecticut: 22 records, May 20 to October 10; 11 records, June 5 to July 28. New York, New Jersey, and Pennsylvania: 18 records, May 21 to August 31; 9 records, June 7 to July 16. Virginia, Kentucky, and Kansas: 18 records, May 14 to September; 9 records, June 2 to July 23. South Carolina and Georgia: 15 records, April 24 to September 16; 8 records, May 21 to June 18. Indiana, Illinois, and Iowa: 31 records, April 28 to October 16; 16 records, June 2 to July 13.

Florida (*floridanus*): 46 records, April 19 to July 26; 23 records, May 13 to June 8.

Texas (*texanus*): 50 records, March 18 to August 19; 25 records, May 11 to June 2.

COLINUS VIRGINIANUS FLORIDANUS (Coues)

FLORIDA BOBWHITE

HABITS

The Florida bird is merely a small, dark variety of the common, northern bobwhite. It is confined entirely to the peninsula of Florida, where it is universally common and generally distributed in all the drier portions of the State, except in the extreme north and the extreme south; in the north it intergrades with the northern form and in the south with the still smaller and darker Cuban form.

If due allowance be made for the different environment in which they live, the Florida birds will be found in similar localities to those chosen by their northern relatives and their habits are essentially the same. Nearly everything I have written about the northern birds applies equally well to the southern. They are equally fond of the society of human beings, where they probably feel more secure from their wild enemies and perhaps find more food. Major Bendire (1892) quotes Doctor Ralph, as saying that "localities they like best are open woods grown up with saw palmettos or low bushes, or fields with woods near them, and they are particularly fond of slovenly cultivated grounds that have bushes and weeds growing thickly along their borders." I have frequently seen them in small villages, in gardens, and about houses; but I have more often found them in the open flat pine woods where there are extensive patches of saw palmettos; these thick clumps of low palmettos are often almost impenetrable and afford them excellent protection from their enemies.

Nesting.—Donald J. Nicholson has sent me the following notes on this subject:

These quail begin to pair off by February, and by March most of them have chosen their mates, but still some will be found in coveys into March. In March the bobwhites begin to think of domestic duties, and the woods and cultivated fields resound with their cheery *bob-bob-white*. This continues until late June, when the calls become very much less frequent. A few pairs breed as early as March or early April, but the height of the nesting season is late May or early June. They also nest up into August, and one nest was found in December by a hunter, which is quite unusual. From 9 to 16 eggs are laid, which are deposited in arched nests of dead grass, in old fields, in pine woods, or on the edges of grassy ponds. The female does not commence to sit until the last egg is deposited. How long it takes to incubate is unknown to me. The birds are quite suspicious if a nest is found and will generally leave it, but if not touched sometimes continue their duties. When a nest is found the bird sits until almost trodden upon, then either flies directly off with great speed, or more often flits off and cackles excitedly, running about close by and feigning lameness.

Eggs.—The number of eggs laid averages less, but the eggs are quite indistinguishable from those of the northern bird. They are not even appreciably smaller. The measurements of 51 eggs average 30.6 by 23.8 millimeters; the eggs showing the four extremes measure **32.6** by 24.1, 31.5 by **24.5**, and **28.9** by **22.8** millimeters.

Young.—Mr. Nicholson says in his notes:

Both parents attend the young and are very courageous in the defense of the little fellows, running about close to the intruder, jumping up to attract attention, and uttering strange notes. When the hiding spot of the young is discovered they scurry for other shelter, cheeping as they run, and if pressed too closely will take flight, when very young. I once saw a bobwhite make a mad rush at the wheels of a passing auto, and after the machine passed the mate also came out. Undoubtedly there were young near by.

Behavior.—In general behavior and habits the Florida bird does not differ materially from its northern relative. Its food does not seem to have been separately studied and is included in reports on the species as a whole. It feeds on such of the seeds, fruits, and insects, mentioned in such reports, as are to be found within its range. Its voice is essentially the same. Once exceedingly abundant in the open country in Florida, it has been greatly reduced in numbers and mainly by its human enemies. It has long been a favorite with sportsmen; its long open season, with little or no protection in many places, has offered them attractive opportunities long after the season in the North has closed. A steadily increasing number of winter visitors have been tempted to spend part of their idle time in hunting, where the shooting is mostly open and the birds easily killed. The quail have shown the effect of this slaughter. They have also suffered greatly from illegal trapping and netting and from shooting out of season, mostly by poor whites and negroes. Fortunately for the future of the Florida bobwhite the laws are now better enforced and there are many large areas where no shooting is allowed and where public sentiment is protecting this and all other desirable birds.

COLINUS VIRGINIANUS TEXANUS (Lawrence)

TEXAS BOBWHITE

HABITS

George N. Lawrence (1853) first called attention to the characters that separate the bobwhite of Texas from the two eastern forms. In general appearance it is decidedly grayer than either of the eastern forms. Mr. Lawrence described it more in detail, as follows:

> This somewhat resembles *O. virginianus*, but is smaller, and differs also in having the lores white, in being without the conspicuous dark markings on the back and wings, and the bright chestnut red so prevalent in the upper plumage of that species; the bill is proportionately longer and narrower, the legs more slender, and the black markings on the abdomen and breast are fully twice as broad.

The Texas bobwhite closely resembles the eastern form in behavior, habits, and haunts. George Finlay Simmons (1925) gives us the following long list of places in which it may be found:

> More or less open country, particularly in mesquite-chaparral-cactus pastures; old plantations, clearings, and cultivated fields; open, semi-arid tree-dotted, bushy pastures in farming country, particularly where such pastures are interspersed with small bodies of woodland; hay, grain, brown stubble, corn, cotton, and open weedy fields; clearings and brushy edges of woodlands; thickets, brush, and briar patches along edges of meadows and creeks; weedy roadsides and fencerows; wooded hillsides; cedar brakes; cultivated fields. Never in bottom woods or open praries. In fall and winter, among dead stalks in cotton and corn fields.

Nesting.—Major Bendire (1892) says:

The favorite nesting site of the Texan Bob White is a bunch of sedge grass. A slight cavity is made in the center, this is lined with a few straws and arched over with similar material. Sometimes a covered way or tunnel leads to the entrance of the nest. Occasionally a nest is placed under a bush and not covered or arched. Two broods are usually raised in a season, and even three at times.

Mr. Simmons (1925) mentions a number of other nesting sites as follows:

Nest on ground along fencerows, in small dewberry thickets, in prickly-pear beds, in brushy mesquite lands or bushy, grassy pastures, in thick tussocks or clumps of big grass, along weedy fence lines or roadsides, or in the middle of cotton or corn fields on the rolling prairies; usually well hidden by a small bush or by weeds and prairie grasses; occasionally placed in a fencerow with overhanging vines or beside a stone wall or a log.

Eggs.—The eggs of the Texas bobwhite are indistinguishable from those of the eastern bird. From 10 to 15 eggs constitute the usual set, but 18 or 19 are occasionally laid, and Bendire (1892) states that H. P. Attwater once found 33 eggs in a nest. Probably this large set was the product of two females. According to Bendire (1892), "two broods are usually raised in a season, and even three at times." The measurements of 59 eggs, in the United States National Museum, average 30 by 24 millimeters; the eggs showing the four extremes measure **32** by 25, 30 by **25.5**, and **27.5** by **22** millimeters.

Plumages.—The molts and plumages correspond to those of the eastern bobwhite, but Dr. Jonathan Dwight (1900) says that "the juvenal plumage is browner than in *virginianus*."

Food.—Mrs. Florence Merriam Bailey (1928) writes:

The Bob White is of special agricultural value because it destroys a large amount of weed seed and a considerable number of insects. Half of its food is weed seed, only a fourth grain—mainly from the stubble fields—and about a tenth wild fruits. Fifteen per cent is composed of insects, including several of the most serious pests of agriculture. It feeds freely upon Colorado potato beetles and chinch bugs, and eats also grasshoppers, cucumber beetles, wireworms, billbugs, clover-leaf weevils, the Mexican cotton-boll weevils, army worms, cotton worms, cutworms, and Rocky Mountain locusts.

Behavior.—Mr. Simmons (1925) says:

Observed singly, in pairs in summer, or by threes and fours; in winter, from middle fall to early spring, in coveys or bevies of from 10 to 30, breaking up for the breeding season and reassembling as soon as it is over. Rather shy and difficult to find after the hunting season has opened. During summer days the birds seek shelter under the bushes which dot the pastures; winter days, in scattered brush heaps and tiny hollows. On spring and summer nights they generally roost in the open fields; on fall and winter nights, roost under cover, usually in lowlands. When frightened, or when preparing to sleep and keep warm, a covey arranges itself in a close-huddled circle, heads out from center; when approach is too close, the birds burst in all directions, making it very difficult to shoot more than one at a time; were they to follow each other, as most gregarious birds do, the hunter would have a better opportunity to exterminate the whole covey. Occasionally, but rarely, takes to the trees when flushed, and remains squatting close to limb and practically invisible.

Voice.—The Texas bobwhite has a range of varied calls similar to those of its eastern relative, but the notes are said to be less loud, clear, and ringing.

Enemies.—Major Bendire (1892) quotes William Lloyd, as follows:

They are very insuspicious, and their low notes, uttered while feeding, attract a good many enemies. I have seen foxes on the watch, and the Marsh Harrier perched on a clump of grass on the lookout, waiting for them to pass. But the many large rattlesnakes found here are their worst enemies. One killed in May had swallowed five of these birds at one meal; another, a female evidently caught on her nest and a half dozen of her eggs; a third, four Bob Whites and a Scaled Partridge. The young are also greatly affected and many killed by heavy rains in June and July; numbers perish then from cold and protracted wet weather. When alarmed by a Hawk sailing overhead they run under the mother for protection, as domestic chickens do.

COLINUS RIDGWAYI Brewster

MASKED BOBWHITE

HABITS

This well-marked species once inhabited a narrow strip of country in southern Arizona, where its range extended for not more than 40 or 50 miles north of the Mexican boundary. It has long since been exterminated in Arizona and has now perhaps disappeared entirely from its more extensive range in Sonora. Herbert Brown was the first to discover the presence of this quail in Arizona, and we are indebted to him for practically all we know of its former distribution and habits. He first met the bird in Sonora, hearing its note and supposing it to be our eastern bobwhite. He says of this incident (1904):

It is not easy to describe the feelings of myself and American companions when we first heard the call *bob white.* It was startling and unexpected, and that night nearly every man in camp had some reminiscence to tell of Bob White and his boyhood days. Just that simple call made many a hardy man heart-sick and homesick. It was to us Americans the one homelike thing in all Sonora, and we felt thousands of miles nearer to our dear old homes in the then far distant States.

In the spring of 1884 a pair of these quail was taken on the eastern slope of the Baboquivari Mountains, Arizona, and brought to Brown as specimens; but because of his absence they were allowed to spoil and were thrown away. Meantime he had reported to a local paper that a pair of bobwhite quail had been taken in Arizona, and the note was republished in Forest and Stream. This aroused the interest of

Robert Ridgway, to whom fragments of the two birds were sent for identification. He pronounced them *Ortyx* (*Colinus*) *graysoni*, a Mexican species.

During that same year Frank Stephens collected in Sonora, Mexico, a male masked bobwhite for William Brewster, which was described by Mr. Brewster (1885) as the type of a new species, which he named *Colinus ridgwayi*. This whole interesting history is given in more detail, with reference to several published articles on the subject, in an excellent paper by Dr. J. A. Allen (1886), to which the reader is referred.

Brown (1904) was told that "in early days they were plentiful in Ramsey's Canyon in the Huachucas, and also on the Babacomori, a valley intervening between the Huachuca and Harahaw ranges." Speaking of conditions prior to 1870, he says:

At that time the valley was heavily grassed and the Apache Indians notoriously bad, a combination that prevented the most sanguine naturalist from getting too close to the ground without taking big chances of permanently slipping under it. For many years Indians, grass, and birds have been gone. The Santa Cruz, to the south and west of the Sonoite, is wider and was more heavily brushed. Those conditions gave the birds a better chance for life and for years they held tenaciously on. Six or seven years ago I was told by a ranchman living near Calabasas, that a small bunch of Bob-white Quail had shortly before entered his barnyard and that he had killed six of them at one shot. It was a grievous thing to do, but the man did not know that he was wiping out of existence the last remnant of a native Arizona game bird. Later I heard of the remaining few having been occasionally seen, but for several years now no word has come of them.

I never found them west of the Baboquivari Mountains, and from my knowledge of the country thereabouts I am inclined to fix the eastern slope of that range as their western limit. Between that and Ramsey's Cañon, in the Huachucas, is a distance of nearly one hundred miles. Their deepest point of penetration into the Territory was probably not more than fifty miles, and that was down the Baboquivari or Altar valley.

As to the causes of the bird's disappearance, he writes:

The causes leading to the extermination of the Arizona Masked Bob-white (*Colinus ridgwayi*) are due to the overstocking of the country with cattle, supplemented by several rainless years. This combination practically stripped the country bare of vegetation. Of their range the *Colinus* occupied only certain restricted portions, and when their food and shelter had been trodden out of existence by thousands of hunger-dying stock, there was nothing left for poor little Bob-white to do but go out with them. As the conditions in Sonora were similar to those in Arizona, birds and cattle suffered in common. The Arizona Bob-white would have thriven well in an agricultural country, in brushy fence corners, tangled thickets and weed-covered fields, but such things were not to be had in their habitat. Unless a few can still be found on the upper Santa Cruz we can, in truth, bid them a final good-bye.

Nesting.—Very little is known about the nesting habits of the masked bobwhite. Brown (1904) offered a reward of $1 an egg for the first nest found for him. A nest containing six eggs was found on the mesa on the eastern side of the Baboquivari Mountains. "Unfortunately these precious things were lost through the cupidity of the finders, whose expectations ran to more eggs, but while waiting for the increase the nest was robbed of the eggs that were then in it. I was, however, notified of the find, but when I reached there I found only an empty nest, a bowl-shaped depression in a bunch of mountain grass."

Major Bendire (1892) quotes from a letter received from Otho C. Poling, relating his fruitless attempt to find a nest and reporting the finding of an egg in the oviduct of a female, which he shot on May 24, 1890.

Col. John E. Thayer has a set of seven eggs in his collection, presented to him by Miss Engel, of Cleveland, Ohio. It was collected in Sonora, Mexico, on May 4, 1903. The nest was "placed in sand under a bunch of dry grass." The female was closely observed, but as the females of this and some other Mexican species are practically indistinguishable, there may be some doubt about the identification.

Eggs.—The eggs of the masked bobwhite are indistinguishable from those of the eastern bobwhite. The single egg in the United States National Museum is pure white and unstained, as it was taken from the oviduct of a bird by Herbert Brown. This egg measures 32.5 by 25 millimeters. The egg recorded by Major Bendire, possibly the Poling egg, measured 31 by 24 millimeters; I do not know just what has become of this egg, or how Major Bendire got the measurements.

Colonel Thayer's eggs are slightly smaller; they average 29.6 by 23.4 millimeters; the largest egg measures **30.5** by **23.6** and the smallest **28.3** by **23.1** millimeters.

Plumages.—The downy young of the masked bobwhite is unknown. In the juvenal plumage, which is worn for only a short time in summer and early in fall, the sexes are alike and closely resemble the similar stage in the Texas bobwhite. Doctor Allen (1889) describes a young male, taken on October 10 and still partly in first plumage, as follows.

The top of the head is blackish, with each feather narrowly bordered with ashy brown. The hind neck, sides of the neck, and jugulum are yellowish white, with each feather barred at the tip with black. The scapulars are brownish, each feather with a rather broad whitish shaft stripe, and barred with yellowish white and black, and the wing coverts have much the same pattern, but the barring is pale cinnamon and brown. The throat is pure white, with new black feathers appearing irregularly along the sides of the chin and upper throat. Breast pale brown, with light shaft stripes and faintly barred with blackish, passing into brownish white with more distinct bars on the upper abdomen. The new feathers along the sides of the breast and flanks are chestnut, tipped with a spot of clear white, which is bordered behind with a more or less V-shaped bar of deep black. The broad yellowish white superciliary stripes extend to the nostrils.

The sexes differentiate during the fall, and young males show continuous progress toward maturity during their first winter. The juvenal plumage does not wholly disappear until December or later; there is more or less white in the black throat and more or less black barring on the breast; but the throat becomes clearer black and the breast and underparts purer "tawny" as the season advances.

Food.—Major Bendire (1892) quotes Brown on this subject, as follows:

Of three stomachs of this species examined, one contained a species of mustard seed, a few chaparral berries, and some six or eight beetles and other insects, ranging in length from a half inch down to the size of a pin head. The second was similarly provided, but contained, in lieu of mustard seed, a grasshopper fully an inch in length. These two were taken on the mesa. The third, from a bird taken in the valley, contained about 20 medium-sized red ants, several crescent-shaped seeds, and a large number of small, fleshy, green leaves.

He also says that Lieut. H. C. Benson, who secured a number of specimens in Sonora, in 1886, wrote him "that they only frequented cultivated fields there, where wheat and barley had been raised." John C. Cahoon, who collected in the same section of Sonora, found these quail abundant there; "several large coveys were seen and 8 specimens shot in one day"; he sent 10 specimens, taken February 5–8, 1887, to William Brewster (1887); they were "haunting patches of weeds in gardens and barren sand wastes, where they fed on the seeds of a plant called red-root."

Behavior.—Brown (1885) says of the habits of the masked bobwhite:

They appear to resemble very closely those of the common quail (*C. virginianus*), only slightly modified by the conditions of their environment. They utter the characteristic call, "Bob White," with bold, full notes, and perch on rocks and bushes when calling. They do not appear to be at all a mountain bird, but live on the mesa, in the valleys, and possibly in the foothills * * *. In addition to their "Bob White" they have a second call of *hoo-we*, articulated and as clean cut as their Bob White. This call of *hoo-we* they use when scattered, and more especially do they use it when separated toward nightfall. At this hour I noted that, although they occasionally called "Bob White" they never repeated the first syllable, as in the daytime they now and then attempted to do * * * I will venture to say that when frightened and scattered they are a hard bird to get. Hear one call, locate it as you may, see one fly and mark it down, and without a dog it is virtually impossible to flush it.

Griffing Bancroft wrote to me in 1928, as follows:

The masked bobwhite, *Colinus ridgwayi*, is virtually extinct. Its former breeding range was confined to a transition zone somewhat oval in shape and lying at an elevation of 3,500 to 4,000 feet and over. The western axial center is situated in Sonora, about 20 miles due south of Nogales. The city of Magdalena appears to mark the southern boundary, while the northern line lies about halfway between Tucson and Nogales. The western swing of the circle naturally is broken and irregular; nevertheless it can be traced quite clearly, as the land falls in every direction but east. As soon as desert conditions begin to intergrade with grass and oak, environments no longer seem favorable for the masked bobwhite.

The eastern limits of the bird's range are also determined by altitude. Just where the mountains begin to get too high for them it is hard to say, though it is not very far east of Nogales. I believe it is about 50 miles and feel quite confident it is not more than 100, though I have not traced that line personally as I have the other.

There have been seen within the past two years two small bands of these quail north and west of Magdalena. Except for this no hunter, sportsman, or observer with whom I have been able to establish contacts has ever seen the bird or heard of its existence. As my inquiries have been carried on for the past six years and have covered a large stretch of country and been thorough, my conclusion that the bird is almost extinct does not seem open to question. The reasons for his disappearance are not so clear. The greatest contributing cause appears to be the habit the Mexican wood cutters have of burning the hillsides in order to get better firewood. The rise in the cattle industry, in the prosperity and population of the country, and civilization in general seems to have wiped out the bobwhite and the turkeys, but not Gambel's or Mearns's quail. *Colinus ridgwayi* can not maintain itself against civilization. The reasons why it can not do so are not wholly obvious.

DISTRIBUTION

Range.—Formerly **north** to southern Arizona (Baboquivari Mountains and Huachuca Mountains). **South** to northern Sonora (Sesabe and Magdalena). The eastern and western limits determined by the extent of grassy plains at altitudes of from 3,500 to 4,000 feet. This bobwhite has now disappeared entirely from Arizona and is nearly, or quite, extinct in Sonora.

Egg dates.—Arizona: 3 records, May 4 to 12.

OREORTYX PICTA PALMERI Oberholser

MOUNTAIN QUAIL

HABITS

The common name, which is fortunately more stable than the scientific name, of this quail remains as we have always known it. But Dr. H. C. Oberholser (1923) has discovered that the bird which Douglas described as *Ortyx picta* was really the lighter-colored bird of the interior, rather than the darker bird of the humid coastal strip, to which we have always heretofore applied the name *picta*. Therefore, the paler bird of the interior must take the name *Oreortyx picta picta*, plumed quail. This relegates the name *plumifera* to synonymy and leaves the bird of the humid coast belt without a name, for which Doctor Oberholser has proposed the name *Oreortyx picta palmeri*, in honor of Dr. T. S. Palmer, who had reached the same conclusion some years ago. The range of this race is restricted to the humid Transition Zone of the Pacific coast, from southwestern Washington south to Monterey County, Calif.

J. H. Bowles (Dawson and Bowles, 1909) says of the status of this quail in the State of Washington:

The Mountain Quail, as it is generally called, and its close relative, the Plumed Quail, are neither of them native to Washington, several crates of living birds having been imported from California between the years 1880 and 1890. So kindly did they take to the conditions they found here, that, at the end of a long season of protection imposed by law, they fairly swarmed in suitable localities. But what a change a few years of persecution have wrought! Where formerly a dozen large coveys could be found within a small area, only an occasional solitary bird, rarely a pair, is now left of this gem among our upland birds. The entire blame cannot be laid at the door of the sportsman, altho modern rapid-fire guns have played their part. By far the worst havoc has been wrought by the treacherous nets, snares, and traps of all descriptions, which unscrupulous persons set in defiance of law. Too lazy to hunt, these human vermin catch the poor birds alive and wring their necks. Before close association with mankind had proved so fatal a mistake, these partridges were among the tamest and most confiding of birds. Utterly unsuspicious of danger, they would run into the yard and eat with the farmer's hens, paying little attention to any passing human being. When flushed from their haunts in the woods, the whole covey would merely fly into the nearest bushes and trees. Now all is changed, for the "fittest" survivors have inherited the knowledge that mankind is their deadliest enemy.

He described its haunts as follows:

Somewhat inclined to high altitudes, as their name implies, the favorite localities for these birds are the large areas in our forests that have been cleared of standing timber. In the course of a year or two these "burns," as they are called, become over-grown with huckle-berry, salal, and occasionally a dense growth of the wild sweet pea. Here is food in abundance at all seasons; for in summer the decayed mold of the fallen trees contains grubs and insects galore as change from fall and winter diet of berries and seeds. Telltale hollows in the soft dry earth, sprinkled with a feather or two, speak of luxurious dust-baths, and a net-work of three-toed tracks in a neighboring wood-road shows where the band has taken its morning constitutional.

Nesting.—Major Bendire (1892) gives the following account of the nesting habits of this subspecies:

Nidification commences about the middle of May, and ordinarily but one brood is raised. The nest is placed on the ground, alongside or under an old log, or on side hills under thick bushes and clumps of ferns, occasionally along the edges of clearings, grain fields, or meadows. A nest found May 27, 1877, near Coquille, Oregon, containing six fresh eggs, was well concealed under a bunch of tall ferns, in a tract of timber killed by a forest fire. Another, taken in Ukiah Valley, Mendocino County, California, June 2, 1883, by Mr. C. Purdy, contained twelve fresh eggs. This nest was found under a bush of poison oak among a lot of dry leaves on a steep hillside. The average number of eggs laid by this Partridge is about ten, most of the sets containing from eight to twelve. An occasional nest contains as many as sixteen, but such large sets are rare.

Eggs.—The eggs of the mountain quail are indistinguishable from those of the plumed quail. The measurements of 61 eggs average 34.1 by 26.7 millimeters; the eggs showing the four extremes measure **35.6** by 27.7, 33.5 by **27.9**, **30.7** by 25.2, and 34 by **25** millimeters.

The plumages, food, and general habits of this quail are so much like those of the plumed quail that I shall not repeat them here. Louise Kellogg (1916) relates the following incident, which illustrates the sagacity of the weasel rather than that of the quail:

On July 8, on north fork of Coffee Creek, the writer caught sight of a weasel in pursuit of a mountain quail. The bird was clucking in a distressed manner and evidently leading the enemy away from where her chicks were. When the weasel got her to a safe distance he ran back, jumped over a log, and was seen to make off with a small victim in his mouth. The whole episode did not occupy two minutes and occurred in a clearing in broad daylight.

DISTRIBUTION

Range.—The Western United States and Lower California; successfully introduced at points in Washington, British Columbia, Montana, and Idaho.

The range of the mountain quail extends **north** to northwestern Oregon (Astoria); southern Washington (Kalama); and northeastern Oregon (Ironside). **East** to northeastern Oregon (Ironside and Vale); western Nevada (Big Creek, Granite Creek, Truckee, Carson City, and Mount Magruder); southeastern California (Willow Creek, Coso, Little Owens Lake, and Thomas Mountain); and Lower California (Laguna Hanson and Mision San Pedro Martir). **South** to Lower California (Mision San Pedro Martir and Valladares). **West** to Lower California (Valladares, La Grulla, Las Cruces, and Los Pozos); western California (San Diego, Frazier Mountain, Mansfield, Big Creek, Monterey, Camp Meeker, Cahto, Stuarts Fork, and Fort Jones); and western Oregon (Port Orford, Coquille, Newport, Netarts, Tillamook, Batterson, and Astoria). The species has been reported as occurring south to Cape St. Lucas, Lower California, but the record is not considered satisfactory.

The range as outlined is for the entire species. *Oreortyx picta palmeri* is confined to the humid northwestern coast region south to the coast ranges of Monterey County, Calif. (Big Creek). *Oreortyx p. picta* occupies the arid and semiarid regions east of the coast ranges and south to southern San Diego County (Campo, Mountain Spring, Cuyamaca, and Volcan Mountains). A third race, *O. p.*

confinis, has been described from the San Pedro Martir Mountains, Lower California.

Migration.—A slight vertical migration down the slopes of the mountains is performed in the autumn. The journey is made almost entirely on foot, the birds following railroads, wagon roads, and trails, sometimes passing close to human dwellings. The start for lower altitudes is made about the first of September, and by October 1 the flocks have entirely abandoned those parts of the summer range above 5,000 feet. The movement has been known to start as early as August 28 (Webber Lake, Calif.). The return trip is made early in spring, but exact dates are not available.

These birds have been introduced with fair success at points in Washington (Whidbey Island, San Juan Island, and others); British Columbia (Vancouver Island and Fraser Valley); Idaho (Nampa, Silver City, and Shoshone); and southeastern and western Montana. Attempts to acclimatize them in the Eastern States and in New Zealand have been failures.

Egg dates.—California (*palmeri*): 8 records, March 10 to June 20. Oregon: 24 records, May 1 to June 15; 12 records, May 29 to June 6. Washington and British Columbia: 4 records, June 10 to 21.

California (*picta*): 50 records, April 7 to August 15; 25 records, May 25 to June 20.

Lower California (*confinis*): 5 records, March 29 to May 28.

OREORTYX PICTA PICTA (Douglas)

PLUMED QUAIL

HABITS

As explained under the preceding subspecies, the scientific names of the two California forms have been changed, but fortunately the English names have shown more stability and will probably stand as they always have. The plumed quail of the semiarid interior ranges was formerly called *Oreortyx picta plumifera*, but will now stand on our new check list as given above. This is the most widely distributed and best known form in California, where it is commonly known as the mountain quail and is so named by Grinnell, Bryant, and Storer (1918).

It is the largest and handsomest of the North American quail, is frequently called "partridge," and has been thought by some to resemble the European partridge, although it is decidedly smaller. It is a shy, retiring species, more often heard than seen. I have hunted for it, where it was common and where it could frequently be heard calling, and been favored with only an occasional glimpse of one walking stealthily away among the underbrush. It prefers to steal away quietly rather than show itself by flying. All the earlier writers speak of it as uncommon or comparatively rare; it was doubtless often overlooked because of its secretive habits. It is even more of a mountain bird than the preceding, ranging up to 10,000 feet in summer. W. Leon Dawson (1923) says of its haunts:

Save in the extreme northwestern and southeastern portions of its range, the Mountain Quail is to be found in summertime somewhere between 2,000 or 3,000 and 9,000 feet elevation, according to local conditions of cover. It inhabits the pine chaparral of the lesser and coastal ranges, but its preference is for mixed cover, a scattering congeries of buck-brush, wild currant, service berry, *Symphoricarpus*, or what not, with a few overshadowing oaks or pines. In the northwestern portion of its range the bird comes down nearly to sea-level and accepts dense cover. In the southeastern portion, namely, on the eastern slopes of the desert ranges overlooking the Colorado Desert, the Mountain Quail, according to Mr. Frank Stephens, ventures down and nests at an altitude of only 500 feet. It is closely dependent here upon certain mountain springs, which it visits in common with *L. c. vallicola* and *L. gambeli*. Under certain conditions, therefore, its breeding range overlaps that of the Valley Quail. There are several instances on record of nests containing eggs of both species, and at least one hybrid has been found, conjectured to be between *O. p. confinis* and *L. c. californica*.

Courtship.—On this subject I can find the brief statement from Bendire (1892) that "the mating season begins in the latter part of March and the beginning of April, according to latitude and altitude. The call note of the male is a clear whistle, like *whu-ié-whu-ié*, usually uttered from an old stump, the top of a rock, or a bush."

Grinnell and Storer (1924) write:

With the coming of the warm days of late spring, and on into early summer, the males perch on fallen logs, open spaces on the ground, or even on branches of black oaks, and announce their amatory feelings by giving utterance to their loud calls with such force and vigor that these resound through the forests for a half-mile or more, commanding the attention of all within hearing. One type of call consists of but a single note, *quee-ark*, and this is repeated at rather long and irregular intervals. One bird timed by the watch, June 3, 1915, gave his calls at intervals of 7, 6, 8, 5, 8, 6, 7, 5, 7, 9, and 9 seconds, respectively, and continued at about the same rate for a long time afterward. This intermittent utterance lends to the call a distinctiveness and attractiveness which would be lost if it were given in quicker time.

Nesting.—Of the nest Bendire (1892) says:

The nest, simply a slight depression in the ground scratched out by the bird, and lined perhaps with a few dry leaves, pine needles, grasses, and usually a few feathers lost by the hen while incubating, is sometimes placed alongside an old log, at other times under low bushes or tufts of weeds, ferns, and, when nesting in the vicinity of a logging camp, a favorite site is under the fallen tops of pine trees that have been left by wood-choppers, the boughs of which afford excellent cover for the nest.

Mr. L. Belding found a deserted nest of this species in a cavity of the trunk of a standing tree near Big Trees, California, but in this locality they nest oftener in thickets of the rock rose or the tar-weed, and according to his observations they do not desert their nests for slight cause, like the Bob White or the California Quail.

H. W. Carriger writes to me:

On June 20, 1914, I found four nests of the Mountain Quail. What struck me as unusual was how close the birds remain on the nest; one was under some brush that a camper had cut the previous year, the leaves were all off and the bunch was a mere handful and, though I stood looking down at the bird, she paid no attention to me and did not get off till I had hit the brush pile several times with a stick. The one with 19 eggs was also a very close sitting bird; I flushed a male and then began a search about, in a radius of about 15 feet from where he flew up; I used a stick and beat all the surrounding bushes and vines (mountain misery), but could not flush the female, which I figured must be in the vicinity; after covering all the near-by ground I sat down near a large tree and accidentally saw a bird move its eye and there, about 6 feet away in a patch of "misery," was a sitting bird; she allowed me to practically touch her before flying. I am sure that I passed this tree and beat this bush before, but not a movement from the bird. The third nest was located without seeing the bird and was at the base of a small tree. The fourth had but one egg and was also at the base of a very large tree.

Grinnell, Bryant, and Storer (1918) mention two cases where nests of this species have been found containing eggs of the California, or valley, quail.

Chester Barlow (1899) describes as follows some nests found in Eldorado County, at an elevation of 3,500 feet in the pine belt:

Three nests of the Plumed Quail were found by us, all built in the tar-weed or "mountain misery" (*Chamaebatia foliolosa*), and all near paths or roads. The one shown in the illustration was built at the foot of a large cedar tree, and was nicely concealed and shaded by the foliage of the weeds. The nesting cavity was about six inches across and three inches deep, lined with feathers from the parent bird. It held ten eggs, in which incubation was well advanced. Several times the bird was flushed in order that we might observe the nest, but she was persistent and always returned. Another nest containing 11 incubated eggs was found on the same day, placed amongst the tar-weed in the shade of large cedars. This nesting cavity was about six inches in depth, and composed of dry leaves from the tar-weed and lined with feathers. From the nests observed it seems certain that the Plumed Quail makes a nest of its own, for the one last mentioned was substantial enough to bring home.

Charles R. Keyes (1905) found six nests of the plumed quail in the heavily timbered portion of the Sierra Nevada at an altitude of 3,000 feet. One nest was "protected under the outer edge of a mass of deer brush (*Ceanothus velutinus*)"; another was "neatly tucked away along the northwest side of a small boulder and partly concealed by dwarf manzanita"; still another "was in rather an open situation under a Murray pine and five feet away from the trunk"; it was "composed entirely of pine needles" and was "partially concealed by low sprigs of manzanita." Of the fifth nest he says:

The fifth nest was found on June 20 by tramping through deer brush near the place where a male had been heard calling for several days. It was the best concealed of any, being under quite a thick mass of ceanothus, though I hardly think I should have overlooked it, even though the female had not flushed with a great whirr of wings when I was three or four feet away from her. The nest was quite well constructed of coarse dry grass, a few small twigs, and many breast feathers from the bird. The measurements were the same as those of the last nest described and the eggs were twenty-two in number, laid in two layers, the lower of the nineteen eggs with three on top in the center.

Eggs.—The plumed quail does not lay so large sets of eggs as the valley quail. Probably the average is not more than 10 or 12, but as many as 19 or even 22 have been found in a nest, probably the product of two females. In shape they vary from ovate to subpyriform; some eggs are quite pointed; the shell is smooth and somewhat glossy. The color varies from pale cream to a reddish buff, or from "pinkish buff" to "pale ochraceous-salmon." They are entirely unspotted. The measurements of 61 eggs average 34.7 by 27 millimeters; the eggs showing the four extremes measure **38** by 28, 35 by **29**, and **33** by **25** millimeters.

Young.—Bendire (1892) says that incubation lasts about 21 days and that "in the higher mountains but a single brood is raised; but in the lower foothills they rear two broods occasionally, the male caring for the first one while the female is busy hatching the second." Probably both sexes share the duties of incubation. Mrs. Irene G. Wheelock (1904) gives the following account of the hatching process:

I stole back alone for a last peep at them, and two had pipped the shells while a third was cuddled down in the split halves of his erstwhile covering. The distress of the mother was pitiful, and I had not the heart to torture the beautiful creature needlessly; so going off a little way, I lay down flat along the "misery," regardless of the discomfort, and awaited developments. Before I could focus my glasses she was on the nest, her anxious little eyes still regarding me suspiciously. In less time than it takes to tell it, the two were out and the mother cuddled them in her fluffed-out feathers. This was too interesting to be left. Even at the risk of being too late to reach my destination, I must see the outcome. Two hours later every egg had hatched and a row of tiny heads poked out from beneath the mother's breast. I started toward her and she flew almost into my face, so closely did she pass me. Then by many wiles she tried in vain to coax me to go another way. I was curious and therefore merciless. Moreover, I had come all the way from the East for just such hours as this. But once more a surprise awaited me. There was the nest, there were the broken shells; but where were the young partridges? Only one of all that ten could I find. For so closely did they blend in coloring with the shadows on the pine needles under the leaves of the "misery" that although I knew they were there, and dared not step for fear of crushing them, I was not sharp enough to discover them.

Bendire (1892) writes:

I met with a brood of young birds, perhaps a week or ten days old, near Jacksonville, Oregon, on June 17, 1883. The male, which had them in charge, performed the usual tactics of feigning lameness, and tried his very best to draw my attention away from the young, uttering in the mean time a shrill sound resembling *Quaih-quaih*, and showed a great deal of distress, seeing I paid no attention to him. The young, already handsome and active little creatures, scattered promptly in all directions, and the majority were most effectually hidden in an instant. As nearly as I was able to judge they numbered eleven. I caught one, but after examining it turned it loose again. The feathers of the crest already showed very plainly.

Dawson (1923) says:

Not less uncanny nor less fascinating are the vocal accompaniments with which a scattered covey of youngsters is coached or reassembled. If the little ones are of a tender age and the need is great, the parent will fling herself down at your feet and go through the familiar decoy motions; but if the retreat has been more orderly, the parents clamber about, instead, over the rocks and brush in wild concern. Once out of sight, the old bird says *querk querk querk querk*, evidently an assembly call, for the youngsters begin scrambling in that direction; while another old bird, presumably the cock, shouts *quee yawk*, with an emphasis which is nothing less than ludicrous.

Plumages.—In the downy young a broad band of deep "chestnut," mixed with and bordered by black, extends the whole length of the upper parts, terminating in a point in the middle of the crown; the rest of the upper parts, including the cheeks, are buffy or buffy white, with large blotches of "chestnut" on the wings, thighs, and flanks and with a dusky line behind the eye; the underparts are grayish white or yellowish white, palest on the chin.

The wings begin to grow almost at once, and the juvenal plumage comes on very fast, while the chick is still small, 2 or 3 inches long. In this plumage the scapulars, which appear with the wings, are "clay color" or "cinnamon-buff," peppered, edged, and partially barred with brownish black; the crown and upper back are "hair brown," the crown barred with dusky and the back mottled with dusky and spotted with white; the crest is brownish black, barred with brown; the wing coverts, tertials, and tail are pale buff, conspicuously patterned with black, washed with bright browns on the wing coverts and tail, and peppered and barred with dusky on the tertials; the breast is "light Quaker drab," with white edgings, becoming whitish on the throat and belly, and brownish on the thighs and crissum.

The juvenal plumage is worn but a short time before it is replaced by the first winter plumage, which is acquired by a complete molt, except that the outer pairs of primaries are retained for a full year. The time of this molt varies greatly with the date of hatching. I have seen birds in this transition stage at various dates from July 10 to September 27, and from less than half grown to fully grown. In first winter plumage young birds are practically indistinguishable from adults, except for the outer juvenal primaries.

Adults have a very limited prenuptial molt in spring, confined to the head and neck, and a complete postnuptial molt late in summer. This species has been known to hybridize with the valley quail, where their ranges meet in the foothills.

Food.—Concerning the food of the plumed quail, Dr. Sylvester D. Judd (1905) says:

Their feeding hours are early in the morning and just before sundown in the evening, when they go to roost in the thick tops of the scrub live oaks. Their feeding habits are similar to those of the domestic hen. They are vigorous scratchers, and will jump a foot or more from the ground to nip off leaves. This bird is especially fond of the leaves of clover and other leguminous plants. It feeds also on flowers, being known to select those of Compositae and blue-eyed grass (*Sisyrinchium*). Flowers, leaves, buds, and other kinds of vegetable matter form the 24.08 per cent marked miscellaneous. The birds probably eat more fruit than these stomach examinations indicate. Lyman Belding says that this quail feeds on service berries, and that during certain seasons it lives almost entirely on grass bulbs (*Melica bulbosa*), which it gets by scratching, for which its large, powerful feet are well adapted. The fruit in its bill of fare includes gooseberries, service berries (*Amelanchier alnifolia*), and grapes (*Vitis californica*). The bird is probably fond also of manzanita berries, for it is often seen among these shrubs. The food of the mountain quail of the arid regions has been studied in the laboratory of the Biological Survey. The stomachs examined, 23 in number, were collected in California. Five were collected in January, 2 in May, 6 in June, 3 in July, 3 in August, and 6 in November. The food consisted of animal matter, 3 per cent, and vegetable matter, 97 per cent. The animal food was made up of grasshoppers, 0.05 per cent; beetles, 0.23 per cent; miscellaneous insects, including ants and lepidopterous pupae, 1.90 per cent; and centipedes and harvest spiders (Phalangidae), 0.82 per cent. The vegetable food consisted of grain, 18.20 per cent; seeds, practically all of weeds or other worthless plants, 46.61 per cent; fruit, 8.11 per cent; and miscellaneous vegetable matter, 24.08 per cent. The grain included wheat, corn, barley, and oats. The legume seeds include seeds of alfalfa, cassia, bush clover, vetch, and lupine. The miscellaneous seeds come from wild carrot (*Daucus carota*), tar-weed (*Madia sativa*), *Collomia* sp., *Amsinckia* sp., labiate plants, dwarf oak, snowbush (*Ceanothus cordulatus*), and thistle.

Behavior.—H. W. Henshaw (1874) writes:

It seems nowhere to be an abundant species. * * * The bevies are very small, and I do not remember to have ever seen more than fifteen together, oftener less. It is a wild, timid bird, haunting the thick chaparral-thickets, and rarely coming into the opening. When a band is surprised they are not easily forced on the wing, but will endeavor to find safety by running and taking refuge in the thickness and impenetrability of their favorite thickets. If forced, however, they rise vigorously and fly swiftly and well, and sometimes to a considerable distance, and then make good their escape by running. During the heat of midday, they will be found reposing under the thick shade of the chaparral, and there they remain till the cooler hours invite them to continue their quest for food.

Grinnell, Bryant, and Storer (1918) say:

When alarmed the Mountain Quail carries its crest feathers erect, bowing backwards towards the tip but not tilted forward as in the case of the Valley Quail. This action gives the bird an alert attitude—consistent with its evident anxiety in case there are young about. Although habitually occupying brushy and forested areas, this quail but seldom perches in trees, and as far as we know the adults never roost in one at night. They stick close to the ground and usually seek safety by running beneath cover rather than by flight. For this reason the Mountain Quail is considered an unsatisfactory bird to hunt. When hunted in the brush they generally run some distance before flying, scattering and finally taking wing like as not behind a bush so as to preclude the probability of a successful shot.

Voice.—Dawson (1923) describes the varied notes of this quail very well, as follows:

The Mountain Quail's is the authentic voice of the foothills, as well as the dominant note of Sierran valleys and of bush-covered ridges. Spring and summer alike, and sometimes in early autumn, one may hear that brooding, mellow, slightly melancholy *too' wook*, sounding forth at intervals of five or six seconds. Now and then it is repeated from a distant hillside where a rival is sounding. This note is easily whistled, and a little practice will enable the bird-student to join in, or else to start a rivalry where all has been silent before. And quite as frequently, in springtime, a sharper note is sounded, although this, I believe, is strictly a mating or a questing call, *queelk* or *queelp*. This has alike a liquid and a penetrating quality which defies imitation, so that the unfeathered suitor is not likely to get very far in milady's affections. Thus, also, I have "witnessed" the progress of courtship and its impending climax in the depths of a bed of ceanothus where not a feather was visible. The *quilk* of the preceding days had evidently taken effect. The lady was there, *somewhere*. The mate was still *quilking*, but his efforts were hurried, breathless. Between the major utterances, ecstatic *took* notes were interjected. As the argument progressed I heard a low-pitched musical series, rapidly uttered, *look look look look look*. (But there was no use in looking.) This series, employed six or eight times, was suddenly terminated by half a dozen *quilks* in swift succession, indicative of an indescribable degree of excitement.

Leslie L. Haskin writes to me:

Their call when quietly feeding is not greatly unlike that of young turkey poults when following their mother. Another call often given is a simple, rapid, chirring thrill, *t-r-r-r-r-r-r-r*, often long continued. Their alarm note when startled but not badly frightened is an exact reproduction in accent, though not in tone, of a hen's cackle. In the partridge the *cut-cut-ca-do* of the barnyard fowl is charmingly altered into a shrill *t-t-t-t-t-tr-r-r-r-r-t* or *tut-tut-tut-tr-r-r-r-tut* all very rapidly delivered, and in sharp crescendo. Because of the difference in tone, and the rapidity of the partridges' delivery, few have noticed this resemblance, but once the ear has grasped the accent the similarity can never be forgotten.

Game.—Although the mountain partridge is a fine, large, plump bird and makes a delicious morsel on the table, it is not highly regarded as a game bird and not fully appreciated by sportsmen. It is a difficult bird to hunt in the dense mountain thickets it frequents. Edwyn Sandys (1904) says:

This comparatively large and exceedingly handsome species is not highly esteemed by sportsmen in general, owing to its true value not being well understood. In certain portions of California, and notably in the Willamette Valley, Oregon, when abundant it affords capital sport, while upon the table it is a delicacy not to be forgotten. As a rule, one, or at most two, broods are found on a favorite ground, the birds seldom, if ever, flocking like some of their relatives. *O. pictus* prefers moist districts and a generous rainfall. It is a runner, and in comparison with Bob-white, by no means so satisfactory a bird for dogs to work on. After the first flush the covey is apt to scatter widely and the beating up of single birds is a slow and frequently a wearying task. On the wing, its size and moderate speed render it a rather easy mark.

Grinnell, Bryant, and Storer (1918) say:

Its flesh is excellent, being declared juicier and more finely flavored than that of the Valley Quail. But its comparatively small numbers, even under normal conditions, the difficulty attendant upon reaching its habitat, and the fact that it does not lie well to dogs, deter many sportsmen from hunting the species. Except when the birds may be out of their natural habitat, as during their fall migration, it takes stiff, hard climbing and a deal of patience to get a limit of ten. In former years Mountain Quail were commonly sold on the markets of San Francisco. In some instances they were trapped along the western flanks of the Sierras and sent to the markets alive. Mr. A. E. Skelton, of El Portal, has reported to us that while shooting for the market near Raymond, Madera County, many years ago he averaged about a dozen and a half Mountain Quail a day. The birds then brought from $2.50 to $4.00 a dozen. At the present time it is illegal to sell quail of any sort, except for propagation and then under permit only.

Enemies.—Predatory animals and birds help to account for the high rate of mortality among the young birds, so that, in spite of the large broods hatched, only a few ever reach maturity. Wildcats and gray foxes seem to be their greatest enemies. These animals are also sufficiently agile to capture the old birds as well, for bunches of their feathers are often found.

Fall.—From the nature of its summer haunts in the mountains, which must be abandoned when the winter snow comes, the plumed quail has developed an interesting migratory habit. Barlow and Price (1901) say:

By the first of September the quail are restless and are beginning their peculiar vertical migration to the west slope of the mountains. Sometimes four to six adults with their young will form a covey of ten to thirty individuals and pursue their way, almost wholly on foot along the ridges to a more congenial winter climate. By October 1 the quail have almost abandoned the elevations above 5,000 feet. In the fall the woodland is full of the disconsolate *peeps* and whistling call notes of the young who have strayed from their coveys.

Lyman Belding (1903) adds:

The fall migration of the mountain quail (*Oreortyx pictus plumiferus*) appears to be influenced but little by the food supply or temperature in its summer habitat in the Sierras, which it appears to leave because the proper time has arrived for its annual tramp down the west slope. The first flocks start about the first of September, or sometimes two or three days sooner. At Webber Lake after three cold cloudy days, they began to move westward August 28, 1900. When they are migrating their whistle is frequently heard, and they do not seek cover for protection but follow a wagon road, railroad, travel in snow sheds, pass near dwelling, and seem to care but little for self preservation.

Winter.—Belding says further (1892):

The mountain quail (*Oreortyx pictus plumiferus*), which are so plentiful in the high mountains in summer, are only summer residents there. They usually spend the winter below the snow line, but as it is not possible to tell just where that is, or rather where it is going to be, they are sometimes caught in snow storms, but I have been astonished at the correctness of their apparent forecast of different winters. A few birds winter high in the mountains, but I think they are parts of flocks which were nearly annihilated, or young birds which got scattered and lost, and a few that were wounded and survived.

OREORTYX PICTA CONFINIS Anthony

SAN PEDRO QUAIL

HABITS

In the mountain ranges of southern California and northern Lower California a grayer race of the mountain quail occurs. This race was discovered and named by A. W. Anthony (1889) from specimens collected in the San Pedro Martir Mountains in Lower California, which he describes as "differing from *Oreortyx picta plumifera* in grayer upper parts and thicker bill."

He says of its haunts and habits:

From an elevation of 6,000 to 10,000 feet above the sea, in the San Pedro Mountains, I found this quail abundant, occurring wherever water and timber afforded it drink and shelter, and only leaving the higher elevations when the frosts of winter make life in the lower valleys desirable. A few pairs bred about my camp at Valladores, 6 miles from the base of the range and 2,500 feet above the sea; but nearly all of the flocks that wintered along the creek at this point were gone in March, leaving only an occasional pair, which sought the shelter of the manzanitas high up on the hill-sides, from whence their clear, mellow notes were heard morning and evening, so suggestive of cool brooks and rustling pines, but so out of place in the hot, barren hills of that region.

CALLIPEPLA SQUAMATA PALLIDA Brewster

ARIZONA SCALED QUAIL

HABITS

The Mexican plateau, with its elevated and arid, or semiarid, plains, extends northward into southern Arizona, New Mexico, and western Texas. Much of this region is dry and barren, except for a scattered growth of creosote bushes, dwarf sagebush, stunted mesquite, catclaw, mimosa, various cactuses, and a few yuccas. In the mesas between the mountain ranges and at the mouths of canyons, where underground streams supply some scanty moisture, there are grassy plains, with a scattered row of sycamores or cottonwoods marking the course of an unseen stream. Such are the haunts of the scaled quail, blue quail, or white topknot, as it is appropriately called in Arizona. Here, as one drives along over the winding desert trails, dodging the thorny shrubs or still more forbidding clumps of cactus, he may surprise a pair or perhaps a little bevy of these birds, invisible at first in their somber gray dress, which matches their surroundings so well. They do not attempt to escape by flight, but scatter in different directions, running with remarkable speed, with their necks upstretched and their white crests erected, dodging in and out among the desert vegetation, like so many rabbits scurrying off to the nearest brier patch. They are soon lost to sight, for they can run faster than we can and will not flush.

George Finlay Simmons (1925) says that, in Texas, it "shuns timbered country," but is "characteristic of the barren plateaus in the mountainous districts of western Texas, usually where the soil is fine, loose, and sandy; broad, dry, arid washes, gulches, and semi-barren plateaus of the hills where hard ground is covered with a few thorny bushes, scattered scrub oak, chaparral, mesquite, sagebrush, and different species of cactus; chaparral and mesquite country, generally in the vicinity of water, but sometimes miles from any stream or pond."

Mrs. Florence Merriam Bailey (1928) says that in parts of New Mexico scaled quail collect about the ranches and huts of the settlers, picking up grain left from feeding the horses, or feeding with the chickens.

Frank C. Willard says in his notes:

Of the three species of quail found in Cochise County, Ariz., the scaled quail is the one most commonly met with on the dry, brush-covered mesas and valleys. Here it frequents the dry washes, or arroyos, with their fringe of mesquites, small desert willows, and an occasional flat area covered with bear grass. The near proximity of living water does not seem to be at all necessary for their existence, as I have frequently found them 7 or 8 miles from any water at all. Such of these quail as do live in the vicinity of water make regular daily trips to it and they congregate more thickly around ranches and water holes than they do away from them.

Along the San Pedro River and the Barbacomari River (a branch of the former), there is a mingling of the scaled quail and Gambel's quail. These two species are also found inhabiting the foothills of the mountains, and the low ranges of hills like those around Tombstone and in the Sulphur Spring Valley. In such localities the two quails occasionally lay their eggs in the same nest.

Nesting.—While going to and from our camp in Ramsey Canyon, in the Huachuca Mountains, Ariz., we frequently saw two or three pairs of scaled quail on a grassy flat where a few sycamores and small bushes were scattered along the course of a dry wash, the bed of an underground stream running out from the canyon. On May 25, 1922, while hunting this flat we flushed a quail from the only nest of this species that I have ever seen. The nest was well concealed under a tuft of grass surrounding a tiny mesquite; the grass was well arched over it, and the hollow in the ground was lined with dry grass and a few feathers. It held 14 fresh eggs.

Nests found by others have been similarly located under the shelter of some low bush, sagebush, creosote bush, mesquite, catclaw, cactus, or yucca, rarely in an open situation among rocks or under a fallen bush. Simmons (1925) says that in Texas the nest is rarely placed in a meadow or grainfield. The hollow is lined with whatever kind of dry grass is available. Willard says in his notes:

> Most of the nesting occurs during the months of June and July. I am inclined to believe that this is because the rainy season in Arizona commences, under normal conditions, early in June. Thereafter there are more or less heavy showers nearly every day. This assures a supply of drinking water within easy reach of the newly hatched young. The nests are usually placed under some tussock of mixed dry and green grass. In the vicinity of gardens, they sometimes build under tomato vines. Where a haystack is available, they are quite likely to work out a hollow near the bottom and lay their eggs there much after the manner of the domestic hen. It is not at all unusual for two scaled quail to lay their eggs in the same nest, if the presence of two distinct types of eggs in the same nest can be considered as evidence. In several instances I have had nests under observation (which did not yet hold complete clutches) and in three of these instances eggs were deposited at the rate of two per day, quite positive proof that two birds were using the same nest.

Eggs.—The scaled quail lays from 9 to 16 eggs, rarely more, and usually from 12 to 14. They are ovate or short ovate in shape and usually quite pointed. The shell is thick and smooth, with little if any gloss. The ground color varies from dull white to creamy white. Some few eggs are thickly, or even heavily speckled with very small spots or minute dots of dull, light browns, "sayal brown" to pale "cinnamon-buff." Most of the eggs are sparingly marked with similar spots. Some are nearly or quite immaculate. Major Bendire (1892) says that "occasionally a set is marked with somewhat more irregular, as well as larger, spots or blotches, resembling certain types of eggs of *Callipepla gambeli*, but these markings are always paler colored and not so pronounced." The measurements of 57 eggs average 32.6 by 25.2 millimeters; the eggs showing the four extremes measure **35.8** by 26, 34 by **27**, **30** by 24.5, and **31.5** by **23.5** millimeters.

Young.—Major Bendire (1892) believed that "two and even three broods are occasionally raised in a season, the male assisting in the care of the young, but not in incubation. This lasts about 21 days."

Mrs. Bailey (1928) writes:

That the downy young are also obliteratively colored is well illustrated by an experience of Major Goldman when climbing the Florida Mountains. At 5,300 feet, among the oaks and junipers, he reports, "I came suddenly on an adult bird and a brood of recently hatched young. The old bird disappeared after giving several sharp cries of alarm, and the young also disappeared in an open patch of short grass. On reaching the place I began looking about carefully and soon saw one young bird flattened down, with not only its little body but its head and neck also pressed close against the ground, its downy plumage blending in well with the color of the ground and the dead grass stems." There it lay, pressed close to the ground until approached within three feet when, "it suddenly started up with sharp peeping cries, and the entire brood which had scattered and hidden in an area about fifteen feet across, half ran, half flew into some thick bushes where they were more securely hidden."

Plumages.—In the downy young scaled quail the forehead, the front half of the crown, in front of a little gray topknot, and the sides of the head are "cinnamon-buff" or "pinkish buff"; there is a broad band of "chestnut" from the middle of the crown, back of the topknot, down to the hind neck, bordered narrowly with black and with broad stripes of buffy white; the auricular spots are dark "chestnut"; the chin and throat are buffy white and the rest of the underparts are pale grayish buff; the back is mottled with pale buff and "russet."

The juvenile plumage starts to grow at an early age, beginning with the wings; on a small downy chick, less than 2 inches long, the wings are well sprouted; the wings grow so fast that the young can fly long before they are half grown. The sexes are alike in the juvenal plumage. In this the crown is "buffy brown" to "wood brown" and the crest is "vinaceous-buff"; the rest of the head, neck, and shoulders shades off gradually to shades of drab; the feathers of the back, scapulars, and wing coverts are from drab to "sayal brown" or "tawny-olive," barred with "sepia" or brownish black, finely sprinkled or peppered with brownish black, with conspicuous median stripes of buffy white and with buffy edgings on the scapulars; the tail is mainly dark drab, but more buffy near the tip, barred and peppered with brownish black; the underparts are buffy white or grayish white, spotted or barred with dusky, most distinctly on the breast, where many feathers are tipped with white arrowheads.

The molt into the first winter plumage takes place during September and October. This is a complete molt, except for the two outer primaries on each wing, which are retained all through the first year. The molt begins on the back, breast, and flanks. Young birds are practically indistinguishable from adults during the first winter and spring except for the retained outer primaries. Both first year and adult birds have a partial prenuptial molt early in spring, restricted mainly to the head and throat, and a complete postnuptial molt in August and September. Albinism occurs occasionally, and this species has been known to hybridize with Gambel's quail and with the bobwhite.

Food.—Mrs. Bailey (1928) sums up the food of this species very well, as follows:

The Scaled Quail apparently eats more insect food than any of the other quails, or more than 29 per cent, as against 70 per cent of vegetable matter. Of this vegetable matter over 50 per cent is weed seeds, among which are thistle, pigweed, and bindweed, a troublesome weed that often throttles other plants. *Dasylirion* seeds almost entirely filled six stomachs examined. Wild fruit, such as prickly pear and the succulent parts of desert plants, together with its larger per cent of insect food, doubtless help it to live with a minimum amount of water. Its insect food includes grasshoppers, ants, and beetles—among them leaf chafers and cucumber beetles—weevils, such as the clover pest and scale insects (several hundred in one stomach) that feed on the roots of plants.

Sylvester D. Judd (1905) says of its vegetable food:

The species resembles the ruffed grouse in its habit of feeding on green leaves and tender shoots. It feeds upon budded twigs, but more often limits its choice to chlorophyll-bearing tissue, often picking green seed pods of various plants. Like domestic fowls, it eats grass blades. Fruit was eaten by only 6 of the 47 birds, and none was taken from cultivated varieties. As might be expected from inhabitants of arid plains, these birds like the fruit of cacti, and have been found feeding on the prickly pear (*Opuntia lindheimeri*). The fruit of *Ibervillea lindheimeri* also is eaten. The blue berries of *Adelia angustifolia*, which furnish many desert birds and mammals with food, are often eaten by the scaled quail. Different kinds of *Rubus* fruits are relished, and the berries of *Koeberlinia spinosa* and *Monisia pallida* also are eaten. The fruit and succulent parts of plants no doubt serve in part in the parched desert as a substitute for water.

Behavior.—The scaled quail is a decidedly terrestrial bird with very powerful legs, which it uses to advantage in the rather open desert growth in which it lives and where it can run very fast in the smooth open spaces among the desert plants. It prefers to escape by running rather than flying; but, if come upon suddenly and surprised, it rises with a whir of wings, flies a short distance, and scales down into cover again, much after the manner of the bobwhite; it then starts running and can not be easily flushed again. If in a flock, they sometimes follow a leader in Indian file, but more often they scatter in several directions and are soon lost to sight. Major Bendire (1892) quotes Dr. E. W. Nelson, as follows:

In many instances I have found them far from water, but they make regular visits to the watering places. On the Jornada del Muerto and on Santa Fé Creek I found them frequenting the open plains, away from the water in the middle of the day, and in the vicinity of the water late in the afternoon. At this time they are often seen in company with Gambel's Quail amongst the bushes and coarse grass or weeds bordering the water courses.

Bendire (1892) quotes William Lloyd as saying:

The Blue Quail loves a sandy table land, where they spend considerable time in taking sand baths. I have often watched them doing so, pecking and chasing each other like a brood of young chickens. Good clear water is a necessity to them. They are local, but travel at least 3 miles for water. In the evenings they retire to the smaller ridges or hillocks and their calls are heard on all sides as the scattered covey collects. Several times I have seen packs numbering sixty to eighty, but coveys from twenty-five to thirty are much oftener noticed. During the middle of the day they frequently alight in trees, usually large oaks, but they roost on the ground at night.

Voice.—Major Bendire (1892) says:

According to Mr. Lloyd their call note sounds something like a lengthened *chip-churr, chip-churr*; the same, only more rapidly repeated, is also given when alarmed, and a guttural *oom-oom-oom* is uttered when worried or chased by a Hawk. The young utter a plaintive *peep-peep*, very much like young chickens. Like the rest of the partridge tribe they are able to run about as soon as hatched.

Mr. Simmons (1925) refers to their notes as: "A single low, long-drawn whistle; a nasal, musical, friendly *pe-cos'*, *pe-cos'*."

Enemies.—Mrs. Bailey (1928) writes:

Although protective coloration and attitudes partly serve their purposes, protective cover is still vitally important, for as Mr. Ligon has found, "Prairie Falcons, Cooper Hawks, Roadrunners, snakes, skunks, wildcats, and coyotes all take their toll of these birds or their eggs"; in the northern part of their range, Magpies destroy both eggs and young; and over much of their range hail, cold rains, and winter storms deplete their numbers.

Mr. Willard says in his notes:

The Gila monster, rattlesnake, and skunk are natural enemies which take a large toll from the nests of the scaled quail. I once observed a female quail fluttering excitedly over a clump of grass and making dashes down at it. On investigating I found a rattlesnake and nine quail eggs in the nest. I dispatched the snake and on opening it found three whole eggs inside. A Gila monster, which I caught and caged, evidently disgorged two scaled quail eggs, as there were two eggs in the box a short time later, and I am sure no one had been near it but myself. In passing, it may be of interest to say that this great lizard will devour a hen's egg by gradually working it far enough into its mouth to be able to clamp down on it with its powerful jaws, crushing it, and then sucking out the contents. They are large enough to swallow easily a quail's egg whole. We occasionally found a mass of loose feathers of this quail scattered on the ground and clinging to near-by bushes. The presence of cat tracks told what was responsible for the tragedy the feathers betrayed. On at least four occasions I have surprised a long-legged Mexican lynx stalking the same game I was after, and was able to collect a cat as well as a quail.

Fall.—Mrs. Bailey (1928) says further:

The entire life of the Scaled Quail is spent in the environment to which it is so well adapted, but in the fall it is sometimes found a few hundred feet higher than in the nesting season. When the young are raised these delightful little Cotton-tops go about in small flocks, visiting water holes and river bottoms. Picking up insects, seeds, and berries as they go, they wander through brushy arroyas, over juniper-clad foothills, cactus flats, and sagebrush or mesquite plains, calling to each other with a nasal *pay-cos, pay-cos,* which by long association comes to take on the charm attaching both to the gentle-eyed birds themselves and to the fascinating arid land in which they make their homes.

Game.—Hunting the blue quail will never figure as one of the major sports, although it is a gamy bird and makes a delicious and plump morsel for the table. The birds are widely scattered over a vast expanse of rough country, on desert plains covered with thorny underbrush, or on stony or rocky foothills where walking is difficult and slow. The hunter must be prepared to do some long, hard tramping, for he is more likely to count the number of miles to a bird than the number of birds to a mile. A dog is useless, for these quail have not yet acquired the habit of lying to a dog. Eastern quail have learned to lie close, a good way to hide from human enemies but a very poor way to escape from the many predatory animals in the West. Scaled quail are shier than Gambel's quail and are generally first seen in the distance running rapidly and dodging around among the bushes. They run faster than a man can walk, and the hunter must make fast progress over the rough ground to catch up with them. By the time he gets within range he will be nearly out of breath and will have to take a quick snap shot at a fleeting glimpse of a small gray bird dodging between bushes. This is far more difficult, under the circumstances, than wing shooting and can not be considered pot shooting. Sometimes, when a large covey has been scattered and rattled, the hunter may surprise single birds and get an occasional wing shot; but they are apt to jump from most unexpected places, ahead of or behind the hunter, and give him a difficult shot. Late in the season they are often found in large packs of 100 or 200 birds, when the chances for good sport are better. Even then the hunter may well feel proud of a hard-earned bag.

DISTRIBUTION

Range.—The Southwestern United States and northern Mexico. Nonmigratory.

The scaled quail is found **north** to southern Arizona (Picacho, Rice, and Clifton); northern New Mexico (Haynes and the Taos Mountains); east-central Colorado (Mattison and Holly); and northeastern Texas (Lipscomb). **East** to Texas (Lipscomb, Mobeetie, Colorado, San Angelo, Fredericksburg, San Diego, Fulfurrias, and Brownsville); Tamaulipas (San Fernando); and San Luis Potosi (Ahualulco). **South** to San Luis Potosi (Ahualulco and Ramos);

Durango (Rancho Baillon); Chihuahua (San Diego); and Sonora (San Pedro and Sesabe). **West** to Sonora (Sesabe); and southern Arizona (Arrivaca, Sierrieta Mountains, and Picacho). It has been detected casually in eastern Texas (Gainesville and Bonham). On August 19, 1926, three specimens were collected at Elkhart, Morton County, Kans. It is a common species across the State line in southeastern Colorado.

The range above outlined is for the entire species. By recognition of the Arizona, New Mexico, and Colorado bird (*C. s. pallida*), the typical subspecies (*C. s. squamata*) is restricted to northwestern Mexico. Another race, the chestnut-bellied scaled quail (*C. s. castanogastris*), inhabits the lower Rio Grande Valley in Texas and adjacent regions in northeastern Mexico.

Attempts to transplant the scaled quail to other regions have generally resulted in failure. Among these may be mentioned introductions in Louisiana, Florida, Georgia, and Washington. So far as is known, the only successful transplantation was made in Colorado at Colorado Springs and probably also at Canyon City. From these points the birds have spread and increased until they are now common in the Arkansas Valley from Pueblo east to the State line, and it appears that the introduced stock has met and blended with the native birds working northward along the Las Animas River.

Egg dates.—Texas (*pallida*): 11 records, May 7 to June 22. Arizona and New Mexico: 37 records, April 16 to September 22; 19 records, June 11 to July 7.

Texas and Mexico (*castanogastris*): 44 records, March 7 to June 28; 22 records, May 3 to June 2.

CALLIPEPLA SQUAMATA CASTANOGASTRIS Brewster

CHESTNUT-BELLIED SCALED QUAIL

HABITS

The scaled quail of the lower Rio Grande Valley in Texas and eastern Mexico is more richly and darker colored than the quail found farther west, and, as its name implies, it has a well-marked patch of dark chestnut in the center of its belly, which is more prominent in the male than in the female. George B. Sennett (1879) thus describes its habitat:

> The foothills of the Rio Grande, about 100 miles back from the coast, are the eastern limits of this bird, as well as of the Cactus Wren and the Yellow-headed Titmouse. The first rise of ground in going up the river occurs at Lomita Ranch, and here we often saw these beautiful birds running about; but although we frequently collected a mile or two below the hill, there we never saw them, and not even in the fertile and heavily wooded lowlands in the vicinity of this hill did we observe them. A few miles up from Lomita and back from the river,

near the water-holes, rises are numerous, covered with thin, poor soil, where cactuses and scrubby, thorny bushes grow, and here the blue quail abounds.

Nesting.—Sennett evidently found only one nest, of which he says:

On the 22d of May, near the buildings of the ranch at San José Lake, Mr. Sanford shot a fine male, which was on the brush fence forming the enclosure. In searching among the weeds where the bird fell, we found a nest and 16 fresh eggs. The nest was under the edge of the fence, and was simply a saucer-like depression in the ground, with leaves for lining.

Three sets of eggs in my collection were taken from hollows in the ground, under cactus plants or bushes, lined with grass, weeds, or trash. Major Bendire (1892) says that "their nests are always placed on the ground; a slight hollow in the sand is scratched out by the bird, usually under a clump of weeds or grass, or a prickly-pear bush. They are very slightly lined with dry grasses."

Eggs.—The eggs of the chestnut-bellied scaled quail are practically indistinguishable from those of the Arizona form, though they may average a little more richly colored or more heavily spotted and a trifle smaller. Major Bendire (1892) says:

Full sets of eggs have been taken near Rio Grande City, and at Camargo on the Mexican side of the river opposite, as early as March 11, and from that time up to July 10. Two broods are unquestionably raised in a season. Mr. Thomas H. Jackson, of West Chester, Pennsylvania, gives the average number of eggs laid by this species as fifteen, based on data taken from twenty-seven sets. The largest number found in one nest was twenty-three.

The measurements of 77 eggs in the United States National Museum average 31 by 24 millimeters; the eggs showing the four extremes measure 34 by 24.5, 33 by 25.5, 28.5 by 23.5, and 30 by 22 millimeters.

Behavior.—In its general habits this quail does not differ materially from its western relative. Its plumage changes, its food, voice, and behavior are all similar. Both forms are resident in their respective ranges, moving about only as the food supply demands.

LOPHORTYX CALIFORNICA CALIFORNICA (Shaw)

CALIFORNIA QUAIL

HABITS

The type race of this species originally inhabited the narrow strip of humid coast region from southwestern Oregon south to southern Monterey County, Calif. It has been introduced on Vancouver Island and in Washington, where it has become well established. It differs from the more widely known valley quail in having the upper parts olive-brown, rather than grayish brown, and the inner margins of the tertials deeper buff. It does not differ materially from the other races in its nesting, food, or general habits. A full account of the habits of the species is given under *L. c. vallicola*, the next form treated. J. Hooper Bowles (Dawson and Bowles, 1909) writes:

This bird and its near relative, the Valley Partridge, are not natives of Washington; but, like the Mountain and the Plumed Partridge, were introduced here from the State of California. Dr. Suckley, one of our pioneer naturalists, tells us that as early as 1857 two shipments of birds were turned out in the vicinity of Puget Sound by Gov. Charles H. Mason and a Mr. Goldsborough. Conditions seem to have proved most suitable for them, since, in the face of constant persecution, they continued to increase in numbers, spreading their ranks over new territory every year. Although often found in dry, bushy uplands, they are much more inclined to damp localities than the Mountain Partridge, their favorite haunts being the low ground of the river valleys. Here they may be found searching for seeds in the weed-patches of the open fields, or gleaning amongst the growing cabbages, beans, and other vegetables of the farmer's garden. Indeed, few birds are so much the friends of the farmer as our partridges, for their food consists almost entirely of weed-seeds, worms, beetles, grass-hoppers, and other insects. What little of the newly-sown crops they may eat is repaid a thousand fold by the vast amount of good they accomplish.

Nesting.—Mr. Bowles, in his notes sent to me, says:

Like most species of introduced game birds these quail lay their eggs in the nests of other varieties of birds. I have in my collection a set of nine eggs of the sooty grouse and three eggs of this quail, personally taken here at Tacoma. All the eggs were heavily and evenly incubated.

I have also a nest of the Nuttall's sparrow containing four eggs of the sparrow and two eggs of this quail, which was taken near the city of Seattle, Wash., on May 8, 1918. Incubation was slight. This set was collected and presented to me by D. E. Brown, of Seattle. His notes say that one of the quail eggs was on end in the nest, the other on top of two of the sparrow eggs. The sparrow was on the nest and showed much anxiety.

Eggs.—The eggs of this quail are indistinguishable from those of the valley quail, but they average slightly larger. The measurements of 60 eggs average 32 by 25 millimeters; the eggs showing the four extremes measure **35.5** by 24.5, 35 by **26, 30** by 24, and **31** by **23** millimeters.

DISTRIBUTION

Range.—Southwestern Oregon, California, extreme western Nevada, and Lower California. Successfully introduced in Washington, Idaho, Utah, Nevada, and British Columbia, as well as Hawaii, New Zealand, Chile, and probably locally in France.

Because of the many attempts to extend the range of this species in the Western States, it is difficult to outline the area to which they are indigenous. It appears, however, that the natural range extends **north** to southwestern Oregon (Anchor and Algoma). **East** to Oregon (Algoma and Klamath Falls); western Nevada (Anaho

Island and Stillwater); south-central California (Fresno, Tehachapi, and San Bernardino); and southeastern Lower California (Pinchalinque Bay, Triunfo, and San Jose del Cabo). **South** to southern Lower California (San Jose del Cabo and Cape St. Lucas). **West** to Lower California (Cape St. Lucas, San Javier, Rosarito, San Andres, San Quintin Bay, and Los Coronados Islands); western California (San Diego, Santa Catalina Island, Santa Barbara, Monterey, Watsonville, Alameda, San Francisco, Marysville, Chico, and Baird); and southwestern Oregon (Grants Pass and Anchor).

The range as above outlined is for the entire species. True *californica* is restricted to the humid coast region from southwestern Oregon south to Monterey County, Calif. The valley quail (*L. c. vallicola*) occupies the rest of the range south to the northwestern corner of Lower California (about latitude 32° N.), including Los Coronados Islands. The birds found on Catalina Island have been described as a distinct race, *Lophortyx c. catalinensis.*

In Lower California, in addition to *vallicola*, which occurs in the northwestern part, the species has been divided into two subspecies. The San Quintin quail (*L. c. plumbea*) is distributed over most of the territory between latitudes 30° and 32° N., while the San Lucas quail (*L. c. achrustera*) is found from latitude 30° N. south to Cape San Lucas. There is more or less intergradation in the areas where these races meet.

As previously indicated, the California quail has been a favorite in attempts at transplantation. The birds on Los Coronados Islands are considered by some to be introduced, although there also is evidence that they fly back and forth to the mainland. They have been successfully introduced into Nevada (Virginia City, McDermitt, Quinn River Valley, Paradise Valley, Lovelock, and probably also Carson City and Reno); Oregon (Willamette Valley and Jackson and Josephine Counties); Utah (Salt Lake County and Ogden); Washington (Olympia, Garfield County, Walla Walla County, Yakima County, and probably many other points, as they are now well distributed over the western part of the State, including the islands in Puget Sound and Bellingham Bay); British Columbia (Vancouver Island, Denman Island, and the Okanagan Valley). Attempts to introduce this species in Illinois, Maryland, Massachusetts, New York, Delaware, Mississippi, and Missouri have not been successful, although in a few instances the birds seemed to thrive during the first season.

California quail also have been successfully transplanted to the Hawaiian Islands (Hawaii, Maui, and Molokai), New Zealand, and Chile. Other foreign experiments apparently have failed.

Egg dates.—California (*californica*): 92 records, January 12 to July 21; 46 records, May 8 to June 8. Washington and British Columbia: 10 records, May 11 to July 3.

California (*vallicola*): 125 records, February 9 to October 29; 63 records, May 1 to June 5.

LOPHORTYX CALIFORNICA VALLICOLA (Ridgway)

VALLEY QUAIL

HABITS

The valley quail being the most widely distributed form, it shall have the most complete life history of the species, as it is the best known of the various subspecies. In Pasadena and vicinity, southern California, it is a common dooryard bird, coming regularly into the city to feed on the lawns and to roost in the trees and shrubbery. On Dr. Louis B. Bishop's lawn, in the thickly settled part of Pasadena, one might see from 10 to 20 of the pretty birds almost any afternoon after 4 o'clock. Although rigidly protected and regularly fed, they seemed very nervous and shy; if they saw us moving, even at a window, they would run or fly into the shrubbery. J. Eugene Law has a flock of 100 to 200 birds, which he feeds every morning during winter on his driveway in Altadena. I was able to photograph some of these birds one morning from a blind, but I found them very nervous; at the slightest noise or movement they would all fly off but would soon return. Mr. Law told me that these quail all bred in the vicinity, nesting commonly in the old abandoned vineyards overgrown with rank grasses and weeds. They travel around in flocks during winter but begin to break up into pairs during March. The latest flocks I saw were two small flocks on April 1. Outside of the cities and towns we saw these quail on the brush-covered hillsides, on the grassy plains in the wider canyons, in cultivated fields and in the fruit orchards, or almost anywhere that they can find a little cover.

Claude T. Barnes writes to me:

In northern Utah the favorite habitat of the valley quail (*L. c. vallicola*) is the patches of scrub oak (*Quercus gambellii*), which grow upon the foothills of the Wasatch Mountains and along the deeper stream-gullies of the valleys. It is very fond, also, of fences along which, in early days, the golden currant (*Ribes aureum*) was planted and permitted to spread along irrigation ditches; in fact, any dense covert adjacent to grainfields and near one of the many crystal streams for which the region is noted suits very well this semidomesticated bird. Many farmers, convinced of not only the aesthetic but also the economic value of the quail, habitually in winter sprinkle grain upon the snow about their barnyards for the quickly responding coveys.

Courtship.—W. Leon Dawson (1923) writes:

The Quail's year begins some time in March or early April, when the coveys begin to break up and, not without some heart-burnings and fierce passages at arms between the cocks, individual preferences begin to hold sway. It is then that the so-called "assembly call," *ku kwak' up, ku kwak' up, ku kwak' u k-k o*, is heard at its best; for this is also a mating call; and if not always directed toward a single listener, it is a notice to all and sundry that the owner is very happy, and may be found at the old stand. Although belonging to a polygamous family, the Valley Quail is very particular in his affections; and indeed, from all that we may learn, is at all times a very perfect model of a husband and father. Even in domestication, with evil examples all about and temptresses in abundance, the male quail is declared to be as devoted to a single mate as in the chaparral, where broad acres may separate him from a rival.

Nesting.—The valley quail is not at all particular about the choice of a nesting site and is not much of a nest builder. A slight hollow in the ground lined with grass or leaves may be well hidden under a bush, hedge, or brush pile, beside a log or rock, in some thick clumps of grass or weeds in an orchard or vineyard, in a clump of cactus or pricklypear, under the base of a haystack in an open field, or even in a cranny in a rock. W. Leon Dawson (1923) shows a photograph of a nest in the latter type of location. Often the nest is near a house, in a garden, or close to a much-traveled path or road. This quail often lays its eggs in other birds' nests. M. L. Wicks (1897) tells an interesting story of a partnership nesting with a long-tailed chat; the quail had laid two eggs in the chat's nest, in which the chat laid four eggs; both birds took turns at incubating. Harold M. Holland (1917) twice found a quail occupying a road runner's nest. John G. Tylor (1913) speaks of a curious habit this quail has of dropping its eggs at random anywhere; this happens early in the season, and he thinks it is due to the fact that the vines under which it wants to nest have not yet developed enough foliage for concealment. Grinnell, Bryant, and Storer (1918) mention a few more odd nesting sites, as follows:

H. R. Taylor records the finding of ten fresh California Quail's eggs in a Spurred Towhee's nest in a cypress hedge about four feet from the ground, and also two eggs of this quail in a Spurred Towhee's nest on the ground, both in Alameda. Near Los Angeles, Wicks found two eggs of the Valley Quail in a Long-tailed Chat's nest. Several cases of tree-nesting of the California Quail came to the attention of W. E. Bryant. The sites which had been chosen were the upright ends of broken or decayed limbs, or the intersections of two large branches. The same observer found a nest in a vine-covered trellis over a much-used doorway, from which the young later successfully reached the ground. Howell found a nest with three fresh eggs four feet above the ground on top of a bale of hay in the shade of an orange tree at Covina, Los Angeles County.

Mr. Dawson (1923) tells of a nest placed "on a horizontal stretch of dense wistaria covering an arbor, at a height of 10 feet from the ground"; when the young were hatched the parent birds called them, and they came tumbling down, stunned at first but not seriously injured. He mentions another nest on the roof of a house.

Eggs.—The valley quail lays ordinarily from 12 to 16 eggs; large sets of more than 20 eggs are sometimes found, but these may be the product of two hens. They are short ovate in shape, and sometimes rather pointed; the shell is thick and hard, with little or no gloss. The ground colors vary from "cream buff" to "ivory yellow" or, rarely, dull white. They are usually heavily marked and show considerable variation. Some eggs are well covered with large blotches, irregularly scattered; others are evenly covered with minute dots; but there are many intermediate variations, and there is generally a mixture of both kinds of markings on the same egg and several types of eggs in the same set. An occasional egg is entirely unmarked. The colors of the markings are dull browns, varying from "snuff brown" or "cinnamon-brown" to "Isabella color." The measurements of 77 eggs average 31 by 24 millimeters; the eggs showing the four extremes measure **34** by 25, 32 by **26**, and **28** by **23** millimeters.

Young.—Mrs. Irene G. Wheelock (1904) writes:

Incubation requires three weeks, and usually the hen alone broods the eggs. after the young are hatched they are kept in the underbrush or heavy stubble and can rarely be discovered, so expert at hiding are they. Like the California partridge they run to cover rather than fly, and they are so swift-footed that it is almost impossible to flush them. When the young are feeding, the adult males constantly call them, either to keep the covey together or to give warning of danger, and they answer each call with a faint piping note. This is not unlike the scatter call of the Eastern Bob White, but consists of two syllables in one tone, or one longer note. It is not unusual to come upon a covey of these when driving through the foothills and valleys of Southern California, but the sensation is simply of something scampering into the brush rather than a definite sight of any bird, unless the cock comes out into view for a moment to sound his warning and draw your attention from the brood to his handsome self.

Bendire (1892) quotes William Proud as saying that only one brood is raised in a season, that incubation lasts about 18 days, and that "as soon as the young are hatched, they immediately leave the nest, keeping under cover as much as possible. Should the brood be disturbed, the old birds will run and flutter along the ground to draw the attention of the dog, or whatever may have frightened them, to themselves and away from the young. In about 10 days these can fly a short distance."

F. X. Holzner (1896) says:

I walked unsuspectingly upon a bevy of Valley Partridges (*Callipepla californica vallicola*), consisting of an old male and female with about 15 young ones. They were in a crevice of a fallen cottonwood-tree. On my stepping almost upon them, the male bird ran out a few feet and raised a loud call of *ca-ra-ho*; while the female uttered short calls, addressed to her brood. Seeing me, she picked up a young one between her legs, beat the ground sharply with her wings, and made towards the bush, in short jumps, holding the little one tightly between her legs, the remainder of the brood following her.

Plumages.—In the small chick of this species the front half of the crown and sides of the head are "ochraceous-tawny"; a broad band of "russet," bordered with black, extends from the center of the crown to the hind neck, and there is an auricular stripe of the same color; the rest of the upper parts are from "ochraceous-buff" to "warm buff," striped, banded, or blotched with black; the chin and throat are white, and the rest of the underparts are grayish white, suffused with buff on the breast.

As with all young quail and grouse, the juvenal plumage comes in while the chick is still very small, the wings and scapulars sprouting first, so that the young birds can fly before they are half grown. In the full juvenal plumage, the forehead is "hair brown," the crown and hind neck "wood brown," and the chin and throat "drab gray"; the feathers of the back, wing coverts, and scapulars are from "hair brown" to "clay color," with median stripes of buffy white, broadest on the scapulars, peppered with black and tipped or banded near the end with black; the tertials are from "sayal brown" to "cinnamon," peppered with black, and bordered on the inner edge with a broad band of black and a broad edge of "pinkish buff"; the rump is grayish buff, barred with dusky and whitish; the tail is from "drab" to grayish buff, tipped with "cinnamon," and peppered and barred with blackish brown; the underparts are grayish white, barred with dusky, more buffy, and marked with triangular whitish spots on the chest; the head crest is "warm sepia." The sexes are alike.

A complete postjuvenal molt, except for the primary coverts and the outer pairs of primaries, begins before the young bird has attained its full growth. The time varies, of course, with the date of hatching, but it takes place between August and October. The last of the juvenal plumage is seen on the head and neck. This molt produces the first winter plumage, which is practically indistinguishable from the adult, except for the outer pairs of juvenal primaries and primary coverts, which are retained until the next postnuptial molt. The sexes are now different.

Young birds and adults have a partial prenuptial molt, confined to the head and neck, early in spring, and a complete molt late in summer and early in fall. Hybrids between this species and *gambeli* and between this and *picta* have been recorded, where the ranges of the species come together.

Food.—These quail are very regular in their feeding habits. When they have found a good feeding place they resort to it day after day, often traveling long distances on foot and not flying unless forced to. They travel in flocks at all times except during the nesting season, when they are paired. Formerly they came to the watering places in immense flocks of hundreds, but now in flocks of 30 or 40, or aggregations of two or three families. Their feeding hours are for an hour or two after sunrise and an hour or two before sunset. During the middle of the day they congregate near the drinking places or rest in the shade of trees or bushes. While feeding, one bird acts as sentinel or guard until relieved by others in turn.

Grinnell, Bryant, and Storer (1918) write:

The Valley and California quails are believed to be more exclusively vegetarian than any other of our game birds, save those of the pigeon family. The United States Bureau of Biological Survey, in an examination of 619 stomachs (representing both subspecies), found that only about 3 per cent of the food consisted of animal matter. The remaining 97 per cent was vegetable material and consisted of 2.3 per cent fruit, 6.4 per cent grain, about 25 per cent grass and other foliage, and 62.5 per cent seeds. The animal food comprised chiefly insects, and of these, ants were most frequently present. Some beetles, bugs, caterpillars, grasshoppers, flies, spiders, "thousand-leggers," and snails were also found in the stomachs examined. A case is cited by Beal of a brood of young quail feeding extensively on black scale.

Fruit evidently does not form any important part of the food of the quail, as it was found in only about one-sixth of the stomachs and then only in very small quantities. Damage is sometimes done to grapes, but this is not shown clearly by examination of stomach contents. Beal mentions two cases where 1,000 and 5,000 quail, respectively, had been seen feeding upon grapes in vineyards. Under such circumstances severe loss was undoubtedly sustained; but these are exceptional instances. Florence A. Merriam states that on the ranch of Major Merriam at Twin Oaks, San Diego County, quail were in 1889 so abundant as to be a severe pest. For several years previously great flocks of them came down the canons to the vineyard, "where they destroyed annually from 20 to 30 tons of fruit." A report comes from the Fresno district to the effect that grape growers are occasionally troubled by the birds scattering the drying raisins from the trays.

Behavior.—The movements and actions of valley quail seem to me strikingly like those of our eastern bobwhites, except that they are less inclined to fly or to hide and more inclined to run. When alarmed or forced to fly they jump into the air with a similar whir of wings and dash away with an equal burst of speed, scaling down into the nearest cover on stiff, down-curved wings. If they alight on the ground, they do not stop, but continue running at terrific speed, their long, strong legs fairly twinkling in a hazy blur; it seems as if they continued to fly along the ground almost as fast as they flew in the air. On the ground their movements are quick, alert, and graceful; their trim and pretty little bodies are held in a semierect attitude, leaning forward a little as they run, with the crest held forward. They are most attractive in appearance and most winning in their confiding ways.

John J. Williams (1903) made some very interesting observations on the use of sentinels by valley quail; his article is well worth reading in detail, but I prefer to quote Grinnell, Bryant, and Storer's (1918) summary of it, as follows:

A flock was heard calling and moving about on a brushy hillside some distance from the observer, but before coming into view a single individual preceded the rest and took his station in the branches of an apple tree, whence he could survey the region round about. After carefully scrutinizing his surroundings for several minutes the *kayrk* note was uttered several times in a low guttural tone. Soon members of the flock were seen coming down the hill in the same direction as taken by the sentinel, but their manner of approach was entirely different; he had exercised great caution and carefully examined the surroundings for possible danger, while they came with their plumed heads held low, searching among the clover roots for seeds and other articles of food. Some preened and fluffed out their feathers; others took dust baths. While so occupied they all kept up a succession of low conversational notes. Meanwhile the sentinel remained on his perch and continued on the alert even after the flock had moved some distance beyond him. Then a second bird mounted a vantage point and took up the sentinel duty and after a few minutes the first relinquished his post. While the flock was still in view, yet a third bird relieved the second. It would seem that by this practice, of establishing sentinels on a basis of divided labor, the flock had increased its individual efficiency in foraging. The same observer also states that he had seen sentinels used when a flock was crossing a road, or when "bathing" in the roadside dust, and that the practice is made general use of in open areas; but he had never observed the habit when the birds were in tree-covered localities. During the breeding season it is known that the male mounts guard while the female is searching for a nesting site, and again when she is incubating the eggs. Sometimes he also performs this guard function after the chicks are out but not fully grown.

Unlike our eastern bobwhite, which roosts on the ground, the valley quail roosts at night in safer places, in bushes or in low, thick-foliaged trees. In the treeless region of Lower California, Laurence M. Huey (1927) found quail roosting in the centers of cactus patches. Dawson (1923) says he has "seen a wounded bird swim and dive with great aplomb."

Voice.—Some of the notes of this quail suggest the familiar *bob-white* of our eastern quail. Grinnell, Bryant, and Storer (1918) have described them very well, as follows:

The Valley Quail has a variety of notes which are used under different conditions and to express various meanings. When anxious or disturbed the members of a flock utter a soft *pit, pit, pit,* or *whit, whit, whit,* in rapid succession, as they run about under the brush or when about to take wing. Then there is a loud call used by the males to assemble the flock when scattered. This has been variously interpreted as *ca-loi'-o, o-hi'-o, tuck-a-hoe', k-woik'-uh, ki-ka-kee', ca-ra'-ho, tuck-ke-teu',* or more simply as *who-are'-you-ah.* However, the easiest and by far the most usual interpretation is *come-right'-here,* or *come-right-home,* with the accent on the second syllable. Sometimes when excited a bird calls *come-right, come-right, come-right-here.* In at least one instance a female bird has been observed to utter this call. The notes of the Valley Quail are less elaborate than those of the Desert Quail, the "crow" lacking the two additional notes which the latter gives at the end; also the Valley Quail lacks much of the conversational twitter of its desert relative.

Enemies.—Quail have numerous enemies; the eggs and young of these and other ground-nesting birds are preyed upon by crows, ravens, jays, snakes, raccoons, weasels, skunks, squirrels, and badgers; the adults also are pursued and killed by hawks, owls, coyotes, foxes, bobcats, and domestic cats and dogs. Gopher snakes are particularly fond of quail's eggs. Joseph Dixon (1930) tells a striking story of a brood of 19 young quail that was entirely destroyed by a pair of California jays, which he says are one of the quail's worst enemies; he saw four chicks carried off by one jay within 15 minutes.

Game.—The California quail, in its two forms, has often been referred to as *the* game bird of California, has been hunted by more sportsmen and market hunters than any other bird, and has been killed in enormous numbers. Its great abundance in former years seems almost unbelievable to-day. Dr. A. K. Fisher (1893) wrote:

Throughout the San Joaquin Valley, Mr. Nelson found it common about ranches, along water courses or near springs. It was excessively abundant at some of the springs in the hills about the Temploa Mountains and Carrizo Plain. In the week following the expiration of the close season, two men, pot hunting for the market, were reported to have killed 8,400 quail at a solitary spring in the Temploa Mountains. The men built a brush blind near the spring, which was the only water within a distance of 20 miles, and as evening approached the quails came to it by thousands. One of Mr. Nelson's informants who saw the birds at this place stated that the ground all about the water was covered by a compact body of quails, so that the hunters mowed them down by the score at every discharge.

Grinnell, Bryant, and Storer (1918) say:

In the days when the Valley Quail was plentiful far beyond its condition to-day, it was a common bird on the markets and could be obtained at practically every hotel and restaurant. Records show that during the season 1895–96 as many as 70,370 quail (mostly Valley Quail) were sold on the markets of San Francisco and Los Angeles; while an earlier report states that full 100,000 were disposed of in a single year in the markets of San Francisco. W. T. Martin, of Pomona, states that in 1881–84 he and a partner hunted Valley Quail in Los Angeles and San Bernardino Counties for the San Francisco markets. Eight to fourteen dozen were secured daily, and in the fall of 1883 the two men secured 300 dozen in 17 days. Martin himself secured 114 birds in one day's hunt. In 1881 and 1882 over 32,000 dozen quail were shipped to San Francisco from Los Angeles and San Bernardino Counties, and brought to the hunters engaged in the business one dollar a dozen. In those days restaurants charged thirty cents for quail-on-toast. By 1885 hunting had become unprofitable because of the reduction in the numbers of quail.

Quoting from T. S. Van Dyke (1892), they say:

At your first advance into the place where the quail last settled in confusion, a dozen or more rise in front of you and as many more on each side anywhere from 5 to 50 yards away. They burst from the brush with rapid flight and whizzing wing, most of them with a sharp, clear, *pit, pit, pit,* which apprizes their comrades of the danger and the course of escape taken. Some dart straight away in a dark blue line, making none too plain a mark against the dull background of brush, and vanish in handsome style, unless you are very quick with the gun. Others wheel off on either side, the scaling of their breasts showing in the sunlight as they turn, and making an altogether beautiful mark as they mount above the skyline. Some swing about and pass almost over your head, so that you can plainly see the black and white around their heads and throats, and the cinnamon shading of their under surfaces.

Although this quail is a splendid game bird and as good on the table as our eastern quail, all sportsmen who have shot both seem to agree that our bobwhite is a far more satisfactory bird to hunt. The valley quail will not lie to a dog, unless thoroughly frightened; it has a most exasperating habit of running, which is quite disheartening to both man and dog. Dwight W. Huntington (1903) referring to Mr. Van Dyke's comments on the former abundance and habits of this quail, says:

He said that when he first came to California, in 1875, quail in flocks now quite incredible soared out of almost every cactus patch, shook almost every hillside with the thunder of a thousand wings, trotted in strings along the roads, wheeled in platoons over the grassy slopes and burst from around almost every spring in a thousand curling lines. The same writer says that the partridges have already deserted many of the valleys and are now more often found in the hills, ready always to run and fly from one hillside to another, and "their leg power, always respectable enough to relieve you from any question of propriety about shooting at one running, they have cultivated to such a fine point that sometimes they never rise at all, and you may chase and chase and chase them and get never a rise." Writing at another time Mr. Van Dyke advises the shooter not to attempt to bag anything at first, but to spend all the time in breaking and scattering the coveys, racing and chasing after them and firing broadsides over their heads and in front of them, until they are in "a state of such alarm that they will trust to hiding." He then advises that the dog (which I presume has been used in coursing the birds) be tied to a shady bush and that the coat be laid aside, that the sportsman may travel fast after the scattered birds.

Occasionally they may behave differently and offer good sport, as in the following account by Henshaw (1874):

As a rule, their ways are not such as to endear them to the sportsman; for they are apt to be wary, and unless under specially favorable circumstances, are not wont to lie closely. I have, however, flushed a large bevy contiguous to a bushy pasture where the scrub was about knee-deep, with cattle-paths through it, and have had glorious sport. The birds lay so close as to enable me to walk almost over them, when they got up by twos and threes, and went off in fine style. The sportsman may now and then stumble upon such chances, but they do not come often. A bevy once up, off they go, scattering but little unless badly scared, the main body keeping well together; and having flown a safe distance they drop, but not to hide and be flushed one after another at the leisure of the sportsman. The moment their feet touch firm ground, off they go like frightened deer, and if, as is often the case, they have been flushed near some rocky hill, they will pause not a moment till they have gained its steep sides, up which it would be worse than useless to follow. Should they, however, be put up hard by trees, they will dive in among the foliage and hide, and there standing perfectly motionless will sometimes permit one to approach to the foot of the tree they are lodged in ere taking wing.

Winter.—What birds are left in the big flocks, after the sportsmen have taken their toll, remain together during fall and winter, formerly in great droves of hundreds, but now more often in flocks of 40 or 60. They are not migratory to any extent. A. C. Lowell, one of Major Bendire's (1892) correspondents in Nevada, told him that they were not able to stand the severe cold, accompanied by a heavy fall of snow in the Warner Valley; 2 feet of snow and 3 nights of 28° below zero killed most of the birds. On the other hand, Major Bendire (1892) tells of a flock that spent the winter successfully near Fort Klamath, Oreg., where the snowfall is quite heavy and the thermometer fell "more than once considerably below zero."

LOPHORTYX CALIFORNICA CATALINENSIS Grinnell

CATALINA QUAIL

HABITS

Based on a series of six specimens Dr. Joseph Grinnell (1906) gave the name *Lophortyx californica catalinensis* to the California quail inhabiting Santa Catalina Island. He characterized it as similar to the valley quail, "but about 9 per cent larger throughout, and coloration somewhat darker; similar to *L. c. californicus*, but larger and much less deeply brownish dorsally." He says further:

The bulkiness of *catalinensis* is at once apparent when one sees it among specimens of the mainland *vallicola*. The tail is particularly long, the rectrices being proportionately broader. The bill is heavier, and the toes and tarsi decidedly stouter These characters hold equally in the males and females. In coloration *catalinensis* shows a deepening of shades especially on the lower surface. In both sexes the flanks and lower tail-coverts are more broadly streaked with brown; the terminal black edgings of the lower breast feathers are broader, and the light markings beneath are suffused with deeper ochraceous. Especially in the female of *catalinensis* is the lower surface darker than in *vallicola*, due to the encroachment of the dark portions of each parti-colored feather upon the light part. The dorsal surface is not however much browner than in *vallicola*—it is decidedly slaty as compared with the deep bright vandyke brown of *californicus* from the vicinity of San Francisco Bay.

It was thought at first that these quail had been introduced from the mainland, but more recent evidence shows that they were probably native on the island, which perhaps was once connected with the mainland. Doctor Grinnell (1906) was assured that they were there at least as early as 1859.

After examining a series of 16 skins, at a later date, Doctor Grinnell (1908) writes:

When compared with a series of the mainland *vallicola* the island birds are distinguished by larger size, especially of the feet, broadness of terminal barring on the posterior lower surface, and broadness of shaft-streaks on lower tail coverts and flanks. An additional character which shows up in the larger series is the averaging more intense and extensive chestnut patch on the hind chest, in the male, of course. This does not seem to be due to the different "make" of the skins. An examination of individual variation in the two series shows that any one character alone is not diagnostic of every single individual. For instance, a small-footed island bird can be duplicated in that respect by an extra large-footed mainland bird. But at the same time the barring and streaking of the former renders it easily recognizable. Then in the matter of barring on the lower surface, a mainland female appears as heavily marked as the average island female. But at the same time the former has a decidedly shorter wing and weaker foot. It is therefore evident that there is a mergence of separate characters thru individual variation; and according to the criterion now apparently most popular, the island form would be given a trinomial appellation. The binomial, however, appears to me most useful, as it signifies complete isolation because of the intervention of a barrier.

LOPHORTYX CALIFORNICA PLUMBEA Grinnell

SAN QUINTIN QUAIL

HABITS

Under the name *Lophortyx californica plumbea* Dr. Joseph Grinnell (1926) has separated the quail found in certain parts of northern Lower California from the subspecies found on either side of it. He describes the new form as "in general characters similar to *Lophortyx californica vallicola* and *L. c. achrustera*, but tone of coloration clearer, less buffy or brownish; gray or lead-color on dorsum, foreparts and sides, and remiges, more slaty than in either." He says further:

It should go without saying that in quail fresh fall plumages should be relied upon chiefly, if not altogether, in seeking color values. When this is done, the quail of the "San Quintin district" show themselves to differ in mass effect appreciably from Valley Quail from anywhere north of the Mexican line. San Diego County birds, even, and those from Riverside and Inyo counties, well east of the desert divides, all are markedly browner dorsally, the remiges browner, the chest less clearly ashy gray, and the "ground" tone of the hinder flanks and crissum more brightly tan. This holds for both sexes. The creamy area on the lower chest of male *plumbea*, while not so pale as in *achrustera*, is not so deep-toned as in average *vallicola*. In females the grayness about the head and on the chest in *plumbea* is almost constantly diagnostic; and in both sexes, the plumbeous tone of the remiges is as a rule strikingly different from the brown tone in *vallicola*. In the dried specimens, the feet and legs of *plumbea* average blacker than in *vallicola*.

Referring to its distribution and haunts, he says in a later publication (1928):

Abundant resident of the northwestern portion of the territory, roughly between latitudes 30° and 32°—practically as comprised in the San Quintin subfaunal district. While the metropolis of the subspecies lies on the Pacific slope of the peninsula, colonies or pairs occur also to the eastward, in canyons at the east base of the Sierra San Pedro Martir, and even at San Felipe, on the Gulf coast (Mus. Vert. Zool.). As regards life-zone, inhabits the Lower Sonoran, Upper Sonoran, and Transition, without any seeming choice. Associationally, adheres to an open or interrupted type of chaparral, especially as adjacent to springs or water-courses. Altitudinally, extends from sea level up to as high as 8,800 feet, on the Sierra San Pedro Martir.

Eggs.—The eggs are probably indistinguishable from those of other races of California quail. Griffing Bancroft (1930) says that the measurements of 150 eggs average 30.6 by 23.3 millimeters.

LOPHORTYX CALIFORNICA ACHRUSTERA Peters

SAN LUCAS QUAIL

HABITS

James L. Peters (1923) is responsible for the name *L. c. achrustera*, which he has applied to the quail of this species inhabiting southern Lower California. Based on the examination of a series of 27 males and 15 females, he says that it is "similar to *Lophortyx californica vallicola* (Ridgway), but slightly paler above; band across breast grayer; the buffy patch on the lower breast of the male much paler; dark feather-edgings on the lower breast, middle and sides of abdomen, narrower; flanks paler."

William Brewster (1902), with much of the same material, noticed that the Lower California specimens were "slightly paler" than California birds and their bills averaged "a little heavier," but he did not consider these differences well marked or constant. But Peters (1923) says that "while the bill character is of no diagnostic value, the color characters are constant and serve to distinguish the valley quail of southern Lower California almost at a glance."

Griffing Bancroft (1930) says of their haunts: "They insist on riparian associations, but they follow these without regard either to altitude or to the character of the country adjoining the stream beds. They definitely do not require the presence of water."

Nesting.—Of the nesting of the San Lucas quail, Bancroft (1930) says:

Our experience with the breeding of these quail was limited to San Ignacio. That was because the nests were too well hidden to be found, except accidentally, and those we saw were shown to us by the natives. Three of the sites were in damp ground in rank grass; one of them, to our surprise, on a tiny islet in a swamp. Two nests were in vineyards, two in natural cavities among the sucker growths of date palms, and one was under a lava rock on the mesa. In all but the last three cases the birds had excavated a cup nearly as deep as it was broad and had lined it with materials brought in, grass, leaves, and feathers. The breeding season commences about the first of June and is hardly well under way until after the middle of that month. The number of eggs in a clutch is rather consistently ten or eleven, sixteen being the most we found in any one nest.

Eggs.—He gives the average measurements of 80 eggs as 32.3 by 24.7 millimeters. The measurements of 15 eggs in P. B. Philipp's collection average 31.8 by 24.3 millimeters; the eggs showing the four extremes measure **33.6** by **25.4**, **29.7** by 23.4, and 31.3 by **23.1** millimeters.

LOPHORTYX GAMBELI GAMBELI Gambel

GAMBEL'S QUAIL

HABITS

Gambel's quail is also very appropriately called the desert quail, for its natural habitat is the hot, dry desert regions of the Southwestern States and a corner of northwestern Mexico. Its center of abundance is in Arizona, but it ranges east to southwestern New Mexico and El Paso, Tex., and west to the Colorado and Mohave Deserts in southeastern California. On the western border of its range it is often associated with the valley quail and has been known to hybridize with it.

This beautiful species was discovered by Dr. William Gambel "on the eastern side of the Californian range of mountains in 1841" and named in his honor, according to John Cassin (1856), who gives us the first account of its distribution and habits, based largely on notes furnished by Col. George A. McCall. He did not meet with it west of the Colorado Desert barrier in California or east of the Pecos River in Texas.

We found Gambel's quail very common in southern Arizona, especially in the lower river valleys, where the dense growth of mesquite (*Acacia glandulosa*) afforded scanty shade, or where they could find shelter under the spreading green branches of the paloverdes, which in springtime presented great masses of yellow blossoms. They were even more abundant in the thickets of willows along the streams or in the denser forests of mesquites, hackberries, and various other thorny trees and shrubs. We occasionally flushed a pair as we drove along the narrow trails, but more often we saw them running off on foot, dodging in and out among the desert underbrush until out of sight. My companion on this trip, Francis

C. Willard, has sent me the following notes, based on his long experience in Arizona:

Gambel's quail is essentially a bird of the areas in southern Arizona where the mesquite abounds. Unlike their neighbor, the scaled quail, they seem to require the close proximity of a water supply. They are, therefore, found principally along the few living streams and close to permanent water holes. I found them swarming in the mesquite forest along the Santa Cruz River south of Tucson and almost as plentiful along the Rillito between Tucson and the mountains. In the valley of the San Pedro River they were also present in large numbers. Between the valley of this last-named river and the various ranges of mountains fringing it are long sloping mesas from 5 to 20 miles wide where the "black topknot" is rarely seen except close to the infrequent water holes. In the foothills of the Dragoon, Huachucas, Whetstones, Chiricahuas, and other less well-known ranges this quail again appears in some numbers but nothing like those in the lower valleys.

Dr. Elliott Coues (1874) has given us the best account of this quail, which I shall quote from quite freely. He says of its haunts:

Gambel's Quail may be looked for in every kind of cover. Where they abound it is almost impossible to miss them, and coveys may often be seen on exposed sand-heaps, along open roads, or in the cleared patches around settlers' cabins. If they have any aversion, it is for thick high pine-woods, without any undergrowth; there they only casually stray. They are particularly fond of the low, tangled brush along creeks, the dense groves of young willows that grow in similar places, and the close-set chaparral of hillocks or mountain ravines. I have often found them, also, among huge granitic boulders and masses of lava, where there was little or no vegetation, except some straggling weeds; and have flushed them from the dryer knolls in the midst of a reedy swamp. Along the Gila and Colorado they live in such brakes as I described in speaking of Abert's Finch; and they frequent the groves of mezquite and mimosa, that form so conspicuous a feature of the scenery in those places. These scrubby trees form dense interlacing copses, only to be penetrated with the utmost difficulty, but beneath their spreading scrawny branches are open intersecting ways, along which the Quail roams at will, enjoying the slight shade. In the most sterile regions they are apt to come together in numbers about the few water-holes or moist spots that may be found and remain in the vicinity, so that they become almost as good indication of the presence of water as the Doves themselves. A noteworthy fact in their history, is their ability to bear, without apparent inconvenience, great extremes of temperature. They are seemingly at ease among the burning sands of the desert, where, for months, the thermometer daily marks a hundred, and may reach a hundred and forty, "in the best shade that could be procured," as Colonel McCall says; and they are equally at home the year round among the mountains, where snow lies on the ground in winter.

In New Mexico, according to Mrs. Florence M. Bailey (1928), it is found

in the Lower Sonoran Zone in quail brush (*Atriplex lentiformis*) and creosote, and in hot mesquite valleys or their brushy slopes, in screw bean and palo verde thickets and among patches of prickly pear. It is not generally found so far from water as the Scaled Quail, which eats more juicy insect food, but at times both are seen in the same landscape.

In inhabited regions, in places where cattle trails lead to water, the Gambel's pretty foot prints call up pleasant pictures of morning procession of thirsty little "black-helmeted" pedestrians, talking cheerfully as they go. For it seems most at home about small farms, such as those cultivated by the Spanish-Americans, which dot the narrow canyons and river valleys.

Courtship.—Springtime in Arizona is most charming as the desert plants burst into bloom with their profusion of many colors. The new fernlike foliage of the mesquite mingles with dangling yellow tassels. The long slender stems of the ocotillo are tipped with vermilion spikes. Even the lowly creosote bush is clouded with yellow haze. The various chollas and the pricklypears are studded with pink, yellow, or crimson flowers, and the little rainbow cactus blooms by the roadside with a wealth of large magenta and yellow blossoms. Even the giant cactus supports a crown of white, and the paloverde is the showiest of all, a great bouquet of brilliant yellow. Then we may look for the trim figure of the cock quail, perched on some low tree, bush, or stump, and listen to his challenging love call.

Major Bendire (1892) has described his courtship very well, as follows:

During the mating and breeding season, the former commencing usually in the latter part of February, the latter about the first week in April and occasionally later, according to the season, the male frequently utters a call like *yuk-kae-ja, yuk-kae-ja*, each syllable distinctly articulated and the last two somewhat drawn out. A trim, handsome, and proud-looking cock, whose more somber-colored mate had a nest close by, used an old mesquite stump, about 4 feet high, and not more than 20 feet from my tent, as his favorite perch, and I had many excellent opportunities to watch him closely. Standing perfectly erect, with his beak straight up in the air, his tail slightly spread and wings somewhat drooping, he uttered this call in a clear strong voice every few minutes for half an hour or so, or until disturbed by something, and this he repeated several times a day. I consider it a call of challenge or of exultation, and it was taken up usually by any other male in the vicinity at the time. During the mating season the males fight each other persistently, and the victor defends his chosen home against intrusion with much valor. It is a pleasing and interesting sight to watch the male courting his mate, uttering at the time some low cooing notes, and strutting around the coy female in the most stately manner possible, bowing his head and making his obeisance to her. While a handsome bird at all times, he certainly looks his best during this love-making period.

Nesting.—My experience with the nesting habits of Gambel's quail is limited to three nests found near Tucson, Ariz., in 1922. On May 19 we were hunting through the mesquite forest, a large tract of once heavy timber that had been much depleted by the raids of Papago Indian woodchoppers. There were only a few large trees left, some very large hackberry trees, which were more or less scattered with many open spaces; but there was plenty of cover left in the extensive thickets of small mesquite and thorny undergrowth, or in the patches of large mesquite, oaks, and hackberry. While walking along the edge of a dry ditch, we flushed a Gambel's quail from its nest under a tiny thorn bush. The nest was a rather deep hollow lined with sticks, straws, leaves, and feathers; it held 10 eggs.

Two days later, while hunting among the giant cactuses, which here were scattered over an open plain, scantily overgrown with low mesquite and greasewood bushes in dry stony soil, we flushed another quail from under a large mesquite bush; way under it, at the base of the trunk and almost beyond reach in the thorny tangle, was its nest with 16 eggs. The next day while investigating a Palmer's thrasher's nest, five feet above the ground in a cholla, we were surprised to find in it three eggs of this quail. I have since learned that it is not unusual for this quail to use old nests of thrashers or cactus wrens. Perhaps the birds have learned by sad experience that ground nests are less safe.

My companion, Frank Willard, who has had much wider experience than I, has sent me the following notes:

From early in May well into July and sometimes even into August nests with eggs may be found. The last week of May and the first week in June seem to be the height of the egg-laying season. The eggs of *gambelii* are laid in more exposed situations than those of *squamata*. The most frequently chosen site is at the foot of a small mesquite or other bush where a slight hollow is scratched in the dry ground. There is one protection, however, which the quail seems to find necessary. There must be something to shade them from the hot midday sun. The scanty shade afforded by the fern leaf of the mesquite is sufficient but there must be some at least.

Two or more females lay their eggs in the same nest very frequently. I have had nests under observation where two or more eggs were added daily to the complement therein. It has occurred to me that this is a wise provision of nature to secure a nest full of eggs with as little delay as possible so that incubation could be undertaken promptly and an even hatching take place without the eggs being exposed to the dry desert heat until one bird could lay a full set, which averages a dozen eggs or more. A few days of exposure to the dry air without the moisture from the body of the sitting bird would make many of the eggs sterile. Nearly every egg in nests where incubation had commenced was fertile and I seldom found more than one unhatched egg among the débris of a nest from which the young had hatched and gone.

On May 14, 1908, I went to collect a set of Gambel's quail eggs which I had been watching for some time. The previous day there had been 18 eggs in the nest, some of them those of the scaled quail. As I looked around the large rock behind which the nest was concealed I found the female quail fluttering above her nest in which was coiled a large rattlesnake. With head uplifted it was striking at the bird which deftly avoided the blows. On my appearance the bird flew away. I prodded the snake, driving it from the nest, and then killed it. Eleven rattles adorned its tail. There were 16 eggs left, all of which were fresh. I foolishly neglected to open the snake and look for the two missing eggs.

I once found a Gambel quail sitting on 16 eggs laid in a Palmer thrasher's nest 5 feet up in a cholla sheltered by a large sycamore. The bird sat very close. On another occasion I found several eggs of this quail in the nest of a Palmer thrasher and the thrasher sitting on them and her own three eggs.

Major Bendire (1892) writes:

The nest of Gambel's Partridge is simply a slight oval-shaped hollow, scratched out in the sandy soil of the bottom lands, usually alongside of a bunch of "sacaton," a species of tall rye grass, the dry stems and blades of last year's growth hanging down on all sides of the new growth and hiding the nest well from view. Others are placed under, or in a pile of, brush or drift brought down from the mountains by freshets and lodged against some old stump, the roots of trees, or other obstructions on some of the numerous islands in the now dry creek beds, refreshing green spots amid a dreary waste of sand. According to my observations only a comparatively small number resort to the cactus and yucca covered foothills and mesas some distance back, where the nests are usually placed under the spreading leaves of one of the latter named plants. If grain fields are near by they nest sometimes amidst the growing grain in these, and should the latter be surrounded by brush fences, these also furnish favorite nesting sites.

Among the nests observed by me two were placed in situations above ground. One of these was found June 2 on top of a good-sized rotten willow stump, about 2½ feet from the ground, in a slight decayed depression in its center, which had, perhaps, been enlarged by the bird. The eggs were laid on a few dry cottonwood leaves, and were partly covered by these. Another pair appropriated an old Road-runner's nest, *Geococcyx californianus*, in a mesquite tree, about 5 feet from the ground, to which apparently a little additional lining had been added by the bird. The nest contained 10 fresh eggs when found on June 27, 1872.

M. French Gilman (1915) found this quail quite tame and confiding, nesting in much-frequented localities, for he says:

Two nests were in the school woodpile, containing 19 and 13 eggs, respectively. Another, in a pile of short boards and kindling about 10 feet from the school woodshed, had 7 eggs in it. The nest out in the fields had 9 eggs, and was at the base of a Lycium bush. About the middle of June I put some straw in an old nail keg, open at one end, and placed it on its side in the forks of a mesquite tree about two feet from the ground. The mesquite had some saplings starting from the trunk that sheltered the keg. June 24, I found that a quail had moved in and had laid two eggs. Later she completed the set, only eight eggs, and successfully hatched all but one. She was quite tame on the nest, and would not be scared off by any mild measures. I tried hammering on the rear of the keg, rolling it gently and talking to her, requesting her to get off and let me count the eggs, but unless I put my hand at the front of the keg she sat pat.

Eggs.—Ten or a dozen eggs constitute the average set, but sets of 18, 19, and 20 have been recorded. These large sets are doubtless the product of two hens, as indicated by Mr. Willard's observations above and by the fact that these sets usually contain two types of eggs. The eggs are short ovate in shape and sometimes somewhat pointed; they are smooth and slightly glossy. The ground colors vary from dull white to "cartridge buff" or "pale pinkish buff." They are irregularly spotted and blotched with a mixture of large blotches, small spots, and fine dots; sometimes the ground color is well covered, but more often not. The markings are in shades of dark or purplish browns, from "warm sepia" or "chocolate" to "snuff brown." One of my sets has 5 eggs of the ordinary type and 11 beautiful eggs with a "pinkish buff" ground color, well covered with small blotches, spots, and fine dots of "deep grayish lavender" and "deep heliotrope gray." This set shows the purplish bloom referred to by Bendire (1892), which turns dark brown, when washed, and dries out to purple again. The measurements of 99 eggs in the United States National Museum average 31.5 by 24 millimeters; the eggs showing the four extremes measure **34** by **26, 28.5** by 24, and 30 by **23** millimeters.

Young.—Incubation seems to be performed by the female alone and is said to require from 21 to 24 days. Both sexes share in the care of the young. Very early and very late dates indicate that at least two broods are raised in a season. Coues (1874) says of the young:

> They run about as soon as they are hatched, though probably not "with half shell on their backs," as some one has said. In a few days they become very nimble, and so expert in hiding that it is difficult either to see or catch them. When the mother bird is surprised with her young brood, she gives a sharp warning cry, that is well understood to mean danger, and then generally flies a little distance to some concealed spot, where she crouches, anxiously watching. The fledglings, by an instinct that seems strange when we consider how short a time they have had any ideas at all, instantly scatter in all directions, and squat to hide as soon as they think they have found a safe place, remaining motionless until the reassuring notes of the mother call them together again, with an intimation that the alarm is over. Then they huddle close around her, and she carefully leads them off to some other spot, where she looks for greater security in the enjoyment of her hopes and pleasing cares. As long as they require the parent's attention they keep close together and are averse to flying. Even after becoming able to use their wings well, they prefer to run and hide, or squat where they may be, when alarmed. If then forced up, the young covey flies off, without separating, to a little distance, often re-alighting on the lower limbs of trees or in bushes, rather than on the ground. As they grow older and stronger of wing, they fly further, separate more readily, and more rarely take to trees; and sometimes, before they are fully grown, they are found to have already become wary and difficult of approach. As one draws near where a covey is feeding, a quick, sharp cry from the bird who first notices the approach alarms the whole, and is quickly repeated by the rest, as they start to run, betraying their course by the rustling of dried leaves. Let him step nearer, and they rise with a whirr, scattering in every direction.

Plumages.—In natal down the young Gambel's quail is much like the California quail at the same age, but the colors are paler and duller. The front half of the crown and the sides of the head are from "clay color" to "pinkish buff"; there is a broad band of "russet," bordered with black, from the tiny topknot to the hind neck, and a dark brownish auricular patch; the rest of the upper parts is light "pinkish buff," banded lengthwise and blotched with "warm sepia"; the underparts are pale grayish buff.

As with all other quail and grouse, the juvenal plumage begins to grow soon after the chick is hatched, appearing first on the wings and scapulars; the topknot, or crest, appears at once, "hazel" at first and then dull brown. In fresh juvenal plumage the feathers of the mantle are variegated with "cinnamon," "pinkish buff," and black, each with a broad, median, white stripe; later these fade to gray, pale buff, and dull brown; the scapulars have buffy edgings; the tertials are at first "pinkish buff," later grayish, barred with dusky and tipped with white; the tail is grayish, barred near the tip with dusky, dull whitish, and dull buffy; the underparts are grayish white, faintly barred with dusky; on the head, which is the last part to be feathered, the forehead is dusky and the crown "mikado brown." In this plumage the sexes are alike, and the birds closely resemble young California quail of the same age.

The birds are hardly fully grown and the juvenal plumage is hardly complete before the change into the first winter plumage begins on the back and wing coverts. This change is accomplished by a complete postjuvenal molt, except that the two outer juvenal primaries, and their coverts, on each wing are retained for a full year. Otherwise the young birds are practically indistinguishable from adults after the molt is completed in October, or later.

Subsequent molts for old and young birds consist of a very limited prenuptial molt in April and May, involving only the head and neck, and a complete postnuptial molt in August and September.

Food.—In the stomachs, collected from January to June, Dr. Sylvester D. Judd (1905) found that less than half of 1 per cent of the food consisted of insects, which included ants, beetles, grasshoppers, leaf hoppers, and stink bugs. Vegetable matter made up 99.52 per cent of the food; 3.89 per cent of this was grain, 31.89 per cent seeds, and the balance, 63.74 per cent, was made up of leaves and shoots of various plants. The grain included corn, wheat, and oats, much of which was probably picked up among the grain shocks, where large flocks have been seen feeding with domestic poultry. The seeds were largely those of leguminous plants such as alfalfa, bur clover, and mesquite, and also of alfilaria, mustard, chickweed, peppergrass, and atriplex. Succulent foliage and shoots form by far the larger percentage of the food. Of this, alfalfa, bur clover, and the foliage of other legumes constitute the greater part. Both the green leaves and pods of alfalfa are freely eaten. In spring this quail shows a fondness for buds, and in some localities its flesh has a distinctly bitter taste due to a diet of willow buds. Certain kinds of fruit also are eaten.

Baird, Brewer, and Ridgway (1905) say that during the summer Gambel's quail feeds extensively on the berries of the nightshade. Grinnell, Bryant, and Storer (1918) say:

Evidence is also at hand that this quail, like many other desert animals, feeds upon the fruit and seeds of certain kinds of cactus. Stomachs of Gambel Quail collected along the Colorado River in the spring of 1910 contained masses of mistletoe berries, and, at the time the mesquites were first coming into leaf, quantities of the tender green foliage of this plant.

It has been said that these quail can not exist very far from water, to which they have to resort twice a day to drink, but Dr. Robert C. Murphy (1917) found them abundant at all hours about his camp in the Colorado Desert, which "was upwards of 20 miles from the river, 7 miles from the miserable hole of the Tres Pozos, 10 miles from the Laguna Salada, and an equal distance from the nearest mountain 'tinaja.' The soil was everywhere sandy and porous; not a suggestion of moisture was to be detected even in the beds of the deepest barrancas." He also says: "The crops of the specimens taken early in April were mostly crammed with caterpillars of the genus *Hemileuca*, assorted sizes of which were at that time marching in legions across the desert."

Behavior.—Gambel's quail is not so persistent a runner as the scaled quail, but it is quite reluctant to fly and prefers to escape by running very swiftly away among the underbrush. It does not often squat and hide, as our bobwhite does, for this would be a poor way to escape from its natural enemies. When it is forced to fly its flight is swift and strong and often protracted to a long distance in the open; in thick brush, which it largely frequents, it flutters rather awkwardly away for a short distance; flocks separated in this way soon begin to call and gather again. We found it not particularly shy, especially among the mesquite thickets, where it probably felt that it was not observed. About the ranches and farmhouses it becomes very tame, often feeding with the domestic poultry. During the heat of the day it rests quietly in the scanty shade of the mesquites, or under denser thickets, or even on the branches of leafy trees. It comes out to feed and drink early in the morning and toward night. It takes to the trees readily at any time and probably roosts in trees at night. When a flock of Gambel's quail is feeding there is usually a sentinel on guard.

Voice.—Cassin (1856), quoting from Colonel McCall's notes, says of the voice of the male in June:

A very good idea may be formed of his cry by slowly pronouncing, in a low tone, the syllables "*kaa-wale*," "*kaa-wale*." These notes when uttered close at hand, are by no means loud; yet it is perfectly astonishing to what a distance they may be heard when the day is calm and still. There was to me something extremely plaintive in this simple love-song, which I heard for the first time during a day of burning heat passed upon the desert.

Again he writes:

Later in the season, when a covey is dispersed, the cry for assembling is "*qua-el*," "*qua-el*." The voice at all seasons bears much resemblance to that of the *California Partridge*—having, in its intonation, no similarity to the whistle of the Virginia or common partridge.

Bendire (1892) gives the mating call of the male as "*yuk-kae-ja, yuk-kae-ja*, each syllable distinctly articulated and the last two somewhat drawn out." Another note, given while moving about in coveys, "resembled the grunting of a sucking pig more than anything else, and it is rather difficult to reproduce the exact sound in print. Any of the following syllables resembles it, *quoit*, *oit*, *woet*, uttered rapidly but in a low tone. The alarm note is a sharp, discordant *craer*, *craer*, several times rapidly repeated, and is usually uttered by the entire covey almost simultaneously."

Dr. Joseph Grinnell (1904) quotes from Joseph Mailliard's notes, as follows:

The notes of the desert quail differ from those of the valley quail in variety, and to a certain extent in character, though they have some notes in common. The "crow" of the latter consists of three notes, varying in length and accent according to the call given, in one case the last note being a falling one. The "crow" of the desert quail, while rather similar to the other, has two additional notes at the end, rendered in a softer tone. Besides the alarm calls the valley quail has a few twittering or conversational notes, while the other species has a lot of these, quite varied and often given in a way that seems remarkably loud to one accustomed only to the notes of the former. Another peculiarity of the desert quail is the queer sound that it makes as it rises from the ground on being surprised into flight—the sort of screeching cackle, on a small scale, that a hen makes when frightened from her nest.

Game.—Although Gambel's quail is a plump and delicious morsel for the table, it is an exasperating bird to hunt. It loves the thickest and thorniest cover and frequents the roughest and hardest country, through which it runs, and keeps on running, faster than a man can follow; often it will take refuge in a rocky creek bed or canyon, where it is hopeless to follow. What few I have shot have required more vigorous leg exercise than they were worth and usually had to be shot on the run. When flushed in the open it flies swiftly and requires good shooting. The birds will not lie to a dog, so the best bird dogs are utterly useless in hunting them, except as retrievers.

In past years these quail were an important item in the market hunter's game bag. Herbert Brown (1900) was informed by an express agent that 3,000 dozen quail were shipped out of Salt River Valley in 1889 and 1890. He says further:

The Mohawk valley, in Yuma county, is probably the most prolific breeding spot in the territory. It was, at one time, a favorite place for trappers and pot-hunters, and it was not until the game law had been amended that their nefarious practices were broken up. In six weeks, in the fall of 1894, no less than 1,300 dozens were shipped to San Francisco and other California markets. The price at first realized, so I was told by the shippers, was $1.12½ per dozen, but later 60 cents only were realized. The Quail were trapped, their throats cut, then sacked and shipped by express. I was told by one of the parties so engaged that he and his partner caught 77 dozens in one day. They used eight traps and baited with barley. Their largest catch in one trap, at one time, was 11 dozens. At the meeting of the next legislature the game law was again amended, and it was made a misdemeanor to trap, snare, or ship Quail or Partridges from the Territory. This effectually stopped the merciless slaughter of the gamiest bird in Arizona—Gambel's Partridge.

Enemies.—Coues (1874) writes:

Man is, I suppose, the Quail's worst enemy; what the White does with dog and gun the Red accomplishes with ingenious snares. The Indians take great numbers alive in this way, for food or to trade with the whites along the Colorado; and they use the crests for a variety of purposes that they consider ornamental. I saw a squaw once who had at least a hundred of them strung on a piece of rope-yarn for a necklace. But the birds have other foes; the larger Hawks prey upon them, so also do the wolves, as I have had good evidence upon one occasion, when hunting in a precipitous, rocky place near Fort Whipple. I heard a covey whispering about me as they started to run off in the weeds, and followed them up to get a shot. They passed around a huge boulder that projected from the hill-side, and then, to my surprise, suddenly scattered on wing in every direction, some flying almost in my face. At the same instant a wolf leaped up from the grass, where he had been hiding, a few feet off, intending to waylay the covey, and looking very much disappointed, not to say disgusted, at the sudden flight.

The quail have numerous other enemies. Coyotes, foxes, wildcats, and various hawks and owls kill the old birds and young; even the little pigmy owl has been known to kill an adult Gambel's quail. Skunks, rats, rock squirrels, snakes, Gila monsters, and even land terrapins eat the eggs. Fortunately these quail are prolific breeders, so they are not exterminated.

DISTRIBUTION

Range.—Southwestern United States and northern Mexico. The range of Gambel's quail extends **north** to southern California (rarely Los Angeles, Hesperia, and Daggett); southern Nevada (Ash Meadows and Pahranagat Valley); southern Utah (Hamblin, Harmony, and Fruita); Arizona (Cedar Ridge, Roosevelt, Nantan Plateau, and Blue); and central Texas (Eagle Springs). **East** to southwestern New Mexico (Socorro and Las Cruces); central

Texas (Eagle Springs and Fort Clark); and southeastern Sonora (Camoa). **South** to southern Sonora (Camoa and Guaymas); and northern Lower California (Laguna Salada). **West** to northern Lower California (Laguna Salada and Signal Mountain); and southern California (Calexico, Pelican Lake, Agua Dulce, Palm Springs, and rarely Los Angeles). Some of the northern localities, such as those in Utah, may possibly be the result of introductions in contiguous areas. Bryant (1889) states that "a few pairs with small young were seen" on the western side of Lower California at about latitude 30° N., but it is likely that the birds seen were San Quintin quail (*L. c. plumbea*). There is one record of the occurrence of *gambeli* at San Diego, Calif., where a female was found (1924) mated with a male valley quail (*L. c. vallicola*).

It is difficult to outline the natural range of this quail, since it has been transplanted extensively. For example, they were found rather commonly at Furnace Creek, Death Valley, having been introduced there from Resting Springs, Calif., by the borax company. They are easily trapped and for this reason probably have been favorites in many ill-advised projects. In the early nineties they were introduced into Massachusetts (Marthas Vineyard). Other eastern experiments were made in Pennsylvania and Kentucky. Naturally, all these were failures, as were also introductions in northern California and Washington.

They were, however, successfully acclimated on San Clemente Island, Calif., and there have been several successful introductions well outside the normal range in Arizona (Snowflake, Holbrook, Vernon, and Colfax Counties) and New Mexico (Cortez, Gallup, San Juan Valley, and Huntington). A successful introduction also is reported from Montana (Blue Creek, near Billings). The most remarkable achievement, however, is the transplantation of these birds to western Colorado. About 1,000 birds from southern California (which locality gave rise to the name "California quail," under which they have appeared in the literature on Colorado birds) were liberated at Montrose in 1885 or 1889. From that point the birds have increased and spread over the Uncompahgre Plateau and the valleys of the Grand and Gunnison Rivers. Because of plumage changes that have taken place since the introduction, they have been described as a subspecies (*L. g. sanus*). There is no definite record of the natural occurrence of Gambel's quail anywhere in Colorado.

Egg dates.—Arizona: 68 records, March 19 to September 20; 34 records, April 20 to May 29. California: 8 records, March 26 to June 2.

LOPHORTYX GAMBELI SANUS Mearns

OLATHE QUAIL

HABITS

The Gambel's quail of southwestern Colorado has been described by Dr. Edgar A. Mearns (1914) as "rather larger than the average" of Arizona birds, from which "it differs in coloration as follows: Adult male with upper parts neutral gray (Ridgway, 1912), unwashed with olive; crown chestnut-brown instead of hazel; chest-patch cartridge buff instead of warm buff or chamois. Adult female with upper parts as in the male, differing from *gambelii* and *fulvipectus* in having the crown darker (sepia instead of cinnamon drab); chin and throat darker and more grayish; chest and abdomen pale olive-buff instead of cream color."

The type specimen came from Montrose County, but it seems to be a rare bird even there. Its habits are probably no different from those of the species elsewhere.

CYRTONYX MONTEZUMAE MEARNSI Nelson

MEARNS'S QUAIL

HABITS

One of the handsomest and certainly the most oddly marked of the North American quails presents a bizarre appearance when closely examined; one look at its conspicuously marked face would brand it as a clown among birds; its dark-colored breast is contrary to the laws of protective coloration and would make it very conspicuous on open ground. But when one tries to find it in its native haunts, squatting close to the ground among thick underbrush, weeds, and grass, one realizes that its dark belly and spotted flanks are completely concealed, that the grotesquely painted face becomes obliterated among the sharp lights and shadows, and that the prettily marked back matches its surroundings so well that the bird is nearly invisible.

Mearns's quail has been called the "fool quail" because it has learned to trust this wonderful protective coloration and lie close, rather than trust to its legs or its wings to escape. How successful it has been nobody knows. The only two I ever saw I almost stepped on before they flew, and I wonder how many more I walked near without knowing it. I question whether it is as much of a fool as it is said to be. It lives in different haunts and has developed different habits from its neighbors, the scaled and Gambel's quails. Its shape and carriage, its white-spotted sides, and its habit of clucking as it walks or feeds have suggested a possible relationship with the guinea fowls.

The haunts of Mearns's quail are generally far removed from the habitations of man. Major Bendire (1892) quotes William Lloyd as saying that, in Texas, "the favorite resorts of the Massena Partridge are the rocky ravines or arroyas that head well up in the mountains. They quickly, however, adapt themselves to changed conditions of life and are now to be seen around the ranches picking up grain and scratching in the fields. In the vicinity of Fort Davis, Texas, they have been exceptionally numerous and may frequently be seen sitting on the stone walls surrounding grainfields in Limpia Canon."

In Arizona we found them in the lower parts of the canyons and in the foothills of the Huachucas and the Chiricahua Mountains, where the ground was rough and more or less rocky, with tall tufts of grass, low bushes, scattered mescals, and small oaks. They range up the sides of rocky ravines and into the mountains up to 9,000 feet in summer and are seldom found below 4,000 feet. In Apache County, according to Major Bendire (1892), "the favorite localities frequented by this species during the breeding season are thick live-oak scrub and patches of rank grass, at an altitude of from 7,000 to 9,000 feet. Here they are summer residents only, descending to much lower altitudes in winter." Henry W. Henshaw (1874) writes: "This beautiful partridge is quite a common resident in the White Mountains, near Apache, Ariz., where, in summer, it seems to shun the open valleys, and keeps in the open pine-woods, evincing a strong preference for the roughest, rockiest localities, where its stout feet and long, curved, strong claws are admirably adapted to enable it to move with ease."

Nesting.—Major Bendire (1892) refers to two nests described as slight hollows, one under a small shin-oak bush, the other alongside a sotol plant. He quotes descriptions of two other nests quite fully. Otho C. Poling wrote:

I was climbing up a steep mountain side on the northeast of the Huachuca Mountains, some 10 miles north of the border, when, at an elevation of about 8,000 feet, I flushed the female almost directly under my feet and shot it. The hillside was covered in places with patches of pines and aspens, as well as with low bushes and grasses. The nest was directly under a dead limb which was grown over with dead grass, and so completely hidden that until I had removed the limb and some of the grass it was not discernible at all. The nest was sunken in the ground, and composed of small grass stems, arched over, and the bird could only enter it by a long tunnel leading to it from under the limb and the grass growing around it. The eggs were eight in number and naturally white, but they were badly stained by the damp ground, their color being now a brownish white. They were almost hatched. The female must have remained on them all the time to have caused such uniform incubation and preserved the eggs from spoiling by the excessive dampness.

G. W. Todd's notes state:

The only nest of this species I have ever seen was situated under the edge of a big bunch of a coarse species of grass, known as "hickory grass." This grass grows out from the center and hangs over on all sides until the blades touch the ground. It is a round, hard-stemmed grass, and only grows on the most sterile soil. According to my observations the Massena partridge is seldom seen in other localities than where this grass grows. I was riding at a walk up the slope of a barren hill when my horse almost stepped on a nest, touching just the rim of it. The bird gave a startled flutter, alighting again within 3 feet of the nest and not over 6 feet from me; thence she walked away with her crest slightly erected, uttering a low chuckling whistle until lost to view behind a Spanish bayonet plant (yucca), about 30 feet off. I was riding a rather unruly horse and had to return about 30 yards to tie him to a yucca before I could examine the nest. This was placed in a slight depression, possibly dug out by some animal, the top of the nest being on a level with the earth around it. It was well lined with fine stalks of wire grass almost exclusively, the cavity being about 5 inches in diameter and 2 inches deep. At the back, next to the grass, it was slightly arched over, and the overhanging blades of grass hid it entirely from sight. The nest was more carefully made than the average bobwhite's nest and very nicely concealed.

There is a set of 13 eggs in Col. John E. Thayer's collection, taken by Virgil W. Owen in Cochise County, Ariz., June 18, 1905. The nest is described as "a slight depression under a bunch of saw-grass, which was growing on a hillside. It was quite compactly built of straws, grass, and leaves, with a tubular entrance extending out about six inches from the nest. The whole nest was roofed over and well concealed by overhanging grass."

Eggs.—Mearns's quail is known to lay from 8 to 14 eggs. These vary in shape from ovate to short ovate or ovate pyriform, usually more elongated than those of the bobwhite. The shell is smooth and somewhat glossy. They are pure white, dull white, or creamy white and unmarked, but often much nest stained. The measurements of 39 eggs average 31.9 by 24.7 millimeters; the eggs showing the four extremes measure **33** by 24.4, 32.3 by **27.3**, **30.5** by 24.6, and 31.5 by **23** millimeters.

Young.—The period of incubation seems to be unknown. Both sexes share in the incubation and in the care of the young. Frank C. Willard, who has seen many nests, tells me that

in about half of the nests examined the male was on the eggs. In two instances both birds were at the nest. In one which I went to see on August 17, 1913, the male was sitting at the entrance with a newly hatched chick poking its head out from under his wing. The female was in the nest, which was well arched over. Around her was a row of little ones, and one was sitting on her back. This charming picture lasted but a few seconds. The two old birds fluttered away, pretending disability to fly. The scarcely dry young could not walk, but crawled away with astonishing rapidity.

Henshaw (1874) describes the following exhibition of parental solicitude:

August 10, while riding with a party through a tract of piny woods, a brood of 8 or 10 young, accompanied by the female, was discovered. The young, though but about a week old, rose up almost from between the feet of the foremost mule, and after flying a few yards dropped down, and in a twinkling were hidden beneath the herbage. At the moment of discovery, the parent bird rose up, and then, tumbling back helplessly to the ground, imitated so successfully the actions of a wounded and disabled bird that, for a moment, I thought she must have been trodden upon by one of the mules. Several of the men, completely deceived, attempted to catch her, when she gradually fluttered off, keeping all the time just beyond the reach of their hands, till she had enticed them a dozen yards away, when she rose and was off like a bullet, much to their amazement.

Plumages.—In the newly hatched chick the upper parts are variegated with "cinnamon-buff," "hazel," "chestnut," and black, mostly hazel and chestnut, fading out to white on the chin and throat and to grayish white on the belly; there is a broad band or patch on the crown, of "hazel," bordered with chestnut, and an auricular stripe of blackish brown.

The juvenal plumage is acquired at an early stage. A small young bird, only about 2½ inches long, is nearly covered above with juvenal plumage; the wings reach beyond the tail, which has not yet started; the sides of the breast and flanks are feathered; but the head, neck, and center of the breast are still downy. This young bird could probably fly.

In this plumage the sexes are much alike, except that in the young male the crissum, lower belly, and flanks are black, and the center of the breast is suffused with brown, whereas in the young female these parts are whitish; these characters are conspicuous in flight. In both sexes the crown is "hazel" or "russet," spotted with black and with some whitish shaft streaks; the sides of the head are buffy white, mottled with black and with a dark brown auricular patch; the mantle is "tawny" to "ochraceous-tawny," barred heavily with black and with broad buffy median stripes; the wing coverts are ashy with rounded black spots; the primaries and secondaries are banded with white spots; the throat is white, and the rest of the underparts are pale buff, or grayish white, barred or spotted with black and white.

The juvenal plumage is worn for only a short time, for the postjuvenal molt begins early in September and is prolonged in some individuals through November. This molt is complete except for the outer two primaries on each wing, which are retained during the first year. The sexes now begin to differentiate rapidly, the brilliant body plumage of the male and the "vinaceous-pink" breast of the female replacing the juvenal plumage.

The molt starts on the upper breast, flanks, and shoulders, working gradually downward; the rich chestnut in the center of the breast of the male is the last of the body plumage to be acquired; but the entire body molt is completed before any change takes place on the head; the conspicuous head markings are not assumed in the young male until December or later. A very limited prenuptial molt takes place in spring and a complete postnuptial molt late in summer and in fall, after which young birds are fully adult.

Food.—Mrs. Bailey (1928) summarizes the food of Mearns's quail as follows:

As far as known, lily bulbs—¾ of the food in 5 specimens and to judge from their large strong digging feet provided with sharp claws perhaps the principal article of their diet—also great numbers of acorns and pinyon nuts, and in addition seeds and spines of prickly pear, acacia, seeds of legumes and spurges, grass blades, berries of mountain laurel, arbutus, and cedar, and such insects as weevils, caterpillars, bugs, crickets, and grasshoppers.

A pair that Mr. Bailey started at the head of the Mimbres at about 8,000 feet had been scratching under the pine trees. "In the freshly scratched ground," he says, "I found a quantity of membranacious shells of a little bulb—probably *Cyperus*—and several of the bulbs. I ate one of these and found it good, starchy, juicy, crisp, and of a nutty flavor. The Quail had dug two or three inches deep in the hard ground and seemed to find plenty of bulbs, but I could not find one by digging new ground, nor could I find the plant which bore them.

Behavior.—Mearns's quail is a gentle, retiring bird of rather sedentary habits. It prefers to walk about slowly and quietly among the rocks, bushes, and clumps of grass on the rough hillsides where it lives. If alarmed, it squats and freezes, immovable, until almost trodden upon or touched, when it rises from almost under foot, flies a short distance, drops into cover and squats again. When greatly alarmed it sometimes flies to a great distance in a very swift and direct flight. Several observers have mentioned coming upon one or more of these birds in the mountain roads, where they are fond of dusting; they showed no alarm, either walking away quietly or squatting and freezing. Captain William L. Carpenter says in his notes:

I once stopped my horse when about to step on one and watched it for some time without creating alarm. After admiring it for several moments squatting close to the ground within a yard of the horse, watching me intently, but apparently without fear, I dismounted and almost caught it with my hat, from under which it fluttered away.

Henshaw witnessed a remarkable exhibition of the confidence that this bird shows in its protective coloration, for he says in some notes sent to Mrs. Bailey (1928):

Of the several quail known to me the "fool quail" of New Mexico and Arizona seems to depend for his safety upon his protective coloration more than any other. As an example I recall one that squatted on a log near the trail our pack train was following, and so closely did the colors of his back and sides harmonize with his surroundings that 12 or 15 pack mules and horsemen passed by him without seeing him or disturbing his equanimity in the least. He seemed so completely petrified by astonishment at the novel sight as to be incapable of motion, and he was so close to us that one might have touched him with a riding whip. While the bird was no doubt actuated to some extent by curiosity, he depended for his safety, I am sure, upon the nice way in which his plumage matched his surroundings and upon his absolute immobility. No one saw the bird but myself, and when the train had passed I had to almost poke him off his perch before he consented to fly. Whoso calls this the "Fool Quail" writes himself down a bigger fool than the bird, who has been taught his lesson of concealment by Mother Nature herself.

Louis A. Fuertes (1903) thus describes his first impression of a Mearns's quail:

I awoke in the cool, just before sun-up, and was lazily dressing, half out of my sleeping bag, when my sleepy eye caught a slight motion in the grass about 20 feet away. I looked and became aware that I was staring at my first Mearns quail. Even as I took in the fact, he apparently framed up his ideas as to *his* vision, and telling himself in a quiet little quail voice that it were perhaps as well to move on and look from a safer distance, he slimmed down his trim little form and ran a few steps. Meanwhile I was clumsily trying to get my gun out from under my sleeping bag, where I had put it to keep it out of the dew. The quail, getting wiser every second, doubled his trot, and with head erect and body trim ran like a plover for a few yards through the short desert grass, and with a true quail *f-r-r-r-r-r-r-r* burst into flight and dropped into the thick brush across the arroyo. The most noticeable thing about him as I watched him running was the curious use of his queer little crest. Instead of elevating it as the mountain quail does his, he raised his painted head on slim neck and spread his flowing crest *laterally*, till it looked like half a mushroom, giving him the most curious appearance imaginable. When he flew I marked him down carefully, hastily drew my boots half on, grabbed my gun and stumped after him with all speed. I got to his point within a short time, but thrash and kick around as I might, I never succeeded in making him flush a second time.

Mr. Willard writes to me:

One morning, as I arrived in front of our store in Tombstone, I found a flock of a dozen or more of these birds running around in the street. Most of them flew up onto the roof of the building, but one male ran into the doorway, stuck his head down into a corner, and waited for me to pick him up.

Voice.—Major Bendire (1892) quotes Mr. Todd as saying:

When scared they utter a kind of whistling sound, a curious combination between a chuckle and a whistle, and while flying they make a noise a good deal like a Prairie Hen, though softer and less loud, like "*chuc-chuc-chuc*," rapidly repeated.

H. S. Swarth (1909) writes:

Their call consists of a series of notes slowly descending the scale, and ending in a long, low trill, the whole being ventriloquial in effect and most difficult to locate. It is easily imitated, however, and the birds readily answer when one whistles; when the flock is scattered they will sometimes even return, calling at intervals as they approach. The only other note I have heard is a quavering whistle uttered as they take flight.

Mrs. Bailey (1928) says that

the low call of the Mearns Quail, suggestive of the quavering cry of a Screech Owl, adds to the fascination of the pursuit of this illusory bird, for it is ventriloquial in quality and leads you such a fruitless chase that you return to camp with an exaggerated interest in this feathered Will-o'-the-wisp.

Fall.—In some notes sent to Major Bendire (1892) Doctor Nelson states that

the birds breeding along the northern limit of their habitat migrate southward in October. In southern Arizona the same result of a warmer winter climate is obtained by descending the flanks of the mountains. The summer range of this species is just above and bordering that of Gambel's Quail in parts of Arizona and New Mexico. The fact that Gambel's Quail changes its range but little in winter results in these birds being found very frequently occupying the same ground at this season. I have never seen the Massena Partridge in coveys larger than would be attributed to a pair of adults with a small brood of young. Frequently a pair raise but three or four, and I do not remember having ever seen more than six or seven of these birds in a covey.

DISTRIBUTION

Range.—Southwestern United States (except California) and northern Mexico. The range of Mearns's quail is extremely circumscribed. It extends **north** to Arizona (Fort Whipple, Camp Verde, Mogollon Ridge, Wilcox, and Marsh Lake); New Mexico (Zuni Mountains, San Mateo Mountains, White Mountains, and Guadalupe Mountains); and central Texas (San Angelo and Mason). **East** to Texas (Mason, Kerrville, Bandera Hills, and San Antonio); Durango (Ramos, El Salto, and Huasamota); and Nayarit (San Blasito). **South** to Nayarit (San Blasito). **West** to Nayarit (San Blasito); western Chihuahua (50 miles northeast of Choix, Sinaloa); central Sonora (La Chumata and Patagonia Mountains); and Arizona (Baboquivari Mountains, Rincon Mountains, Mount Turnbull, and Fort Whipple).

The species has a vertical range from 4,000 to 9,000 feet, the birds moving to the higher altitudes early in fall and retiring upon the approach of winter. They apparently do not, however, descend below the lowest parts of the breeding range.

Egg dates.—New Mexico, Arizona, and Mexico: 29 records, April 24 to September 5; 15 records, June 23 to August 16.

Family TETRAONIDAE, Grouse

DENDRAGAPUS OBSCURUS OBSCURUS (Say)

DUSKY GROUSE

HABITS

The big "blue grouse" of the southern Rocky Mountains ranges from southern Idaho and northern Colorado to Arizona and New Mexico. Two other subspecies of *obscurus* occupy the more northern parts of the Rocky Mountains. Its home is in the mountains from about 2,000 feet in the foothills, or as low as quaking aspens grow, up to timber line at 11,000 or 12,000 feet. It was formerly very common in the coniferous forests, but it has now disappeared from many of its former haunts and is becoming scarce in others. Coues (1874) quotes from Mr. Trippe's notes, regarding its haunts in Colorado, as follows:

The "Gray Grouse," as this species is universally called, is a rather common bird throughout the mountains, from the foot-hills up to timberline, and, during summer, wanders at times above the woods as high as the summit of the range. Excepting for a brief period in August and September, it rarely approaches the vicinity of clearings, frequenting the dense pine forests, and showing a preference for the tops of rocky and inaccessible mountains. In its nature, in short, it is the exact counterpart of the Ruffed Grouse, having the same roving, restless disposition; living upon the same diet of buds and berries; frequenting the same rugged, craggy mountain haunts; and, like that bird, is more or less solitary in its habits, and constantly moving from place to place on foot.

Mrs. Florence M. Bailey (1928) writes of it in New Mexico:

In the depths of the coniferous forest you may suddenly discover a Dusky Grouse with its small pointed head and henlike body sitting quietly on a log facing you, as if secure in its disguise—a dusky bird in the dusky woods surrounded by shadowy tree trunks. From a forest trail you may flush one that has been dusting itself in the soft earth, or hearing a muffled ventriloquial hooting may creep up within sight of the lordly cock at the foot of a conifer, with purplish red neck pouches dilated. Again in the open, you may be startled by a loud whir and look up to see great dark forms with a wide spread of wing disappearing over your head; or, on a steep mountain side, catch sight of a big Grouse sailing off below you with stiff outstretched wings and a spread tail, whose gray band makes a striking mark to follow among the branches. If still more fortunate, you may surprise a family in a mountain meadow, for strawberries are evidently one of their favorite summer foods.

Courtship.—Bendire (1892) quotes from some notes received from Denis Gale, as follows:

During the mating season if you are anywhere near the haunts of a pair you will surely hear the male and most likely see him. He may interview you on foot, strutting along before you, in short hurried tacks alternating from right to left, with widespread tail tipped forward, head drawn in and back and wings dragging along the ground, much in the style of a turkey gobbler. At other times you may hear his mimic thunder overhead again and again in his flight from tree to tree. As you walk along he leads, and this reconnoitering on his part, if you are not familiar with it, may cause you to suppose that the trees are alive with these Grouse. He then takes his stand upon a rock, stump, or log, and in the manner already described distends the lower part of his neck, opens his frill of white, edged with the darker feather tips, showing in its center a pink narrow line describing somewhat the segment of a circle, then with very little apparent motion he performs his growling or groaning, I don't know which to call it, having the strange peculiarity of seeming quite distant when quite near, and near when distant; in fact, appearing to come from every direction but the true one. The first time I heard the sound I concluded it was the distant laboring of one of our small mountain sawmills wrestling in agony with some cross-grained saw-log.

Nesting.—I have never seen a nest of this grouse, and very little seems to have been published about it. W. L. Sclater (1912) says: "The nest, which has been described by Henshaw and Burnett, is placed on the ground, generally in an open glade but sheltered or somewhat concealed; that found by Burnett was placed in a hollow under two old logs, it being a simple structure of dried grass or pine needles." Mrs. Bailey (1928) says that the nest is "a shallow depression beside a log or under grass or bushes, slightly lined with a few pine needles or a little grass."

Bendire (1892) describes a nest found by Henshaw, as follows:

A nest found June 16 contained seven eggs on the point of hatching. The nesting site was a peculiar one, being in an open glade, where the grass had been recently burned off. The nest proper was a slight collection of dried grass placed in a depression between two tussocks, there apparently having been no attempt made at concealment.

Eggs.—From 7 to 10 eggs are usually laid by the dusky grouse, and as many as 12 have been found in a nest. Bendire (1892) says that "an egg is deposited daily, and incubation does not commence till the set is completed." The eggs are ovate to elongate ovate in shape. The shell is smooth, with little or no gloss. The ground colors vary from "pinkish buff" to "pale pinkish buff" or to "cartridge buff." They are usually evenly covered with very small spots or minute dots, generally quite thickly; some eggs are less thickly and more irregularly covered with somewhat larger spots; very rarely an egg is immaculate. The colors of the spots are "sayal brown," "clay color," or "cinnamon-brown," rarely darker. The measurements of 54 eggs average 49.7 by 34.9 millimeters; the eggs showing the four extremes measure **52.3** by 34.4, 50.6 by **37.1**, **46.2** by 34, and 50 by **31.8** millimeters.

Young.—Bendire (1892) quotes Denis Gale as follows:

Upon one occasion I met with a covey which had just been hatched; they were quite nimble, and with the exception of one which I caught they hid themselves with great address. Until I released the little prisoner the female showed great distress, clucking in the most beseeching manner, accompanied with suitable gestures, similar to but more tender and graceful than those of our domestic hen. She stood within 6 or 7 feet of me pleading her cause and easily won it. In her beautiful summer dress of brown, handsomely plumed as she was, she looked very interesting.

In a single instance only, with a brood about ten days old, have I noted the presence of both parents. Perched upon a fallen tree the male seemed to be on the lookout, while the female and young were feeding close by. This seeming indifference of the male while the brood is very young, allowing his mate to protect them, if he really is always near at hand, looks very strange, and yet it may be the case, since he is generally with the covey when the young are well grown. Directly the young are able to travel, the hen Grouse leads them to some desirable opening, skirting the timber or gulch, where bearberries, wild raspberries, gooseberries, and currants, as well as grasshoppers, worms, and grubs are abundant, managing them just as the domestic hen does her brood. The young grow rapidly, and when about two weeks old can do a little with their wings; then, instead of hiding on the ground, they flush and endeavor to conceal themselves in the standing timber. Until almost fully grown they are very foolish; flushed, they will tree at once, in the silly belief that they are out of danger, and will quietly suffer themselves to be pelted with clubs and stones till they are struck down one after another. With a shotgun, of course the whole covey is bagged without much trouble, and as they are, in my opinion, the most delicious of all Grouse for the table, they are gathered up unsparingly.

Plumages.—The sequence of molts and plumages is the same as in the sooty grouse, of which I have found more material for study. The racial characters are apparent in immature birds. Richardson's grouse has been known to hybridize with the Columbian sharp-tailed grouse (Allan Brooks, 1907).

Food.—Mrs. Bailey (1928) summarizes the food of the dusky grouse very well, as follows:

In 45 crops and stomachs examined, the food consists of 6.73 per cent animal matter—5.73 per cent grasshoppers, and the rest beetles, ants, and caterpillars—and 93.27 per cent vegetable matter—seeds, fruits, and leaves, coniferous foliage amounting to 54.02 per cent, fruits 20.09 per cent, including manzanita berries, mountain ash, service-berry, currant, gooseberry, etc. One cock shot at 11,600 feet on Pecos Baldy in a strawberry patch had both crop and gizzard filled mainly with strawberries. The crop of another shot between 8,000 and 9,000 feet contained 27 strawberries, 28 bear-berries, 12 Canadian buffalo-berries, flowers of Indian paint brush and milk vetch, leaves of vetch and buffalo-berry, and a few ants and caterpillars, while its gizzard was filled with seeds of bear-berry, Canadian buffalo-berry, and strawberry, a few green leaves, and a number of ants, beetles, and other insects. Grasshoppers, the green leaves of blue-berry and vetch, salal, and other berry seeds, needles of Douglas spruce and fir, together with gravel and hard quartz grinding stones were among the items that the field examination of other specimens revealed. The quartz grinding stones were found in gizzards apparently filled with hard coniferous needles. These needles seem to be the regular winter food as under a winter roosting tree on Pecos Baldy the winter dung was composed entirely of spruce needles. (Three birds taken in September near Golden, Colorado, had their "crops crammed with the berries of kinnikinick.")

Behavior.—Coues (1874) quotes from Mr. Trippe's notes, as follows:

On being suddenly startled, this bird takes wing with great rapidity, sometimes uttering a loud cackling note, very much like that of the Prairie Hen on similar occasions, frequently alighting on the lower limb of a tree after flying a little way, and watching the intruder with outstretched neck. Sometimes they will fly up to the top of a tall pine and remain hidden in the thick foliage for a long time; nor will they move or betray their position, although sticks and stones are thrown into the tree, or even a shot fired. Late in summer many of them ascend to the upper woods to feed upon the multitudes of grasshoppers that swarm there in August and September, in the pursuit of which they wander above timber-line, and may sometimes be met in great numbers among the copses of willows and juniper that lie above the forests.

Edwyn Sandys (1904) says that this grouse

is most difficult to locate even when perched upon a limb only a few yards away. In its native woods the light is baffling and there is a confusion of shade, amid which the general slaty tone of the plumage is barely distinguishable. A coat of feathers especially designed with a view to protective coloration could not better serve the purpose, and the bird appears to be perfectly aware of this. Indeed, its habit of trusting to its trick of treeing and remaining motionless has earned for it the names of "fool-grouse," which I believe should be applied only to *young* birds. These unquestionably will tree and foolishly maintain their positions while their comrades are being shot or clubbed down, but the older birds, except in seldom disturbed localities, are wiser.

But fool grouse or no, when once the bird concludes to start there is no more foolishness. With a nerve-shaking whirring it promptly gets to top speed, and usually darts downhill, a maneuver which greatly adds to the difficulty of the shot. When taking wing it cackles like a scared fowl.

Game.—Referring to it as game bird, Sandys (1904) writes:

Among western sportsmen it is termed the "blue," or "gray," grouse, and those who have enjoyed the pleasure of shooting and later eating it have yet to be heard from in the line of adverse criticism. Its sole fault as a game bird consists in its seldom being found in cover which affords a fair chance to the gun. In fact, it is such an inveterate lover of trees that it takes to the branches as naturally as a duck takes to water. Like the ruffed grouse, it will tree, and remain motionless until it fancies it has been observed; then it at once departs with a sounding rush, which may only be stopped by the quickest and most skilled of shots. I have flushed it when it seemed to do hardly anything more than leap from the ground to a convenient limb, and more than once, while seeking to trim off its head, it has left the perch so suddenly that the gun could not be shifted in time to prevent the wasting of a shell—and this little joke at the expense of a notoriously quick shot.

Only those familiar with the western cover can understand how easy it is to fail to bag at short range a bird about as large as a common barnyard hen—to be accurate, of between three and three and one-half pounds' weight. The tenderfoot would imagine such a bird, rising close at hand, to be an easy, perhaps too easy, mark. Let the tenderfoot climb the steeps and try a few blue grouse as they leave the trees, and his song may take on an undertone suggestive of blasted hopes and trust betrayed. In the first place, the cover usually is standing timber big enough to stop a locomotive, to say nothing of small shot. This timber, as I found it, is about as close as it can stand, thereby forming something closely akin to a gigantic stockade with extremely narrow interspaces.

In spite of the bird's penchant for timber it frequently is found in the open and in grain fields. In such places the sportsman may enjoy "blue grouse" shooting as it should be, and sport of a very high order. Then the full strength and speed of the game becomes apparent, and the man who makes uniformly good scores has no reason to fear any ordinary company.

Fall.—Prof. Wells W. Cooke (1897) says:

In August they begin to gather into flocks of 10 to 15 individuals and visit the grain fields or the more open gulches and foothills for berries. In September they wander above timber-line to feed on grasshoppers, reaching 12,500 feet. In winter they come down into the thick woods during the severest weather, but many remain the whole year close to timber-line.

DISTRIBUTION

Range.—The Rocky Mountain region of the United States and Canada.

The range of the dusky grouse extends **north** to east-central Yukon (probably the south fork of the Macmillan River); and southern Mackenzie (Mount Tha-on-tha). **East** to southern Mackenzie (Mount Tha-on-tha); Alberta (Stony Plain and 60 miles west of Calgary); Montana (Zortman, Judith Mountains, Big Snowy Mountains, and Fort Custer); Wyoming (Sheridan, Trappers Creek, Guernsey, Wheatland, Pole Mountain, and Sherman); Colorado (Estes Park, Boulder, Golden, Jefferson, Wet Mountains, Sangre de Cristo Mountains, and Fort Garland); and New Mexico (Culebra Mountains, Halls Peak, Pecos Baldy, San Mateo Mountains, Eagle Peak, and Mogollon Mountains). **South** to southwestern New Mexico (Mogollon Mountains and San Francisco Mountains); Arizona (Blue Mountains, Mount Thomas, Mount Ord, and the Kaibab Plateau); and central Nevada (Arc Dome). **West** to central Nevada (Arc Dome, Toiyabe Mountains, Monitor Mountains, Ruby Valley, Clover Mountains, and East Humboldt Mountains); eastern Oregon (Fort Harney, Strawberry Mountain, and Turtle Cove); eastern Washington (Butte Creek, Hompeg Falls, Okanogan, and Haig Creek); eastern British Columbia (Princeton, Nicola, Ashcroft, Bonaparte, Babine Mountains, Nine-mile Mountain, Fort Connolly, Groundhog Mountain, Tset-ee-yeh River, Second South Fork, Doch-da-on Creek, and Telegraph Creek); and Yukon (Teslin Lake and probably the south fork of the Macmillan River).

Three subspecies of *Dendragapus obscurus* are recognized. True *obscurus* is found from central Arizona and southwestern New Mexico north to northern Colorado, northern Utah, and southeastern Idaho, extending west to central Nevada. *D. o. richardsoni* occupies the range from Wyoming and south-central Idaho north to western Alberta and central British Columbia and west to eastern Oregon and Washington. *D. o. flemingi* is found in northern British Columbia, southern Yukon, and southwestern Mackenzie.

Although technically nonmigratory, both the dusky and the sooty grouse have the curious habit of performing a postbreeding vertical migration. The cocks will sometimes move to the higher elevations while the hens are still incubating their eggs, usually in July or early in August. The females and half-grown young follow, so that by the last of September the species has deserted the breeding grounds. The winter months are usually spent in the upper spruce forests, the birds living entirely in the trees. By May or June they have descended again to their breeding areas.

Egg dates.—Colorado and Utah (*obscurus*): 18 records, April 17 to July 1; 9 records, May 21 to June 13. British Columbia and Alberta to Idaho and Wyoming (*richardsoni*): 15 records, May 3 to August 20; 8 records, May 13 to June 23.

DENDRAGAPUS OBSCURUS RICHARDSONI (Douglas)

RICHARDSON'S GROUSE

HABITS

The name *D. o. richardsoni* was applied to all the "blue grouse" of the northern Rocky Mountain region until, in 1914, the bird of the extreme north was separated under the name *flemingi*, the next form. Richardson's grouse is now restricted to the intermediate territory, from central British Columbia and western Alberta to eastern Oregon, south-central Idaho, and Wyoming. It differs from the southern bird, *obscurus*, in having no very distinct, terminal, gray band on the tail, and from the northern bird, *flemingi*, in being lighter colored. It does not differ materially from either in its habits, except as hereinafter mentioned.

M. P. Skinner (1927) writes:

Here in the Yellowstone National Park, we are in the borderland occupied by both the southern and northern forms. It is very difficult to place these Park birds under either *obscurus* or *richardsoni*, for there are birds of each form present and all degrees of graduation between them. I once found a bird dead near the Buffalo Ranch that had no band at all on the end of its tail. Twice I have seen very dark birds with only a slight amount of gray tipping their tails, so that they were more typical of the Sierra Mountain form. Another time, I found a bird so tame that I could approach close enough to determine positively that he had more than three-quarters of an inch of gray on the ends of the middle tail-feathers. At other times, I have found feathers with terminal bands more than a half inch wide. There does not seem to be any segregation of the two forms within the Park, but, so far as I know, the different forms are found in all sections. Both *obscurus* and *richardsoni* are larger than Ruffed Grouse, but the Yellowstone specimens are even larger than the average.

Richardson's Grouse live in all kinds of brush and forest from the willows as low as 5,500 feet altitude, through the service-berry (*Amelanchier alnifolia* Nutt.) areas, aspen groves, Douglas firs, limber pines, lodgepole pines, spruce and white-bark pines to the stunted spruces at timber-line (about 9,500 feet). I have seen them out on the open plateaus and high meadows above timber-line, almost up to the 10,000-foot level. These timber-line birds were seen from July to October, but they apparently went down into the forests during the winter. I have seen these grouse in thick-growing lodgepole saplings where the small trees were so thick it appeared impossible for the grouse to walk or fly through them; and I have also seen them out in burned forests, especially where there were berries to be had.

While Richardson's Grouse prefer the forests, I have found them out in the sage-brush (*Artemisia tridentata* Nutt.), sometimes as much as five hundred yards from the nearest tree. Usually, they are wild in such localities, and soon fly to the nearest trees for protection; but on May 31, 1921, one was found hiding under a two-foot sage-bush in the open. I have also seen single birds in open grasslands without brush and as much as two hundred yards from the nearest tree; and once I rode my horse past one in full view in the grass and within twenty feet of me. No doubt this tameness is largely due to the fact that they are absolutely protected and have nothing to fear from man within the Park. They often come boldly about the buildings of the little village at Mammoth, and even walk across the lawns, both in winter, and when the tourists are numerous in summertime. I have sometimes thought the birds liked the open in winter, because of the sunny warmth there. Still, they are also out in the open on the warmest of summer days.

Spring.—A. W. Anthony (1903) gives a very interesting account of the extensive spring migration of Richardson's grouse in eastern Oregon. The flight was mainly southward from the higher ridges, heavily timbered with pine and fir, to the sage-covered benches and ridges where they nest. He writes:

On the first of March, 1902, when the first of the migrating grouse made their appearance along the edge of the timber north of Sparta, the snow was from 2 to 4 feet in depth, though the lower slopes near Powder River were bare and had begun to show the first signs of sprouting grass. Snow squalls and rough weather seemed to check the southward flight until about the 10th, although a few birds were passing over daily. The tracks on the snow bore ample testimony as to the manner in which the migration was made.

From the higher slopes north of Eagle Canon, the birds sailed until the rising ground brought them to the surface of the snow on the south side of the creek, usually well above the canon. From this time until the highest point of the ridge south was reached the journey was performed on foot. Immediately north of Sparta lies a conical peak known as Baldy, some 700 feet above camp, the highest point in the ridge south of Eagle Creek. From the top of Baldy, and in an area not to exceed 100 feet square, I think fully 85 per cent of the grouse passing over Sparta take their departure. From east, north and west up the steep, snowy slopes hundreds of trails led toward the top and not one could be found leading downward. The flight from the top of the peak was almost invariably undertaken at about sunrise or sunset. It is only when birds are disturbed and driven from the peak that they will attempt to cross to the southern ridge during the middle of the day. Throughout the day grouse are arriving along the upper slopes of Baldy, singly, in pairs, and small flocks that have perhaps formed since the southward march began, as I think they do not winter in company, but the flight from the peak is usually in flocks of from a dozen to a hundred birds. Though the ridge south of Sparta is 400 feet or more lower than the top of Baldy, it is fully a mile and a half distant in an air line, and the flight is seldom sustained to carry the birds to the top. Usually they alight on the snow half way up the slope, and after a few moments' rest, continue the journey on foot; those passing over in the evening spend the night, I think, in the pines, the last of which are seen along this divide; but those arriving in the morning soon pass on, walking down any of the small ridges leading toward Powder River.

Courtship.—Maj. Allan Brooks (1926) has published a full account, with illustrations, of the courtship display of this grouse, from which I quote, as follows:

His first position was a crouching one, the tail spread to extreme extension cocked right over the back and a little to one side, the neck feathers showing as a snowy mass, with the red gular sac looking like a small oyster on a large shell.

He maintained this attitude for several minutes, then the head was raised, the neck swelled, and he turned towards me and commenced to nod his head; the gular sacs were a deep purple-red, the "combs" over each eye changed from yellow to a dusky orange and were inflated to the extent that they almost met on the crown, and the inversion of the neck feathers showed as a huge blaze of white on each side. After six or eight nods the head was lowered to within two inches of the ground and with the neck inflated until the sacs showed a diameter of three inches, the tail still elevated and spread to its full extent, the feathers of the lower back standing on end, the wings trailing on the ground, the bird made a short quick run of six or eight steps curving to the right and emitted the deep *Oop!*

Brooks compares the hooting of Richardson's grouse with that of the sooty grouse, as follows:

Often in later years a male Richardson's Grouse has been seen uttering his low hooting, similar to the resonant hooting of the Sooty Grouse (*Dendragapus obscurus fuliginosus*) but with only a small fraction of its volume. In my experience Richardson's Grouse always utters this hooting from the ground. The tempo is the same as in the Sooty Grouse, five or rarely six deliberate evenly spaced hoots or grunts—*Humph–humph–humph–ma-humph–humph*—but the sound is barely audible up to 75 yards or so. In the Sooty and Sierra Grouse this chant assumes a dominant character, ventriloquial to a degree it sounds far off when quite near and yet has a carrying power of at least 2 miles. In neither species is the tail very widely spread when uttering, nor is there any special posture or action excepting a crouching attitude, high up in a coniferous tree in the Sooty Grouse, and on the ground, usually on some rocky ridge, in Richardson's. The Sooty Grouse also has the single *Oop!* although I have rarely heard it and then only late in the breeding season.

A very pronounced distinction in the two species, however, is the nature of the gular air sacs. In the Sooty Grouse and allied races these in the breeding season become cellular, gelatinous masses, capable of great distention, and the exterior surface is velvety, deeply corrugated, and of a deep yellow color. In the fall this specialized character is largely lost, the skin loses most of its corrugations but still retains a yellow color.

In Richardson's Grouse and its sub-species *flemingi* very little change from normal in the character of the neck-skin occurs in the breeding season, the exterior is flesh-colored tinged with purple, deepening to purple-red when temporarily surcharged with blood.

Charles de B. Green (1928) has noticed a curious fluttering habit of the male grouse, which he describes as follows:

Years afterward I saw it done for the first time in my experience. I was lying in bed looking out of the door at a female grouse walking on the ground some 20 yards away. Near her was a small pine, the lowest branch of which was about 10 feet from the ground. On this branch was a male bird and, just as I caught a glimpse of him, he fell as straight as a string to the ground, tumbling over as he came, in what looked like a ball of feathers. This took only a second or two, but as he fell the fluttering sound was made. Then he made short runs at the female, stopping abruptly each time he got near, and giving one hoot. This running and hooting went on until they were lost to view, but there was no more fluttering.

In the last few years, while in the sheep business, it has been my custom in March to be with the flock before dawn. Every day for a month this very same fluttering is the first sound to be heard in the dusk, before things are at all clearly seen at 50 yards away. In fact, before the faintest sign of daylight, from all around comes in fluttering sound. Only once was I able to see the bird: I was watching a grouse about 30 or 40 yards away in the growing light, when he sprang about 3 feet into the air, turned over, and came down with the fluttering sound on the same spot from which he sprang.

Nesting.—Anthony (1903) writes:

A few birds undoubtedly remain and nest throughout the timbered region of Powder River Mountains, but the percentage is small indeed compared with those that nest on the bare sage plains along Powder and Burnt Rivers. Many of the nests are placed in the shelter of the scattered growth of chokecherry, aspen, or cottonwood that fringes the water courses tributary to the river; and a few of these nests may produce young that reach maturity, but fully as many birds lay in the shelter of a bare rock, or scanty sage brush in the open plain, in company with Sage Grouse; and fortunate indeed is the bird, nesting in such location, that raises its young. In a circuit of not over 6 miles from my camp on Powder River the past May, were ranged not less than 20,000 sheep which tramped out the nests so completely, that, while finding dozens of broken nests, I saw not one that had not been destroyed, of either Richardson's or Sage Grouse, and only one young bird. Nevertheless, many of them do escape, as their numbers testify, although I am told, on good authority, that there are very few in comparison with their former numbers.

Skinner (1927) says that, in Yellowstone Park, "nesting takes place in May, and the nests are usually placed at the foot of forest trees at any altitude from 6,300 to 8,000 feet, and perhaps even above the last-named height. The Richardson's nests are shallow depressions lined with grass, pine needles and leaves, and contain from seven to ten creamy eggs speckled and blotched with brown,

Fresh eggs are laid from May 10 until almost the first of July in belated cases."

Eggs.—The eggs of Richardson's grouse are similar to those of the dusky grouse. The measurements of 32 eggs average 47.9 by 33.3 millimeters; the eggs showing the four extremes measure **51.7** by 34, 47.5 by **36, 43** by 33.5, and 45 by **32** millimeters.

Young.—Bendire (1892) mentions, in some notes from Robert S. Williams, the behavior of some young grouse that were probably too young to run away or hide; he writes:

On June 21, 1885, while crossing over the almost bare summit of a small knoll in the foothills of the Belt Mountains, I suddenly almost ran into a brood of young Richardson's Grouse, which had evidently been hatched out but a very short time. The young, about ten in number, were closely huddled together, the old bird standing by their side, with head up, and eyes fairly blazing at the unexpected intruder. I was almost within reach of them, but neither old nor young made a single motion or uttered a sound while I stood watching them for several moments; and I left them in the same position.

Mrs. Florence M. Bailey (1918) writes:

As we rode out of the dark woods the peeping voices of young were heard, and as the first horse shied a big mother grouse flew conspicuously into the top of a low evergreen, while her brood, circling out on widespread curving wings like young quail, disappeared under cover. Early in August, on the Swiftcurrent, an old grouse and seven half-grown young, finding our camp nearly deserted, walked calmly past the tents and under the kitchen awning on their way to the creek. On reaching it the mother flew across, calling the brood till they followed, when they all walked off toward the blueberry patch in the pine woods.

Food.—In Yellowstone Park, according to Mr. Skinner (1927), their food consists largely of berries, such as bear berries (*Arctostaphylos uva-ursi* L.), huckleberries (*Vaccinium scorparium* Leiberg), high bush blueberries (*Vaccinium membranaceum* Dougl.), service-berries (*Amelanchier alnifolia* Nutt.), false buffalo berries (*Shepherdia canadensis* L.), raspberries (*Rubus strigosus* Michx.), gooseberries (*Ribes saxosum* Hook. and *Ribes parvulum* Gray), and strawberries (*Fragaria americana* Porter). In addition they eat many insects, especially grasshoppers. When other food becomes scarce they eat fir, pine, and spruce needles.

Aretas A. Saunders has sent me the following interesting notes:

In August and September the food is mainly grasshoppers obtained in the mountain meadows. A method of obtaining them is interesting. A small flock of 8 or 10 birds, apparently a female and her fully grown young, stand in a circle some 30 feet or more in diameter, the birds 10 or 12 feet apart. Each bird faces toward the center of the circle, and they slowly move inward, scaring up the grasshoppers as they go. As the circle grows smaller the grasshoppers are more and more concentrated in the center, and the birds capture a great many. A grasshopper jumps and flies from a bird on one side only to fall victim of one on the opposite side. I have watched this performance once in the Sun River country, and flushed birds that were in this formation several other times, so that I believe it is a common habit.

Enemies.—The same observer says:

Although not here hunted by man, these grouse have plenty of other enemies, such as wolves, coyotes, mink, weasels, and all the other predacious fur bearers, as well as the duck hawks, Cooper's hawks, and western horned owls. The big western red-tailed and Swainson's hawks do not bother the grouse even when they are small. I remember once watching a redtail circling overhead. I could not see what it was hunting, but apparently it was not Richardson's grouse, for almost at the same time an undisturbed grouse and seven little fluffy youngsters were noted in the grass under the circling hawk. Soon, the mother grouse flew up into a small sapling and from there clucked loudly to her brood, while the little grouse tried their best to fly up to her, one at a time, on very shaky little wings. Sometimes I see a Richardson's grouse with many tail-feathers missing, as if it had just escaped an enemy by the sacrifice of a few feathers.

Fall.—The fall migration to the higher elevations is thus described by Anthony (1903):

The return migration is less pronounced in its beginning, and more gradual in its progress. Toward the last of July the broods of well grown young, attended by the adults, begin to appear along the ridges, returning as they came by walking invariably up to the tops of the hills and ridges and as invariably flying as near to the top of the next as their gradually descending flight will carry them. Before the middle of August, the migration is in full swing, and flocks are seen each evening, passing over Sparta. Frequently they alight in the streets and on the house-tops. I recall with a smile the memory of a flock of a dozen or more which lit one evening in front of the hotel. For a time pistol bullets and bird shot made an accident policy in some safe company a thing to be desired, but strange to relate none of the regular residents of the town were injured. The same may be said of most of the grouse, though one, in the confusion, ran into the livery stable and took refuge in a stall, where it was killed with a stick.

Straggling flocks from south of Powder River prolong the fall migration until near the first of October, after which none are seen below the high elevations north of Eagle Creek.

Winter.—Skinner (1927) says of the winter habits of these grouse in Yellowstone National Park:

Normally in summer, these grouse are on the ground, or on low logs and boulders; and they live mainly in the evergreen trees while the snow covers the ground. But, when skiing through the winter forests in December and January, I have had roosting birds burst out from under snow drifts. At other times, most of these birds roost in heavy coniferous trees. If not disturbed, they may stay in a small grove of trees, and not descend to the ground for several successive days. At such times, they eat needles for food and use the snow instead of water.

Mrs. Bailey (1918) was told that in Glacier National Park where the snow is from 1 to 25 feet deep, "they roost in holes in the snow."

DENDRAGAPUS OBSCURUS FLEMINGI (Taverner)

FLEMING'S GROUSE

HABITS

The range of Fleming's grouse, as given in the new A. O. U. Check List, includes northern British Columbia, southern Yukon, and southwestern Mackenzie. The bird was described by P. A. Taverner (1914), based on a small series of "specimens taken within 30 miles of Teslin Lake, on the boundary between British Columbia and Yukon Territory, longitude 130° 30′, at the west base of the Cassiar Mountains. The range of the form cannot therefore be defined."

It was named after J. H. Fleming, of Toronto, Canada, and is said to be "like *Dendragapus obscurus richardsoni*, without terminal tail band, but darker in general coloration even than *D. o. fuliginosus*."

The subspecies is based on the characters of the male, and Mr. Taverner (1914) says:

> The characters of the female are less marked than of the male and without series for comparison may be difficult of recognition. However, they average in the same directions as the male, being bluer underneath than *richardsoni* and darker dorsally, with the rufous or rusty markings bolder and more decided in character.

Maj. Allan Brooks (1927) adds the following suggestion:

> The best character for separating *flemingi* from typical *richardsoni* is not the darker coloration of the males; any large series will show dark colored *richardsoni* and, as in Swarth's Teslin Lake bird, light colored *flemingi;* but the blacker under tail coverts of the last named form with small white tips, instead of the white tip covering almost the whole exposed portion of the feather as in *richardsoni*. But birds from Revelstoke, Selkirk range, in southern British Columbia show the extreme of blackness of the lower tail coverts. It is possible that a wedge running southward to or near the British Columbian southern boundary splits the range of *richardsoni*.

Harry S. Swarth (1926) says of the haunts and habits of Fleming's grouse:

> In the Atlin region the "blue grouse" is resident and fairly common at high altitudes. It is a favorite game bird of the region, both from its large size and from the excellent quality of its flesh. Its habitat is about timber line, where there is open country interspersed with clumps of balsam firs. The dense thickets of these stunted trees, with their gnarled and spreading branches, afford shelter from enemies and from inclement weather, and in the foliage food also is furnished when other sources fail.
>
> The broods are cared for solely by the hen. The old cock is usually solitary during the summer, though males of the previous year sometimes form small coveys, together with non-breeding females. Such gatherings were encountered on several occasions. The hen with a brood is sometimes tame to the verge of stupidity; I found several that were, literally, as indifferent to

approach as any barn-yard fowl. I have, however, seen an occasional covey of young birds that was extremely hard to approach. The broods often feed over open meadows, where they are exposed to attack by hawks and other enemies, and there must be a heavy mortality from such causes. That this is so is borne out by the small size of most of the broods encountered, and by the number of hens seen with no broods at all.

DENDRAGAPUS FULIGINOSUS FULIGINOSUS (Ridgway)

SOOTY GROUSE

HABITS

In the heavily timbered, humid coast ranges, from southern Alaska to extreme northern California, we find one of the most widely distributed and best-known races of the "blue grouse" group. It is the darkest of all the races, except the extremely dark race of the closely related species *D. obscurus flemingi*, which is found in northern British Columbia. All the "blue grouse" (genus *Dendragapus*) have until recently been regarded as subspecies of *Dendragapus obscurus*, but the studies and suggestions of Allan Brooks (1912, 1926, and 1929) and of Harry S. Swarth (1926) have resulted in separating the various races into two groups as subspecies of two distinct species. The new A. O. U. Check List, therefore, gives three races of the Rocky Mountain, or eastern, group, *obscurus, richardsoni*, and *flemingi*, all as subspecies of *obscurus;* and this leaves the four coastal, or western, races, *fuliginosus, sierrae, sitkensis*, and *howardi*, as subspecies of a new species, *fuliginosus*. Major Brooks (1926) gives the following distinctive characters of the two groups:

Group I, including *richardsoni* and *flemingi.*

1. Air sacs. Skin not conspicuously thickened or corrugated even in the mating season, color flesh, changing to purple red under the influence of excitement.
2. Voice. "Hooting of five or six notes audible for less than 100 yards, uttered from the ground. Note: The single hoot when in full display is alike and common to both groups.
3. Tail. In adult males, squarer, the feathers truncate at the tips; terminal band of gray darker, sometimes (rarely) absent or but faintly indicated.

Group II, including *fuliginosus, sitkensis, sierrae* and *howardi.*

1. Air sacs. Skin highly specialized in the mating season, thick, gelatinous, the surface deeply corrugated into a series of tubercles of a velvety texture and of a deep yellow color. This condition is reduced when the mating period is over.
2. Voice. "Hooting" of five or six notes of great power, audible for several miles. Always (?) uttered from high up in a tree.
3. Tail. In adult males rounded, the feathers rounded at the tips; terminal band of light gray averaging narrower than in group No. I.

For a further, detailed discussion of the question the reader is referred to Mr. Swarth's (1926) remarks under *flemingi*, with a map showing the distribution of the seven forms and a study of the plumage changes.

Major Bendire (1892), who had a wide experience with this species, writes:

The favorite locations to look for the Sooty Grouse during the spring and summer are the sunny, upper parts of the foothills, bordering on the heavier timbered portions of the mountains, among the scattered pines and the various berry-bearing bushes found in such situations, and along the sides of canons. According to my observations these birds are scarcely ever found any distance within the really heavy timber. In the middle of the day they can usually be looked for with success amongst the deciduous trees and shrubbery found along the mountain streams in canons, especially if there is an occasional pine or fir tree mixed amongst the former.

In the vicinity of Seattle, Wash., in 1911, we found sooty grouse still quite common almost within the city limits. Where much of the virgin forest had been cut off, where some of the land had been cultivated, and where scattered houses were being built, the grouse still clung to the remnants of the coniferous forests and the brushy clearings near them. W. Leon Dawson (1909) has expressed it very well, as follows:

Indeed, the Blue Grouse and the Douglas fir are nearly inseparable. In the sheltering branches of this tree the bird takes refuge in time of danger; from its commanding elevation he most frequently sends forth the challenges of springtime; and in its somber depths he hides himself thruout the winter season. So great is this devotion on the part of the bird that it is found indifferently at sea level or at the timber line of the highest mountains; and it will not willingly quit a favorite piece of woodland even tho the supporting forests be cut away on every side.

Referring to its summer haunts on Mount Rainier, Wash., R. A. Johnson (1929) says:

The summer range of this species is restricted to the alpine meadow zone, a habitat extending around the mountain approximately between the elevations of 5,000 and 6,500 feet. The ground cover of this habitat consists largely of heather meadows, small mountain ash, and mountain willow thickets interspersed with clumps and individual trees of alpine fir (*Abies lasiocarpa*) and mountain hemlock (*Tsuga mertensiana*), the branches of which usually droop so that the lowest ones touch the ground.

Courtship.—Early in March the grouse begin to migrate downward from their winter resorts in the heavily forested mountains to their summer homes in the more open foothills and valleys. From that time until the last of May their haunts resound with the loud, deep-toned hootings of the male, his challenge to his rivals, or his courtship love notes. I once had a good opportunity to watch a fine old cock grouse hooting and displaying under very favorable circumstances. It was on Mercer Island in Lake Washington, within the city limits of Seattle, where Samuel F. Rathbun had taken us to see some of these grouse. After some difficulty in following up the ventriloquial notes, we located the performer on a horizontal limb, close to the trunk of an enormous Douglas fir, fully 50 feet up in the densest part of the tree, but in plain sight; we watched him for some time through powerful glasses and could plainly see every detail. He turned about occasionally on the branch, facing first one way and then the other, with drooping wings, lifting and spreading his tail. When ready to hoot he stretched out his neck, on which two large, white, rosettes appeared, swelling open and showing the naked sacks of dull yellow skin, which puffed out to semiglobular shape with each of the four or five hoots; the plumage of the whole neck and throat swelled out with each note and the bill opened slightly. The hooting notes were much like the soft, low notes of the great horned owl; when near at hand they seemed to be softer and less powerful, but they really have great carrying power and can be heard for a long distance. They have been likened to the noise made by blowing into the bung-hole of an empty barrel or by swiftly swinging a rattan cane. They were given in groups of four to six notes each; I wrote down the full group as *hoooo*, *hoot*, *hoot*, *hoot*, *a-hoot*, *hoot*, or sometimes as four, five, or six straight *hoots*, or as different combinations of the above notes. I recorded the intervals between the hootings as varying from 12 to 36 seconds; the following series was noted: 12, 18, 22, 23, 35, 14, 22, 22, 16, 17, 19, 14, and 32 seconds. During the intervals of silence the bird assumed a normal pose, or strutted about. Dawson (1909) says: "The hooting, or grunting notes, of this Grouse are among the lowest tones of Nature's thorobase, being usually about C of the First Octave, but ranging from E Flat down to B Flat of the Contra Octave."

Leslie L. Haskin has sent me the following notes:

Early in March the males begin to manifest their presence by their muffled hooting, which proceeds from high up in the trees. At this time the birds will usually be found sitting close to the bole of the fir, their bodies hard to discern against the general grays and browns of the surrounding limbs. The "hooting" of the grouse is one of the most distinctive and peculiar bird notes of the Douglas-fir region, as characteristic in its way as the drumming of its relative, the ruffed grouse. This call can be best imitated by closing the lips tightly, puffing out the cheeks, and then articulating the sound *oo* in a deep tone, low in the throat. The hooting of the grouse has a muffled, ventriloquial quality that makes it exceedingly hard to locate—now seeming far, now near, now high, now low—and as a dozen birds may be calling at the same time it becomes very confusing.

As March advances the hooting becomes more pronounced, and the birds begin to move about more freely, and when mating takes place they descend to the ground. Now, where no grouse were visible only a few days previous, the whole forest may be alive with them. The females have a cackling call to which, early in spring, the males respond very readily. An experienced hunter, by imitating this cackle, can call the birds to him and thus pot them with the least possible effort.

Nesting.—My first and only nest of this species was found on May 7, 1911, on the island in Lake Washington referred to above. Mr. Rathbun guided us to an old clearing, which he said was the breeding ground of these grouse; it was an open, sunny hillside, surrounded by large fir trees and covered with large fallen logs, clumps of brakes, shrubs, and small trees. We had been hunting here less than 10 minutes when a big grouse flushed from under a small fir tree just ahead of me, and there was the nest with seven fresh eggs. The nest was on a little knoll under the small fir, which was only about 7 feet high, and was only partially concealed; the nest cavity was about 7 inches in diameter and 3 inches deep, with a well-built rim of green mosses and ferns, and was lined with dry leaves, ferns, bits of rubbish, and plenty of grouse feathers.

F. Seymour Hersey collected a set of seven eggs for me, near Tacoma, Wash., on May 3, 1914. The nest was in an open situation in a fir forest and was plainly visible at a distance of 25 feet; it was under the end of a fallen log, a hollow 4 inches deep and 7 inches in diameter, well shaped, and lined with twigs and moss, with which the ground was carpeted, and with a few feathers of the bird.

J. Hooper Bowles has sent me his records of some 23 nests found near Tacoma between April 30 and May 22. He says in his notes:

In the selection of a nesting site this grouse, as a rule, prefers a very dry, well-wooded locality, where it scratches out a considerable depression in the ground at the base of a tree, under a fallen branch or other shelter. I have found nests under old boxes, a large tangle of wire netting from a henyard, etc. However, at times they will nest as far as a hundred yards from trees of any kind and with very little concealment. The nesting material is the same as a barnyard hen would use under similar conditions, being mostly what it can reach when the bird is on the nest and what falls into it in the course of incubation. A goodly number of feathers also come out of the bird, so that the nests are often objects of decided beauty.

The period of deposition of the eggs is extremely variable, and it is doubtful if they ever lay an egg a day regularly. To give what may be considered an extreme case was a nest found on May 3, 1928, when it contained two eggs. This nest was built in the side of a huge ant hill about 4 feet in diameter and against a large fir tree, the whole place being swarming with large and ferocious red ants. On May 9, the nest contained only five eggs, and on the 12th it still had only five eggs and had them half buried with hill material, so I concluded that the ants had made the bird desert. However, on passing the place on May 22, I walked over to see if they were all covered over and was astounded to see the female grouse sitting peacefully on her treasures. She was completely surrounded with ant hill and hundreds of ants were running all around her busy with their duties, but not an ant went on her, although a friend and I watched her for some 20 minutes. We flushed her and found that she had cleaned out the nest and lined it with green fir sprays.

Why the ants did not molest her is beyond comprehension, as they certainly nearly ate us alive while we were examining and photographing. The eggs varied from addled to slightly advanced in incubation, so it would be difficult to judge when the last one was laid.

Bendire (1892) gives an interesting account of a "most exposed nest, without any attempt at concealment whatever," at an elevation of 6,800 feet in Oregon; it was in "a beautiful oval-shaped mountain meadow of about an acre in extent, near the summit of which stood a solitary young fir tree." There were no other trees within 30 yards. He lay down to rest under the shade of this tree, when his setter dog came running up and pointed a grouse sitting on her nest within 3 feet of him. She allowed him almost to touch her before she fluttered off; the nest held two chicks and seven eggs on the point of hatching.

Eggs.—The sooty grouse lays from 6 to 10 eggs, but 7 or 8 are the commonest numbers; as many as 16 have been found in a nest, but evidently the product of two females. The eggs are indistinguishable from those of the dusky grouse, which I have already described. Bendire (1892) describes them as follows:

The eggs are ovate in shape, and the ground color varies from pale cream to a cream-buff, the latter being more common. In a single set before me it is a pale cinnamon. The eggs are more or less spotted over their entire surface with fine dots of chocolate or chestnut brown; these spots vary considerably in size in different sets, ranging from the size of No. 3 shot to that of mustard seed. These markings are generally well rounded, regular in shape, and pretty evenly distributed over the entire egg. An egg is usually deposited daily and incubation does not begin until the set is completed, the male taking apparently no part in this duty nor in the care of the young after they are hatched.

The measurements of 92 eggs average 48.5 by 35 millimeters; the eggs showing the four extremes measure **53** by 34.5, 52 by **37**, and **45** by **32.5** millimeters.

Young.—But one brood is raised in a season. Various observers have reported the incubation period, as from 18 to 24 days; this duty is performed by the female alone. Bendire (1892) writes:

The cocks separate from the hens after incubation has commenced, I believe, and keep in little companies, of from four to six, by themselves, joining the young broods again in the early fall. At any rate, I have more than once come upon several cocks in June and July without seeing a single hen amongst them. High rocky points near the edges of the main timber, amongst juniper and mountain mahogany thickets, are their favorite abiding places at that time of the year. The young chicks are kept by the hen for the first week or two in close proximity to the place where they were hatched, and not until they have attained two weeks' growth will they be found along the willows and thickets bordering the mountain streams. Their food consists at first principally of grasshoppers, insects, and tender plant tops, and later in the season of various species of berries found then in abundance everywhere, as well as the seeds of a species of wild sunflower, of which they seem to be very fond. It is astonishing how soon the young chicks learn to fly, and well, too, and how quickly they can hide and scatter at the first alarm note of the mother bird, which invariably tries by various devices to draw the attention of the intruder to herself and away from her young. A comparatively small leaf, a bunch of grass, anything in fact will answer their purpose; you will scarcely be able to notice them before they are all securely hidden, and unless you should have a well-trained dog to assist you, the chances are that you will fail to find a single one, even when the immediate surroundings are comparatively open. After the young broods are about half grown, they spend the greater portion of the day, and I believe the night as well, among the shrubbery in the creek bottoms, feeding along the side hills in the early hours of the morning and evening. During the heat of the day they keep close to the water, in shady trees and the heavy undergrowth. They walk to their feeding grounds, but in going to water they usually fly down from the side hills.

Plumages.—In the downy chick the head and underparts vary from "cream color" to "ivory yellow"; the crown is mottled with black and a little "hazel," and the auriculars are spotted with black; the upperparts are variegated with "hazel," "chestnut," dusky, and pale buff. The wings begin to grow soon after the chick is hatched; in a chick 3 inches long they already reach beyond the tail. These first wing feathers and their greater coverts are broadly tipped with white and have white shaft streaks.

The juvenal plumage comes on very rapidly and is fully acquired before the young bird is half grown, the last of the natal down disappearing on the belly and head. In the full juvenal plumage, in which the sexes are alike but in which the racial characteristics begin to show, the crown is mottled with "amber brown" or "hazel" and black; the feathers of the mantle are variously patterned with dull browns, "hazel" or "tawny," and black or dusky, with conspicuous white shaft streaks, broadening at the tip; the primaries, secondaries, and tertials are barred, notched, or mottled with pale buff on the outer web; the chin and throat are yellowish white and the belly dull white; the breast and flanks are spotted with dusky and pale buff; the rectrices are narrow and pointed, banded, and mottled, much like the plumage of the back. The juvenal flight feathers are molted during July and August; the molt begins as soon as the last of these feathers are fully grown, or even before that; and the body molt into the first winter plumage is continuous from August to October. The postjuvenal molt is complete, except that the outer pairs of primaries are retained for a full year. A. J. van Rossem (1925) says of the tail molt:

The juvenal rectrices are shed at a very early age. The lateral pairs go first, followed soon after by the central pairs. The chicks can be scarcely more than two or three weeks old when the tail-feathers are dropped, and the characteristic post-juvenal tail begins to appear beyond the tips of the coverts. These new tail-feathers are comparatively slow in growing, and reach maturity when the first winter plumage is fully acquired. Their shape and size, as well as the shape of the tail itself, is diagnostic of birds of the year. As compared to that of the adult, the tail is much rounder and more fan-shaped, because of the greater proportionate shortness of the lateral feathers. The individual feathers are shorter, narrower, and have rounder tips. Most of these rectrices are replaced during the following winter, spring, and summer, but some (usually the outer pairs) are apparently always retained until the following fall. The replacement-feathers are similar to those of the adult. However, the whole tail (replacement and first winter feathers alike) is cast at the second fall moult, at which time the longer and broader tail-feathers of maturity are acquired. Replacement is so invariable as to preclude the possibility of accidental or fortuitous moult. These differences are most easily seen in males. They are present in females in lesser degree.

The first winter plumage is much like the adult and the sexes are quite unlike; but young birds, during the first year, can always be recognized by the outer primaries, by the shape of the tail, and by the narrow tail feathers. There is little evidence of even a very limited prenuptial molt, but a complete postnuptial molt takes place the next summer, which produces the fully adult plumage, when the bird is more than a year old. Of this Harry S. Swarth (1926) says:

The first post-nuptial molt begins about the middle of July of the second year and lasts until about the middle of September. The change in character of rectrices is the one conspicuous feature of the mature plumage. I cannot find that there is any renewal of rectrices (except sporadically, presumably as the result of accidental feather loss) until this molt regularly begins. This, I believe, is the only point in which I disagree with van Rossem in the conclusions drawn by him regarding molt in this genus. The fully adult tail, now acquired, is square ended, the feathers broad and truncate. Minor color differences are a clearer gray coloration below and less white spotting on breast and sides, while the mottling on dorsal surface of wings and on interscapulars is less in extent, and gray instead of brown.

According to A. W. Anthony (1899) the sooty grouse has been known to hybridize with the ring-necked pheasant.

Food.—Dr. Sylvester D. Judd (1905a) sums up the food of the "blue grouse" (various races), as follows:

The food consisted of 6.73 percent animal matter—insects, with an occasional spider—and 93.27 percent of vegetable matter—seeds, fruit, and leaves. Grasshoppers constitute the bulk of the animal food, amounting to 5.73 percent. Beetles, ants, and caterpillars form the rest of the insect food. One stomach contained the common land snail (*Polygyra* sp.).

Of the vegetable food, he says:

Browse is eaten by the blue grouse to the extent of 68.19 percent of its annual food, and is distributed as follows: Buds and twigs, 5.28 percent; coniferous foliage, 54.02 percent; other leaves 8.89 percent. The species spends most of its time in pine forests feeding on needles, buds, and flowers. The yellow pine (*Pinus ponderosa*)—male flowers, the white fir (*Abies concolor*), *Abies magnifica*, the Douglas fir (*Pseudotsuga mucronata*), the western hemlock (*Tsuga heterophylla*), and the black hemlock (*Tsuga mer-*

tensiana) are among the trees that afford it subsistence. Plants other than conifers furnish 14.17 percent of the annual food of the species. This material includes red clover leaves, willow leaves, blueberry leaves, miterwort (*Mitella breweri*), birch shoots, and poplar flower buds. During July, in Montana and Utah, field agents of the Biological Survey have seen the bird feeding on the leaves, buds, and flowers of the Mariposa lily (*Calochortus*). It eats also the blossoms of lupine, columbine, and the Indian paint brush (*Castilleja*).

At times it visits fields for oats and other grain. It feeds also on pine seeds (*Pinus flexilis* and other species). It picks up polygonum seeds (*P. polymorphum* and others), is fond of wild sunflower seeds, and has been known to sample false sunflower (*Wyethia mollis*), caraway (*Glycosma occidentalis*), and the capsules of *Pentstemon gracilis*. It picks up also the seeds of various species of lupine, and is fond of acorns, including those of the canyon live oak (*Quercus chrysolepis*).

Among the berries eaten he includes manzanita berries (13.48 per cent of the whole), bear berries, gooseberries, huckleberries, serviceberries, salmon berries, and the fruits of red elder, honeysuckle, cherries, mountain-ash, salal, and currants.

Mr. Bowles writes to me:

In the examination of a great many stomachs of this grouse I have invariably found the contents to be 100 per cent vegetable, with no signs of animal food of any kind. Green leaves of different kinds form at least 75 per cent of their diet, such as grass, ferns, kinnikinnick, etc., with different small berries next, while one was packed with dry green peas. This diet continues from about the first of April until the end of October, when the only food of any kind that I have found consists of the tips of green fir sprouts of the Douglas fir. As with the rest of the family, small gravel is swallowed to aid in grinding up the food.

Behavior.—When rising from the ground in the open clearings in its summer haunts the sooty grouse flushes with a loud whirring of wings, like other grouse, and flies directly away across the clearing to the nearest timber, with a strong steady flight. It usually alights in a tall, thick, coniferous tree, where it stands so still that it is not easily seen; it crouches down on a limb lengthwise or huddles against the trunk, where its colors match its surroundings so perfectly that it is easily overlooked; only the keenest, practiced eyes can discover it. If flushed from a tree on the mountainside, where it is usually found in fall and winter, it sails downward on silent, extended wings and disappears in the forest below.

In regions where these grouse have not been hunted much they are tame, almost to stupidity, the young birds particularly; it has been said that, when a number are perched in a tree, several of them may be killed, if the lowest one is shot first, before the others take flight; this must be unusual, however. Even where they have been hunted, they are not nearly so wary or so resourceful as the ruffed grouse. They seem to feel particularly safe in the tops of the tall fir trees, where it is sometimes difficult to flush them even by shooting at them. The males during the mating season are quite unconcerned, even in the presence of quite an audience; they seem wholly absorbed in their hooting and keep up their display in plain sight. How different from the behavior of the shy ruffed grouse!

R. A. Johnson (1929) writes:

On September 6, I observed a brood of Sooty Grouse going to roost. I located them by their low cackling noise which is often heard at such times. This cackle is a subdued one and is somewhat similar to that of the domestic fowl, except for the absence of the outbursting notes with which the domestic hen ends her performance. In going to roost each bird settled upon a thick mat of branches about eight feet from the trunk of the tree and from twelve to thirty feet from the ground. In this position their dusky color blended somewhat with the moss and lichen-covered branches. The birds settle singly, usually with the head turned toward the tip of the broad frondlike branch upon which they perch, yet concealed for the most part by overhanging branches.

Voice.—In addition to what has been written above, I quote the following observations by John M. Edson (1925), which describe the hooting of this grouse, as well as its tameness:

The voice of the grouse has an almost ventriloquial carrying power. Although in reality his notes are not loud, they often may be heard for very considerable distances. Still, it happens as frequently that the supposedly distant hooter is in fact close at hand. The hooting appeared to commence about 4 A. M., or perhaps a bit earlier, and was heard off and on till after 9 P. M., and sometimes even as late as 9:45. The notes of the hooting Sooty Grouse may perhaps be described as deep bass, but soft in quality, expressed as: *Oot, oot, oot, oot, t-oot,* the second and fifth notes being noticeably subdued. Different individuals vary this slightly. Frequently the last note is omitted, and occasionally but three notes are given. The watch was held on one bird that had been vociferating steadily for some time, one afternoon about 3 o'clock, and it was found that his performances numbered just seven per minute for each of four minutes, following which he became silent for a time. This bird gave only four notes.

The most interesting incident in connection with our grouse acquaintanceship occurred on the morning of the 11th. The weather was pleasant, with a few fleecy clouds and a southwest breeze. Somewhere down the rocky ridge that pitched to the north from our camp, a grouse had been hooting since 4 o'clock. As we sat by our camp-fire at breakfast the hooting seemed to draw nearer, and eventually as if the bird might be out in the open space that commenced not far below the camp. I rose to peep over a near-by rock that obstructed my view, wondering if the bird might not be in sight. And indeed he was. Standing upon another small, flat rock immediately behind the first and just eight paces from where I stood, was our performer all posed for his act. I moved out cautiously till he was in full view, then stood motionless. The bird seemed not the least disturbed by my presence, and after giving me an inquisitive glance, soon started the hooting ceremony once more. Standing with his side toward me, his body pitched at an angle of about 45 degrees, the tail slightly drooping, head well up and neck and breast feathers somewhat puffed out, he began by drawing down his head and further inflating his feathers till the bill and head, except from the eyes up, were concealed. Then throwing open the pocket of his neck feathers he showed a horizontally elongated patch of white lining, in the midst of which was distinguishable the yellowish air-sacs peculiar to the genus.

It was not greatly distended or conspicuous, being partially concealed by the feathers. With his effort of giving voice to his feelings, the bird's whole form pulsated with each note, the half-spread tail vibrating vertically. After the concluding note, he would raise his head, close the neck pocket and calmly look about. He repeated this numerous times for six or seven minutes. Occasionally a whiff of smoke from the camp-fire would sweep past, and this would take his instant attention but caused no serious alarm.

Fall.—Writing of conditions on Vancouver Island, where the migratory habits of the sooty grouse are evidently the same as elsewhere, Swarth (1912) says:

The species is locally migratory, descending into the valleys during breeding season, and retreating into the higher mountains at the end of the summer. The old males go first, beginning to leave about the time the females are bringing their young from the nest. At Beaver Creek a few still lingered through June and could occasionally be heard hooting. In the mountains south of Alberni, in July, no old males were seen at the bottoms of the basins, or in the canons, where females with young were frequently met with, but on the higher slopes and the summits of the surrounding ridges they were quite abundant. At the top of Mount Douglas (altitude about 4,200 feet) several were heard hooting July 14 to 16.

At Errington, early in September, sooty grouse were abundant and gathered in flocks, usually of from six to ten individuals, though as many as fifteen were seen in one gathering. At this time there were no males in the lowlands, these flocks being in all probability composed usually each of a female with her brood; but a trip to the summit of Mount Arrowsmith, September 6 to 8, disclosed the presence of the cock birds in numbers everywhere on the higher slopes of the mountains. About the second week in September the others began to follow, and they soon became quite scarce in the lowlands. By the end of the month but very few remained.

Game.—The "blue grouse," or "gray grouse," or "mountain grouse," as it is variously called, has been the finest game bird of the northwest coast region; but it is becoming very scarce near the centers of population. It is a fine, large, heavy bird, weighing 3 or 4 pounds. When the bird is feeding on berries in summer or fall, its flesh has a delicious flavor and it is often very fat; but in winter, when it feeds almost exclusively on coniferous browse, it has a resinous flavor which is not so good. It lies well to a dog and, when flushed in open clearings or in the mountain pastures or berry patches, it flies away with a straight, steady flight, making very pretty shooting. In the heavily timbered mountains it is not so easy a mark, as it glides silently out of the top of some tall fir tree and goes scaling downhill at a swift pace. It is a difficult mark too, as it flies away among the maze of tree trunks in a heavy stand of Douglas fir, where many a charge of shot finds a tree trunk instead of the bird.

Winter.—Reversing the habit of most other birds, the "blue grouse" spends the summer in the lowlands and retires to the higher mountain forests to spend the winter. Mr. Haskin says in his notes:

It is a remarkable bird, with habits all its own. Throughout the winter months the birds are seldom seen, even in places where they are most abundant, for at this season they retire to the heavy fir timber, and spend their time very quietly high up in the trees. During this season they feed almost exclusively on fir buds, and do not even descend to the ground to drink, for the abundant rainfall makes it possible for them to quench their thirst in the treetops. Personally, I do not think that the grouse ever voluntarily comes to the ground during these months of retirement. Only when by accident they are disturbed, as when woodsmen are felling trees, are they likely to be seen at all. In sections where the grouse are very abundant you may pass through the woods day after day, and unless you understand their ways, never suspect that such a bird is present.

DISTRIBUTION

Range.—Pacific coast mountain ranges of the United States, Canada, and southeastern Alaska.

The range of the sooty grouse extends **north** to southeastern Alaska (Glacier Bay and Portage) and northwestern British Columbia (Wilson Creek). **East** to British Columbia (Wilson Creek, Hastings Arm, and Westminster); central Washington (Barron, Buck Creek Pass, Cascade Tunnel, Ellensburg, and Signal Peak); central Oregon (Wapinita, Fort Harney, and Drews Creek); northeastern California (Warner Mountains and Honey Lake); western Nevada (Truckee and Marlette Lake); and southeastern California (Bishop Creek, Mount Whitney, and Piute Mountains). **South** to southern California (Piute Mountains and Mount Pinos). **West** to California (Mount Pinos, Dunlap, Yosemite Valley, Big Trees, Mount Sanhedrin, summit of the Yolla Bolly Mountains, Summerville, and the White Mountains); western Oregon (Glendale, Roseburg, Spencers Butte, Newport, and Tillamook); western Washington (Cape Disappointment, Grays Harbor, Hoh River, and Crescent Lake); western British Columbia (Victoria, Mount Douglas, probably Della Lake, Haida Mountain, and Massett); and the western part of southeastern Alaska (Coronation Island, St. Lazaria Island, Sitka, Cross Sound, Skagway, and Glacier Bay).

The sooty grouse, formerly considered as a subspecies of the dusky grouse (*Dendragapus o. obscurus*), is raised to full specific rank in the 1931 edition of the A. O. U. Check List. Three additional subspecies are recognized, the distribution as above outlined being for the entire species. True *fuliginosus* is confined to the mountains of the northwest coast from northwestern California and Oregon, north to Alaska (Skagway) and northwestern British Columbia. *D. f. sitkensis* occupies the islands of the southeastern Alaskan coast (except Prince of Wales Island), Queen Charlotte Islands, and Porcher Island; *D. f. sierrae* is found from central southern Washington and Fort Klamath, Oreg., south on the inner side of the coast range to Mount Sanhedrin, Calif.; and *D. f. howardi* is confined to an area in southern California from Mount Pinos east through the Tehachapi Range and north in the Sierra Nevada to about latitude 36° N.

As stated under *Dendragapus o. obscurus*, strictly speaking the sooty grouse is nonmigratory, but locally it has a curious vertical migration as it descends into the valleys during the breeding season and retreats to the higher mountains at the end of summer.

Egg dates.—Washington and Oregon (*fuliginosus*): 60 records, April 16 to July 12; 30 records, April 30 to May 22. California (*sierrae*): 5 records, April 27 to June 9. California (*howardi*): 1 record, May 21. Alaska (*sitkensis*): 1 record, June 2.

DENDRAGAPUS FULIGINOSUS SIERRAE Chapman

SIERRA GROUSE

HABITS

The sooty, or "blue," grouse of the Sierra Nevada and the inner side of the coast ranges, from southern Washington to about latitude 31° N. in California, was described by Dr. Frank M. Chapman (1904) as "most nearly related to *Dendragapus obscurus*, but the nuchal region often browner and usually vermiculated with black, the whole dorsal region less black and more heavily vermiculated with brown and gray; terminal tail band narrower and more speckled with blackish; the median tail-feathers more heavily marked with gray or brownish; the scapulars and tertials with the terminal white wedge less developed or entirely wanting; the basally white neck-tufts practically absent; the throat averaging duskier and the feathers of the sides, flanks, and under tail-coverts with much less white. Differs from *Dendragapus obscurus fuliginosus* in much paler coloration above, in the heavier vermiculation of the entire upper surface, practical absence of neck-tufts, whiter throat, and paler underparts." He says further:

In spite of the fact that the Sierra Grouse more nearly resembles *obscurus* than it does *fuliginosus* it apparently has been derived from the latter rather than from the former. That is, it represents a southern extension of the northwest coast form and not a westward extension of the Rocky Mountain form.

This theory is supported by the apparent continuity of range of *sierrae* and *fuliginosus* and by their evident intergradation in the vicinity of Klamath, Oregon. Several of the specimens, in an admirable series collected by Major Bendire, at Fort Klamath, are referable to *sierrae* rather than to *fuliginosus*, though not typical of the former. Other examples in this series, however, are much nearer to *fuliginosus*. On the other hand, I have no material proving continuity of range in *sierrae* and *obscurus*, and the character of the country intervening between the nearest known portions of their respective ranges would lead one to suppose that they do not intergrade geographically.

It lives chiefly in the Canadian Zone and only locally in the Upper Transition, ranging "upward into the Hudsonian Zone during late summer," according to Grinnell and Storer (1924), who write:

Acquaintance with the Sierra Grouse may begin in several ways, but rarely does it come in the conventional manner through which we learn to know most birds. Upon entering the Jeffrey pine and red fir forests of the Canadian Zone in spring and early summer, one may often hear a very un-bird-like, dull sodden series of booming notes that have a ventriloquial quality. These are the courting notes of the male grouse. Less often, whatever the time of the year, the introduction may come suddenly and much more impressively when, close at hand, a heavy-bodied "blue grouse" rises quickly from the ground and makes off through the forest on loudly whirring wings, and showing an expanse of square-ended gray-banded tail. When a small flock of the birds get up, as they often do, in rapid succession, or even simultaneously, the aggregate effect is bewildering, to say the least.

The Sierra Grouse lives in the high country throughout the year, never migrating to lower levels as does the Mountain Quail. The thick heavy plumage and legs feathered clear down to the toes enable the grouse to withstand the cold of the midwinter months; while their ability to subsist on pine and fir needles assures them at any season an abundance of food to be easily obtained without seeking the ground.

A. B. Howell (1917) says of its haunts:

Although most of the published information pertaining to the Sierra Grouse gives one the impression that these birds haunt the pines and associations of scant undergrowth, my experience has been that they seldom resort to the larger conifers except to roost, and to escape their enemies by remaining motionless in the upper branches. At least in the locality under consideration, their favorite habitat is in the vicinity of dense aspen thickets, and the tangles of manzanita, hazel and other brush on the dry hillsides and benches of the high Transition Zone, from which they flush to the timbered ravines.

Courtship.—Grinnell and Storer (1924) have described the booming courtship of this grouse very well, as follows:

During the spring and early summer, the males are in the habit of taking solitary positions near the tops of pines or firs, sixty or more feet above the ground, where they stand on horizontal limbs close to the trunk. They hold such positions continuously for hours, one day after another, and send forth at intervals their reverberant booming. With different birds the series of notes comprising this booming consists of from five to seven syllables, six on an average. The quality of the sound can be likened to that produced by beating on a water-logged tub, *boont, boont, boont', boont', boont, boont*, crescendo at the first, diminuendo toward the end of the series. As each note is uttered the tail of the bird is depressed an inch or two—perhaps an index to the effort involved. The separate series of notes in two instances were uttered at intervals of 40, 20, 25, 45, 12, 21, and 29 seconds, and again 10, 10, 20, 26, 14, 15, 17, 12, 11, 15, 13, 28, 17, and 11 seconds, respectively. These two birds had been heard booming for a long time before we began to pay special attention to them, and they continued long after we finished this record. The ventriloquial quality is discovered when one attempts to locate the producer, a difficult feat as a rule. The observer may succeed in locating the proper tree, but is likely to circle it many times, peering upward with painfully aching neck, and still utterly failing to locate the avian performer amid the foliage high overhead. The notes are commonly supposed to be produced by the bird's inflating and exhausting the glandular air sacs on the sides of the neck. These sacs are covered by unfeathered yellow skin, and we think it more likely that they serve only as resonators, being kept continually inflated, while the air actually producing the sound passes to and from the lungs along the regular air passage. It rests with some one gifted with patience for long-continued observation to determine exactly how the notes are produced.

W. Leon Dawson (1923) also says:

As the hooter becomes vehement he struts like a turkey-cock, spreading the tail in fan-shape, dropping the wings till they scrape the ground, and inflating his throat to such an extent as to disclose a considerable space of bare orange-colored skin on either side of the neck. This last certainly makes a stunning feature of the gallant's attire, for Nature has contrived that the feathers immediately surrounding the bald area should have white bases beneath their sooty tips. During excitement, then, as the concealing feathers are raised and reversed, a brilliant white circlet, some five inches in diameter, suddenly flares forth on each side of the neck, to the great admiration, no doubt, of the observant hen.

These more emphatic demonstrations are probably reserved for such time as the hen is known to be close at hand, for I have never frightened a strutting cock without finding a female hard by, at least at no greater distance than the lower branches of a neighboring tree. She has responded to the earlier calls of the male by a single musical toot note, uttered at intervals of approach; but once arrived at the trysting place she has become very shy, and will take no part in the celebration, save by a few tell-tale clucks and many coy evasions. On these occasions, also, the cock works himself up into such a transport that he becomes oblivious to danger, so that he may be narrowly observed or even captured by a sudden rush.

Nesting.—The nesting habits of the Sierra grouse are similar to those of other forms of the species. Mrs. Irene G. Wheelock (1904) writes:

In May or June, according to location, the wooing begins, and soon the mother is brooding on her eight buffy eggs in the shade of a fern tangle, near a log, or in a clump of manzanita. No part does the father take in the three weeks of patient incubation, but the mother can seldom be surprised away from the nest. It would be far easier to discover the eggs were she not covering them, for so protective is her coloring that you may be looking directly at her and never suspect it, although at that very moment you are searching for a nest. Her food is all about her—buds, berries, and insects. If she leaves the eggs, it is only to stretch her tired little legs and pick up a few dainties close by.

Eggs.—The eggs of this grouse are indistinguishable from those of other races of the species. The measurements of 23 eggs average 48.7 by 35.2 millimeters; the eggs showing the four extremes measure **45.1** by 34.3, 49 by **37.5**, **51.2** by 34.6, and 46.1 by **32.8** millimeters.

Young.—Mrs. Wheelock (1904) says that as soon as the young are out of the shell and dry, away goes the mother,

proud as a peacock, with them at her heels. And now the father is introduced to family cares, and he scratches for bugs, calling the young with imperative little chucks to come. He is the drill-master of the little flock, teaching them with infinite patience all that they need to know of wood lore. He stands on guard at every suspicious noise, and whistles his warning when danger threatens. When their wing-feathers have developed and they can flutter up to a low branch in the bush, they roost there instead of cuddling under the mother's broad wings at night. But they remain with the parents and evidently under discipline throughout the first six or eight months of their existence. In the wintry weather, when their mountain homes are covered deep with snow, they often sleep huddled together deep in a drift, waking to feed upon the buds of the coniferous trees, but seldom seeking a lower level. They are the hardy mountaineers, the children of the forest ranges.

Grinnell and Storer (1924) write:

By early July the new broods of grouse are to be looked for in the brush-bordered glades of the forests. When the chicks have been partly reared the males desert their mates, and, forming in flocks of 6 or 8, work higher in the mountains. The females remain with, and continue to care for, their offspring, these family units remaining separate for the time being. Finally, as the summer wanes, they, too, work up into the Hudsonian Zone. Thus, while the Mountain Quail go down-hill in the fall, the grouse go up-hill.

Food.—The same observers say on this subject:

One of the above-mentioned male birds was shot, and its crop was found to contain 1,520 needle tips of the lodgepole pine. The bitten-off ends of needles varied from one-fourth to one inch in length. The crop also contained a few fragments of very young pistillate cones. The bill of this bird was smeared with pitch. The crop of an adult female grouse obtained at Walker Lake held eleven ripe rose hips, and the gizzard was filled with the hard seeds of the rose, together with grains of quartz which of course had served to grind the resistant portions of the bird's food.

Grinnell, Dixon, and Linsdale (1930), in their report on the Lassen Peak region, say:

The crop of a female trapped on July 6, 1924, in a rolled-oat baited steel trap contained a quantity of unripe manzanita berries. On September 16, 1923, at 8,200 feet on Warner Creek, a bird was discovered as it was eating manzanita berries. At the same station a grouse was watched as it stalked grasshoppers. The bird would stretch out its neck and slowly approach the insect until within one-half meter when it would make a quick rush forward and capture the hopper in its bill.

DENDRAGAPUS FULIGINOSUS HOWARDI Dickey and van Rossem

MOUNT PINOS GROUSE

HABITS

The southernmost race of the *fuliginosus* group of "blue grouse" has been recently discovered on Mount Pinos in Kern County, Calif.

It was described by Dickey and van Rossem (1923) and named in honor of O. W. Howard. The characters given are:

Nearest to *Dendragapus obscurus sierrae*, but differing from that form in paler dorsal coloration, and in coarser and more conspicuous vermiculation and barring. Underparts darker, a brownish suffusion replacing the clearer gray of *sierrae*. The white median shafting and terminal pattern of the feathers of flanks and sides reduced in area and entirely lacking on anterior part of body, whereas in *sierrae* traces of this pattern extend forward to the shoulders. Wing slightly longer; tail decidedly longer and much more graduated, with terminal band averaging wider. Culmen, tarsus, and middle toe averaging slightly longer and decidedly heavier.

Throughout the range of *Dendragapus obscurus* in California there is a gradual geographic variation which particularly affects the length and graduation of the tail. These characters increase steadily from north to south. Birds from Mount Pinos express in ultra-typical form this lengthening of the tail itself, as well as the greater ratio between the length of the lateral and median rectrices, a truly striking character which the writers have termed "graduation" in the above description. In this same region, the variation in color and pattern from typical *sierrae* is also most pronounced.

J. R. Pemberton (1928) says of the haunts of this grouse:

These birds inhabit the crests of some of the higher mountains from the southern extremities of the Sierra Nevada through the Tehachapi Range to Mount Pinos in Southern Kern County, California. Mount Pinos is the southernmost recorded station. This high peak reaches an altitude of 8,826 feet and is beautifully wooded with several species of pines and the silver fir. The grouse live only on the higher portions of the mountain and I believe have not been observed below 7,800 feet, which is the elevation of the old sawmill. In a sense their range coincides with the areas where the silver fir reaches its best development. The upper part of Mount Pinos consists of a rather gently rolling table-land. The automobile road ends at an altitude of 7,800 feet, and in a walk of two miles the summit, 8,826 feet, is reached. The mountain is really a broad ridge with an exceedingly steep north slope which falls 3,800 feet in a distance of three miles to San Emigdio Creek. This creek runs in an east and west valley paralleling the longer diameter of Mount Pinos.

Nesting.—Mr. Pemberton (1928) was the first, and so far as I know is the only one, to find the nest of this grouse. Describing his experience with it, he writes:

This year, Dudley DeGroot and the writer spent May 21 looking for eggs, being unable to make the trip earlier. The interesting discovery was made that at that time no birds could be located on the higher part of the mountain, while well down on the cliff-like north slope many hooters could be heard. We believed that the hooters were near the sitting females, so we spent our time clambering about on this steep slope. Many tons of rocks were rolled down but no birds could be flushed. Finally, as I was about ready to give it up and about 200 feet below the rim of the steep slope at about the 8,200 foot level, I flushed a female at a distance of about 50 feet and immediately saw the eggs. The bird left with a great whirr, lit on the lowest branch of a large pine about 100 feet distant, clucked a few times as she walked to the end of the limb, and then flew noiselessly downhill. The location was near a point where a hooter had been circling all day and, although he moved his location many times, it was now evident that he had been in sight of the sitting female all the time. The nest was in clear open ground and without the slightest cover for the eggs. A depression less than an inch in depth seemed to have been scratched out of the dry, sandy soil and lined rudely with bits of pine bark, a few needles and vegetable trash. Many feathers lay loosely with the eggs. It was a poor excuse for a built nest and was rather a simple resting place for the eggs. I believe the following generalizations can be made. Howard's Grouse nest on Mount Pinos during the first week in May, and full sets will be found before the 15th. The nests are fairly well down on the steep north slope and placed in entirely open ground in sunny spots well covered from the distance by observation trees. Nests ought to be found by search near where hooters are active. In early May the snow banks will eliminate all unlikely ground. As soon as the young are able to walk they are led to the flatter upper slopes of the mountain where there is good cover and more food. It is obviously unsafe to attempt a statement concerning the number of birds which live on Mount Pinos, but one can say that there are not many and I believe that the number is less than one hundred.

Eggs.—I saw these eggs in Mr. Pemberton's collection and, as I remember them, they are like certain types of sooty-grouse eggs He (1928) describes them as follows:

The five eggs were nearly ready to hatch and the embryos had feathers an inch long. They resemble miniature turkey eggs but with larger spots. The ground color is light buff while the spots are auburn, using Ridgway's *Color Standards and Nomenclature* (1912). The more prominent spots are 2 and 3 millimeters in diameter and one egg has two spots 8 and 10 millimeters in diameter. The measurements are 49 by 36, 49 by 37, 50 by 36, 51 by 36, 51 by 37; the average is 50 by 36.5.

Behavior.—The summit and steep slope of Mount Pinos have proved difficult country to hunt on, and late snowstorms, cold rains, and high winds make it very uncomfortable for the collector, even late in spring. Several good collectors have made repeated attempts to get specimens of this grouse without success. The birds seem to be very wild and unusually crafty in avoiding capture. Dickey and van Rossem (1923) say:

The birds of Mount Pinos display a sagacity in eluding capture that is utterly beyond anything observed by the authors in birds from the central or northern Sierra Nevada. One "hooting" site, in a Jeffrey Pine, was carefully watched on several different occasions during a period of two years, before the bird was located and secured. By contrast, the species in like season in the Sierras is often lacking in suspicion to the point of actual stupidity.

DENDRAGAPUS FULIGINOSUS SITKENSIS Swarth

SITKA GROUSE

HABITS

On some of the islands of southeastern Alaska there occurs a race of the sooty grouse in which the females are conspicuously reddish in color. Harry S. Swarth (1921) has described this race under the name *sitkensis*, characterizing it as

most nearly like *Dendragapus obscurus fuliginosus*. Adult male not appreciably different from the male of *D. o. fuliginosus*. Adult female and immature of both sexes, as compared with those of *fuliginosus*, much more reddish in general coloration. This color feature affects practically all the plumage except some limited areas, as the slaty colored abdominal tract, the chin and throat, and the unmarked and generally concealed portions of the remiges and rectrices. The predominant color dorsally is close to pecan brown. Individual feathers are barred with black and brown, and are brown tipped. On head and neck brown predominates, the narrow black bands being almost entirely hidden. Upper tail coverts and central rectrices are conspicuously of this reddish brown color. Breast and sides are mostly pecan brown and black. There are conspicuous white spots on sides of breast and flanks. Tarsus brown.

Female *fuliginosus*, in comparison, is colored as follows: The upper parts are a duller brown with a great deal of black showing through and with the brown everywhere sprinkled with black or gray. There are no pure reddish brown areas as in *sitkensis*. The neck above is predominantly grayish; upper tail coverts and remiges are mostly grayish. Breast and sides are mostly gray and black, with very little reddish. Feathers on sides of breast are dull brownish, mottled with black and tipped with white. Flanks are mostly grayish. Tarsus gray.

There are four eggs of the Sitka grouse in the collection of P. B. Phillip, which measure 52.7 by 35.3, 52.7 by 35.1, 52.7 by 35.1, and 49.7 by 35.1 millimeters.

Alfred M. Bailey (1927) says of their habits:

The birds are often so high they will not flush with the discharge of a gun; in fact, I have seen a Grouse sit within twenty feet of a gunner, who fired a dozen shots with a high power rifle in an attempt to shoot the bird in the head, without the bird seeming the least alarmed. In the early fall, many repair to the mountain tops with their broods, where they find ample cover among the dwarfed pines and dense alder thickets; then they drop to sea level during the cold winter months, and one will often see them below snow line, where the tide has cleaned the beaches. They feed on the hillsides, among the dead devil-club and berry bushes, and rarely fly when one passes. Although strongly tainted with the spruce, which makes up a great part of their food, the flesh of these birds affords a welcome addition to the camp-fare when one is afield.

CANACHITES CANADENSIS CANADENSIS (Linnaeus)

HUDSONIAN SPRUCE GROUSE

HABITS

CONTRIBUTED BY CHARLES WENDELL TOWNSEND

The Hudsonian spruce grouse thrives best in regions where man is absent. In fact it remains so woefully ignorant of the destructive nature of the human animal that, unlike its cousin, the ruffed grouse, it rarely learns to run or fly away, but allows itself to be shot, clubbed, or noosed, and, in consequence, has earned for itself the proud title of "fool hen." As a result, wherever man appears, the spruce grouse rapidly diminishes in numbers, and, in the vicinity of villages or outlying posts, is not to be found. It is a bird of the northern wilderness, of thick and tangled swamps, and of spruce forests, where the ground is deep in moss and where the delicate vines of the snowberry and twinflower clamber over moss-covered stubs and fallen, long-decayed tree trunks.

Although spruce grouse are resident wherever found even to the northern limit of their range, a certain extent of movement occurs among them in winter, dependent probably on the food supply and not on the severity of the cold.

Courtship.—As with the ruffed grouse, "drumming" by the wings is an important feature of the courtship, but in this the spruce grouse has not reached so high a degree of evolution. It appears to be at a stage midway between the bird that in courtship flies with rapidly and noisily fluttering wings and the bird that stands still and flutters or "drums" with its wings.

J. L. Devany (1921), writing of the courtship of the spruce grouse, says:

His favorite location at such a time is between two trees standing apart some 20 or 30 feet, and with the lower branches large and horizontal. Perched on one of these branches he pitches downward, pausing midway to beat and flutter his wings, and ascend to a branch of the opposite tree. After a short interval this manoeuver is repeated and so continued by the hour, swinging back and forth from tree to tree, the time between each swing being as exact as if measured by a watch. If such an ideal situation is not at hand, the fact does not prevent the "fool hen" from giving vent to his exuberance. Selecting a small open space among the bushes, he takes his stand in the center and, like a jack-in-the-box, pops up a few feet in the air and, giving his triumphant fluttering, drops again to earth * * *. The sound produced by the drumming of the Canada grouse can in no-wise compare with that of the ruffed grouse; it has neither the roll nor the volume. It is in fact little more than a flutter, such as might be made by birds forcing their way through thick branches after buds or berries. Unlike the ruffed grouse, however, he seems to have no very strong objections to an audience. The performance of a ruffed grouse can only be witnessed by the exercise of stealth and caution. Our little spruce partridge on the other hand will peer and look at the intruder, and then, as if suddenly remembering, go through his evolutions with a gusto that excites our startled amusement. Though the drumming of the grouse is peculiar to the male, its practice is not confined to the nesting season alone, but may be heard in any month of the year, and occasionally at any hour of the day or night.

Everett Smith (1883) thus describes the performance:

The Canada Grouse performs its "drumming" upon the trunk of a standing tree of rather small size, preferably one that is inclined from the perpendicular, and in the following manner: Commencing near the base of the tree selected, the bird flutters upward with somewhat slow progress, but rapidly beating wings, which produce the drumming sound. Having thus ascended 15 or 20 feet it glides quietly on the wing to the ground and repeats the manoeuvre. Favorite places are resorted to habitually, and these "drumming trees" are well known to observant woodsmen. I have seen one that was so well worn upon the bark as to lead to the belief that it had been used for this purpose for many years. This tree was a spruce of 6 inches in diameter, with an inclination of about 15 degrees from the perpendicular, and was known to have been used as a "drumming tree" for several seasons. The upper surface and sides of the trunk were so worn by the feet and wings of the bird or birds using it for drumming, that for a distance of 12 or 15 feet the bark had become quite smooth and red as if rubbed.

Bendire (1892) quotes another description of the drumming by James Lingley:

After strutting back and forth for a few minutes, the male flew straight up, as high as the surrounding trees, about 14 feet; here he remained stationary an instant, and while on suspended wing did the drumming with the wings, resembling distant thunder, meanwhile dropping down slowly to the spot from where he started, to repeat the same thing over and over again.

Bendire also quotes this from Manly Hardy: "My father, who has had opportunities to see them drum, told me they drummed in the air while descending from a tree."

Nesting.—The nest of the spruce grouse is difficult to find, as it is generally placed under the low protecting branch of a spruce or in deep moss and concealed in a tangle of bushes. As the mother bird so perfectly matches the dead leaves and twigs of the forest floor, and as she does not move except in imminent danger of being stepped upon, the difficulty of discovery is increased, and it not infrequently happens that the intruder steps in the nest and breaks some of the eggs before he realizes that it is there. This method of finding a nest occurred when Mr. Bent and I were cruising along the Canadian Labrador coast. Mr. Bent had offered a reward to anyone who would bring him a set of the eggs of this bird, as our own search had hitherto been fruitless. While we were anchored behind Little St. Charles Island, a fisherman came on board with eight of these beautiful eggs in his hat. He explained somewhat ruefully that he had stepped into the nest almost on the sitting bird, and crushed four of a set of 12 eggs before he knew they were there.

The nest is generally a slight depression in the moss, lined with dead grass and leaves. Lucien M. Turner describes a nest found in the neighborhood of Fort Chimo, Ungava, as "merely a few grass stalks and blades loosely arranged among the moss of a higher spot under the drooping limbs of a spruce situated in a swamp. A few feathers of the parent bird were also in the nest."

A. D. Henderson describes a nest as follows: "It was in a muskeg, a slight hollow in the moss scantily lined with a few twigs and leaves of the Labrador tea. It was under a moss-covered, dead, fallen spruce branch beneath a low-branching green spruce. The sitting bird was very reluctant to leave the nest." Another "nest was in a hollow lined with dry leaves and spruce needles under a small spruce bush, about 2 feet high, on the edge of a muskeg."

Eggs.—[AUTHOR'S NOTE: The spruce grouse and its near relative, Franklin's grouse, lay the handsomest eggs of any of the grouse. Ten or a dozen eggs usually make up the set, but as many as 14 or even 16 have been found in a nest. Sets of less than eight are probably incomplete. The eggs vary in shape from ovate to elliptical ovate. The shell is smooth with a very slight gloss. The ground colors vary from "cinnamon" to "pinkish buff," or from "cream-buff" to "cartridge buff." They are usually boldly and handsomely marked with large spots and blotches of rich browns, sometimes more sparingly marked and sometimes thickly and evenly covered with small spots and dots. The colors of the markings vary from "chestnut-brown" or "chocolate" to "hazel" or "russet." One odd set in my collection has a "cartridge buff" ground; one egg is nearly immaculate and the others are sparingly, or only slightly, spotted with "bone brown." The measurements of 54 eggs average 43.5 by 31.7 millimeters; the eggs showing the four extremes measure **47.1** by **34**, **40** by 30.4, and 40.1 by **29.8** millimeters.]

Young.—The duration of incubation is about 17 days, according to Lucien M. Turner. Incubation is performed by the female and she alone looks after the young. Bendire (1892) states that "an egg is deposited every other day, and incubation does not begin until the clutch is completed." Turner, however, states that "laying begins about the fifth of June and incubation about the twelfth" in Ungava.

The young are able to run about and follow the mother almost as soon as their feathers dry, and they are able to fly vigorously at an early age, when they appear about a quarter of the size of the adults. One of these young, a good flyer, that I collected at Shecatica Inlet, Canadian Labrador, on July 23, 1915, measured 5 inches in length, and its wing measured 3.5. The adult's length is about 13 inches and the wing 6.5. On this occasion the brood of young startled me by flying up with a slight whirring sound almost from under my feet. They flew to the branches of a low spruce, while the mother appeared most conspicuously, standing in a bed of curlew-berry vines and reindeer lichen, with head up and tail erected. As a rule the young fly off and conceal themselves so thoroughly that it is difficult to flush them again, while the mother, clucking and ruffling her feathers, flies to a spruce tree or remains on the ground, in both cases allowing an approach to within a few feet. On one occasion, when I was in the marshes about the mouth of the St. Paul River, Canadian Labrador, the female stood on a small rock and crooned, while the young, one after the other, until seven in all, flew and joined her on the rock.

Mr. Bent contributes the following note on the behavior of mother and young when the latter are still unable to fly; this was near Hopedale, Labrador:

> We caught one of the young and had an interesting time watching the mother bird in her solicitude. Her boldness was remarkable and she showed no fear whatever; we could walk right up to within a few feet of her, on the ground or in small trees. We tied the young one with a string and, as soon as the mother heard its peeping notes, she came right up to it, clucking and scolding, with her feathers all ruffled up and her tail spread like a turkey's; she strutted around over the logs and rocks near us; her soft clucking notes sounded like *kruk, kruk, kruk,* with an occasional rolling note like *krrrrruk,* soft and low. The young evidently understood it as a danger signal, for they remained so well hidden that we found only two of them. While I was photographing the captive little one, the mother came almost near enough to touch and even ran between the legs of the tripod.

Plumages.—[AUTHOR'S NOTE: In the small, downy chick the general ground color is yellowish buff, varying from "chamois" above to "colonial buff" below; there is a black spot at the base of the culmen, a larger one in the middle of the forehead, one on each of the lores, and a broken stripe on the auriculars; there is a large patch of "hazel," bordered with black, on the crown and occiput; the back and rump are washed with "hazel" and "tawny" and indistinctly spotted with black; the underparts are unmarked.

The juvenal plumage appears at an early age, beginning with the wings; these start to grow within the first 5 days and, at the age of 10 or 12 days, the wings reach beyond the tail and the young bird can make short flights. By the time the young bird is half grown it is fully feathered. In this juvenal plumage the sexes are alike and resemble the adult female, but are browner above, rustier on the head and neck, and whiter on the chin. The crown is "cinnamon-rufous" or "hazel," spotted with black; the back, scapulars, and wing coverts are "ochraceous-tawny" or "tawny," boldly patterned with black blotches or bars, and with broad, median, buffy stripes with triangular white tips; the remiges are sepia, the primaries narrowly notched with buff, the secondaries edged with buff, and the tertials barred and spotted with "ochraceous-tawny"; the pointed rectrices are "ochraceous-tawny," heavily barred and peppered with black; the breast is "ochraceous-tawny," spotted with black, the belly grayish or yellowish white, faintly spotted with dusky, and the chin and throat yellowish white.

Beginning early in August and lasting through September, the postjuvenal molt takes place, during which the sexes begin to differentiate, the young males showing patches of black feathers in the breast. This molt is complete except for the two outer primaries on each wing. It begins on the breast, extends to the flanks and back, and is finally completed on the throat and crown. When this molt is completed in October, young birds can hardly be distinguished from adults, though there is more white in young birds and the outer primaries are diagnostic.

Adults probably have a very limited prenuptial molt about the head and neck in spring; and they have a complete postnuptial molt in August and September. The Hudsonian spruce grouse is a grayer bird than the Canadian, with rather more white and purer gray in the male; the difference is even more pronounced in the female, which is much more purely black and gray, with much less buffy or ochraceous.]

Food.—The spruce grouse not only lives in spruce woods but depends upon the buds, tips, and needles of the spruce, as well as of the fir and larch, for a considerable part of its diet. This is particularly the case in winter when snow and ice cover the ground, concealing many berries, which it enjoys eating in summer. In the latter season, I have found in Canadian Labrador the following stomach contents of this bird: A young able to fly had eaten 5 red spiders, 10 green snowberries, and 75 achenes of a bulrush (*Scirpus caespitosus*); an adult had eaten 25 snowberries, 20 crowberries (*Empetrum nigrum*), and many leaf tips of a dwarf bilberry (*Vaccinium ovafolia*). Another adult had been feeding entirely on crowberries and a third had been eating the leather woodfern (*Dryopteris marginalis*). Bearberries (*Arctostaphylos uva-ursi*), as well as grass and weed seeds and various insects, including grasshoppers, are also eaten in summer, although the regular diet of spruce is not entirely given up.

E. A. Preble (1908) found nothing but spruce needles in the stomachs of four spruce grouse. Another, taken on the shore of an inlet had in its crop several mollusks (*Lymnoea palustris*). The crops of a number taken late in fall and in winter contained only the needles of the jack pine (*Pinus banksiana*). A young bird just ready to fly had eaten bits of the American rockbrake fern (*Cryptogramma acrostichoides*), blueberries (*Vaccinium uliginosum*), and mountain cranberries (*V. vitisidaea*).

Lucien M. Turner says:

> The food of the spruce partridge consists of the tender, terminal buds of spruce; and this, in winter, seems to be their only food * * * mixed with, at times, an astonishing quantity of gravel. I was surprised to find these stones of such uniformity of size and material. Crystallized quartz fragments, in certain instances, formed alone the triturating substance.
>
> If a bird be opened when just killed the contents of the gizzard has a powerful terebinthine odor which quickly pervades the flesh and renders it uneatable to a white person. In the spring and summer months these birds

consume quantities of berries of *Empetrum* and *Vaccinium* and in the fall the flesh of the young * * * has a fine flavor, and, as the meat is white, it is very acceptable.

In some instances the flesh of adults in summer is also free from any taste of turpentine. According to A. L. Adams (1873) it is said that the flesh is sometimes poisonous when the birds have been eating mountain-laurel berries.

Behavior.—The chief characteristic of the Hudsonian spruce grouse is its unsuspicious character, which amounts, indeed, to stupidity. This is illustrated by an experience of Lucien M. Turner, who says:

I once shot 11 and did not move a yard in distance to do so. The people of Labrador employ a method which they term "slipping," i. e., a slip noose on a long pole which enables the holder to slip the noose over the heads of the birds and jerk them down. One who is expert in this method rarely fails to obtain all the birds within reach.

D. G. Elliot (1897) says: "I have seen birds push this noose aside with their bills without changing their position, when through awkwardness, or unsteadiness of hand on account of a long reach, the noose had touched the bird's head but had not slipped over it." I have known a botanist to kill an adult grouse by throwing his short-handled collecting pick at it.

The plumage of spruce grouse often makes them difficult to distinguish from their surroundings, and if their tameness depends on this protective coloration, they are overconfident, for, in a setting of reindeer lichen or snow, or an open branch of a spruce, they are very conspicuous. When flushed they generally fly only a few yards or even feet, and, alighting in trees, they continually thrust the head and neck now this way, now that, and appear to be blindly trying to discover what has disturbed them. As a rule the flight is noiseless, or a slight sound only is heard, but at times they rise with a loud whir of wing beats.

I have already mentioned the tameness or boldness of the female bird with her brood. This boldness is also shown at the nest containing eggs. J. Fletcher Street writes:

The nest was somewhat hidden under a dense spruce shrub, and while I was cutting away some of the inclosing branches to obtain a better view, the bird left and at first charged toward me. Then she withdrew and kept retreating as I approached toward her, keeping about 10 feet between us. She exhibited but little concern after having left the nest but would not return to it while I remained nearby. I left the locality for three definite periods and upon each return found the bird sitting upon the eggs, yet becoming more wary at each successive disturbance.

Referring to this bird, Joseph Grinnell (1900) says: "After the snow came, grouse were seldom found for they remained continually in the trees. I saw but few tracks on the snow all winter, though in the fall their tracks were numerous on the sand-dunes and among willows along the river." According to A. Leith Adams (1873), they do not dive under the snow like the ruffed grouse. Lucien M. Turner says: "I have reason to suspect that some of these birds retain their mates for more than one season as I have frequently found a pair together in the depth of winter and these two being the only ones of the kind to be found in the vicinity." Audubon (1840) says that "the males leave the females whenever incubation has commenced, and do not join them again until late in autumn; indeed they remove to different woods, where they are more shy and wary than during the love season or in winter." He also imparts the following curious information:

All the species of this genus indicate the approach of rainy weather or a snow storm, with far more precision than the best barometer; for on the afternoon previous to such weather, they all resort to their roosting places earlier by several hours than they do during a continuation of fine weather. I have seen groups of grouse flying up to their roosts at mid-day, or as soon as the weather felt heavy, and have observed that it generally rained in the course of the afternoon. When, on the contrary, the same flock would remain busily engaged in search of food until sunset, I found the night and the following morning fresh and clear.

D. G. Elliot (1897) says:

The spruce grouse is found usually in small flocks consisting generally of one family, but also old males are met with alone, and I have always regarded it as a bird that was rather fond of solitude. Frequently, even in autumn, when the nights were becoming frosty, and snow flurries would hide the sun by day, heralding the coming winter, I have seen an old male, in the recesses of a swamp, strut about with ruffled feathers and trailing wings, as if the air were balmy and mild and spring were at hand.

Tales are told of immense numbers of these birds collecting in great flocks, or "packs," but such collections have probably not occurred for many years.

Voice.—The spruce grouse is a silent bird except when disturbed. His courtship "song" is instrumental, made by the rapid "drumming" of the wings striking the air as already described. I have heard slight clucking sounds from young birds and somewhat similar cluckings from adult males. Adult females when disturbed cluck incessantly, a sound described by Mr. Bent as "*kruk, kruk, kruk,* with an occasional rolling note like *krrrrruk,* soft and low." Street says: "The only note that the grouse uttered at any time was a low *chuck chuck* upon the occasion of her first leaving the nest." Forbush (1927) records the voice of the immature male as "a low wailing whistle, *weeo-weeo-weeo.*"

Field marks.—The male is a handsome bird distinguished by its compact form, its jet-black breast contrasting sharply with white, its red combs over the eyes, and its yellow-tipped tail. The female is a plain brown bird barred with black above, in this way differing from the ruffed grouse, which is spotted. It is smaller than the ruffed grouse and has a shorter tail.

DISTRIBUTION

Range.—Northeastern United States, Canada, and Alaska.

The spruce partridge is nonmigratory. Its range extends **north** to Alaska (Noatak River, Coldfoot, Fort Yukon, and Circle); Yukon (latitude 66° 40′ N.); Mackenzie (Mackenzie River, Fort Franklin, Lake Hardisty, Gros Cape, and Fort Simpson); northern Saskatchewan (Cochrane River); northern Manitoba (Lac du Brochet, Fort Churchill, and York Factory); northern Ontario (Fort Severn); northern Quebec (Fort George, Great Whale River, Fort Chimo, and Whale River); and Labrador (Okkak). **East** to Labrador (Okkak); eastern Quebec (Rigolet, Groswater Bay, head of the Magdalen River, and Mount Albert); northeastern New Brunswick (Bathhurst); Nova Scotia (Baddeck, Canso, Halifax, and Shelburne); eastern Maine (Fort Fairfield, Mount Katahdin, Houlton, Kingman, Calais, Orono, and North Livermore); and southern New Hampshire (Dublin). **South** to southern New Hampshire (Dublin); northern New York (Raquette Lake); southern Ontario (Kingston, Peterboro, and Bradford); northern Michigan (Au Sable River, Vans Harbor, and Palmer); northern Wisconsin (Mamie Lake); northern Minnesota (Northern Pacific Junction, Leech Lake, and Hallock); southern Saskatchewan (Fort Pelly and Osler); central Alberta (Mundare, Blueberry Hills, and Simpson Pass); southeastern British Columbia (Goat Mountain); and northern Washington (Chopaka Mountain and Barron). **West** to northwestern Washington (Barron); northwestern British Columbia (Flood Glacier, Glenora, and Atlin); and Alaska (Chilcat, Kodiac, Nushagak Lake, Aleknagik, Bethel, Russian Mission, Nulato, Kowak River, Kotzebue, and Noatak River).

Spruce partridges are of casual occurrence in Massachusetts (Gloucester, in 1851, and Roxbury, about 1865).

The 1931 edition of the American Ornithologists' Union Check List of North American Birds recognizes four races of *Canachites canadensis*, all of which are included in the foregoing ranges. True *canadensis* is found from the Labrador Peninsula west to the eastern base of the Rocky Mountains west of Edmonton, Alberta. *Canachites c. canace* ranges over the Maritime Provinces of Canada (New Brunswick and Nova Scotia); northern New England and

New York; southern Quebec, Ontario, and Manitoba; and northern Michigan, Wisconsin, and Minnesota. It is now largely extinct in the southern part of its range. *Canachites c. osgoodi* is found from Great Slave Lake and Athabaska Lake west to the Yukon region and the Mount McKinley Range of Alaska, while *Canachites c. atratus* occupies the coast region of southeastern Alaska.

Egg dates.—Central Canada (*canadensis*): 8 records, May 23 to June 14. Labrador Peninsula: 11 records, June 1 to July 4.

Alaska (*osgoodi*): 9 records, May 11 to June 25.

Quebec to Nova Scotia and Maine (*canace*): 21 records, May 5 to June 24; 11 records, May 24 to June 2.

CANACHITES CANADENSIS OSGOODI Bishop

ALASKA SPRUCE GROUSE

HABITS

The Alaskan form of the spruce grouse, or spruce partridge, was discovered and described by Dr. Louis B. Bishop (1900) and named in honor of his companion on the Yukon River trip, Dr. Wilfred H. Osgood. He described it as "similar to *Canachites canadensis* but with the ochraceous buff bars replaced everywhere by cream-buff and grayish white. On the upper parts the gray tips are paler, the ochraceous buff replaced by cream-buff and whitish, and the pale bars of the cervix grayish white instead of buff; below the white tips are larger, the pale bars whitish and cream color instead of buff, becoming cream-buff only on the jugulum."

Doctor Osgood (1904) says of its habitat:

> The range of the spruce grouse is practically coextensive with that of the spruce tree. We traveled much of the time near the western limit of the timber, and found grouse fairly common, even up to the edge of the tundra, where the spruce was considerably scattered. The last one seen was a fine cock, which was started very early on the morning of September 10, from a small beach on the Nushagak River about 25 miles above its mouth. The grouse are said to occur within a very few miles of Nushagak, however.

Herbert W. Brandt contributes the following notes on his experience with this grouse in Alaska:

> The Alaska spruce grouse proved to be a common bird throughout the wooded area that we traversed while en route to Hooper Bay. We first met with it about 60 miles west of Nenana, and from that time thereafter, when we were in the spruce areas, we were continually coming upon it. This noble fowl was common in the spruce timber right up to the highest pine growth in the Beaver Mountains, but its apparent preference is for the densely grown spruce river bottoms. The "fool hen's" noted lack of fear was often in evidence, and its retreating from ahead of our caravan often quickened the pace of the chase-loving dogs. It proved to be much more arboreal in habits than the other Alaskan gallinaceous birds we encountered, for we seldom saw it on the ground,

and its snow tracks are rarely observed, which contrasts markedly with the network of telling ptarmigan trails that everywhere enliven the barren snow wastes. In the heart of the dense spruces, such as it frequents, the beautiful dark plumage pattern gives it almost complete coloration protection, and if it did not reveal itself by movement, this bird would seldom be observed.

Nesting.—According to A. H. Twitchell the Alaskan spruce grouse nests regularly in the vicinity of the Beaver Mountains at the head of the Distna River, and here he has noted nests containing from five to eight eggs. One nest found by him on June 10, 1924, was placed out in the open in winter-dried grass near a small, dead spruce, which afforded but scant concealment. This bird chose a site about 100 yards from the reindeer corral in a scattered growth of small spruce, and in an area where there was often considerable activity. The male bird was frequently seen flying about, but the female was very wary. The nest was sunken 5 inches in the moss and made of circular-formed dry grass and a few dead spruce twigs, and contained a number of feathers of the sitting bird. The nest was found on June 1, when it contained one egg, and on June 10 the clutch numbered eight eggs, all of which proved to be fresh. During the egg-laying period, when the bird was off the nest the eggs were hidden beneath a covering of surrounding dead vegetation, artfully arranged before the bird departed.

Eggs.—In shape, the egg of the Alaskan spruce grouse is elongate-ovate, and the surface reflects a noticeable luster. The shell is somewhat greasy and quite sturdy. The ground color is prominent, as the spots occupy less than one-third of the surface, and the egg, on account of the bold richness of its markings, is quite handsome. The ground color varies from "salmon color" to "chamois" and "cream buff." The spots are flecked over the entire surface, but are sparsest on the larger end. In size these markings range from dots to those the size of a pea. There are only a few of the larger spots on each egg and these are well scattered over the surface, seldom exhibiting confluence. In contour the spots tend to be circular, with their rims well defined. When the pigment is thin the color is "chestnut" to "chestnut-brown," but the usual shade is "haematite red." When the egg is newly laid it is evident that both the ground color and markings are soft and moist, like those on the egg of the willow ptarmigan, as each egg somewhere on the surface is streaked with feather scratches, which often show distinctly the individual feather barbs. Occasionally a considerable area is so rubbed, exposing the ground color. When the egg dries, however, the markings are very durable.

Food.—Dr. Frank M. Chapman (1902) quotes from J. D. Figgins's notes, as follows:

In all the timber region I visited, the Canada Grouse was found common and breeding. Their chief food during early summer is the leaves of various deciduous bushes and spruce needles. About the 1st of August they repair to the edge of the barren grounds for berries which are then ripening. These are their food until September, when they return to the timber where raspberries and currants are abundant. During winter and spring their food consists entirely of spruce needles. Both adults and young appreciate their protective coloration, and when approached remain perfectly motionless until the danger is past. During the winter their color is to their disadvantage, and they become very shy and will not allow a close approach.

Near the base of the Alaska Peninsula, Doctor Osgood (1904) found these grouse in abundance about Lake Clark, "more common there than" he had "ever found them elsewhere in Alaska." He says:

They feed largely on berries in the summer time, being particularly fond of those of *Vaccinium vitis-idaea*, which they eat almost exclusively from the time the little green berry first begins to swell until it is dead ripe. At this time the flesh of the birds is sweeter than in the early winter, when a diet of spruce needles has made them fatter but less palatable. In the spruce forest, which is their ordinary habitat, they are unable to obtain on the moss-covered ground the grit necessary for a gallinaceous bird, so they make daily excursions to the shores of the rivers and lakes where fine gravel is to be had in abundance. Early morning before sunrise is the time for this; then they may often be seen on the beaches, singly, in pairs, or in small flocks. Doubtless they also come to the rivers to drink, though pools are common enough in the swampy openings in the timber. On the Chulitna River one was caught in a steel trap which had been set for a possible mink or weasel in the marsh grass at the water's edge.

CANACHITES CANADENSIS CANACE (Linnaeus)

CANADA SPRUCE GROUSE

HABITS

This is the form of the spruce grouse found in extreme southern Canada and the extreme Northern States, east of the Rocky Mountains. The male is practically indistinguishable from the male of the Hudsonian spruce grouse, but the female is decidedly more rufous or rusty, both above and below. The haunts and habits of the two are practically identical.

Edward H. Forbush (1927) has given us the following attractive description of its haunts:

In the dense spruce, fir, cedar and tamarack swamps of the great Maine woods the Spruce Grouse dwells. Where giant, moss grown logs and stumps of the virgin forest of long ago cumber the ground, where tall, blasted stubs of others still project far above the tree-tops of to-day, where the thick carpet of green sphagnum moss deadens every footfall, where tiny-leaved vinelets radiate over their mossy beds, there we may find this wild bird as tame as a barn-yard fowl. In the uplands round about, there still remain some tall primeval woods of birch and beech and rock maple where the moose and bear have set their marks upon the trees. In winter the deer gather in the swamps, and there their many trails wind hither and yon. Gnarled, stunted trees of arbor vitae, some dead or dying, defy the blasts of winter, while the long, bearded Usnea droops streaming from their branches.

An equally satisfactory account comes from the facile pen of William Brewster (1925):

For the most part the birds frequent dense, matted growths of cedar (i. e. arbor vitae), black spruce, and hackmatack (American larch), overspreading, low-lying, flat, and more or less swampy lands bordering on sluggish streams or on semiopen bogs similar to those known as Muskegs in the far North. From such coverts they wander not infrequently up neighboring hillsides to evergreen forests on still higher ground beyond, or perhaps into neglected pastures choked with intermingling young balsams, red spruces, and white spruces no more than eight or ten feet tall. Nor are they unknown to appear well out in rather wide upland clearings, where the only available cover consists of thickets of raspberry bushes, or even in river—or brook—meadows, where it is furnished solely by rank grass. Ramblings, thus venturesome, are exceptional, of course, and undertaken, I believe, at no seasons other than late summer and early autumn, when the lowly vegetation that clothes such perfectly treeless ground is most luxuriant, and also best supplied with berries or insects of various kinds; these Spruce Partridges devour eagerly whenever, and wherever, they can obtain them readily, although subsisting during the greater part of the year on a nearly unmixed diet of spruce and balsam spills (leaves), plucked mostly from branches at least fifteen or twenty feet above the ground.

Courtship.—William Brewster (1925) gives a slightly different account of this from what others have given; he relates Luman Sargent's experience with it, as follows:

Many years ago he was skirting a dense swamp, when his attention was attracted by a peculiar whirring sound that came from it. Advancing cautiously he soon perceived two Spruce Partridges, cock and hen, together on the ground. The cock left it presently, and vibrating his wings with great rapidity began mounting upward in a spiral course around the trunk of a large balsam, producing all the while a continuous drumming sound. After rising to a height of about 20 feet, and making three or four complete turns around the stem of the tree, he alighted on one of its branches where he rested for a moment or two and then flew down just as he had risen, that is by circling spirally around the trunk, with the same uninterrupted sound of wings. On reaching the spot where he had left his mate, he strutted about her like a Turkey cock, with wide-spread tail. Luman saw all this repeated fifteen or twenty times. For the first 10 feet above its base the trunk of the balsam was smooth and bare, but above that the Partridge had to conduct his drumming flights, both upward and downward, through numerous stiff branches. The sound of his drumming was distinctly audible at least 50 yards away.

Watson L. Bishop's account, quoted by Bendire (1892), of the display of a male bird in captivity is as follows:

The tail stands almost erect, the wings are slightly raised from the body and a little drooped, the head is still well up, and the feathers of the breast and throat are raised and standing out in regular rows which press the feathers of the nape and hind neck well back, forming a smooth kind of cape on the back of the neck. This smooth cape contrasts beautifully with the ruffled black and white feathers of the throat and fore breast. The red comb over each eye is enlarged until the two nearly meet over the top of the head. This comb the bird is able to enlarge or reduce at will, and while he is strutting the expanded tail is moved from side to side. The two center feathers do not move, but each side expands and contracts alternately with each step as the bird walks. This movement of the tail produces a peculiar rustling, like that of silk. This attitude gives him a very dignified and even conceited air. He tries to attract attention in every possible way, by flying from the ground up on a perch, and back to the ground, making all the noise he can in doing so. Then he will thump some hard substance with his bill. I have had him fly up on my shoulder and thump my collar. At this season he is very bold, and will scarcely keep enough out of the way to avoid being stepped on. He will sometimes sit with his breast almost touching the earth, his feathers erect as in strutting, and making peculiar nodding and circular motions of the head from side to side; he will remain in this position two or three minutes at a time. He is a most beautiful bird, and he shows by his actions that he is perfectly aware of the fact.

Nesting.—The nesting habits of this grouse are not essentially different from those of its more northern relative. Brewster (1925) records a nest found by Aldana Brooks near Richardson Lake, Me., "where the land was low and wooded with black ash, birch, alder, and a few larches. It was sunk, he said, in the top of a little mound with no rock, log, or even tree-trunk very near it. There were nine eggs. The bird did not leave them until almost stepped on, when she fluttered off over the ground for a few yards, and then stopped to watch Brooks who finally continued on his way without molesting her, or taking any of the eggs, which he never saw again."

Watson L. Bishop (1890), who was succeeded in domesticating this bird, says:

As the nesting season approaches I prepare suitable places for them by placing spruce boughs in such a way as to form cozy little shelters, where the birds will be pretty well concealed from view. I then gather up some old dry leaves and grass and scatter it about on the ground near where I have prepared a place for the nest. She will then select one of these places, and, after scratching a deep, cup-shaped place in the ground, deposit in it her eggs; * * * if there should be sufficient material within easy reach of the nest the bird will sometimes cover the eggs up, but not in all cases.

No nesting material is taken to the nest until after three or four eggs are laid. After this number has been deposited, the hen, after laying an egg, and while leaving the nest, will pick up straw, grass, and leaves, or whatever suitable material is at hand, and will throw it backward over her back as she leaves the nest, and by the time the set is complete, quite a quantity of this litter is collected about the nest. She will then sit in her nest and reach out and gather in the nesting material and place it about her, and when completed the nest is very deep and nicely bordered with grass and leaves.

So strong is the habit, or instinct, of throwing the nesting materials over the back, that they will frequently throw it away from the nest, instead of toward it, as the hen will sometimes follow a trail of material that will turn her "right about" so that her head is toward the nest, but all the time she will continue to throw what she picks up over her back. This, of course, is throwing the material away from the nest. Discovering her mistake, she will then "right about face" and pick up the same material that an instant before was being thrown away, and throw it over her back again toward the nest.

Eggs.—The eggs of the Canada spruce grouse are indistinguishable from those of the Hudsonian race, already described. The measurements of 53 eggs average 43.2 by 31.1 millimeters; the eggs showing the four extremes measure **47.5** by 31.5, 46.5 by **32.5**, **39.9** by 31, and 40.4 by **29** millimeters.

Food.—On dissecting some young birds shot on a meadow, on September 11, Brewster (1925) "found in their crops very many grasshoppers of various kinds and sizes, numerous ripe raspberries, a few leaves of *Spiraea tomentosa* and (in one crop only) a few larch spills." Evidently these birds had wandered out of their usual haunts for a change of diet. He says further:

In this connection it may be well to add that in the crops of two young Spruce Grouse only about half-grown and killed in the Tyler Bog on August 13, 1873, I found raspberries, blueberries, checkerberries, and balsam *buds* as well as needles; that from the crop of an adult female shot near Mollidgewauk Stream on September 28, 1890, I took 51 berries of *Viburnum lentago?*, some fragments of small mushrooms, and a few spills of the black spruce; and that a young male found and killed in company with the old female just mentioned had in his crop 13 Viburnum berries, uncounted pieces of mushrooms, and a few larch spills. Hence it will appear that food of various kinds other than that supplied by the foliage of coniferous trees is partaken of rather freely by Spruce Grouse in late summer and early autumn.

Behavior.—Although they usually flutter awkwardly away or silently fly for very short distances in the thick woods, when flushed in the open by Brewster (1925) they behaved quite differently, "rising all at once like Quail, from within a space no more than two yards square, with what seemed a deafening roar of wings, they sped straight for the woods, flying precisely like Ruffed Grouse and quite as swiftly."

Forbush (1927), while walking on a trail, almost stumbled over a male spruce grouse. He says:

The bird was somewhat startled and flew heavily up into a near-by spruce, alighting near the tip of a little limb about 20 feet from the ground. As the limb drooped under his weight, he walked up it to the trunk, hopped up a branch or two higher, and immediately began to feed on the foliage. After a few minutes of this, he moved a little into another tree and continued feeding. Pounding on the trunk with an axe did not alarm him, and it was only after several sticks had been thrown and one had hit the very limb on which he sat that he was induced to fly.

Edwyn Sandys (1904) writes:

The writer has twice caught mature specimens with his bare hands, and it is a common trick of woodsmen to decapitate a bird with a switch, or noose it with a bit of twine. Once the writer came precious near hooking one with a trout fly, at which the grouse had pecked. Only a dislike to needless cruelty, and a respect for a fine rod, saved this particular bird. Quite often the brood is met with in the trail, when they will sedately step aside about sufficiently far to make room for the intruder's boots, meanwhile regarding him with a laughable air of affectionate interest. No doubt this grouse could fly rapidly should it choose to exert its powers, but it is content with more leisurely movements.

Game.—The spruce grouse is not much esteemed as a game bird, as it lacks most of the qualities that appeal to the sportsman. Its haunts are usually too difficult to hunt in, it is too tame and stupid to make its pursuit interesting, and, except when on rare occasions it is found in open clearings or on meadows, it seldom offers a flying shot. It is not highly regarded as a table bird, for its flesh is said to be unpalatable. This is probably so in winter, when it has been feeding on spruce and balsam leaves; its flesh is then dark and decidedly resinous in flavor. But during fall, when it feeds largely on berries, green herbage, and insects, its flesh has a very different color and flavor. Brewster (1925) says that some young birds, shot in September, "proved delicious eating, their flesh being much sweeter and finer flavoured than that of any Ruffed Grouse. Both before and after cooking it was nearly as white as the Ruffed Grouse's, where as the fully-matured Spruce Grouse has invariably dull reddish flesh somewhat too redolent of spruce foliage to be relished by everyone, although I do not dislike it. The flesh of at least some of the young becomes, almost, if not quite, as dark as that of the adults, by the last of September."

These birds are killed for food all through the fall and winter by hunters of large and small game, by lumbermen, and by trappers and others. They are so easily killed that they are disappearing very rapidly and are now very scarce in northern New England in any but the most inaccessible regions.

Winter.—Walter H. Rich (1907) writes:

During a snowstorm the Spruce Grouse usually flies up into the densest clump of spruce or fir trees in the neighborhood, and, under their thick, arching branches, snow-laden and bending, he finds shelter from the weather and food in abundance. He may not leave the tree for several days if undisturbed and the storm continues. The question of temperature troubles him little, and with his wants all provided for, the Spruce Grouse is more independent in his mode of life than any of his feathered neighbors, for when other birds are scurrying about for something to eat and perhaps going hungry, this gentleman finds plenty of food in his shelter, and sits in comfort, "at ease in his own inn."

CANACHITES CANADENSIS ATRATUS Grinnell

VALDEZ SPRUCE GROUSE

HABITS

Dr. Joseph Grinnell (1910), in describing the spruce grouse of the coast region of southeastern Alaska, says that it "resembles *Canachites canadensis osgoodi* of the interior of Alaska (Yukon and Kowak Valleys), but general tone of coloration darker: white markings less in extent; black areas more extended; and grays less ashy, more olivaceous."

Of its distribution he says: "The indications are that this form is generally distributed in the humid coast belt from the eastern side of the Kenai Peninsula southwestwardly at least as far as Hawkins Island, and probably beyond."

Referring to its haunts and food, he writes:

Spruce grouse were not abundant in the Prince William Sound region, but appeared to be generally distributed. Two shot on Hawkins Island were both in heavy timber near the beach. Their crops were filled with the fresh green leaf-buds of spruce. On Hinchinbrook Island a male was shot on the mountain side near timber-line. The example secured by Heller on Hoodoo Island was flushed from a rank growth of salmonberry bushes; its crop contained berries, some fern fronds and a few seed pods of the devils-club. Grouse sign was noted on Chenega Island; and, as previously noted, skins were secured at Knight Island and at the head of Port Nell Juan.

CANACHITES FRANKLINI (Douglas)

FRANKLIN'S GROUSE

HABITS

This handsome species might well have been named the western spruce grouse, for it is the western counterpart of the well-known spruce grouse of eastern and northern Canada. It lives in similar haunts, has similar habits, and is so closely related to the spruce grouse that it may eventually be shown to intergrade with some of the western races of *canadensis* and be reduced to subspecific rank. It occupies a comparatively limited range in the mountainous interior of the Northwestern States and southwestern Canada.

While stationed in Idaho, Major Bendire (1892) found these grouse quite common

along the edges of wet or swampy mountain valleys, the so-called "Camas prairies," or the borders of the numerous little streams found in such regions among groves or thickets of spruce and tamarack. Few naturalists have as yet been sufficiently interested to invade their favorite haunts. In the summer of 1881 I found a single covey, numbering about ten birds, in the low flat and densely timbered region between the southern end of Pend d'Oreille Lake (the old steamboat landing) and Lake Coeur d'Alene, Idaho, at an altitude not exceeding 3,500 feet, I should think. I bagged three of these birds, and was quite surprised to find them in such a locality. As far as I have been able to learn, they usually occurred only at altitudes from 5,000 to 9,000 feet, and scarcely ever left the higher mountains. They were scratching in the dust on the trail I was following, and simply ran into the thick underbrush on each side, where they were quickly hidden.

Courtship.—Thomas T. McCabe has sent me the following interesting notes on the courtship of Franklin's grouse, as observed by him and Mrs. McCabe in British Columbia on May 28, 1929:

In the course of a morning's nest hunting in second-growth spruce and balsam, carpeted with deep green moss, we had found the cock in his usual locality, sitting quietly on a tussock, showing no erection of the crimson combs, no inclination to display, and typically indifferent to us. An hour later and a little before noon we were about 200 yards from this point, when he appeared above us, flying through the tree tops, and lit in a spruce about 15 feet from the ground and close to a pack trail. He still displayed no unusual excitement, and we left him again for 15 minutes. When we came back the hen had appeared and was squatting flat on the ground in the center of the beaten trail. Her appearance was normal and remained so through the ensuing episode. Perhaps 20 feet away the cock was walking down the trail toward her in a typical attitude of display—head drawn up and back, tail spread through two-thirds of a circle and vertical (not bent forward over the back), the fine undercoverts falling back from it like the sticks of a fan, the wing points slightly dropped, the combs bulged upward into elongate crimson rolls, which met in the center of the cranium. In this guise he strutted very slowly, with a statuesque pause of about 8 seconds every 2 or 3 feet.

At a distance of about 10 feet the whole bird was transformed with the suddenness of a conjuring trick, and the similarity to the courting ruffed grouse disappeared. The tail snapped together and sank nearly to the ground. The head was lowered and extended far forward. The plumage was flattened so that a hard sleekness replaced the fluffy rotundity, and the size of the bird diminished by half. The attitude was like that often assumed by *Bonasa* or *Lagopus* when luring an intruder away from nest or young, but in place of their whining note a low guttural was produced, vibrant and threatening, in from 5 to 7 periods, the first two distinct and slow, the remainder losing interval and less sonorous, trailing off to silence. This was accompanied by a slight movement of the tail in the vertical plane, rather a periodic trembling than a snapping like that of the courting *Bonasa*. One of us thought that with this movement the rectrices were slightly opened. Between these utterances the bird moved 3 to 6 feet, very slowly, and not in short runs after the manner of *Bonasa*. The movements were in various directions, a few perhaps directly toward the hen, but for the most part oblique and keeping 5 or 6 feet away.

After about 4 minutes of this the hen took wing, silently, but with amazing suddenness and speed for so phlegmatic a bird, flashed down the trail, and made a quick turn into the woods. Her sudden start scarcely gained a foot on the eager male, and both disappeared together. Perhaps a final scene was enacted near by. Perhaps many more were set, with many variations, before the subtle interplay of impulse and reaction rose to its climax. The details of the primitive drama, with its suggestions of threatening, beseeching, lamenting, lie beyond our power of interpretation, but, except in the opening movement, the element of simple and lavish *display*, so widespread among other genera, was absent.

Nesting.—The nesting habits of Franklin's grouse are similar to those of the spruce grouse. The eggs are very rare in collections, as the nests have seldom been found, and very little has been published about them. Major Bendire (1892) writes:

> Through the kindness of Mr. W. E. Traill, in charge of one of the Hudson Bay Company posts in British Columbia, parts of three sets of these rare eggs, fifteen in number, were collected during the season of 1890; taken on May 20, 27, and 30, respectively. The nests were shallow depressions in the moss-covered ground, lined with bits of dry grass, and were placed at the borders of spruce thickets. The eggs were fresh when found.

A set of six eggs, fresh when taken on May 27, 1906, is in my collection; it was taken by E. C. Bryant in Flathead County, Mont. The nest was at the end of an uprooted tree among some lodgepole pines; the hollow in the ground was lined with pine needles, weeds, and other material that came handy.

William L. Dawson (1896) describes a nest that he found in Okanogan County, Wash., as follows:

On the 28th of April, 1896, I found a nest of this bird at an altitude of about a thousand feet above Lake Chelan. It was placed in the tall grass, which clothed the side of an inconspicuous "draw" bottom, and although the plough had recently turned up the soil within five feet of her, the mother bird clung to her post. I took several "snap shots" of her at close range, and she allowed me to advance my hand to within a foot of her, when she stepped quietly off the eggs and stood looking back at me over her shoulder. The nest was a depression in the gravel-filled soil, lined with grass and dry corn leaves, besides a few stray feathers; depth 3 inches, width 7 inches.

Eggs.—The eggs of Franklin's grouse are similar to those of the spruce grouse; what few I have seen average more finely and more evenly spotted with smaller spots, but practically all types can be matched in a series of either. They are beautiful eggs and greatly in demand by collectors. The measurements of 33 eggs average 42.7 by 31.2 millimeters; the eggs showing the four extremes measure **45.1** by 30.2, 44.5 by **33**, and **39** by **30** millimeters.

Young.—Mrs. Florence M. Bailey (1918) gives the following interesting account of her experience with a brood of young in Glacier National Park:

A brood of three half-grown buffy-breasted and tailless young were seen in the Waterton Valley about the middle of August, wandering around enjoying themselves in deep, soft-carpeted woods of spruce and fir, where they jumped up to pick black honeysuckle berries from the low bushes, or answered their mother's call to come and eat thimbleberries. One of them, which flew up on a branch, also passed the time eating fir needles. When surprised by our appearance the little fellows ran crouching down the trail showing a keen hiding instinct, but their mother had little sense of danger. When the young were approached she merely turned her head over and called mildly in soft remonstrance. She was the genuine fool hen of Montana, we were told, whom the Flatheads and the mountain Indians never kill except when in great need of food, as the birds are so tame they can be snared at will, without ammunition; as the Indians say, with string from a moccasin.

The same brood, we supposed, was met with a few days later on the same trail. One of the young was in the trail and the mother was sitting on a log when we came up, but on seeing us she called the little ones into the bushes. When driven out for a better view she climbed a bank adorned with bear grass, dwarf brake, and linnaea carpet, and, stopping under a long drooping spray of Streptopus—under whose light-green leaves hung beautiful bright red berries—she jumped up again and again to pick off the berries. Then, flying up on a fallen tree trunk almost over my head, she sat there looking very plump and matronly and entirely self-possessed, while I admired the white and tawny pattern of her plumage. She sat there calmly overlooking the brushy cover where the young were hidden and showed no disapproval when the three came out and walked a log by the trail. She called to them in soft, soothing tones and they answered back in sprightly fashion. It would have been so easy to win their confidence completely and to watch their engaging ways that it was trying to have to leave them and pass on up the trail.

Plumages.—The Franklin's grouse chick is beautifully colored. The central crown patch, which is bordered with black, and the upper parts in general are rich brown, from "Sanford's brown" to "amber brown"; the colors of the forehead, sides of the head, and underparts vary from "mustard yellow" to "Naples yellow," deepest and tinged with brownish on the forehead and flanks, and palest on the sides of the head and belly; there are black spots below the eyes, on the lores and auriculars, on the lower forehead, and on the rump; and there is a black ring around the neck.

The juvenal plumage comes in first on the wings, when the chick is only a few days old, then on the scapulars, back, flanks, and breast, in that order; the tail appears next and the head and neck are the last to be feathered. By the time the young bird is half grown it is fully clothed in juvenal plumage. The feathers of the upper parts are beautifully patterned in rich browns, black, and white; those of the crown are barred with black and white; those of the back, scapulars, wing coverts, and flanks are barred and patterned with "tawny" and "ochraceous-tawny," and have narrow bars or large areas of black separated by bars of creamy white; many of these have central shaft stripes of creamy white, broadening on some into a white tip; the breast is "cinnamon-buff" or yellowish white, with large black spots; the chin and throat are white and the belly grayish white; the tail is barred with sepia and grayish buff and tipped with white. During the latter part of August the molt begins from the juvenal plumage, in which the sexes are alike, into the first winter plumage. This is a complete molt, except that the two outer primaries on each wing are retained for a year. The sexes now differentiate and look much like adults, though the black areas are less purely black and there is more white spotting.

Adults may have a very limited prenuptial molt in spring, but they have a complete postnuptial molt in summer and early in fall. J. H. Riley (1912), who collected a fine series of these grouse in British Columbia, says:

The males were never found with the females and young, but always by themselves and in full molt, July 18th to 21st, while at this time it had barely begun in the females. All the males taken had molted the tail and the new feathers were just appearing, while the only female taken that the molt had progressed so far was shot August 27th. This seems to show that while the female is brooding and bringing up the young, which she does unaided by her spouse, he goes on by himself and moults, while the process in the female is delayed until her young are able to shift for themselves.

In the series of females collected tnere are two phases of plumage; one of which I shall call the red phase and the other the gray phase. In the red phase the lower parts, down to the abdomen, are tawny ochraceous with the sub-terminal black bars on the feathers often interrupted, giving to these parts a beautiful "spangled" appearance; in the upper parts the tawny ochraceous barring is very prominent; the middle tail feathers and upper tail-coverts are black crossed by irregular narrow ochraceous bars and tipped with white. In the gray phase the lower parts only as far as the upper breast are ochraceous-buff, the feather of the breast being broadly tipped with white, and the black bars are not interrupted, giving to the breast and abdomen the appearance of being black and white, entirely different from the red phase; the light bars on the neck and upper back are ochraceous-buff and the tawny ochraceous of the red phase in the rest of the plumage of the upper parts is replaced by wood or hair brown; the central tail feathers are black, barred irregularly or stippled with wood brown and tipped with white; the upper tail-coverts lack the white tips. Both phases were taken at the same locality with young; the gray phase is the rarer and in our series there are intermediate stages.

Food.—The same writer says of the summer food: "The food contents of the crops of the adults was either spruce leaves or the green berries of a low-growing plant, while that of the young was the blossoms of the red heather *Phyllodoce empetriformis*, and a few insects."

Aretas A. Saunders has sent me the following notes:

Late in summer this bird seems to feed mainly on berries, particularly the "huckleberries" (*Vaccinium*). Up to the middle of October their crops contained a blue-colored berry of this genus that grows along the edges of spruce forests. There were sometimes the smaller red-berried ones, another *Vaccinium* that forms undergrowth in the lodgepole-pine forest. All these are called "huckleberry" in Montana, though allied to the blueberries. I do not think the true huckleberry (*Gaylussacia*) ever grows there. The only other food I have noted is spruce and balsam needles, and late in the season these alone are found in the crops. Apparently these needles are the main winter food, though I have never shot and examined a bird later than late November.

Behavior.—Mr. Saunders says in his notes:

All through Montana this bird is known as "fool hen" because of its lack of fear of man. It will sit still, even when close to the ground and allow one to approach very near. They are often killed with sticks or stones. When a dog approaches they fly up into the trees, and sit there. By shooting the lowest one first, I have shot several in a flock, the others sitting and waiting their turns. In Jefferson County, Montana, we had a small brown spaniel that would put them up a tree, and then stand beneath and yelp till we came. In Lewis and Clark Counties, on the upper waters of the Sun River, I once climbed a small pine, and grasped a cock Franklin grouse by the foot, just to see if I could do it. The bird moved to a higher limb when I let go, but did not fly away. The male, even in fall, is fond of puffing out its black breast, and opening and shutting the red "comb" over its eye, apparently by a sort of lifting of its "eyebrows."

John O. Snyder (1900) says that "one sat sedately on a limb while a revolver was emptied at her. The shots having missed, roots and stones were thrown, which she avoided by stiff bows or occasional steps to the side."

Fall.—Mr. McCabe writes to me:

The male, of course, takes no part in rearing the young and is never seen near the broods in the summer. Yet in the late summer or early autumn the birds gather into mixed groups, young males and females and adult females, to the number of six or seven, under the leadership of a single adult male. The latter when they are disturbed, assumes his display attitude, just as we have described it at courting time, utters a rapid clucking sound, and approaches the intruder, a magnificent creature, while the rest either draw off quietly or squat in supposed concealment. There would be nothing peculiar about this were it not for the fact that these groups only hold together for the two or three autumn months and then dissolve. The birds are very hard to find in winter as they remain sluggishly in the hearts of the big balsams, eating the needles (we counted 5,500 in one crop) or sometimes on the ground in dense masses of small balsams whose lower branches are weighed down into the snow, but when found (we collected three females last January) they are invariably single.

Mr. Saunders tells me that though he has traveled on snowshoes many times in midwinter, through forests where Franklin's grouse are known to occur, he has never seen one at that season. Evidently they remain well hidden.

DISTRIBUTION

Range.—Northwestern United States and southwestern Canada. It is nonmigratory.

The range of Franklin's grouse extends **north** to southeastern Alaska (Kasaan Bay); northern British Columbia (Tatletuey Lake, Thudade Lake, Ingenika River, and Hudsons Hope); and central Alberta (Edmonton). **East** to central Alberta (Edmonton, Pipestone River and Banff); western Montana (St. Marys Lake, Belton, Paola, Mount McDonald, and Belt Mountains); and central Idaho (Baker Creek). **South** to central Idaho (Baker Creek, Sawtooth City, and Resort); and northern Oregon (Mount Hood). **West** to northern Oregon (Mount Hood); Washington (Cowlitz Pass, Bumping Lake, Yakima Pass, Lake Chelan, and Pasaytens River); British Columbia (probably Chilliwack, Alpha Lake, Fort George, Fort St. James, Stewart Lake, Babine Lake, and Nine-mile Mountain); and southeastern Alaska (Kasaan Bay).

Egg dates.—British Columbia, Alberta, and Montana: 13 records, May 18 to July 29; 7 records, May 27 to June 9.

BONASA UMBELLUS UMBELLUS (Linneaus)

RUFFED GROUSE

HABITS

Spring.—During the first warm days of early spring the wanderer in our New England woods is gladdened and thrilled by one of the sweetest sounds of that delightful season, the throbbing heart, as it were, of awakening spring. On the soft, warm, still air there comes to his eager ears the sound of distant, muffled drumming, slow and deliberate at first, but accelerating gradually until it ends in a prolonged, rolling hum. The sun is shining with all its genial warmth through the leafless woods, thawing out the woodland pools, where the hylas are already peeping, and warming the carpet of fallen leaves, from which the mourning cloak butterflies are rising from their winter sleep. Other insects are awing, the early spring flowers are lifting up their heads, and all nature is awakening. The breast of the sturdy ruffed grouse swells with the springtime urge, as he seeks some moss-covered log, a fallen monarch of the forest, or perhaps a rock on which to mount and drum out his challenge to all rivals and his love call to his prospective mate. If we are fortunate enough to find his throne, on which he has left many a sign of previous occupancy, we may see the monarch of all he surveys in all his proud glory.

Courtship.—Dr. Arthur A. Allen, who has made some careful studies of the display and drumming of the ruffed grouse and shown some wonderful photographs of them, has contributed, at my request, a very full account of the whole performance, with some quotations from other, earlier observers. I have had to condense it somewhat, but it is substantially as follows:

"In a species as well known as this familiar game bird, which has claimed the attention of naturalists and sportsmen for nearly 200 years, and whose courtship performances have been watched and described by many observers, one would not expect many discrepancies in the accounts—at least among those recent observers who have had the benefit of the arguments of the earlier naturalists. Such is not the case, however, and it seems worth while, therefore, to summarize, here, the descriptions of the plumage display and the varied explanations of the drumming performance, before concluding with the writer's personal experience. Published records of the courtship performances of the ruffed grouse date back to the year 1755 when a communication from George Edwards on the pheasant of Pennsylvania was printed in the Philosophical Transactions of the Royal Society of London (Edwards, 1755). In this classic communication Edwards quotes largely from a letter received from that famous naturalist, John Bartram. He says in part:

> When living, they erect their tails like turkey-cocks, and raise a ring of feathers round their necks, and walk very stately, making a noise a little like a turkey, when the hunter must fire. They thump in a very remarkable manner, by clapping their wings against their sides, as is supposed, standing on a fallen tree. They begin their strokes at about two seconds of time distant from each other, and repeat them quicker and quicker, until they sound like thunder at a distance, which lasts about a minute, then ceases for 6 or 8

minutes, and begins again. They may be heard near half a mile, by which the hunters find them. They exercise their thumping in a morning and evening in the spring and fall of the year.

Edwards likewise quotes a Mr. Brooke, surgeon of Maryland, who says:

> The beating of the pheasant, as we term it, is a noise chiefly made in the spring by the cock birds. It may be distinctly heard a mile in calm weather. They swell their breasts, like a pouting pigeon, and beat with their wings, which sounds not unlike a drum.

Edwards then goes on to quote from La Hontan (1703), who in his New Voyages to North America, vol. 1, p. 67, in speaking about the grouse says:

> By flapping one Wing against the other, they mean to call their Mates, and the humming noise that issues thereupon, may be heard half a quarter of a League off.

"There is the argument in a nut shell; it is a problem of long standing. Bartram says that the grouse beats its body with its wings; Brooke intimates that it merely fans the air; La Hontan reports that it hits one wing against the other. A further complication is advanced by Hodge (1905) when he tells us that the grouse was called 'the carpenter bird' by the Indians because they believed that it beat upon a log with its wings to produce the drumming sound.

"Let me quote from some of the apparently more authentic descriptions and explanations of the act. Audubon (1840) states that—

> * * * the drumming is performed in the following manner. The male bird, standing erect on a prostrate decayed trunk, raises the feathers of its body, in the manner of a Turkey-cock, draws its head towards its tail, erecting the feathers of the latter at the same time, and raising its ruff around the neck, suffers its wings to droop, and struts about on the log. A few moments elapse, when the bird draws the whole of its feathers close to its body, and stretching itself out, beats its sides with its wings in the manner of the domestic Cock, but more loudly, and with such rapidity of motion, after a few of the first strokes, as to cause a tremor in the air not unlike the rumbling of distant thunder.

"Between 1842 and 1874 one finds numerous references to the drumming of the grouse, but which of the four beliefs the respective authors hold as to the method of its production seems to depend upon whom they quote. William Brewster (1874), however, writing in the American Sportsman, describes the drumming of a grouse as actually watched by him from a distance of 12 feet:

> Suddenly he paused, and sitting down on his rump and tarsi, crossways on the log, with tail slightly expanded and hanging down loosely over the edge behind, with body exactly perpendicular, neck stretched to its full length and feathers drawn closely to the body, he stretched out his wings stiffly at nearly right angles with the body. In this attitude he remained several seconds,

and I was instantly reminded most forcibly of the pictures one sees of that singular family of birds, the penguins. Now the wings were drawn slightly back, a quick stroke given forward, at the air, and a pulsating throb entirely different from any sound I have ever heard, struck my ear, producing at such short range an almost painful sensation on the drum; the wings were immediately recovered, and another stroke, a trifle quicker than the first, was succeeded by another still quicker, until the wings vibrated too fast to be followed by the eye, producing the well-known terminal "roll of muffled thunder," and not till then the "semicircular haze." I say not till then, for the first two or three strokes could be distinctly followed by the eye. This over, the bird immediately rose to its feet, shook its feathers with an air of relief, and resumed its attitude of repose. * * *

I think the drumming of the ruffed grouse is produced by the forward beats of the stiffened wings on the air, the planes of their motion being nearly horizontal, about four inches in length, with the initial ends represented by the points of a wire passed through the center of the erect body from side to side.

"Brewster's explanation of the drumming seems to have been rather generally accepted until 1905. At that time Prof. C. F. Hodge, of Clark University, was experimenting in the rearing of grouse in captivity and enjoyed exceptional opportunities for watching the drumming performance at a distance of but a few feet and made 40 different photographs of the drummer. He published quite an extended account of this drumming with his conclusions and a number of his photographs in the Country Calendar (Hodge, 1905). He states:

As to the matter of interpretation, I can not entirely agree with Professor Brewster. The appearance to the eye, however, supports his theory that the wing strikes nothing but air. But I am convinced that at just the critical moment, when the sound is produced, the wing moves with too lightning-like rapidity, even in the first slow strokes, for the eye to follow it. The wing, consequently, disappears from sight as it approaches the contour surfaces of the feathers of the sides. We must defer here to the eye of the camera, and some of the photographs certainly show the blur of the rapidly vibrating wings coming up and touching the tips of the feathers along the sides. It is the impact of the stiffly held concave wing on the feather cushions of the sides that causes all the sound. In fact, the sound, so far as quality goes, can be best imitated by striking with a wing properly stretched, or even a concave fan, on an extremely light eider-down cushion.

"The next account which should be included is that of Frederick K. Vreeland reporting in Forest and Stream, for April, 1918, and reprinted in the Bulletin of the American Game Protective Association. After watching a grouse drum at a distance of 6 feet and after taking a remarkable series of photographs of the drumming bird, he came to the conclusion that the thumping sound was produced by the wings striking behind the back, and he introduces a photograph that he says 'will prove to the most skeptical that they (the wings) did actually strike behind the drummer's back.'

"E. J. Sawyer, after watching from a distance of a dozen feet the beginning, progress, and ending of at least a hundred drummings during the spring of 1921, concludes in the Roosevelt Wild Life Bulletin for March, 1923:

1. The outward and upward motion [of the wings] is chiefly responsible for the drumming sound, particularly during the first half of the performance, the inward and forward motion of the wings being for the most part silent or nearly so.

2. The striking of the air alone with the wings is practically the sole cause of the sound.

"With this framework to build upon, the writer here submits pictures (pls. 26, 27) of a grouse drumming, taken from the rear, which are made from enlarged motion pictures. Unfortunately it is not practicable to reproduce all the 123 frames making up the drum but a careful analysis of the pictures printed should satisfy the most skeptical as to how the sound is produced. The film was made in the spring of 1929 on Connecticut Hill, 17 miles southwest of Ithaca, N. Y., after four nights spent in a blind about a dozen feet from the drumming log. The first night, April 26, the bird came to the log at 1.40 a. m. and drummed every 5 minutes until 5 o'clock. He then moved to another log about 100 feet away and drummed until 6 o'clock.

"The night of April 27 was cloudy and windy but warmer and the grouse did not arrive until 4 a. m. He drummed every 3 minutes until 5 a. m. and then every 5 minutes until 5.30, and then walked off the log not to return that morning.

"Stormy weather prohibited returning to the blind at the drumming log until May 4, when we resolved to try an experiment that might hold the grouse on the log later in the morning and at the same time throw some light on the object of the drumming as well as the method. We hoped it might also give us an indication as to the polygamous proclivities of the grouse. Accordingly we took with us a captive female grouse in a crate with 2-inch wire netting over the top. This we set in front of the log, concealing the sides with boughs so that the female would be most visible from the log. The experiment was entirely a success.

"The night was cloudy, and it rained intermittently with occasional snowflakes. At 4.45 we were awakened by a fluttering in the crate and peering through the peephole beheld the male bird in full display—tail up, ruffs extended, wings drooping but pressed to the body and not touching the log. He moved along the log ever so slowly until near the female. Then he lowered his head, extending his neck and shaking out the ruffs still more, and made a few pecks at the log in front of him, though not always hitting it, with his bill. Next he started shaking his head and ruffs with a rotating motion and commenced a series of short hisses, each one sounding like drawing the palm of one's hand rapidly backward, forward, and backward again over the sleeve or the trouser leg. He shook his head more and more rapidly, the hisses corresponding, and finally with a quick little run forward and a prolonged hiss, he struck a pose and held it for several seconds. In this pose the tail was swung over more to one side, the rump feathers on the offside lifted, and the extended head and ruff turned toward the female. The male continued this strutting and posing, usually on the log but sometimes in front of the crate, until 5.30, when he attempted to mate with the female reaching his head far down through the wire and apparently seizing hold of her, for he pulled out at least one feather. He went through all the motions of actual mating though the wire separated them by several inches. He then proceeded down the log to his accustomed place and drummed. This seems rather significant as indicating that the drum is probably a challenge to other males even more than an announcement of his presence to the female.

"Audubon states that 'the female, which never drums, flies directly to the place where the male is thus engaged,' but so far as I know, no one has ever seen a female grouse come to the male on or near his drumming log, although a great many hours have been spent by different observers watching the drumming bird. I think we are justified in concluding that while she may do so occasionally, she does not do so with any regularity and that the male must find her sometime during the day when he is not drumming. The drumming, therefore, resolves itself primarily into a challenge to other males to keep out of the drummer's territory.

"Let us next analyze how the sound is produced. The ordinary drum, such as the one filmed, requires almost exactly 8 seconds from the first wing beat until the last. With the motion-picture camera taking 16 pictures per second, the performance is registered on 123 frames. The first one or two wing beats are almost silent and are given while the bird is in a nearly normal horizontal position, the wings striking downward and inward. The bird's tail is being lowered against the log during this preliminary beat or beats. Then abruptly he stands erect with his tail against the log, wings drooping at his sides and appears to throw his 'shoulders' back. This might give the impression that the wings were struck behind the back, because the forward stroke of the wing follows so instantaneously that the eye scarcely perceives it, and it is given with such force and the wings come back to the normal position so quickly that the entire action registers on only one frame of the motion-picture film having an exposure of approximately one-fiftieth of a second. Between the ' thumps ' the wings of the bird register on the film with scarcely a blur representing the intervals between thumps. The varying tempo of the intervals between thumps has been noticed by all observers and as registered on the film is as follows, each number being the number of pictures or the number of sixteenths of a second between thumps:

5–6–8–8–6–5–5–4–4–3–3–3–2–2–1–2–1–1–1–1–1–1–000000000000000000000–1.

"If one now examines the series of pictures he will see that not once is the back blurred, as it would be if the wings struck behind the back, and that wherever the wings have moved with sufficient rapidity to cause a compression of the air and resulting sound, they are registered forward and upward. This then is the effective sound-producing stroke of the wing—*forward and upward*—not outward and upward as stated by Sawyer—more like his inward and forward, which he says is silent or nearly so.

"Moreover, if one watches the tail of the grouse during the drumming performance, he will see it become more and more flattened against the log, for ' action and reaction are equal and opposite in direction ' and the forward-upward stroke of the wings tends to drive the bird backward and downward on its tail. The reaction that follows cessation of drumming is even more clear to the observer, for always, upon the completion of the drum, the bird pitches slightly forward and the tail lifts from the log as if it were a spring under compression; when the pressure is suddenly released by the cessation of drumming, the tail throws the bird forward and upward and is itself carried upward by the impetus given the bird.

"A single weak thump heard at the conclusion of the drum registers on the film in frame 122 after an interval of one frame where the wings are quiet. It corresponds to the beginning ' thump ' and is given as the bird pitches forward and is in a more horizontal position. This stroke is forward and downward rather than upward and perhaps helps the bird to regain its balance.

"During the four nights and mornings spent in the blind the grouse drummed approximately one hundred times. I am frank to confess that I did not watch every performance, for the strain of keeping one's eyes at a peephole is considerable in the small hours of the morning. I did watch most of the performances, however, until I was absolutely convinced that the sound was produced as here set forth."

Nesting.—The ruffed grouse is a woodland bird, and its nest is almost always in thick woods or under dense cover, though I once found a nest in a fairly open situation; it was placed at the base of a small white birch in a clearing, with only a few small trees and bushes near it. Most of the nests I have seen in Massachusetts have been found by flushing the bird while hunting through heavy woods in search of hawk's nests. The commonest location is at the base of a tree; this may be a large oak among heavy deciduous timber, or a birch or other small tree in lighter, mixed woods; several nests have been in dense white-pine groves at the base of a large or a small pine. One nest was beside a rock in mixed woods, one was partially hidden under a corner of a woodpile in open pine woods, and others have been well concealed under fallen dead pine boughs or under old piles of brush. Several have been within 50 or 100 yards of a red-shouldered hawk's nest. One that I was watching was near a crow's nest; all the eggs but one were taken from this nest, probably by the crows. Edward H. Forbush (1927) tells of a nest, found by J. A. Farley, that was directly under a sharp-shinned hawk's nest. The nests are merely deep hollows in the ground, lined with whatever material is at hand, usually oak or other hardwood leaves; I have seen nests in pine groves that were lined with nothing but pine needles. There are usually a few feathers of the grouse mixed with the leaves. The female is a close sitter and often does not leave the nest until the intruder is close at hand; but, when approached on a second visit, she is more apt to flush wild. In leaving she flies directly from the nest, if close pressed, with a great whir of wings, which makes the leaves fly and thus partially covers the eggs. Perhaps, on a more leisurely departure, she may cover the eggs more carefully.

All the nests that I have seen have been on the ground in perfectly dry situations. But Major Bendire (1892) says: "Mr. Lynds Jones, of Grinnell, Iowa, found a nest of the Ruffed Grouse in a hollow stump, and Mr. C. M. Jones, of Eastford, Connecticut, found one in a swamp, on a little cradle knoll, surrounded by water. Mr. William N. Colton, of Biddeford, Maine, records a nest found between the stems of three young birches, fully 8 inches from the ground." George M. Sutton (1928) reports two nests found in a sphagnum bog near Hartstown, Pa., one of which was "sheltered by leaves of skunk cabbage." E. A. Samuels (1883) records two instances where this grouse has nested in an abandoned crow's nest in a tree.

Eggs.—From nine to a dozen eggs constitute the usual set, occasionally fewer and often as many as 14. Lester W. Smith sent me a photograph of a set of 23 eggs, found in Connecticut, and he told me that every egg hatched. This was perhaps the product of two females. I believe that ordinarily an egg is laid each day until the set is completed; but often, especially if bad weather occurs, an interval of a day or two may intervene. E. P. Warner (1911) reported that, in a nest he had under observation, he found 3 eggs on April 17, 4 on the 20th, 6 on the 24th, 10 on the 30th, and 14 on May 7, all of which hatched.

The eggs are ovate in shape, with variations toward short-ovate or elongate-ovate. The shell is smooth with a very slight gloss. The ground colors vary from "chamois" to "cream-buff" or "cartridge buff," or, more rarely, from "pinkish buff" to "cinnamon-buff." About half of the eggs, perhaps more, are entirely immaculate; others are more or less spotted with a few small spots or dots of "sayal brown," "clay color," or duller buffs. The measurements of 73 eggs average 38.9 by 29.6 millimeters; the eggs showing the four extremes measure 42.7 by 28.3, 40 by 32, and 33 by 25 millimeters.

Young.—The incubation period has been variously recorded by different observers at 21, 24, or 28 days; probably 21 days is the normal period under favorable circumstances, which may be lengthened by cold or wet weather or by interrupted incubation. The female alone performs this duty and raises only one brood in a season. The young leave the nest soon after they are hatched, or as soon as the down is dry, leaving the empty eggshells in the nest neatly split into halves.

The female ruffed grouse is a model mother, assuming full care of the young, leading them away from the nest, teaching them to scratch and hunt for insects and seeds among the fallen leaves, and showing them where the best berries are to be found and what green food is good to eat. The young are at first very sensitive to dampness and must not be allowed to wander in wet grass or herbage; she broods them under her wings, keeping them dry and warm during wet weather, and she broods them also at night. When they are older she leads them to bare places in the woodland roads, where they are taught to dust themselves and free their plumage of vermin. A species of wood tick causes the death of many very young chicks by burrowing into the skin of the neck or back—a situation with which the mother seems unable to cope. William Brewster (1925) obtained some evidence that the ticks kill the small chicks by boring through the tender skull into the brain.

But her devotion is shown to the best advantage in her heroic defense of the young against their enemies. While walking quietly through the woods we may be startled by a shrill, whining cry and see the infuriated bird, bristling with rage, rushing toward us, her tail spread and all her plumage extended; she seems twice her natural size and imposing enough to cause any enemy to pause. Sometimes she is less aggressive and merely flutters away, feigning lameness, or skulks away, crouching close to the ground and uttering the same whining cry, which is the signal to the young to hide. When the young are older a clucking note is given as a signal to fly. The behavior of the mother is so startling that we have lost sight of the young; they have disappeared completely; and search as we may, our chances of finding any are small. We had better not hunt for them, as we may step on them. But, if we conceal ourselves and wait patiently, we shall see a pretty sight, which is well described by Edmund J. Sawyer (1923) as follows:

There follows perhaps ten minutes of silence. Then comes a low, mewing note, *pe-e-e-e-u-u-r-r-r*. The note can be imitated by trying to pronounce the word "pure" in a strained, tremulous way with the mouth nearly closed. Soon there is an entirely different note like the low clucking of a hen or turkey; this grows louder and more confident and I catch a glimpse now and then of the watchful hen picking her cautious way back among the low plants. *Tsee—tsee—tsee-e-e-e*, answers a chick here and there about me, all unseen. *Puck-puk-puk*, from the mother; *tsee-tsee-tsee-e-e-e*, from the chicks, and one of the latter comes flying down from some leafy lower branch; *tsee-tsee-tsee*—and another appears from around a stump or log. There follows more calling back and forth, more chicks come out of hiding and already the *puk-puk-puks* have begun to grow faint in the distance as the mother quickly leads the brood off under cover of the ferns. I have on two or more occasions discovered one of the chicks in his hiding place on the leafy ground. In each case he was merely squatting there, his coat of mottled down perfectly matching the browns and grays of the forest floor.

The wings start to grow soon after the young are hatched, and before they are half grown they are able to fly, or, at least, to flutter up into the lower branches of a tree. They are zealously guarded by their mother all through the period of growth, and in the fall their father joins the family group, which keeps together during winter in a loose flock. Edwyn Sandys (1904) had an interesting experience with a pointer pup, which was attacked by a hen grouse, the guardian of a brood about as large as quail. He describes the incident as follows:

A sudden tremendous uproar attracted my attention, and, to my astonishment, I saw an old hen grouse vigorously belaboring the bewildered pup with her wings and giving him a piece of her mind in a torrent of cacklings, such as I had never dreamed a grouse capable of uttering. The poor pup, after first trying to make a point, and then to grab her, finally bolted in dismay. She followed him for about a dozen yards, beating him about the rump with her wings, which kept up a thunderous whirring. She acted exactly like a wrathful old fowl, and the pup like a condemned fool.

Edward H. Forbush (1927) relates the following incident:

Once I saw a fracas between the ordinary inoffensive rabbit and a grouse hen, defending her chicks. She "bristled up" and struck at bunny, but he apparently tried to leap upon her. In the ensuing running fight he drove her about a rod. Her chicks having hidden in the meantime, she then flew away. Very rarely, when the young are in danger, the male bird appears and takes his turn at running toward and strutting near the intruder, and he has been known to care for a brood after the death of the mother bird.

The following extract is quoted from the journal of Dr. Edgar A. Mearns:

We surprised an old hen pheasant (*Bonasa umbellus*). She gave a loud squall, and whirred loudly as she beat her wings upon the ground in front of us. The little chicks, only 2 or 3 days old and about 15 in number, at first piped out most lustily in their first surprise and bewilderment; but soon they recovered, and piled over each others' backs pell-mell in their efforts to escape and hide. Several of them rushed directly into the water of the brook close beside which we discovered them. One swam clear across, another was carried a distance down the stream and then crawled under a stone beside the bank. Others hid under stones and débris in the brook; and one fellow was actually drowned in its effort to hide. We found it floating dead upon the water when we returned to the spot sometime afterward. The old bird dragged herself over the ground with a great clucking; but when I ran rapidly after her, she took to wing and flew a little way off, and for a few minutes remained silent. We had captured three of the little chicks, and now examined to see what sort of hiding places the balance of the flock had chosen. One was packed like a sardine between two stones in the brook, with its head and streak of its back visible; another was wedged tightly between a stone and some herbage growing beside it. Two cute little fellows were found under shelving stones in the brook, running imminent risk of drowning. The drops of water were trickling off of one bird's head in rapid succession, and it was found fairly drenched. I verily believe that, had I not rescued it from this perilous situation it would surely have died rather than be discovered.

Plumages.—In the ruffed grouse chick the entire crown and back are "tawny" or "russet," darkest on the back and rump, shading off to "pale ochraceous-buff" on the sides of the head, chest, and flanks; the underparts are pale yellow, shading off to yellowish white on the chin and belly; there is a black auricular patch, but no other spotting on the head. The wings begin to grow soon after the chick is hatched and reach beyond the sprouting tail before the chick has grown much. The juvenal wings are fully developed and the young bird has reached the flight stage before it is half grown. The juvenal remiges are "light vinaceous-cinnamon," unmarked except for a very fine sprinkling of a slightly darker shade, somewhat lighter tips, and darker bases.

The juvenal plumage is at its height when the bird is about three-quarters grown, for soon after that the remiges are molted and the first winter plumage begins to appear. In full juvenal plumage the sexes are alike and closely resemble the adult female, but they are much browner above and below, less distinctly barred on the underparts, and more distinctly barred on the tails, which show both red and gray phases. The mantle is variegated with bright and dull browns, heavily barred and finely sprinkled with black, and has many broad buff or whitish shaft streaks; the chin is whitish instead of buff; and the chest is washed with "hazel."

The first winter plumage is acquired by a complete postjuvenal molt, except that the two outer juvenal primaries on each wing are retained throughout the first year. This molt begins in summer before the bird is fully grown and is generally completed before October. This plumage is practically adult and the sexes are now distinguishable. The new ruffs are a duller, more brownish black in the young male and are at first tipped with "hazel."

Adults have a very limited prenuptial molt, confined to the head and chin, and a complete postnuptial molt in August and September. The red and gray phases, most conspicuous in the tails, are present in this and in all other races of this grouse, though one or the other phase is supposed to predominate in each one of the races. In the northern races gray tails predominate; from Pennsylvania southward "silver tails," as they are called, are rare; in the western and northwestern races red tails are rare; New England birds, as a whole, are about halfway between the extremes. "Red ruffs," birds with brownish ruffs, tinged with a coppery red sheen, are occasionally seen in many of the races.

Food.—Forbush (1927) has published the most complete and condensed list of the vegetable food of the ruffed grouse that I have seen, based largely on Dr. Sylvester D. Judd's report (1905a). Following is his list in full:

Nuts or *Seeds*: Hazelnuts, beachnuts, chestnuts, acorns. Seeds of tick trefoil, hornbeam, vetch, hemlock, pitch pine, maple, blackberry, lily, beggar's ticks, chickweed, sheep sorrel, sedges, violet, witch-hazel, beech drops, avens, persicaria, frost weed, jewel weed. *Buds, Blossoms* or *Foliage*: Poplar, birch, willow, apple, pear, peach, alder, hazel, beech, ironwood, hornbeam, blackberry, blueberry, spruce, arbor vitae, Mayflower, laurel, maple, spicebush, partridge berry, sheep sorrel, aster, green ovary of bloodroot, clover, purslane, wood sorrel, yellow sorrel, heuchera, chickweed, catnip, cinquefoil, buttercup, speedwell, saxifrage, liveforever, meadow rue, smilax, horsetail rush, azalea, false goat's beard, dandelion, cudweed. *Fruit*: Rose hips, grapes, smooth sumac, dwarf sumac, staghorn sumac, scarlet sumac, poison ivy, partridge berry, thorn apple, cockspur thorn, scarlet thorn, mountain ash, wintergreen, bayberry, blackberry, huckleberry, blueberry, cranberry, sarsaparilla berries, greenbrier, hairy Solomon's seal, smooth Solomon's seal, black raspberry, raspberry, domestic cherry, cultivated plum, wild black cherry, wild red cherry, elder, red elder, black haw, nannyberry, withc rod, maple-leaved arrow wood, high-bush cranberry, mountain cranberry, snowberry, feverwort, black huckleberry, black alder, flowering dogwood, bunchberry, cornel, silky cornel, pepperidge, mulberry, bittersweet, manzanita, barberry, Virginia creeper.

Doctor Judd's analysis (1905a) showed 89.08 per cent of vegetable matter and 10.92 per cent of animal matter in the crops and stomachs of 208 grouse, collected in every month of the year in Canada and in 14 States. He says:

The animal food is almost all insects. The vegetable food consists of seeds, 11.79 percent; fruit, 28.32 percent; leaves and buds, 48.11 percent, and miscellaneous vegetable matter, 0.86 percent. The insect food proper includes grasshoppers, 0.78 percent; caterpillars, 1.15 percent; beetles, 4.57 percent, and miscellaneous insects, 3.86 percent. Some miscellaneous animal matter, made up of spiders and snails, is also eaten. The ruffed grouse eats a somewhat smaller proportion of insects than the bobwhite, but, like it, feeds on them to a large extent in the breeding season.

Judd lists among the animal food mainly insects, various grasshoppers, crickets, various caterpillars, cutworms, army worms, cotton worms, apple worms, various beetles and their larvae, clover weevil, potato beetle, various flies, bugs, ants, spiders, oak galls made by insects, snails, and slugs.

The foregoing lists are probably not complete, for the grouse will eat, at different seasons, a great variety of food. In spring they are fond of the catkins, blossoms, and tender leaves of many of the plants named above, the fresh blades of new grass, and the wild strawberries, when they come. Forbush (1927) adds:

Perhaps the plant most sought after in the New England coastal region is the cow-wheat, a low growing plant with small white blossoms which thrives almost everywhere that this bird is found. Ruffed Grouse in confinement are so fond of it that they eagerly eat quantities of it, consuming the entire plant, root and branch. Edible mushrooms are taken eagerly. Fern leaves which remain green in swamps under the snow of winter are eaten then as well as at other seasons.

During summer, when the birds find their food on or near the ground, insects begin to form an important part of their food, about 30 per cent of the adult food, according to Doctor Judd (1905a). He says that the newly hatched chicks are nearly, or wholly, insectivorous, feeding on cutworms, grasshoppers, beetles, ants, wasps, spiders, and caterpillars. The old birds, too, like to wander out into the fields and meadows near the woods in search of grasshoppers and crickets and to scratch among the woodland leaves for other insects and grubs. All kinds of berries and fruits claim their attention during summer and fall; I have found them frequenting regularly the edges of cranberry bogs near the woods, as well as wild-apple trees in secluded spots.

In winter, when their ground food is buried under the snow, they have to resort to trees and bushes for what fruits and berries are left, for leaves that remain green, and for dried catkins and buds. They are said to feed largely on leaves of sheep-laurel and mountain-laurel; and people have been poisoned by eating birds that had fed on such diet. I wonder if the poisoning was not due to berries of poison sumac and poison ivy, which are easily obtained in winter.

The ruffed grouse has a bad habit of budding cultivated apple trees, quite extensively when other food crops fail. Forbush (1927) gives us the following surprising figures:

Mr. Charles Hayward reports that he found in the crop of a grouse 140 apple buds, 134 pieces of laurel leaves, 28 wintergreen leaves, 69 birch buds, 205 blueberry buds, 201 cherry buds and 109 blueberry stems. Another bird had 610 apple buds in its crop and a third had more than 300. Weed and Dearborn found in the crop of a female ruffed grouse 347 apple buds, 88 maple buds and 12 leaves of sheep laurel.

This damage may be serious during certain winters, especially in orchards close to woods where the grouse are numerous; he mentions a case where a tree has been denuded of buds and killed. But, if not overdone, budding may be more beneficial than injurious, amounting to merely healthful pruning, for he says:

For twenty years one or two birds customarily "budded" on an apple tree near my farmhouse window. This tree seemed to be their favorite, but notwithstanding the "budding" or because of it, the tree bore a good crop of large apples nearly every year, while other trees not "budded" by the grouse often bore none. Apparently the thinning of the buds by the birds was a benefit to the crop.

Doctor Judd (1905a) quotes from a letter from Miss M. E. Paine, as follows:

The ruffed grouse eats the buds of apple trees, but it is a help rather than a damage. Last year a wild apple tree on top of a hill, between pasture and mowing, was almost entirely budded. I thought entirely at first, but the terminal buds were almost always left uninjured, also many minute buds on each limb. The result was the terminal buds were pushed out and grew rapidly and strongly. The tree blossomed abundantly and the fruit hung in clusters toward the ends of the branches. The tree is of medium size and the branches droop to the ground. In the fall the golden apples occupied fully as much room as the green leaves, and as one looked at the tree a few rods away—a perfect picture, barrels of apples on it, all nearly perfect and fair, just the result of a vigorous trimming.

William Brewster (1925) describes their method of budding as follows:

At six o'clock this morning my assistant, R. A. Gilbert, called me to see some Ruffed Grouse budding in a large wild apple tree that stands within sixty yards of our old farm-house, from one of the eastern windows of which I was able to watch them very satisfactorily through my field-glass. Five or six were noticed at once and before many minutes had elapsed I counted no less than *nine* scattered all over the tree, a few being low down on stout limbs close to its main trunk and hence inconspicuous, but the greater number near the ends of its longer upper branches, where they could be plainly seen, while one or two were perched on the very topmost twigs, boldly outlined against the grey sky and looking as big as Hen-hawks. They were busily engaged in budding, an operation which I have never before witnessed to such good advantage. It was not less surprising than interesting to see birds ordinarily so shy and retiring, and so very stately and dignified of bearing, assembled thus numerously in an isolated, leafless tree not far from a house, hopping and fluttering almost ceaselessly amid its branches, thereby displaying unwonted activity and sprightliness, as well as apparent fearlessness. At times, however, they would all stand erect and motionless for a few moments, evidently looking and listening intently.

Those feeding near the ends of long and slender branches had some difficulty in keeping their foothold and were constantly obliged to jerk up their tails, and flutter their wings in order to preserve their balance, especially when as often happened, they stretched forward or even for a moment bent almost straight downward after the manner of Redpolls or Pine Siskins similarly engaged. They picked off and swallowed the buds in rapid succession, with much the same quick, bobbing motion of the head as that of a domestic fowl feasting on corn. The supply of such buds as they chose, within reach of the most favouring perch, seldom lasted more than a minute or two. When it became exhausted the partridge either moved still farther out among the terminal twigs, or flew to another part of the tree. Birds at work not far from the trunk behaved somewhat differently, and with decidedly more dignity and deliberation, doubtless because the buds they were obtaining grew on short twigs within easy reach of thick and perfectly rigid branches on which they could stand or work as easily as on level ground.

Behavior.—There is a striking difference in behavior between the unsophisticated grouse of the primitive wilderness and that of the wise and wary birds of thickly settled regions. Birds that have never heard the roar of a gun and have not learned to know their dangerous human enemies are often absurdly, almost stupidly, tame; whereas the birds that have been persistently hunted have developed such a high degree of wariness and strategy as to make it difficult to outwit them. Formerly in much of New England and eastern Canada the ruffed grouse well deserved the name of "fool hen," and was one of the easiest of birds to shoot. It would either walk quietly away or fly up into the branches of a tree and stare stupidly at the intruder. It was an easy matter for a good shot to pick off its head with a rifle, and it was considered unsportsmanlike to shoot grouse in any other way. It has often been said that, when a number of grouse are perched in one tree, if the lowest one is shot first and then the next lowest one, the others will remain until the last one is killed. I doubt, however, if this has often happened; it hardly seems credible; and even in Audubon's time it was doubted. Even now, in the wilder portions of Canada, in the southern Alleghenies, and in some of the Western States, the grouse are absurdly tame and take but little notice of human beings.

The normal behavior of sophisticated ruffed grouse will be referred to later, but, while we are on the subject of tameness, we must consider the numerous cases recorded in print of abnormal tameness of individual grouse in regions where their fellows are the wildest. Space will not permit reference to all of more than a dozen such cases of peculiar behavior that I have heard or read about; one or two samples must suffice. In all these cases an individual grouse, sometimes a male and sometimes a female, showed a strong attachment for, or a decided interest in, one or more human beings, with the element of fear entirely eliminated. Carleton D. Howe (1904) published a full account of the behavior of a hen grouse that developed a strong friendship for a farmer and even allowed herself to be handled by other people. The friendship lasted through at least two seasons. "When Mr. Rand called 'Chickee,' 'Chickee,' the bird would come out of the woods and sit upon his knee. From his knee she would fly to his shoulder, and then to the ground. The bird would repeat this performance a half dozen times, clucking contentedly the while."

Howard H. Cleaves (1920) tells an interesting story of a belligerent cock grouse called "Billy" that "went forth to battle" with a motor tractor; he was evidently attracted by the noise it made, was not in the least afraid of it, and would even ride on it in motion. When first seen he was

25 yards up the road, his ruff extended and his head lowered and jerking nervously, after the manner of a rooster about to make battle with his foe. Billy took the middle of the lane and, following a peculiar, sinuous course, came steadily on to meet us with reckless abandon. The contrast was absurd. On the one side was a wild bird not larger than a bantam, and on the other were five adult humans led by a mobile mass of several thousand pounds of steel from which emanated a loud noise: a feathered David and a mechanical Goliath.

At the instant when it seemed that further advance by either side would mean annihilation for the eccentric Grouse, the pilot brought his tractor to a stop and descended to the ground, whereupon began one of the most remarkable of exhibitions. Billy darted toward Mr. Armstrong's feet and pecked at his trousers, and when Mr. Armstrong walked away the bird ran after him with the greatest agility, striking with wing or beak on coming within range. If a hand were extended toward him, Billy would peck it also and, most extraordinary of all, he would permit himself to be picked up and freely handled, perching on finger, wrist, or shoulder. When on the latter he was invariably prompted to investigate one's eyes and nose with his sharp beak!

A number of other published accounts illustrate similar traits. These abnormal birds are usually resident in some restricted area where they can generally be called by the human voice or whistle or come to the sound of a moving vehicle, a woodchopper's ax, or a stick rustling among the leaves. They follow their human friends about like pet dogs, can be coaxed to eat out of human hands, will often peck at them in a possibly playful manner, and will eventually allow themselves to be handled. Rae T. Hadzor (1923) tells of a hen grouse that flew into the yard one fall, possibly to escape from some enemy, and lived there about a year, mingling with the chickens but roosting by herself in an open shed. She became tame enough to eat out of the hand and even laid a set of eggs in the orchard. Of course, they did not hatch.

When a ruffed grouse is suddenly flushed it springs into the air with a loud whirring of wings, which is quite startling and disconcerting to a novice, and goes hurtling off through the trees or bushes at terrific speed, gaining momentum very quickly. Evidently it depends largely on its feet for the initial spring from solid ground, for it has difficulty in rising from soft snow, where it leaves the imprint of its whole body and wings in its struggle to rise. But it does not always make a noisy "getaway"; I have often seen one flit softly and silently up and over a stone wall, fence, or bush when it was not frightened or thought it was not observed. Again, when flying from a tree, it usually launches downward and flies away almost silently. The roar of a rising grouse, often too far away to be seen, is a common sound in regions where the birds are wild. Its flight is strong, exceedingly swift, and usually quite direct, but not, as a rule, prolonged for more than 150 or 200 yards, unless the bird is crossing a river or an open space between tracts of woods. A common habit is to fly low and straightaway along a woodland road or path, but its usual method is to rise above entangling undergrowth and then fly away through the trees, soon setting its wings and scaling down into thick cover. I have always thought it particularly skillful in dodging the branches of trees in its swift flight through thick woods, but evidently it is not always successful in this, for Forbush (1927) says:

It does not, like the Wood Duck, so control its movements as to avoid the twigs and branches of trees, but dashes through them. I have seen one in such a case strike bodily against a limb and fall to the ground. This bird had been fired at in a neighboring wood, and had crossed the open with tremendous speed to another wood where it struck the limb. Aside from the shock the bird was unhurt. Mr. Albert A. Cross of Huntington sent me a Ruffed Grouse that in full flight had collided with the forked and broken end of a dead limb, driving one of the prongs three inches into its breast and the other into its vitals, and tearing the head and neck from the body.

Forbush also speaks of a habit I have never noted:

A hard-pressed bird has been known to go into shoal water, apparently for concealment. Mr. L. Barber tells us that a grouse that was startled by his dog alighted in the water. She was entirely under water except her head which was covered by a projecting bush. Mr. W. L. Bishop writes that he killed a Goshawk near a brook, and afterward discovered by traces on the snow that the hawk had been pursuing a Ruffed Grouse. He found the frightened bird in the brook entirely submerged with the exception of its head. Though the Ruffed Grouse seems to drink mostly the dew and raindrops from the leaves, it is not afraid of water, and if winged over water can swim fairly well.

Grouse are much given to dusting themselves in soft, dusty places in woods roads, in country highways or on old rotted logs or stumps. They have favorite dusting places in which a few telltale feathers may be found.

Voice.—The vocal accomplishments of the ruffed grouse are quite simple. The commonest note heard when the grouse is slightly alarmed is a sharp *quit–quit*, usually given while walking on the ground and indicating nervousness. The squealing or whining note uttered while defending its young is probably also a signal to them to hide, as a clucking note is a signal to older young to fly. Then there is the call of the female to her young, *crut–crut*, *car-r-r*, and various soft cooing notes and chatterings.

Enemies.—Besides its archenemy, man, who has shot and snared it almost to extinction in many places, the ruffed grouse has many natural enemies and is subject to many diseases. It always managed to survive, however, until man came on the scene; its large broods have helped it to come back to normal numbers after periods of scarcity. Foxes destroy large numbers of grouse, as well as their eggs and young; feathers scattered about their burrows and tracks in the snow tell the story. Forbush (1927) says:

> Mr. C. E. Ingalls, writing of an experience at Templeton, stated that he saw a fox approaching the nest of a Ruffed Grouse near the edge of the woods. "A big ball of feathers," writes Mr. Ingalls, "flew out at that fox and drove him some distance into the grassland." The fox, nevertheless, returned to the attack only to die in his tracks by a well-directed bullet from the rifle of the watcher, not, however, until the brute had filled both mouth and throat with egg contents from the nest of the devoted mother.

Wandering dogs, stray cats (of which we have too many in our woods), lynxes, and perhaps raccoons and weasels kill many old and young grouse, the former being probably mostly caught on their nests. Skunks, opossums, raccoons, and squirrels undoubtedly rob the nests. The goshawk, also called "partridge hawk," levies heavy toll during periods of its abundance; it is often named as one of the chief causes of the periodic scarcity of grouse. The Cooper's, red-tailed, and red-shouldered hawks probably kill a few. Great horned owls pounce on them in their night roosts and are very destructive. The evidence against the screech owl and the long-eared owl, both of which have been seen eating grouse, does not seem conclusive. Crows are canny nest hunters and doubtless break up many nests; I feel confident that a nest that I was watching was robbed by a family of crows that had a nest near by. Dr. Charles W. Townsend (1912) reports a crow seen flying off with a freshly killed grouse in its claws; examination of the body of the grouse, which the crow was seen to drop, led to the conclusion that the crow had killed it.

Fall.—Audubon (1840) refers to short migratory flights of grouse in October across the Ohio and Susquehanna Rivers. These are probably nothing more than autumn wanderings in search of food. But there is much evidence of an incipient or suppressed migratory instinct in the erratic short flights of ruffed grouse during the so-called "crazy season" in fall. At such times they certainly do behave queerly. I have repeatedly known them to appear in my yard in the center of the city, or to kill themselves by flying against buildings or through windows. Once one flew through a window into our machine shop, where the machinery was running and scores of men were working. One of my neighbors once found one inside her house with no visible means of entrance except through a chimney. Forbush (1927) says that—

> Some have been known to go through the glass of moving motor cars or trolley cars and even into locomotive headlights. So careless are they of obstructions that a high wire fence around a covert is likely to kill all the Ruffed Grouse within its confines. Dr. A. O. Gross found that three birds which had been killed by flying against obstructions were infected by internal parasites, and he suggests the possibility that the irritation caused by such parasites may be the initial cause of the "crazy" behavior commonly observed.

Game.—Although the bobwhite may be more universally popular, for the reasons stated elsewhere, I think most sportsmen will agree that the ruffed grouse, known in the South as "pheasant," in the North as "partridge" or "patridge," and in Canada as "birch partridge," is the unrivaled king of North American upland game birds. Shooting into a flock of whirring quail gives a thrill, but it is comparatively easy, and shooting the straight-flying prairie chicken in the open is child's play compared with stopping the swift rush of the wily grouse through the treetops. The thundering roar of the rising bird, the flash of nitro at a vanishing glimpse of brown feathers, the dull thud of a plump partridge falling to earth, and the whir of wings among dry leaves as it beats its final tattoo, combine to produce the thrill of thrills for the successful sportsman. And with the freshly killed monarch of the woods in hand come admiring thoughts, so well expressed by Charles B. Morss (1923):

> In no other game bird do the tones of gray, black, cinnamon, and white shade and blend with such quiet harmony. Child of the wilderness that he is, in the full dark pupil of that eye surrounded by an iris of October's own brown, seem always to dwell the brooding shadows of the great forest he loves so well. And in the moulding of him Nature seems to have embodied all of the beauty, all of the charm, all of the inexplicable strangeness and romance of the autumnal woods and produced her feathered masterpiece—the perfect game bird. * * * And wherever you chance to find him—in the still shadow of ravine and glen where the climbing bittersweet twines its orange offering about old stumps and windfalls—on rocky hillsides clad with second growth where the wild barberry fruits in crimson racemes and berries of the wintergreen flash among the leaves—or in the grass-grown tangles of birch meadow and maple swamp where glows the steady flame of the black alder—always is he the woodland's pride, alert, instinct with life, and filled with a spirit and dash that furnishes, when in such mixed cover as we were hunting this day, the very climax of shooting with the shotgun.

A *good* partridge dog adds much to the pleasure and success of the hunt, but good partridge dogs are scarce, and a poor one is worse than none. I once had one that showed real "bird sense," knew where to hunt for the birds, would not run too far away, was careful about flushing birds, and was a fine retriever. But I love to hunt alone, with nothing to distract my attention from the beauties of the autumn woods, to watch and study the interesting habits of the other wild creatures, to learn the haunts of the wily grouse and match my wits against his. Then, if I can, unaided, outwit this wizard of the woodland glades, learn to beat him at his many clever tricks, I feel that I have earned my prize. Well he knows the trick of putting the trunk of a big tree or a thick tangle of leaves and branches between the hunter and himself in his headlong flight; or running off to one side, he will rise behind the gunner and get away safely; perhaps he will alight in the thick top of a pine tree and slip away on the farther side of it on silent wings, giving the hunter an unexpected and difficult shot. Usually, if we miss, we can watch his distant flight to mark him down and flush him again; but he may run a long distance or fly out across an opening to another bit of cover and escape. Occasionally we get a pleasant surprise by killing a bird we could not see, shooting in the direction it has taken behind thick brush. Or we may think we have made a clean miss, as we see the bird keep on and on rising up and up into the sky until it appears as a small speck; but if we watch it, we may see it drop like a stone, shot in the head; then we need a good dog to find it. There is no bird that so tests the skill, patience, and endurance of a good wing-shot as the ruffed grouse and no shooting that calls for so much experience and intelligent study. There are very few who can make a respectable ratio of birds killed to shots fired; if one takes his shots as they come, there are very few that can put a bird in the pocket for every three empty shells. One must know where to look for his birds and study their food habits in the localities he hunts. One of the surest places to find them in my section is where an old apple orchard has been abandoned and overgrown with grapevines, briers, and birches, or where old apple trees grow along the edges of the woods, or where tangled thickets of berry-bearing shrubs, junipers, cedars, pines, and other forest growths are encroaching on deserted pasture lands. When the beechnut crop is good they may be found in sheltered spots on the sunny side of the woods or brushy hillsides. In the mountains farther south they frequent the rhododendron clumps, the ravines lined with laurel thickets, or the dense undergrowth along the streams, where shooting is difficult. In closing this chapter I am tempted to quote the appreciative words of one who has substituted the camera for the gun; Edmund J. Sawyer (1923) says:

And now there is little enough satisfaction in the reflection that that gun shot many a grouse, albeit all of them on the wing and not one over a dog. I have, after all, never taken a Grouse except through the immense advantage of my infernal powder and lead. I never outwitted him fairly; I have never held his limp form in my hand without feeling the rebuke of his matchless wings. I found no just ground to glory over the dead body of that perfect product of the wild outdoors, that past master of woodcraft with his wings, which so immeasurably outmatched the best my limbs could do; those wings with their damning, rebuking evidence—a drop of lead-tinctured blood. The triumph was all his.

Winter.—If not too much disturbed the young birds remain in the family group with the female during fall and winter; and the male joins them late in fall. They fatten on the abundant crop of berries, fruits, and nuts in preparation for winter, their legs are now warmly feathered, their thick new plumage protects them against the increasing cold, and the comblike scales on their feet grow out to help them walk on the snow. They now seek their winter quarters in thick woods, where they can find shelter from cold winds and a good food supply within easy reach. They lead comparatively inactive lives, spending much of the daytime roosting in evergreen trees, in vine-clad thickets, on sunny borders, or even under the snow. They feed early in the morning and at dusk, mainly on the buds of poplars, birches, and apple trees. Their tracks in the snow are easily recognized, as the toes are widely spread and the tracks are in a direct line, one directly in front of another. In severe weather, when the snow is deep, they often dive into the soft drifts and find a snug, well-protected bed a foot or two below the surface. Unless the snow becomes very heavily frosted they can easily burst a way out in the morning, or unless an enemy finds their hiding place. A good shelter is sometimes found under a low branch or pile of brush covered with snow, or under a log banked with snow. Such places serve as either day or night roosts. When there is little or no snow the birds gather for the night in the thickest groves of pines or other conifers, roosting on the branches near the trunk, often a number together in the same tree, where the cover is dense enough to protect them.

Mortality.—The periodic scarcity of ruffed grouse, with subsequent recovery to more normal numbers again, has long been a fruitful subject for discussion and study. Their natural enemies have probably served only to keep their increase within check. Their decrease is due to many other causes, mostly chargeable to man. Trapping, snaring, and smoking out roosting places have all been stopped, and shooting has been reduced and periodically stopped. But clearing and cultivating land still goes on, and sportsmen are still increasing. Severe winters and unfavorable breeding seasons have their temporary effects, as do the periodic scarcity of certain insects, on which the young depend for food, and the occasional inroads of heavy flights of goshawks or great horned owls.

But the principal causes of decrease and of excessive mortality are the various diseases to which grouse are subject and the numerous parasites that attack them. At my request, Dr. Alfred O. Gross, who has been making an extensive study of this subject for several years, has contributed a condensed, but quite complete, report on the diseases and parasites of the ruffed grouse, which I consider important enough to quote almost verbatim. At least 6 infectious diseases and about 25 parasites have been discovered in the examination of more than 2,000 birds received from a wide range extending from Quebec to Virginia and westward to Minnesota.

Among the infectious diseases he mentions enlargement of the spleen, enteritis (unknown etiology), hepatitis (diseased liver, unknown etiology), enterohepatitis (blackhead), aspergillosis (a fungus disease chiefly of the respiratory system), and bird pox (an integumentary disease, producing tumorlike growths, generally in the region of the head and mouth). Following is Doctor Gross's report on external parasites:

"Bird lice are wingless ectoparasitic insects with mouth parts adapted to biting, not sucking. They feed on parts of the feathers and epidermal scales and when present in large numbers cause severe irritation and considerable annoyance to the host. Lice of the genus *Gonicotes* have been taken most frequently from New England specimens. Grouse collected during the summer months in southeastern New York were heavily infested with the louse *Esthiopterum perplexum*.

"Ticks of the species *Haemaphysalis leporis-pallustris* have been taken from birds collected throughout the entire range of the ruffed grouse. The heaviest infestations of the parasite were found among grouse living in the heavily wooded sections of northern Maine and along the Canadian Labrador coast.

"The ticks generally attach themselves in the region of the head, usually on the chin and skin around the eyes or on the nape, all places where it is difficult for the host to dislodge them. The ticks vary in size, depending on the relative quantity of sucked blood they contain. Even a few ticks are a discomfort to the birds, and in cases of heavy infestation they become a serious menace, decreasing the vitality and resistance of the grouse and in certain rare cases resulting in the death of the bird. Observers have reported that entire broods of young grouse have been destroyed by ticks.

"It has been demonstrated that the tick may serve as a carrier of such diseases as tularemia, an infectious disease of rabbits, and since the tick *Haemaphysalis leporis-pallustris* is a common ectoparasite of both the rabbit and the grouse it may thus eventually prove to be of considerable importance in the life of the grouse. The ticks *Haemaphysalis cinnabarina* and *H. punctata punctata* have also been taken from ruffed grouse.

"Live birds received from Canada and others collected in northern Maine were found to be heavily infested with the northern fowl mite

Lyponyssus sylvarium. Several individuals of a flock of ruffed grouse kept in captivity at Brunswick, Me., were killed by an infestation of myriads of these blood-sucking mites.

"The hippoboscid fly *Ornithoponus americanus* is light brown and about the size of a common house fly but with longer wings and with a radically different life history and behavior. This fly is very active and difficult to secure since it is quick to disappear among the feathers or it may leave the bird entirely. The flies attack the birds in the region of the head, where they suck the blood of the host and are thus capable of bearing blood parasites and diseases."

The internal parasites are more numerous, of greater importance, and apparently more often prove fatal. Of these Doctor Gross writes:

"The internal parasites are chiefly in the alimentary tract, where they are introduced along with the food, chiefly in the form of eggs and larvae. The life histories of some of the parasites involve a secondary host.

"The crop worm *Thominx annulata* is an extremely slender worm found beneath the epithelial lining of the crop and gullet. It is from 2 to 3 inches long, but less than the diameter of a hair. The walls of the crop and gullet are very much thickened and thrown into folds and ridges by the presence of this parasite.

"The stomach worm *Dispharynx spiralis* is one of the most important of the internal parasites. In southern New England and New York State a large percentage of the birds found dead and examined, died from the effects of this nematode. *Dispharynx spiralis* is between one-fourth and one-half of an inch in length, pointed at both ends, and always rolled up in a characteristic spiral form. It becomes established in the glandular walls of the proventriculus and in advance cases of infection this region is so excessively swollen as to equal the size of the gizzard. The worms become sexually mature in the grouse, and great numbers of eggs are passed through the alimentary tract and eliminated with the droppings. The eggs will not develop if taken directly into the body of another grouse, but the life cycle involves an intermediate host, which Dr. Eloise B. Cram, of the United States Bureau of Animal Industry, working with the New England ruffed grouse investigation, has found to be the pillbug.

"The gizzard worm *Cheilospirura spinosa* is distributed over a wide range from Maine to Pennsylvania and west to Wisconsin. In Wisconsin it is second only to *Ascaridia lineata* in the percentages of cases of infection. Gizzard worms are slender, the female obtaining a length of 1½ inches. They are usually found between the chitinous lining and muscular walls of the gizzard. Like *Dis-*

pharynx the life cycle of this nematode worm involves a secondary host, which Doctor Cram has discovered to be a grasshopper.

"The intestinal worm *Ascaridia lineata* is the commonest of all the internal parasites and in certain sections of the range of the ruffed grouse as many as 75 per cent of the birds examined were found to be infected. This worm is comparatively large, ranging from 2 to 4 inches in length. It is yellowish white and pointed at both ends. Because of its size and abundance it is reported by sportsmen more frequently than all other parasites combined. Ascarids normally reside in the intestine and while there do comparatively little harm aside from absorbing a certain quantity of food. A number of cases have been found, however, where the worms have been outside the intestine among the vital organs and in the body cavity where there was no evidence of shot wounds or scars to indicate that the worms had made their way through artificial openings.

"The hatching of the eggs occurs normally after the eggs are swallowed, and the life cycle does not involve a secondary host. In the larval stages this parasite is capable of doing serious injury to the mucous lining of the alimentary tract, especially in young birds. Some of the larvae may even penetrate the mucous lining and thus be transported by the blood stream to other parts of the body. Furthermore, wounds made by the larvae may open the way to infections of a serious nature.

"The cecal worm *Heterakis gallinae* is a small nematode parasite usually found in the ceca. Rarely it occurs in the small intestine, colon, and rectum. It is white and very rigid, and the head is bent dorsally, a character that serves to distinguish it from young *Ascaridia*. The eggs of *Heterakis* pass in the feces and under favorable conditions of temperature and moisture develop in about 10 days to the point where the eggs contain infective embryos. When the eggs are swallowed by the grouse the embryos are released and then develop into adult worms in the region of the ceca. This parasite is known to have an important relation to the dreaded disease enterohepatitis commonly known among poultrymen and game breeders as blackhead. The eggs of *Heterakis* may carry for great lengths of time the blackhead germs and thus facilitate the spread of this disease from bird to bird. A species of *Contracaecum* has been taken from the ruffed grouse.

"Tapeworms, designated also as cestodes, constitute important parasites in the intestines of certain groups of birds. Three species of minor importance have thus far been found in the ruffed grouse.

"A large number of the ruffed grouse examined have been found to be infected with flagellate protozoan parasites. Thus far six species have been identified, of which *Histomonas meleagridis* (blackhead) is of the greatest pathological importance. Blackhead is a common and serious disease of poultry and of captive game birds and has also been found in a number of cases of ruffed grouse collected in their natural covers.

"The group Sporozoa is represented by two species of *Eimeria*, which are responsible for the disease coccidiosis, which is not only important in grouse raised in captivity, but, as in the case of blackhead, is also found in grouse killed in a natural habitat."

DISTRIBUTION

Range.—United States, Canada, and Alaska; north to the limit of trees. The species is nonmigratory.

The range of the ruffed grouse extends **north** to Alaska (Nulato, Tanana, Rampart, Fort Yukon, and the Porcupine River); Mackenzie (Roche Trempe l'Eau, Fort Simpson, Fort Providence, Fort Rae, and probably the Slave River Delta); Manitoba (Hairy Lake, Oxford House, Knee Lake, and York Factory); northern Ontario (Martin Falls, probably Fort Albany, and Moose Factory); Quebec (Lake Mistassini); and Labrador (Hamilton Inlet). **East** to Labrador (Hamilton Inlet and Sandwich Bay); southeastern Quebec (Wolf Bay, Natashquan, and Perce); Nova Scotia (Baddeck, James River, Musquodoboit, and Halifax); Maine (Calais, Columbia Falls, Mount Desert, and Portland); Massachusetts (Boston and Cape Cod); New York (Shelter Island); New Jersey (New Brunswick and Vineland); Maryland (Laurel); Virginia (Bush Hill and Surrey); western North Carolina (Roan Mountain, Mount Mitchell, and Black Mountain); western South Carolina (Caesars Head); and northern Georgia (Brasstown Bald and Grassy Mountain). **South** to northern Georgia (Grassy Mountain and Cloudland); northeastern Alabama (De Kalb County and Long Island); western Tennessee (Danville); southern Missouri (Holcombe and Current River); formerly northwestern Arkansas (Fayetteville); formerly eastern Kansas (Manhattan); rarely Colorado (Estes Park and Sweetwater Lake); southwestern Wyoming (Fort Bridger); northern Utah (Parleys Park and Barclay); central Idaho (Lardo); and northern California (Oak Bar and Eureka). **West** to northwestern California (Eureka); Oregon (Anchor, Empire, Eugene, and Tillamook); Washington (Vancouver, Cougar, Olympia, Elkhorn Ranger Station, Elwha River, Ozette Lake, and Neah Bay); British Columbia (Alberni, Parksville, Nootka Sound, Fort Rupert, Port Simpson, and Hastings Arm); and Alaska (Juneau, Tanana Crossing, Lake Minchumina, Tacotna, Akiak, Russian Mission, and Nulato).

74564—32——12

Ruffed grouse have been extirpated from much of the south-central parts of their range, as in Iowa, Nebraska, Kansas, Missouri, and Arkansas. Casual occurrences outside the normal range are not numerous, probably the most unusual being a specimen that was taken near Camden, S. C., on December 27, 1904.

The range as above described is for the entire species, which has, however, been divided into six subspecies. *Bonasa u. umbellus* occupies the territory from Massachusetts, Virginia, and Georgia west to southern Minnesota, eastern Kansas, and northern Arkansas; *B. u. togata* is found north of *umbellus*, meeting it on the south in Massachusetts, central New York, northern Michigan, northern Minnesota, and eastern North Dakota, while northward its range extends to Maine, northern Quebec, and northern Ontario. The race *B. u. thayeri* is confined to Nova Scotia and probably eastern New Brunswick.

The gray ruffed grouse (*Bonasa u. umbelloides*) is found from western South Dakota west to the coast and Cascade Ranges in British Columbia, north to west-central Mackenzie, and south to northern Colorado and northern Utah; the Oregon ruffed grouse (*B. u. sabini*) occupies the Pacific coast region from Humboldt County, Calif., north through Vancouver Island and the adjacent mainland of British Columbia; and the Yukon ruffed grouse (*B. u. yukonensis*) is found in the interior of Yukon and in Alaska.

Egg dates.—Southern New England and New York (*umbellus*): 72 records, April 5 to June 10; 36 records, May 9 to 24. New Jersey and Pennsylvania: 15 records, April 20 to June 12; 8 records, May 9 to 27. Maryland to Iowa: 13 records, April 28 to June 4; 7 records, May 6 to 27. Michigan to Minnesota: 11 records, May 2 to June 10.

Northern States and southern Canada (*togata* and *thayeri*): 52 records, April 14 to June 26; 26 records, May 15 to 28.

Mountain regions of Western States and Canada (*umbelloides*): 45 records, May 1 to July 10; 23 records, May 18 to 29.

Oregon and Washington (*sabini*): 40 records, April 19 to June 18; 20 records, May 5 to 14.

BONASA UMBELLUS TOGATA (Linnaeus)

CANADA RUFFED GROUSE

HABITS

The ruffed grouse of northern New England and eastern Canada is a grayer bird, with more conspicuous and darker markings below, than typical *umbellus*. Its range is roughly the same as the spruce grouse, or "spruce partridge," from which it is commonly distinguished as the "birch partridge."

William Brewster (1925) well describes its haunts in Maine, as follows:

> Ranging from the lowest levels to the crests of the higher mountains it occurs practically everywhere, although seen comparatively seldom among heavy, unmixed spruce timber, and still more rarely in the larch and arbor vitae swamps, so beloved by Spruce Grouse. It likes best to dwell in woods composed of intermingling evergreen and deciduous trees. Far back in these it is found oftenest about deserted lumber camps, and along old logging roads, where enough sunlight has been let in to stimulate a vigorous growth of underbrush; or along the courses of alder-fringed brooks or runs, where ferns flourish in rich, moist soil; or on river banks freely exposed to the sun, but densely overgrown with cornels, Viburnums, and other berry-bearing shrubs. Near permanent settlements it is given to frequenting wood edges, neglected pastures, and the outskirts of crudely tilled farms, where young spruces, balsams, birches, maples, and alders have been permitted to spring up in crowded thickets about sunny little openings filled with tall bracken. The birds are here reasonably safe from human molestation, except in autumn when everyone possessed of a gun bags as many of them as he possibly can.

Nesting.—The nesting habits of this grouse are similar to those of other ruffed grouse. Brewster (1925) describes three nests found in May, 1896, near Lake Umbagog, Me. Of one found on May 14, he says: "This nest was directly under the main stem of a fallen poplar, on a dry knoll wooded with second-growth poplars and birches among which were interspersed a few balsams and spruces. It was 30 yards back from a public road and within 10 yards of an open pasture." The next day he found one "at the edge of a thicket of alders covering rather wet ground, between two large, buttressed roots of an old stump. Overspreading branches of a small arbor vitae and Viburnum, growing close beside it, screened it so perfectly that the brooding bird could be seen only from the direction whence we happened to approach." On May 16, another "was in a very exposed situation, quite outside of the border of wild cherry, mountain maple, and other undergrowth that fringed an extensive forest, half-encircling an upland mowing field, and in the field rather than in the forest, although but a yard or two from where the latter ended. Here it was sheltered from observation and from blazing sunlight by only a few dead sticks, the remains of a disintegrated brush fence."

J. W. Banks, of St. John, New Brunswick, wrote to Major Bendire (1892):

> Here with us a very common nesting place is what is called a fallow. This is a piece of woods chopped down in the fall, to be burned when sufficiently dry, usually in the latter part of May or early in June. Being composed chiefly of spruce and fir, it burns very rapidly. I found two nests (or rather the remains, for the eggs were badly scorched) in one of these burnt fallows,

and a few feet from each nest the bones of the mother Grouse. A farmer acquaintance told me of finding a nest of this bird, which contained ten eggs, in a fallow he was about to burn, and knowing of another nest with an equal number of eggs, the thought occurred to him to put the eggs in the nest of the other bird that would not be endangered by the fire, and watch developments. He had the satisfaction of knowing that the eggs were hatched.

Eggs.—The eggs of the Canada ruffed grouse are similar to and indistinguishable from the eggs of the more southern race. The measurements of 71 eggs average 39.2 by 30.3 millimeters; the eggs showing the four extremes measure **44** by 31, 42.2 by **31.5**, **36.6** by 28.7, and 38.6 by **28.2** millimeters.

Young.—Bendire (1892) quotes the following from Ernest Thompson Seton's notes:

Every field man must be acquainted with the simulation of lameness, by which many birds decoy or try to decoy intruders from their nests. This is an invariable device of the Partridge, and I have no doubt that it is quite successful with the natural foes of the bird; indeed, it is often so with man. A dog, as I have often seen, is certain to be misled and duped, and there is little doubt that a mink, skunk, raccoon, fox, coyote, or wolf, would fare no better. Imagine the effect of the bird's tactics on a prowling fox; he has scented her as she sets, he is almost upon her, but she has been watching him, and suddenly with a loud "whirr" she springs up and tumbles a few yards before him. The suddenness and noise with which the bird appears causes the fox to be totally carried away; he forgets all his former experience, he never thinks of the eggs, his mind is filled with the thought of the wounded bird almost within his reach; a few more bounds and his meal will be secured. So he springs and springs, and very nearly catches her, and in his excitement he is led on, and away, till finally the bird flies off, leaving him a quarter of a mile or more from the nest.

Plumages.—The molts and plumages are the same as in the other races. There are also both color phases, gray and red. As mentioned elsewhere, these two color phases occur in all the races, but in this and in other gray races the red phase is rarer and less pronounced, the reverse being the case in the red races.

Food.—The long list of food given for the ruffed grouse (*umbellus*) would apply equally well for this race, with due allowance for the different species of plants and insects available. Manly Hardy, in some notes sent to Major Bendire (1892), says that it feeds not only on the poplar buds but also on the hard old leaves. He writes:

I have killed one with its crop filled with such leaves on the 20th of August, and they eat them continuously, until the last have fallen in late October. They do this when other food is abundant. Buds of willow, yellow and white birch, hophornbeam, thorn plums, rosehips, leaves of tame sorrel, of the rock polypod, fungus from birch trees, the seeds of touch-me-nots (*Impatiens fulva*), wild raisins, and highland cranberries (both species of *Viburnum*) form also a part of their bill of fare. They seem to be especially fond of beechnuts. I have a record of finding seventy-six in one bird's crop and over sixty in another.

Ora W. Knight (1908) says:

In the winter they "bud" seeming to prefer the yellow and white birches, and the poplar, but also eating spruce, fir, pine, maple and in fact many other buds.

Brewster (1925) writes:

To the best of my knowledge the Birch Partridges of the Umbagog Region never eat the spills of coniferous trees, although subsisting almost wholly on the buds of deciduous ones during rather more than half of the year. In late spring and early summer their food is gleaned mostly from the surface of forest-shaded ground and consists largely of insects and low-growing herbaceous foliage of various kinds. Even where it is most plentiful the birds seldom linger anywhere to seek it, but continue to advance, picking up a leaf or grub now here, next there, so daintily and infrequently that they often ramble on slowly for a quarter of a mile or more before filling their crops. They are somewhat less dainty and fastidious when feasting in late August on the fruit of low blueberry bushes, while in September I have often seen them alight at sunset in cornel or Viburnum (especially *V. opulus*) bushes on river-banks and literally gorge themselves in the course of a few minutes, almost without change of foothold, on the berries which these shrubs commonly bear in such profuse and crowded clusters. Later in the season the pale orange fruit of the mountain ash is similarly dealt with whenever it can be had plentifully, which is not oftener than every other year. The glowing red berries of the black alder are also eaten freely in late autumn. The birds seem to have little or no liking for oats, but are exceedingly fond of buckwheat, and to obtain it will venture out fifty yards or more from neighboring coverts into stubble fields where it has recently been grown and harvested.

Behavior.—Illustrating the tameness of ruffed grouse under primitive conditions, Brewster (1925) relates his experience with them in the Umbagog region early in the seventies, as follows:

At first I undertook to hunt them with a setter, and to shoot them only on wing, as had been my practice in coverts nearer home. My good dog found and pointed them readily, but was evidently not a little puzzled to comprehend why they should stand conspicuously upright in open ground, or on mossy logs, regarding him with seeming indifference from a distance of only a few yards, instead of rising far in advance, or crouching unseen in dense brush, as had been the unvarying habit of all birds of their kind with which he had had previous experience. When I stepped in ahead of the staunch setter with the intention of flushing the Grouse, their behavior was still more surprising especially if, as often happened, there were as many as five or six together. For instead of rising promptly on wing as I wished, and expected them to do, they would begin a snickering outcry almost precisely like that of a Red Squirrel, nod their heads slightly a few times, and then start off at a slow walk with crests erect and perhaps also widespread tails, shaking their heads and necks, and twitching their expanded ruffs at each deliberate step, and continuing unceasingly to utter their derisive and unseemly snickering. This was most likely to happen in a narrow cart-path tunneled through the forest, or on the outskirts of some woodland opening. In either case the birds had seldom far to go before reaching fallen tree-tops, or dense evergreen thickets, from which it was difficult if not impossible to dislodge them, at least by the aid of a dog, who would never flush his birds. Into such sanctuaries they commonly skulked on foot, if not too closely pressed. Even when I forced them to take to wing by running after them, they rarely went more than a few yards before alighting in a tree, or dropping again to the ground, over which they might continue to hasten, if much alarmed, until it was useless to follow them farther.

After this and several similar experiences, he learned that the only way to secure any of these unsophisticated grouse was to "pot" them in the "time-honored fashion of local hunters" by shooting off their heads as they sat on trees, bushes, or logs, or on the ground. More than once he was able to bag all the birds in a flock, shooting one at a time, while the survivors refused to fly. While hunting for ducks one day in a boat, he saw a grouse sitting on a log. "When I finally shot her," he says, "the report of the gun started six others, hitherto unseen, although close about her. Flying only a few yards, they alighted in trees and bushes within plain view, and remained there gazing at us while, sitting in the bow of the boat, I loaded and fired until the last bird had fallen."

Such shooting was unsportsmanlike, of course, unless the hunter gave the birds a chance by using a rifle and shooting at the head. It was the only method in those days, but is no longer necessary now in any but the more remote regions, for the birds have become educated and have learned to give the sportsman a chance to show his skill. Good sport may now be had with dog and gun in these same woods.

Fall.—Brewster's (1925) very full and exceedingly interesting account of the Canada ruffed grouse contains several references to a migratory movement or an erratic autumn wandering, similar, perhaps, to their queer behavior during the restless "crazy season" that we have noted elsewhere.. I quote a story, one of three similar incidents, told him by Luman Sargent:

After spending an October night (before 1870) in the old log-camp of Middle Dam and leaving it at about sunrise the next morning, he had gone only a short distance into neighbouring woods when he saw a Partridge on a log and shot it. Startled by the report of his gun so many birds of the same species rose far and near on wing, one after another, that for half a minute the air seemed full of them as far as the eye could range through trees and brushwood. There could not have been less than one hundred of them, he said. When followed they all kept moving on by successive flights in the same direction, and to the southward. They were so restless and shy that he had difficulty in getting near them and was able to kill only a few more before losing track of them altogether.

Winter.—Ernest Thompson Seton (Ernest E. Thompson, 1890), referring to the habits of the ruffed grouse in Manitoba, says:

It seems to be the normal habit of this bird to roost in a snow drift during the coldest weather. The wonderful non-conductivity of the snow is well known, but may be forcibly illustrated by the fact that although the thermometer registers 35° below zero, the 10 inches of snow which fell before the severe frost came, has effectually kept the wet earth in the woods from freezing, although the temperature has been at or below zero for over a week. In view of these facts it is easy to understand that the grouse in the snow drift are quite comfortable during the coldest nights. In general the bird will be found to run about before burrowing into the drift; each makes its own bed, usually 10 or 20 feet from its neighbor; they usually go down a foot or so and along 2 feet; they pass the whole night in one bed if undisturbed, as the large amount of dung left behind would indicate. They do not come out at the ingress, but burst through the roof of their cot at one side; they do not usually go straight ahead and out, because their breath during the whole night has been freezing into an icy wall just before their nostrils.

BONASA UMBELLUS UMBELLOIDES (Douglas)

GRAY RUFFED GROUSE

HABITS

The name gray ruffed grouse describes very well the characters on which this race of the ruffed grouse is based; it is a decidedly gray bird. Its range extends from Mackenzie south through the Rocky Mountain valleys and east to the western edge of the Great Plains. The birds found in Manitoba, of which we collected a good series, are intermediate between this and *togata*. It intergrades with *sabini* where their ranges meet.

Harry S. Swarth (1924) found it "abundant throughout the poplar woods of the lowlands" in the Skeena River region of northern British Columbia, where the birds are "less grayish, more brownish" than typical *umbelloides*. In the Stikine River region he (1922) found the birds "relatively gray colored, but not so ashy" as the birds from the Yukon region (*yukonensis*); they are practically the same as *umbelloides* from Alberta.

In the Glacier National Park, Mont., Mrs. Florence M. Bailey (1918) found this grouse "in the pines and aspen thickets of the eastern slope and also in the dense hemlock woods of the western slope of the mountains." M. P. Skinner has sent me the following notes on its haunts in the Yellowstone National Park, Wyo.:

This grouse is more likely to be in the aspen groves than Richardson's grouse is. In fact, it is quite apt to prefer the aspen groves, but it is also found in forests of fir, mixed aspen and fir, lodgepole pine, mixed aspen and lodgepole, and in the spruce forests. Although this grouse is rather scarce in Yellowstone National Park, it frequents the forest in all sections between 6,500 and 8,500 feet elevation, but I do not find them in the timber-line forests. It seems to be in heavy, thick standing forests, open forests, and thick sapling growth. Occasionally it is found out on the grasslands, but not so much as Richardson's grouse. Still, I think the gray ruffed grouse rather prefers an open stand of trees, especially if berries are present.

Courtship.—Skinner says that "as early as March 20 the males begin to strut and court the females. One day I found a pair near

Tower Falls. The female sat as if roosting on a horizontal pine branch, 4 feet from the ground, in a clump of thick-growing saplings; a male, with ruffs spread and tail spread in an upright fan, strutted about below. He may have been drumming, although I did not actually hear him."

J. H. Riley (1912) heard it drumming in the fall in Alberta; he says: "When we first made camp, September 5th, and for about a week thereafter, we seldom heard this bird drum, but before we departed, September 22nd, one used to drum at intervals throughout the day, though the weather was cloudy with rain and later snow."

Thomas T. McCabe writes to me that at Indian Point Lake, near Barkerville, British Columbia, he heard one drumming for many weeks, not on the usual log, but on the large sloping root of a big spruce on the lake shore, 175 yards away from a nest he had found; he watched him drum at the same spot 10 days after the nest had been robbed.

Nesting.—Mr. McCabe very kindly collected and sent to me this nest and the 10 very handsome eggs it contained. He took it on May 27, 1929, at Barkerville, British Columbia, and he writes to me that

> the nest was in heavy, open, mixed, second-growth spruce and balsam (*Abies lasiocarpa*), which had nearly the dignity of primeval forest, and was carpeted with our universal green moss. It was wedged between two roots of a 9-inch balsam, and sunken to the extent that the top was about level with the moss. There was absolutely no bushy or herbaceous concealment. The brooding bird left in each case when we were from 15 to 20 feet away, but it did not approach us. It kept at a distance of about 40 feet, running, crouched low, whining like an eager hound, and then flew away. The leaves lining the nest are *Populus tremuloides*, seldom far to seek here, even in the deep timber.

Major Bendire (1892) mentions a nest found in Montana, "under the trunk of a fallen cottonwood tree, which rested about a foot from the ground. Otherwise the nest was not concealed in any way." Another nest near Nulato, Alaska, was "found in an old willow stump." In Yellowstone National Park, Mr. Skinner records two nests in his notes; one was in a grove of quaking aspens, under a fallen tree; the other was at the foot of a lodgepole pine, a hollow in the pine needles.

Eggs.—The eggs of the gray ruffed grouse are practically indistinguishable from those of its eastern relatives. The measurements of 55 eggs average 40.3 by 29.8 millimeters; the eggs showing the four extremes measure **43** by **31.5**, **38** by 30, and 40.7 by **28.7** millimeters.

Young.—Swarth (1922) found two broods of young in the Stikine River region, of which he writes:

> The young of one brood were still unable to fly. Our first knowledge of their presence was derived from the mother bird, who burst forth from

the bushes and charged us furiously. She kept tail and ruff widely spread, the head crest depressed. She was mewling in a very catlike fashion, and also hissing from time to time. There was an occasional faint *peep* from the grass nearby, and once I caught a glimpse of a yellow chick slipping away through the shubbery, but the young were too agile to be captured.

The young of the second brood were somewhat larger and able to fly. This second mother tried to toll us away from the chicks by feigning a broken wing; the noise she made was not unlike the whining of a small puppy. Her actions, all together, gave the impression that she was frightened rather than angry. However, if frightened, she still did not desert her trust, but remained nearby, dragging herself back and forth across the road, with wings dropping and all her feathers pressed closely against her body. Her tail was not spread nor were her ruffs displayed at any time, all in striking contrast to the behavior of the first bird that morning.

Plumages.—The molts and plumages are similar to those of the eastern bird. It has a red phase, which apparently occurs about as often as the gray phase occurs in the eastern birds. Swarth (1924) collected two red-tailed birds in a series of 14 in the Skeena River region. Ernest Thompson Seton (1885) has described the red phase, as it occurs in Manitoba, as follows: "In general appearance this bird differs but little from the well known *Bonasa umbellus umbelloides*, but it is distinguished by being more decidedly marked—thus the bars on the belly are complete and nearly black—and by having copper-colored touches on the back, the subterminal tail band and the *ruff* a rich, iridescent, coppery red." He says that about 10 per cent of the birds have copper ruffs and only about 20 per cent are pure *umbelloides.*

Food.—Skinner says of its food in the Yellowstone: "In spring, they eat the buds of aspens. In summer, they frequent berry patches and sometimes visit small openings for grasshoppers. In winter, they are said to eat mistletoe berries." In the Stikine region Swarth (1922) found that the food was practically all vegetable matter, consisting of leaves and stems of *Populus*, *Galium*, *Artemisia*, and *Viburnum*, with a few berries.

Behavior.—Skinner contributes the following notes on the behavior of the gray ruffed grouse:

These are resident birds, remaining throughout the year in one locality, and I see no evidence that they even move up and down the mountains, as other resident species do. But in winter they live mostly in the treetops, and we do not see them on the ground much before March 15 or April 1. Although these birds have the same habits and probably the same disposition toward their environment as the eastern ruffed grouse originally did, they have retained their comparative tameness toward man under the protection accorded in Yellowstone National Park. My notes are full of references to them as "quite tame" and "very tame," but occasionally I find one that is wild. The tame ones are recorded in all places, at every elevation within the bird's range, and at all seasons. This unsuspicious attitude extends also to men on horseback, the grouse sitting or remaining in their tracks while the horseman rides by, although sometimes a grouse will move around to the opposite side of a tree from the intruder. I have even been permitted to ride past a grouse within 5 feet. At times, especially in autumn when they frequent the roads, the grouse often run across in front of a horse and even apparently under his nose. I have seen my horse almost step on a grouse. Sometimes they run ahead of my horse down the trail. Even a mother with young is often extraordinarily tame, and she does not make use of the broken-wing tactics nearly so often as unprotected grouse do. Often a bird springs directly from the ground to a tree overhead and allows me to pass directly under. Still, the gray ruffed grouse is not so apt to come about houses and hotels as Richardson's grouse.

BONASA UMBELLUS SABINI (Douglas)

OREGON RUFFED GROUSE

HABITS

In the humid coast belt, west of the Cascades, in Oregon, Washington, and British Columbia, we have the darkest, most richly colored, and one of the handsomest races of the ruffed grouse. The grouse of the southern Alleghenies are quite richly colored, but they will not compare in this respect with these western birds, which Baird, Brewer, and Ridgway (1905) describe as follows:

> The upper parts are dark orange chestnut, mottled with black, the cordate light spots very distinct. The feathers of the breast are strongly tinged with reddish-yellow; those of the sides marked with broad and conspicuous bars of black, instead of the obsolete brown. The under tail-coverts are orange-chestnut, within distinct bars of black, and an angular terminal blotch of white. All the light brown blotches and edgings of the eastern variety are here dark brown or black. The jugular band between the ruffles is very conspicuously black.

They say of its haunts, quoting J. K. Lord:

> Dr. Cooper also speaks of this grouse as very abundant everywhere about the borders of woods and clearings. It was common near the forests east of the Cascade Mountains up to the 49th degree. In the spring their favorite haunt is in the vicinity of stagnant pools, or in the brush around a marsh in which the wild swamp-crab, the black birch, and the alder grow.

William L. Finley (1896) writes:

> They are generally found on low land, a river bottom or along some small creek, but in times of high water, they will go to higher ground. I have often seen them when the water is high, in some small tree or bush, when the water was several feet deep under them, and around them for a half mile. In a boat, at such times, one can row right under the bird, or within a few feet of it. A great many are killed along the river bottoms in this way by hunters.

Courtship.—The drumming performance of this grouse is apparently the same as with the eastern birds, but "Mr. Lord also states that he has seen the males of this species fighting furiously during the pairing season. Ruffing up their necks, with their heads and backs almost in a straight line, and with wings dropped, they circle round and round each other, striking and pecking until the vanquished gives in, and the victor mounts upon a log and proceeds to drum furiously."

Nesting.—W. Leon Dawson (1909) says: "At the foot of a maple in some swampy thicket, or close beside a fallen log, the female scrapes a slight depression in the earth, lining it roughly with dead leaves and a few small twigs."

In some notes sent to me by D. E. Brown, he describes four Washington nests. One was at the foot of an elder tree in an old river bed, in the dead leaves, with no cover at all. Two nests, found April 19, were in open woods, under the ends of logs. And one, found May 6, was at the side of a log at the edge of large woods. In all cases the eggs were laid on dead leaves. Bendire (1892) describes a nest found on Vancouver Island as "a slight hollow in the ground scratched out by the bird, placed under the fallen branches of a spruce tree. The cavity was lined with dead leaves and spruce needles, as well as a few feathers. This nest was found close to a small creek and was well concealed."

William L. Finley (1896) reports a nest found by G. D. Peck in Oregon that contained eight eggs of the sooty grouse and seven eggs of the ruffed grouse; the sooty grouse was flushed from the nest. He has also sent me a photograph of a nest containing eight eggs of the ruffed grouse and three eggs of the ring-necked pheasant.

Eggs.—The eggs are indistinguishable from those of other ruffed grouse, though they may average slightly darker in color and somewhat more often spotted. The sets will average smaller in number. The measurements of 58 eggs average 41 by 30.2 millimeters; the eggs showing the four extremes measure **44** by 31.5, 43.5 by **32**, **38** by 29, and 40.9 by **28.2** millimeters.

Plumages.—Confirming what I have said elsewhere about the presence of both color phases in all races of the ruffed grouse, H. S. Swarth (1912), referring to a series collected on Vancouver Island, says:

The dichromatism of the species is very apparent in the series collected, the gray and the red-colored birds being conspicuously different. Those in the gray phase are quite uniform in color and markings, but the reddish birds show considerable variation. The former all have black ruffs, and gray tails with a black subterminal band. Of the reddish birds some have red ruffs, some black, and others are variously intermediate. Some have a gray tail with a red band, some a red tail with a dark band, and one a red tail with a darker red band. These different styles of coloration are not indicative of age, sex, or season, for both phases are represented among adults and immatures of both sexes.

Food.—The food of the Oregon ruffed grouse is similar to that of its eastern relatives, differing only with the available supply of berries, seeds, leaves, fruits, and insects. William H. Kobbé (1900) writes: "They are extremely fond of the small wild crab apples

(*Pyrus rivularis*) which grow in the low, damp woods. The birds visit these trees very early in the morning and late in the evening, at which times they may be found silently perched upon the branches."

Mr. Dawson (1909) adds: "They are fond of the fruit of the Cascara, which they gather from the ground; and wild crab-apples are favorites in season. These last ripen about the middle of October, and from that time until the alders bud again these Grouse are often to be found in evergreen trees."

Behavior.—In general the habits of the Oregon ruffed grouse do not differ from those of the rest of the species, except so far as they are affected by differences in environment. In the wilder sections they are quite unsophisticated.

Enemies.—J. H. Bowles (1901) tells the following remarkable story:

The carnivorous habits of chipmunks as related in the recent issues of The Condor were very interesting to me, though I believe mice are far more guilty. Mice are a perfect pest to ground-builders in this country, as they burrow into the ground several yards away from the nest and then tunnel until they reach the bottom of the nest.

They then dig upward into the nest and carry the eggs into their tunnel to eat. I have often found broken and unbroken eggs several feet from the nest in a burrow. I have never actually seen mice do this, but the tunnels are much too small for anything else. The Ruffed Grouse (*Bonasa umbellus sabini*) are the worst sufferers that I have yet found, and their eggs are the largest that I have seen destroyed in this manner. All the small ground-builders suffer more or less.

Game.—Dawson (1909) writes:

From the point of view of the sportsman, this bird is not to be compared with the Ruffed Grouse of the Eastern States. Its cover is too abundant, and it does not take the discipline which has educated the wily "partridge." It seldom allows the dog to come to a correct point, usually flushing into the nearest small tree, where it sits peeping and perking like an overgrown chicken, regarding now the dog and now the hunter. Pot-shooting the birds under these circumstances can hardly be called sport, but their fondness for dense thickets often makes it the only way in which they can be obtained.

Edwyn Sandys (1904) evidently agrees with Dawson, for he says:

In British Columbia the sport, as found, could not compare with that of the East. Those who know the wonderful western province will readily guess why. In many places the trees almost rival the famous big conifers of California, and they are crowded together as thickly as it is possible for such mighty trunks to stand. Frequently the lower spaces are filled with ferns of such size and luxuriance as to suggest semitropic lands rather than a portion of Canada. In such cover the keenest of guns can do little or nothing. The writer is over six feet tall, but in that cover he felt like a veritable babe in the wood. The size of the firs was almost oppressive—but the ferns—ye gods! such ferns. In places they grow like the big western corn, close and rank, towering a yard or more above one's head. Among them, grouse after grouse can buzz away unseen, while, in addition, the tremendous fronds combine to form a most baffling light.

BONASA UMBELLUS THAYERI Bangs

NOVA SCOTIA RUFFED GROUSE

HABITS

The ruffed grouse of Nova Scotia had long been recognized and was finally described by Outram Bangs (1912) under the name *B. u. thayeri*, given in honor of Col. John E. Thayer. He designates it as

similar to *Bonasa umbellus togata* (Linn.) but general color of upper parts darker, more dusky or sooty, less grayish; the whole underparts (except throat) heavily and regularly banded with dusky, the dark bands much blacker and much more boldly contrasted against the ground color—less blended.

B. umbellus thayeri presents two phases of coloration, which are both very dark, and not very different; a phase in which there is much dull chestnut or burnt sienna in the upper parts and tail and another in which the tail is wholly dull gray and black and the upper parts are but little varied with dark chestnut markings. The color and markings of the underparts is not different in the two phases, except that very reddish birds sometimes have the bases of the feathers of the upper chest dull chestnut instead of dusky.

Bangs says further in regard to it:

Some years ago I was accustomed to go shooting every autumn in Nova Scotia, and each season I was more and more impressed by the very dark coloration of the Ruffed Grouse killed there. I therefore made into skins during my last two shooting trips to this province a series of sixteen Grouse.

It is probable that the new form is confined to the almost insular province of Nova Scotia, although I cannot be sure about the bird from the coast of New Brunswick as the specimens I have before me are in worn midsummer plumage, and not comparable with the Nova Scotia specimens, all of which were taken in October.

We have no reason to think that the habits of this grouse, or any of the chapters in its life history, are essentially different from those of its neighbors in other parts of Canada where the environment is similar.

Eggs.—The eggs are indistinguishable from those of other ruffed grouse. The measurements of 47 eggs average 40.5 by 30.5 millimeters; the eggs showing the four extremes measure **44** by **33**, **37.9** by 30, and 40 by **29.2** millimeters.

BONASA UMBELLUS YUKONENSIS Grinnell

YUKON RUFFED GROUSE

HABITS

The ruffed grouse of the interior of Alaska and Yukon is, according to Dr. Joseph Grinnell (1916), the "largest and palest of the races of *Bonasa umbellus;* nearest like *B. u. umbelloides*, but general coloration of light-colored parts of plumage more ashy, and pattern of dark markings finer." He says also that it—

occurs along the Yukon River Valley down nearly to its mouth, as also in adjacent wooded areas west even into the Seward Peninsula.

As with the other subspecies of the Ruffed Grouse, *yukonensis* shows two color phases. Three out of the eleven specimens at hand have pale rusty tails; but even in this "red" phase the race is distinguishable from the corresponding phase in the other subspecies by paler tone of coloration. Typical *umbelloides* is still a *gray* bird, but its grayness is more leaden, and its browns and blacks are deeper. The extreme fineness of the intricate pattern of barring and mottling on the plumage is in *yukonensis* an appreciable character.

Dr. E. W. Nelson (1887) writes of its haunts:

Like the Spruce Grouse, and sharing with the latter its range in Northern Alaska, this bird is found everywhere where wooded land occurs, reaching the head of Norton Sound and vicinity of Bering Straits, following the belts of timber as they approach the sea in this portion of the Territory. It is not uncommon in the vicinity of Nulato, where it frequents the deep spruce growths, and feeds exclusively upon the buds of these trees, its flesh being tainted in consequence. Dall found it nesting there in May, and a set of eggs was found in a willow stump. Like the Spruce Grouse, this bird is found wherever spruces occur, and both species range well into the Kaviak Peninsula, so that they are found within a very short distance of Bering Straits.

LAGOPUS LAGOPUS ALBUS (Gmelin)

WILLOW PTARMIGAN

HABITS

The willow ptarmigan, with its various so-called subspecies, is of circumpolar distribution, inhabiting the arctic or subarctic regions of both the North American and Eurasian Continents. European writers know it as the willow grouse, applying the name ptarmigan to only the rock ptarmigan and its races. It seems more logical to apply the name ptarmigan to both species that assume a white plumage in winter. The British red grouse is the only species of the genus *Lagopus* that does not have a white winter plumage; it does, however, have the feathered toes, one of the principal characters of the genus. The following account includes the so-called Alaska ptarmigan.

Spring.—As soon as the first bare spaces appear on the sunny slopes of the tundra the ptarmigan begin their spring migration from their winter quarters, among the willows in the sheltered valleys of the rivers and creeks of the interior, to their breeding grounds on the open hills or tundra. H. B. Conover says in some notes he sent to me:

On our sled trip from Nenana to Hooper Bay, this grouse was not encountered until we reached the Kuskokwim Mountains, where we found it very plentiful. It was generally encountered in flocks of 15 to 100 feeding among the dwarf willows. A few males shot on April 5 had an occasional brown feather showing on the head and neck. The morning of April 24 we left Mountain Village on the Yukon and cut across the tundra for Hooper Bay, a journey which took us four days. For the first two of these we encountered ptarmigan everywhere along the willow-bordered sloughs and creeks. The hens were still white, but the heads and necks of the cocks were about a third into the red spring plumage. As we approached the coast and the willows became scarcer, these birds were no longer seen. On May 9 the ground about Point Dall was beginning to show in spots through the snow and the first ptarmigan made their appearance. Two days later they were common. Each male now took possession of a little spot of bare ground, whence he sent out his challenges, *com-ere, com-ere, go-bec, go-bec.* Between calls they would bob their heads as if they were pecking at the ground, or, jumping about 6 feet into the air, glide down to the earth, cackling as they descended. The hens seemed to have but one call, a cackle similar to that of a tame chicken. Often two cocks were seen chasing each other around over the tundra, but only rarely would they seem to stand and fight it out. In the evenings and early mornings these birds were especially noisy, and often it was no great stretch of the imagination, what with the calls of the water-fowl, to imagine oneself in some great barnyard. About this time the Eskimo boys began to range the tundra with their bows and arrows, and many an unwary cock and sometimes a hen were killed by the blunt shafts of these 8 and 10 year olds.

Courtship.—Edwyn Sandys (1904) gives the following very good account of this performance:

The love-making of the ptarmigan is not unlike that of the Canada Grouse, or "spruce-partridge." The males, with their plumage changing from white to the handsome summer dress, strut with all the pomposity of their kind. The red combs over the eyes are swollen and very conspicuous, as the bird struts with head thrown far back, tail raised and spread, and wings trailing. Presently he leaps into the air, raises himself higher and higher with a vigorous flapping, then sails on set wings through a descending spiral, which brings him back to his starting-point. While thus a-wing, he utters a curt, gruff challenge oft repeated, a defiance to all rivals. Again he struts, and again goes into the air, frequently to see male after male arise from near-by stations. While so occupied the birds make considerable noise, the bark-like challenge of other calls being heard for some distance. Meanwhile, the females loiter about in the cover, admiring the efforts of the males, and gradually acknowledging their charms. The inevitable battles follow—spirited encounters, in which many hard knocks are given, and much pretty plumage marred, until the weaker have been well whipped.

Nesting.—Herbert W. Brandt has sent me some elaborate notes on this species. He says of its nesting:

The willow ptarmigan at Hooper Bay is not at all particular as to the location of its home site, for it dwells impartially from the drift-strewn sea beach to the higher altitudes on the mountains. Down under the protection of a drift log, a clump of grass, a small bush, a mossy hummock, or any screeny object, she scrapes out a cavity to fit her requirements. This she lines more or less with any material at hand, and here she deposits daily her rich crimson egg. During the period of egg accumulation, we found the nest to be covered with surrounding material, because the bird does not begin to incubate until the full complement is satisfied. When the first egg is laid there is but little form to the nest, but as the set progresses, the birds mold it into the proper shape, and by the time incubation has progressed, the eggs snuggle together in a well-cupped basin. The brooding female spreads her feathers as does the sitting domestic hen, and when her frail body is examined in hand, it seems almost incredible that she can cover so large a clutch of eggs. The extreme measurements of 12 nests examined are: Total height, 5 to 7 inches; inside diameter, 6 to 8 inches; depth of cavity, 3 to 6½ inches.

F. Seymour Hersey collected some 10 sets of eggs for me between St. Michael and the mouth of the Yukon. They were mostly in fairly open situations on the tundra. One nest was a depression in the tundra moss at the base of a small clump of grass; it was lined with dry grass, leaves, and a few feathers; it measured 7 inches across and 5 inches deep. Another was in a hollow under a dwarf willow on a raised mound on the tundra. Another was in a deep hollow in a wet place on the border of a marsh, a very open situation; it had a heavy lining of dry grass and a few feathers. Some of the hollows were very shallow, not more than 2 inches deep. There were three sets of 7, two of 8, one of 9, three of 10, and one of 11 eggs.

George G. Cantwell writes to me of a nest he found on Copper River, Alaska, that was "placed on an open river bar among a light growth of willows, close to a growth of mountain spruce." Stanton Warburton, jr., tells me of a nest he found late at night, saying:

To preserve them intact for a photo in the morning, I put my khaki coat over the nest and eggs, completely covering them. Then (as an experiment) I wrinkled up the collar by the eggs so that it formed an opening not over 2 inches high. This slight opening did not expose the eggs to view from any angle, as they were still completely covered. Early the next morning when I approached, the female flushed from right beside my coat. All 11 eggs were now in a new nest just outside my coat, not over a foot from the original location. During the night the female had made a new nest, moved all 11 eggs into it, and recommenced incubation.

MacFarlane mentions in his notes another case where the eggs were probably removed by the birds. He "had reason to know that some, at least, of the nests were used by ptarmigan several seasons in succession."

Joseph Dixon (1927) writes:

The male ptarmigan spends the day hiding in little thickets, keeping within 50 or 100 feet of the nest. He has a definite form or nest of his own which he occupies when roosting. One reason for his staying so close is the danger of Short-billed Gulls finding the ptarmigan's nest. These egg thieves work in organized gangs, usually three together. One will swoop down at the female, trying to make her shift about on the nest so as to expose the eggs. The second or third gull following tries to slip in and grab an egg. As soon as the gulls appear, the hen ptarmigan gives a peculiar call for help. Upon hearing this the cock ptarmigan bursts forth like a rocket and charges the thieving gulls. He doesn't beat around the bush but flies directly at the intruders, knocking them down with the impact of his body. An average cock Willow Ptarmigan at this season weighs 507 grams, while one on the Short-billed Gulls which was shot weighed 358.2 grams. In addition to being one-third heavier than the gull the cock ptarmigan flies much the faster of the two, and when he hits a gull it is almost like a Duck Hawk striking a duck.

Eggs.—The willow ptarmigan lays commonly 7 to 10 eggs; as many as 17 have been found in a nest, and in late or second sets there may be only 5 or 6. An egg is laid each day, and incubation does not begin until the last egg is laid. For a description of the beautiful eggs, I can not do better than to quote Mr. Brandt's remarks, as follows:

Eggs in the same set follow the same type of ground color, the same style of markings, and are nearly uniform in size; and as each bird seems to lay a type of egg individually its own, it may be noted that scarcely any two sets of eggs of this interesting species are exactly alike. In shape the egg is almost always ovate, but in rare instances tend to short ovate with stubby ends, and nearly elongate ovate when more slender in shape. The surface of the shell is smooth and often somewhat greasy like that of a duck egg; the texture is hard; and the egg sturdy like its parents. The luster is rather shiny and this apparently increases somewhat as incubation progresses.

A study of the coloration of the egg of the willow ptarmigan is of special interest because of its change of hue after it is laid. When the egg first appears, the markings are from "ox-blood red" to "scarlet red," and the whole surface is vivid and moist, and it appears as if it has been dipped in fresh, red paint. This undried pigment is very easily rubbed off, and in consequence, it is unusual that an egg does not show the sign somewhere on the surface of being brushed by the feathers of the parent bird. In fact, these rubbed spots may show distinctly the individual barbs of the ptarmigan feather that scratched it when it was wet. As the moisture from the egg dries, however, the pigment sets rapidly, and at the same time so darkens, like congealed blood, that by the time the eggs are a few hours old, the brilliant reds turn to blackish brown. Once the pigment becomes dry and sets, it is very durable, and egg-shells that have lain out in the weather from the previous year still retain their bold markings.

The ground color is usually inconspicuous, because it is seldom that more than half of it is visible, and often it is all practically hidden by the overlying spots. The only place that the ground color is prominent is on those areas where there is an aforementioned rubbed spot. This ground color exhibits many variations of the pale creamy tints, such as "ivory yellow" and "sea-foam yellow," but a few sets are further decorated with paler reds, making the ground color "orange-crimson" to "vinaceous-tawny" and "pecan brown," while the ground color of one egg is even "ocher red."

The markings are the richest of any egg we collected at Hooper Bay and are irregularly flecked in profusion all over the surface. In size, these spots range from the finest pepperings to blotches thumb-nail in size, and are all more or less confluent, some so much so that they cover the surface and almost envelope the ground color. When the spots are large, the ground color is often well defined, and then the most handsome effects are produced. If the ground color is distinctly reddish, the surface markings are usually not nearly so numerous, and because of their sparse distribution, the egg then approaches in appearance that of the spruce grouse (*Canachites canadensis*). The markings show almost uniformly blackish brown with a reddish suggestion, yet where the pigment has been scratched very thin, the color is often as light as "maroon" and even "garnet brown," while the deeper colors are "warm blackish brown" and "blackish brown (1)."

The measurements of 250 eggs, in the United States National Museum, average 43 by 31 millimeters; the eggs showing the four extremes measure **48** by 31.5, 46.5 by **34**, **39** by 30, and 39.5 by **28** millimeters.

Young.—Mr. Conover says in his notes:

Newly hatched young were first found on June 22. The incubation period seems to last about 22 days. A nest found on June 2 with 11 eggs had 12 on June 3, and on being visited on June 25, it was found to be empty. The chicks are very precocious. One day a hen was flushed from a nest containing two eggs and eight youngsters still damp. Hardly had she left when every downy chick scrambled weakly from the nest and attempted to hide in the grass. The minute they were replaced, out they would go again, until finally they became tired out and stayed in the nest. Toward the end of June broods were constantly encountered about the tundra. Both parents were always with them and the cock was especially combative, although discretion always got the better part of valor. The young after running a few feet would suddenly disappear, whereupon the hen would join the male in threats and attempt to lead one off. It was amusing to imitate the peeping of a chick and watch the cock go into a frenzy, ruffling himself up, making short dashes here and there, and in unmistakable language telling you just what he was going to do if you didn't get away from his children. After a few minutes of this, both birds would be worn out and would retire a short way to watch for your next move. By July 22 the young were about a third grown and had begun to shed their first brown primaries and grow their new white ones. The adults were then in the midst of shedding their toenails.

Dixon (1927) noticed that the family party traveled as follows:

First came two or three chicks in the thick grass, then the mother surrounded by the other chicks; the cock sometimes led and at other times brought up the rear. I timed them and found that they covered a lineal distance of 45 feet in five minutes. Following this there came a period of rest of five minutes, during which the mother hovered her brood of young. We never saw the cock hover the young; but when one of the chicks became entangled in a network of twigs he was right there and helped it get free. By noon the ptarmigan family had wandered out in the low bushes 100 yards from where they had hatched. The chicks were now nearly 24 hours old, and all of them were strong and lusty, each able to run about with agility and to secure food for itself. At Copper Mountain, about 4 o'clock in the afternoon of July 12, a family of Willow Ptarmigan came feeding along through the dwarf willows near camp. There were six young about the size of quail. The cock kept a lookout for enemies from elevated positions while the hen herded the young along through the willows. The hen kept up a running conversation with the young as did also the cock. This liaison note was a loud *ke-ouck*, repeated at intervals of from five to ten seconds. The cock's call was somewhat coarser than that of the hen. I had difficulty in hearing the thin peeping of the chicks at a distance of fifty feet, but it served to keep them together. The young were very active, jumping up into the willows and catching insects over a foot off the ground.

Plumages.—In natal down the young willow ptarmigan has a large patch of "burnt sienna," bordered with black, on the center of the crown and occiput; the rest of the head and the underparts are "colonial buff"; the upper parts are variegated with "colonial buff," black, and "cinnamon-rufous," the last mainly in the center of the back and rump, bordered by broad black bands; a small spot on the lores, a larger auricular spot, and a narrow line behind the eye are black.

The juvenal plumage comes in first on the wings, then on the scapulars and back, while the chick is very small; the juvenal remiges are "sepia," bordered with "cinnamon-buff" or buffy white, and their coverts are tipped with buff; the last of the down to disappear is on the chin, neck, and belly. The juvenal plumage is at its height before the young bird is half grown, at which time the postjuvenal molt begins, early in August. In full juvenal plumage the feathers of the mantle are black, edged, notched, or barred with "ochraceous-tawny," and with triangular white tips; the breast and flanks are "ochraceous-tawny," heavily barred or spotted with black or dusky; the belly is buffy or grayish white; the remiges are as stated above, except the outer two, which are the last to appear and are white; the rectrices are black, edged, spotted, or barred with "ochraceous-tawny."

When the young bird is nearly half grown the postjuvenal molt begins by shedding the juvenal remiges, which are replaced by white ones. This is a more or less incomplete molt, involving most of the wings (except the two pairs of outer white primaries), the tail, and a varying extent of the body plumage. In this intermediate, or preliminary, winter plumage, the sexes are still practically alike. There may be only a few scattered, reddish brown, finely vermiculated, or mottled feathers, or the renewal may be nearly complete; the belly, flanks, and legs become white.

This intermediate plumage is worn for a very short time while the young bird is getting its growth; for, during the last of August and through September, a supplementary, partial molt takes place, which completes the change into the white winter plumage. Young birds can be distinguished during their first winter by the outer primaries, which are more worn and often speckled at the tips.

Both young and old birds have a partial prenuptial molt in spring, the date varying greatly with the latitude and season. The reddish spring plumage begins to appear on the head and neck of the male at about the time that bare ground begins to appear in its summer home, from the first to the middle of April in Alaska. The females molt a month or more later. Dr. Jonathan Dwight (1900) says:

Females may now be distinguished with certainty from males for the first time by plumage characters, the barring being coarser and extending to the head, throat, and breast, the feathers of which in the male are reddish brown, chiefly with narrow, dusky terminal bands, and often tipped, on the chin especially, with white. It should be observed that parti-colored feathers basally or terminally white, may be assumed at this moult on the internal borders of the sternal band, just as in juvenal dress, the abdominal wedge, flanks, legs, and feet, retaining as a rule the white feathers of the winter plumage. The white remiges and their coverts are always retained and often much of the rest of the wing plumage, the median rows of coverts being the ones renewed if any are. The tail-coverts may be renewed, but the fourteen black rectrices remain.

Dwight says of the next molt:

Even before the nuptial dress is fully acquired the postnuptial moult sets in, beginning a little prior to the postjuvenal and resulting in an intermediate plumage partly white and partly reddish brown which may hardly be told from that of young birds at the same season. It should be observed that the moult of the remiges now includes the two distal primaries which are retained in young birds. Adults, however, seem to be somewhat grayer with finer mottling or vermiculation, the throat being of a deeper red-brown with less barring than that of young birds. Practically young and old, both males and females, are all indistinguishable except by inconstant differences when clothed by the preliminary winter dress, but their age and sex may usually be told by the left over tell-tale feathers of an earlier plumage.

A supplementary molt early in fall, September and October in Alaska, completes the change into adult winter plumage. Females are indistinguishable from first winter birds, the feathers of the crown being basally gray, whereas in the adult male these are basally black. Winter adults in high plumage often have a decidedly rosy tint, which soon fades in the dry skin.

Food.—During summer ptarmigan feed on the tender leaves and flower buds of the willows, birches, and alders, with a fair percentage of berries, such as mountain cranberry, crowberry, blueberries, arbutus, and kinnikinnick. They also eat what insects they can find. Turner (1886) writes:

During the winter these birds subsist on the past year's twigs of the willow and alder or other bushes. I have cut open the crops of many of these winter-killed birds and found them to contain only pieces of twigs about one-third of an inch long, or just about the width of the gape of the posterior, horny part of the bill, as though this has been the means of measurement in cutting them off. The flesh at this time is dry and of a peculiar taste. In the spring the Ptarmigans congregate in great numbers on the willow-bushes and eat the tender, swelling buds. The flesh then acquires a bitter, but not unpleasant, taste. As open weather advances they find berries that have remained frozen the entire winter, and tender grass shoots, and later, insects. The young are insectivorous to a great degree in their youngest days. They consume great numbers of spiders that are to be found on the warm hillsides.

Dixon (1927) sent the stomach of a 5-day-old chick to Washington for examination; it contained 17 yellow caterpillars, 1 spider, 15

Thysanura, and 15 other insects and larvae, but no vegetable matter. He says:

The eight chicks foraged in a loose flock covering an area about five feet wide and six feet long. They pursue small insects and mosquitoes which they run down or reach up for and pick off the grass. I watched one chick catch a cranefly and after hammering and pecking at it awhile he concluded that it was too tough, gave it up as a bad job, and left it.

Dr. Joseph Grinnell (1900) writes:

Occasionally a few spruce needles were also found. The gizzards of the birds obtained, invariably contained a quantity of small polished pieces of clear quartz, this probably being the hardest substance for the purpose obtainable by the birds. A bare place on a sand-bar in the river, kept clear of snow by the wind, was wont to be frequently visited by the ptarmigan and I have seen them scratching over the gravel in such places, even in the coldest mid-winter weather.

In addition to many of the above items, E. A. Preble (1908) lists mushrooms, tops and seeds of grasses, and leaves, seeds, and berries of various other plants.

Behavior.—Ptarmigan rise from the ground with a loud whir of wings and harsh cackling notes. When scattered about feeding among the willows they do not all rise at once but jump up singly or a few at a time, and there are generally a few laggards. When well under way their flight is strong, swift, direct, and often prolonged for a long distance. They alight readily on trees and bushes, where they are skillful at balancing. In winter it is difficult to see them on the snow-laden branches, where they look much like balls of snow. On bright, sunny days their shadows show up plainly on the snow-covered ground; but on dull, hazy days they are very hard to see, unless the black bill or eye is in motion. The black tail is entirely concealed, except in flight, when it serves as a very good direction mark for other members of the flock to follow. They seem to understand the value of protective coloration, for, if the ground remains bare after the white plumage is assumed, they are very shy; but, after the snow comes, they become very tame. Grace A. Hill (1922) calls attention to the fact that, while the ptarmigan are molting into the white winter plumage, they frequent the open tundra at the time the cottongrass (*Eriophorum polystachion*) is bearing its white cottony plumes, which aid the birds in their protective coloration.

Ptarmigan are evidently monogamous and make quite devoted couples. Dixon (1927) tells of a male, perched on the top of a spruce, standing guard over his mate while she was feeding. Mr. Dixon writes:

The bird gave a couple of warning calls as I approached the tree, and then it dawned on me that he was probably standing guard while his nesting mate fed. So I hunted around; and sure enough, I found the female ptarmigan feeding in some dwarf willows about twenty feet from where I stood. As soon as I started after the female the male ptarmigan flew down from the tree top and ran off ahead of me, trying in various ways to decoy me away from his mate.

I had been told of an instance where a cock willow ptarmigan had attacked and routed a large grizzly bear that happened to stumble upon his nest. But even after seeing the ptarmigan drive off the gulls I did not fully appreciate the furiousness of the attack until June 23, when I came across an old hen ptarmigan with her brood of small young which were just able to fly. I rushed after the young, trying to catch one. Just as I was about to grab a chick, a willow bush in front of me exploded and the cock ptarmigan flew directly into my face, knocking my glasses to one side as he slapped my face with his beating wings. He then dropped to the ground, but instead of retreating flew directly into my face again; but this time I was ready for him and caught him with my bare hand when he became mixed up with my mosquito head net. The bird then tried to bite and to flap his way to freedom. As I started off with the cock under my arm the hen ptarmigan left her young and came rushing at me and then crawled feebly about at my feet as though in mortal agony. When I started away she rushed frantically about flapping my heels with her wings at every step. Every time she rushed at me she hissed. When the male found he could not escape he uttered a few croaking notes and the hen left me at once and went back to her chicks.

Voice.—Dixon (1927) says that the warning cackle of the male "sounds like running a nail over a stiff comb." He mentions three notes of the female: A harsh *ke-ouk, ke-ouk* is a warning of danger; a soft purring *keer-er-eerk* is a hush-a-bye, when hovering young; and a clucking *cuck-cuck* is a note used to call the chicks to her.

The downy chicks give a soft *cheep-cheep-cheep* when in distress or when separated.

Enemies.—Gulls, jaegers, hawks, owls, foxes, wolverenes, and other predatory birds and animals levy heavy toll on the ptarmigan and their eggs and young. Ptarmigan are so plentiful that they furnish the principal food supply for many of these creatures, as well as for the human natives. Dixon (1927) writes:

After the young ptarmigan are out of the shell they are menaced by Black-billed Magpies as well as by the foxes. Thus on June 24 a family of four young and two adult magpies was found systematically working the willows in the Savage River bottom for ptarmigan chicks. When these magpies located a pair of adult ptarmigan they would retire stealthily and hide in the willows near by, until the ptarmigan chicks began to run about. Then the magpies swooped down and grabbed the chicks before they could hide, and then carried them off and ate them. A cock ptarmigan that I watched put one magpie to flight, but where there were six and in another case *nine* magpies working together against two adult ptarmigan the odds were overwhelming. As a result of this persecution by the magpies we found that by July 10 many families of young ptarmigan had been reduced to only one or two individuals. Gyrfalcons also levy continuous toll on ptarmigan; and since these large falcons are relatively numerous in the Mount McKinley district, the aggregate number of ptarmigan killed by them is considerable. It is thus easy to see why the hen ptarmigan lays from 6 to 12 eggs. If only one or two eggs were laid each season the species would soon become extinct.

Game.—Were its haunts not so far removed from the centers of civilization, the willow ptarmigan would be a popular game bird. Our experience with the Aleutian ptarmigan taught us that these birds possess excellent game qualities. Except during the breeding season, when they are very tame, they are wild and wary enough to give good sport. Their flight is strong and swift and sufficiently prolonged to give one all the exercise he wants. Their thick winter plumage is somewhat shot resisting, so that they have to be hard hit, with a close-shooting gun; it often requires considerable chasing to get within effective range. Edwyn Sandys (1904) gives a thrilling account of a winter ptarmigan hunt, with its hardships and dangers. The flesh of the old birds in winter is apt to be dark and dry and to have a bitter flavor, as a result of a steady diet of willow buds, but that of young birds in fall, fed on fresh foliage, berries, and insects, is lighter colored and very good to eat.

Dr. E. W. Nelson (1887) writes:

Among the Alaskan natives, both Eskimo and Indian, especially those in the northern two-thirds of the Territory, this bird is one of the most important sources of food supply, and through the entire winter it is snared and shot in great abundance, and many times it is the only defense they possess against the ever-recurring periods of scarcity and famine.

The Eskimo of the Kaviak Peninsula have a curious way of taking advantage of the peculiarities of this bird in their migrating season. Taking a long and medium fine-meshed fishing-net they spread it by fastening cross-pieces to it at certain distances; then taking their places just at sunset in early November or the last of October, on a low open valley or "swale," extending north and south, they stretch the net across the middle of this highway, with a man and sometimes two at each cross-piece, while the women and children conceal themselves behind the neighboring clumps of bushes. As twilight advances the net is raised and held upright. Ere long the flocks of Ptarmigan are seen approaching skimming along close to the snow-covered earth in the dim twilight, and a moment later, as the first birds come in contact with the obstacle, the men press the net down upon the snow sometimes securing 50 to 60 birds. While the men throw themselves upon the net and hold it down, the women and children rush forward and kill the birds by wringing their necks or by biting their heads. On some evenings several flocks are thus intercepted, and the party of natives return to their houses heavily laden with spoils. In winter the birds are snared in their haunts by placing fine nooses attached to low bushes close to the ground. Sometimes small brush fences are built with snares at the passage-ways purposely left open. In spring, as the snow begins to leave the mossy knolls here and there, the natives shoot a male bird and stuff it roughly with straw, and, mounting it on a small stake, place this effigy upon one of the bare knolls in a conspicuous position; then they surround it with a fine sinew net held in place by slender stakes. The hunter then conceals himself close by and imitates the challenge cry of the male. All around can be heard the loud cries of the pugnacious birds, and attracted by the decoy notes of the native some of them are almost certain to bestow their attention upon the decoy; they approach swiftly, and either fly directly at their supposed rival or alight and run at him in blind rage. In either case their jealousy is fatal, as they are at once hopelessly entangled in the net of the hunter, who disposes of them and repeats the maneuver indefinitely, generally returning home well laden.

Ptarmigan are subject to great fluctuation in numbers from year to year, and during periods of scarcity they may be nearly or quite absent from regions where they were once abundant. Alfred M. Bailey (1926) says:

In 1919, ptarmigan were very scarce throughout the territory; in December, on a trip to within a short distance of the source of the Copper River, we saw but one bird. In 1920 the birds began to return, and in 1921 they were reported abundant at all points where they usually occur. I am unable to explain the cause of this scarcity at intervals, for, so far as I know, no disease has been reported among them.

Mr. Warburton, who spent the summer of 1929 about the mouth of the Yukon, writes to me:

The scarcity of these birds was a great disappointment. I had expected to find them as plentiful as I had seen them about Nome and Teller, Alaska, in the summers of 1924 and 1925. At that time on the Seward Peninsula they were most plentiful, together with many rock ptarmigan. This year the exact reverse was true, at least about the Yukon mouth.

Fall.—Turner (1886) says that the ptarmigan migrate to the interior late in the fall. He writes:

When the snow has pretty well covered the ground in late November the Ptarmigans assemble in immense flocks, often numbering thousands. I was once out on the higher grounds just south of the Crooked Canal. I ascended a slight hill and came, unexpectedly, on one of these large flocks that covered acres of ground. I was among them before either was aware of it. They flew, and made both the air and earth tremble. There must have been over five thousand birds in this one flock. They flew beyond a neighboring hill-range. Approaching night and a heavy snow falling prevented me from following them.

Winter.—When the snow covers the tundra so deeply that no food or shelter can be found there, the ptarmigan are forced to migrate into the interior valleys, river bottoms, and creek beds, where they can find shelter among the willows, alders, or spruces and can feed on the bugs and twigs of these trees or such berries and fruit as still remain above the snow. They often congregate at that season, and, particularly when migrating, in enormous numbers. Mr. Brandt says that, where the railroad crosses the Continental Divide in Alaska, he saw large flocks arising "like snow drifts in motion, alongside the snorting engines, and whirling away over the great white hills."

In winter the ptarmigan's feet are thickly covered with long, hairlike feathers, resembling the foot of a hare, which serve as snowshoes and enable the bird to walk on soft snow. Sandys (1904) writes:

And Nature, as if realizing the perils of the ptarmigan asleep, has taught it to plunge beneath the cold drifts to escape the cold, and to *fly at*, not *walk to*, the chosen drift, so that there will be no telltale trail for some keen nose to follow to the sleeping-place. And this the bird invariably does, going at speed and butting its way into the snow, leaving never a print to betray its retreat, from which it *flies* in the morning. The game of life and death is interestingly played up North—where the weak white snow-shoers are ever hiding from the strong white snow-shoers forever searching over a field of baffling ice-bound white.

Doctor Nelson (1887) describes another method of roosting:

On November 25, 1877, they were numerous, in large and small flocks, along the bushy gullies and hill slopes on the shore of Norton Sound, but were shy. In many places where they had stopped the night before, their sleeping-place was well marked. In each instance they had occupied a small clear spot in the midst of a dense thicket, and in no case had the birds approached on foot, but had flown in over the top and plumped down into the soft snow, where they had remained during the night, each bird thus making a mold of itself in the snow. In some instances there were fifteen to twenty of these molds in the snow in an area of a few feet. In leaving their stopping-place the birds arose and flew directly from their "forms," as was shown by the marks of the wings on each side as they touched the snow in rising, so there were no tell-tale tracks to or from these places; the open places were undoubtedly chosen to allow the birds an unobstructed escape in case they were surprised by prowling foxes, which hunt these thickets for food.

DISTRIBUTION

Range.—The range of the willow ptarmigan as a species is circumpolar, extending in the Old World from Scandinavia, Russia, and Siberia south to Turkestan, and in North America from Greenland, Newfoundland, the Arctic Archipelago, and Alaska south through Canada. The species is casual in winter in the Northern United States.

Breeding range.—The North American breeding range extends **north** to Alaska (Cape Lisburne, Wainwright, Point Barrow, Smith Bay, Delta of Colville River, Camden Bay, Humphrey Point, and Demarcation Point); Franklin (Bay of Mercy, Port Kennedy, Felix Harbor, and Igloolik); northern Quebec (Fort Chimo); and Labrador (Okkak). **East** to Labrador (Hamilton Inlet) and Newfoundland (Raleigh and St. Johns). **South** to Newfoundland (St. Johns and south coast); west-central Quebec (Carey Island); northern Ontario (40 miles south of Cape Henrietta Maria); northern Manitoba (50 miles north of York Factory and Fort Churchill); southern Mackenzie (Artillery Lake and Fort Resolution); central British Columbia (Moose Pass, Icha Mountains, and Ninemile Mountain); and southeastern Alaska (San Juan Island). **West** to Alaska (San Juan Island, Kruzof Island, Glacier Bay, Nushagak, Nelson Island, Igiak Bay, Askinuk Range, Pastolik, St. Michael, Nome, Mint River, Cape Blossom, and Cape Lisburne).

Winter range.—Although willow ptarmigan perform a very definite migration, individual birds are frequently found in winter almost at the northern limits of the breeding range. At this season they have been recorded north to Alaska (Nunivak Island, Nulato, Kutuk River, and Miller Creek); Mackenzie (mouth of the Dease River, Fort Rae, and Fort Reliance); northern Manitoba (Fort du Brochet); Franklin (Igloolik); Quebec (Great Whale River, and Piashti); and Newfoundland. Normally they are found south to southern Quebec (Lake St. John); central Ontario (east of Cochrane and Martin Falls); central Manitoba (Norway House and Grand Rapids); Saskatchewan (Cumberland House and Fort Carleton); and southwestern Mackenzie (Fort Simpson).

Migration.—Though the bulk of the willow ptarmigan move south each fall or winter, and return north in spring, the movement has no regularity and is directly correlated with the food supply and, in consequence, with the fall of snow.

First fall arrivals are: Quebec, Lake Mistassini, October 25, and Godbout, November 2; Ontario, Martin Falls, about October 20; Manitoba, Lac du Brochet, November 4, Grand Rapids, November 12, and Winnipeg, January 12; Mackenzie, Fort Rae, October 1; and Alberta, Fort Chippewyan, October 11.

In spring, late departures for the north are: Quebec, Lake Mistassini, about May 1; Ontario, Cochrane, March 20; Manitoba, Winnipeg, March 21; and southwestern Mackenzie, Fort Simpson, March 12. They have been observed to arrive at Fort Resolution, Mackenzie, on June 28, and at Demarcation Point, Alaska, as early as April 6.

Casual records.—The willow ptarmigan has been reported a few times in the Northern United States. Among these records are: Maine, one at Kenduskeag, April 23, 1892; New York, one at Watson, May 22, 1876; Wisconsin, two taken near Racine in December, 1846; Minnesota, one taken at Sandy Island, Lake of the Woods, April 20, 1914; and Montana, three taken in Glacier National Park in the winter of 1913–14. Prof. W. B. Barrows believed that they occasionally occurred on Keweenaw Point, Mich.

The range as above described is for the entire species, which in the 1931 edition of the American Ornithologists' Union Check List is subdivided into five subspecies. According to this authority the subspecies *albus* is found in North America from the eastern Aleutian Islands, northern Mackenzie, northern Banks Island, and central Greenland south to central British Columbia, central Alberta, central Ontario, and southern Quebec. The Ungava ptarmigan (*L. l. ungavus*) occurs from northern Quebec west probably to the eastern shore of Hudson Bay. Allen's ptarmigan (*L. l. alleni*) is confined to Newfoundland. The Alaska ptarmigan (*L. l. alascensis*) is found on the Alaskan mainland (except the southeastern coast), northern Yukon, and eastward for a distance not yet determined. Alexander's ptarmigan (*L. l. alexandrae*) occurs on Baranof Island, Alaska, and adjacent islands, west to the Shumagin Islands, and south to Porcher Island. This race also may occupy a narrow strip on the mainland from Glacier Bay to central British Columbia.

Egg dates.—Northern Alaska: 68 records, May 25 to July 10; 34 records, June 6 to 25. Arctic Canada: 37 records, June 2 to July 7; 19 records, June 10 to 21.

Labrador Peninsula: 18 records, June 1 to 30; 9 records, June 6 to 23.

Newfoundland (*alleni*): 11 records, May 12 to June 30; 6 records, June 8 to 12.

Southern Alaska and British Columbia (*alexandrae*): 3 records, May 28, June 25 and 26.

LAGOPUS LAGOPUS ALLENI Stejneger

ALLEN'S PTARMIGAN

HABITS

The willow ptarmigan of Newfoundland was originally described by Dr. Leonhard Stejneger (1884) as similar to the common willow ptarmigan, "but distinguished by having the shafts of both primaries and secondaries black, and by having the wing-feathers, even some of the coverts, marked and mottled with blackish." He examined only 14 specimens in all, all of which presented the above characters. In a large series (I have 75 in my own collection and have examined a great many in other collections) these characters appear to be none too constant in Newfoundland birds and to crop out occasionally in willow ptarmigan elsewhere, with a great range of individual variation. In this connection it is interesting to note what Harry S. Swarth (1924) has to say about the willow ptarmigan of southern Alaska:

> It is of interest to note in *alexandrae* the frequent presence of black shafts on the primaries, sometimes on secondaries and greater coverts. This character has been considered an important feature of the Newfoundland subspecies (*L. l. alleni*), but obviously it can not be used as a feature characteristic of that race alone. In an immature female from Prince of Wales Island, which has acquired the winter flight feathers, not only are primaries and secondaries with distinct black shafts, but there are large, tear-shaped spots of black near the tips of all the primaries and most of the secondaries. Furthermore, the primaries have a black "freckling" over much of their surface, and the greater coverts are also marked with black, though to a lesser degree.

Dr. Hart Merriam (1885) has called attention to the great variation in the extent of black in a large series of wings sent to him by Napoleon A. Comeau from Godbout, Quebec. Perhaps the range of *alleni* should be extended to include eastern Quebec.

Allen's ptarmigan are very abundant and widely distributed in Newfoundland, on the upland tundra in the central and northern parts of the island and on the mountains. J. R. Whitaker tells me that from about the middle of April to September 20 they live on the high tundras and rocky ridges; he has never seen a bird on the lower ground during summer. F. S. Hersey, who went to Newfoundland in September and October, 1913, to collect ptarmigan for me, found these birds more abundant than he ever found willow ptarmigan anywhere in northern Alaska. He found them on the hills a few miles back from Cape Ray, and some were seen in the Lewis Hills near the west coast; but they were most numerous about Gafftopsail in the interior of the island. This is a rough, barren, open region, tending to run into low, rocky ridges, separated by lower marshy areas, carpeted with thick mosses, often knee-deep, and dotted with small ponds. There is a little cover here, although a few sheltered spots may have low shrubs and nearly prostrate blueberry bushes, and in a few places there are small areas of exceedingly tough, thick, dwarf spruces.

Courtship.—Referring to the courtship of this ptarmigan, Mr. Whitaker says in his notes:

The cock bird nearly always mounts a rock to utter his challenge; having gained this vantage point he looks all around, then with head erect and breast expanded bleats out his notes; they will start a little before sunrise and continue for a short time and are seldom heard during the day. After the young are hatched crowing appears to cease entirely. The crow of this bird is very like the British red grouse; it is quite easy to imitate, but to me, very difficult to write. The first part sounds like the low trembling bleat of a nanny goat, or, if one can imagine it, a very coarse drum of a snipe; there are usually five quickly uttered tremulous notes run off *er-er-er-er*, followed by three or four *gobeck-gobeck-gobeck;* these latter may be copied perfectly by partly closing your throat and uttering *gobeck* harshly through the nose.

Nesting.—Whitaker says that

nesting begins in May and the young are usually full grown by the end of August. The nest is usually placed at the base of some bowlder and is well hidden; the hen bird sits very close indeed. Some years ago a nest was found close to a watershoot by the railway; several trains a week stopped here for water; the train hands used to take passengers to see the bird sitting, which did not upset her in the least.

A set of eight fresh eggs, collected for me near Gafftopsail on June 16, 1912, was taken from a hollow in the tundra moss, under a little bush, but in plain sight, between two hummocks. Mr. Hersey found a nest from which the young had hatched, which he describes in his notes, as follows:

The site was on the side of a ridge some 20 feet above the marshy ground at its base. A few bushes grew at this point, and under the overhanging branch of one of these the nest had been made. Even at this season the hollow that had been the nest was well defined. There was no indication of lining beyond the accumulation of fine dry leaves, bits of sticks, moss, and other vegetation that made up the general ground cover on the whole ridge. The eggshells from which the young had hatched were still in the nest.

Eggs.—The eggs of Allen's ptarmigan are indistinguishable from those of its mainland relative. The measurements of 54 eggs average 42.5 by 33 millimeters; the eggs showing the four extremes measure **45.6** by 32.3, 44.1 by **37.6**, **40.1** by 30.2, and 43.4 by **28.7** millimeters.

Plumages.—The molts and plumages of this race are similar to those of the willow ptarmigan. It is interesting to note, however, that many of the wings sent by Mr. Comeau to Doctor Merriam (1885) taken from birds killed in November, were "deeply tinged with a delicate and very beautiful shade of rose-pink which is more pronounced than in a freshly killed Roseate Tern." Mr. Whitaker has also noted this, as he says in his notes: "They are in full white winter plumage by the end of October; I say white, but there is a beautiful faint pink flush; I first noticed this when having shot several one winter day and laid them on a block of pale blue ice, this showed up the pink to advantage; the pink quickly fades after death."

Food.—In his notes on food, Whitaker writes:

As the snow melts off the high ground many berries that have lain snug all winter provide plenty of food; after these are shriveled up by the sun and before the new crop matures, the birds feed on the tips of a low-growing plant, which looks like a very dwarf heather. In August the blueberries and partridgeberries (*Mitchella repens*) ripen and on these they feed almost exclusively. When the snow begins to pile up on the highlands during October the bulk of the birds move down to lower levels and begin feeding principally on buds of scrub birch; they also eat buds of blueberry and pussy willow.

The birds shot by Mr. Hersey had their crops filled with blueberries and their leaves, the small green berries from a dwarf evergreen resembling a cedar, or with small white seeds.

Behavior.—Mr. Whitaker says in his notes:

These ptarmigan are very tame in the wilder districts, where they are not disturbed or shot, and seem loath to take wing; they will run in front of you stopping every few yards and spread their tails with a quick jerky motion; they often utter a suppressed little grunting note, possibly a protest against your intrusion in their midst.

Mr. Hersey's notes contain the following observations:

> The birds were usually found in parties of 3 or 4 to as many as 11 and 13 and, as many were in a plumage that indicated they were birds of the year, it was assumed that these flocks usually represented a pair and their grown-up young. They were very erratic in the matter of taking flight, on some occasions not flying until nearly stepped on and again taking wing before a person was within 100 yards of them. As a general thing they were wildest on sharp frosty days or when rather windy and would lie closest when the day was warm and calm. I do not believe the ptarmigan range over any large area at this season. Birds were seen day after day on the same ridges and were believed to be the same flocks. On bright clear days a flock often spends several hours crouched on the ground in the sunlight, particularly in rocky places, where one or two males will take up a position on the rocks. These sentinels perform their duty poorly, as it is usually rather easy to approach and shoot such birds and then secure one or more from the flock with the remaining barrel as the birds take wing. At times the birds on the ground will not flush, even after one or more has been shot. I remember one very warm day when I came upon three birds crouched on the ground a few feet apart. I was unable to flush any of them and was obliged to back away and shoot them on the ground one at a time, the remaining birds paying no attention to the sound of the discharge. When flushed they are frequently silent, but some birds give voice to a harsh *cack-cack-cack-cack-cack*. On foggy mornings birds, presumably males, may be heard "crowing," at least the notes sound very similar to the spring call of the male. They may be a signal or flock call.

Game.—Allen's ptarmigan, known also as "willow grouse" or "partridge," is by far the most important game bird in Newfoundland. It is jealously guarded during the close season, only to be shot in enormous numbers during the open season. Many parties of gunners go out every fall from St. Johns, as well as visiting sportsmen from Canada and the United States, camp along the line of the railway, and kill hundreds of the birds. Fortunately for the birds, there are immense tracts of uninhabited country, inaccessible by branch lines or by roads, where they are not disturbed, so that they do not seem to be in any immediate danger of extermination.

Winter.—Mr. Whitaker says that "quite a number perish during the winter; they make a shallow scratching in the snow in which to roost and are frequently buried by drifts and are imprisoned; I have often found them dead in the spring when the snows are melting, and once saw seven within a few feet of each other that had met this fate."

LAGOPUS LAGOPUS ALEXANDRAE Grinnell

ALEXANDER'S PTARMIGAN

HABITS

The willow ptarmigan of the humid coast region of southern Alaska and northern British Columbia is darker and more richly colored than birds from northern Alaska and Labrador, with a smaller and narrower bill, the general color above being deep, rich chestnut thickly vermiculated with black. The most typical birds of this race are found on the islands along the coast of southern Alaska; the type came from Baranof Island. It was described by Dr. Joseph Grinnell (1909) and was named in honor of Miss Annie M. Alexander. Alfred M. Bailey (1927) first saw them on Willoughby Island in Glacier Bay, of which he says:

This island is about five miles in length, with scant vegetation, other than alder and stunted spruce. The knobs are devoid of plant life, and it is only on the terraces that soil can hold. The southern and eastern slopes are rather densely clothed with alders, however, and form ideal cover for Ptarmigan. Seven other birds were seen by Young, one of which was still in the winter white. The pair collected were breeding birds.

He suggests a good reason why they should prefer to live on the islands:

The lack of predatory animals was very noticeable, for with the exception of two Eagles, none was noted, and no signs of predatory animals were seen on the outer Beardslee Island, although those nearer shore must have some carnivorous mammals. We saw a fox in June and August on the mainland shore, and wolves are abundant. It is rather apparent then that the birds are free from molestation, in direct contrast to their life upon the mountain slopes of the mainland. Ravaged by fur-bearers, it is possible the Ptarmigan first used the islands for protection, and having found both food and comparative safety, have continued to live under such conditions. On the other hand, the islands might be considered as being alpine in nature, with timber-line conditions, as the spruce are small and willow and alder predominate, with the characteristic profusion of small growths. The soil is scant, the glacial sands and moraine débris being exposed, while the windward shores of the outer Beardslee are precipitous; the glacial winds sweep down channel, icebergs line the shores, and taking all into consideration, the region is probably the coldest of southeastern Alaska.

Harry S. Swarth (1924) noted that the birds he collected in the Skeena River region of northern British Columbia "are intermediate in color between *lagopus* of the interior and *alexandrae* from the islands; the average is nearer *alexandrae*." He says of their haunts there:

Ptarmigan are said to occur occasionally in the lowlands of the Hazelton region in midwinter, but during most of the year they are restricted to the Alpine-Arctic mountain tops. We found them in limited numbers on the timberless summit of Nine-mile Mountain. There are miles of open country on the two converging ridges that form the top of this mountain, barren of trees save for occasional thickets of dwarfed or prostrate Alpine conifers, and here, at long intervals, we encountered ptarmigan.

Nesting.—We know very little about the nesting habits of Alexander's Ptarmigan. According to Doctor Grinnell (1909), Joseph Dixon "records that at Coppermine Cove, Glacier Bay, July 10 to 20, the feathers and bones of a ptarmigan were found near a nest of broken eggs on the summit of the mountain, 2,100 feet. The nest was under a stunted hemlock. All the feathers were white, so the ptarmigan must have laid early."

There is a set of 11 eggs in Col. John E. Thayer's collection, taken by John Koren on Kodiak Island, Alaska, on June 25, 1911. The eggs were fresh and "were placed on a bed of moss in a two-foot groove in an elevated part of the tundra"; both parents were present. The eggs are not distinguishable from those of the willow ptarmigan. The measurements of 20 eggs average 42.8 by 30.9 millimeters; the eggs showing the four extremes measure **45.9** by 31.5, 43.6 by **32**, **41.5** by 30.8, and 42 by **30.2** millimeters.

Young.—Swarth (1924) writes:

In all, ten broods of willow and rock ptarmigan were encountered (the species were not always to be differentiated) and about five or six single birds in addition. The broods ranged from three to twelve in number; the aggregate of young birds seen was about fifty. The chicks grew rapidly. Some seen on July 25, and a day or two later, were down-covered and unable to fly. At that time they were accompanied by the female parent only, and the male birds were flushed separately. By August 10 the young ptarmigan were the size of quail and larger, and were strong on the wing. The old males were then associated with the families. In some of the larger broods seen the difference in size among the young was so marked as to suggest the junction of two families. It might happen that upon the death of a hen her offspring would seek the companionship of another family.

Food.—These ptarmigan probably feed on as varied a diet as other ptarmigan, but the following note by Alfred M. Bailey (1927) is all that I can find in print on the subject:

A flock was flushed from a bed of wild strawberries, at an altitude of scarcely thirty feet, when I had expected to find them above timber line. There were several pairs of adults, as well as many young, and a good series was taken. An examination of crop contents proved the birds had been feeding entirely on strawberries and pea-vine, no alder or willow buds being found. After finding ptarmigan in such a low altitude, I searched all the points along the east mainland shore, and did not fail once to find them, where there were berries. It was noticeable they preferred the points where they could feed close to a fringe of alder.

Behavior.—Mr. Bailey (1927) says also:

I returned to Glacier Bay again from October 10–14, and observed the Ptarmigan under still different conditions. A stop was first made on the outer Beardslee, upon which the young bird had been taken. I was interested to see if the Ptarmigan had left the island, as it was now drab and dry looking, and the birds were assuming their winter's white. I found them very abundant, and over forty were flushed from the dense alders, and a few taken. They were extremely wild, as they should have been, for their changing plumage—entirely white below with many white feathers in head and neck, made them extremely conspicuous among the leafless alders. I had little chance to observe them, due to their wildness. Stops were made at two other points, not visited in August, and ptarmigan were noted in both places. At Sandy Cove, October 14, I collected another series from the point they were first found in August. These were as conspicuous as the others, but when the flock was broken up, the birds did not flush so readily, and I had a chance to study them at close range. They ran over the ground the same as in August, evidently believing themselves inconspicuous, and when closely pressed, crouched quail-like and depended upon their "protective coloration." When the birds became scattered several climbed into alders, about a foot from the ground, and sat hunched in some convenient crotch, where they were more evident than ever. I was surprised to find strawberries still abundant here, hidden under the mosses in shady places, and each of the specimens taken had its crop full.

Doctor Grinnell (1909) quotes from Chase Littlejohn's notebook as follows:

While searching for eggs of the glaucous-winged gull on one of the small islands on the east side of Glacier Bay on July 14, I suddenly came upon a flock of ptarmigan in a little opening among some spruce, hemlock, and alders, which covered the ground in dense masses in spots; the remainder of the area supported a thick growth of grass interspersed with patches of moss and low-growing flowering plants. There were about eighteen birds all told, young and old, and as near as I could determine there were four or five old birds present. They would not fly after they were first flushed, but kept dodging about on the ground, sheltered by the thick cover; several times I saw them, but so near that a shot would have ruined them as specimens.

LAGOPUS LAGOPUS UNGAVUS Riley

UNGAVA PTARMIGAN

HABITS

The *ungavus* race of the willow ptarmigan was described by J. H. Riley (1911) as "like *Lagopus lagopus albus*, but with a heavier bill," based on a series of 20 birds collected at Fort Chimo, Ungava, in the northwestern portion of the Labrador Peninsula. He gives as its probable range "from Ungava and probably the eastern shore of Hudson Bay south." Birds that I have seen from the eastern and southern parts of the Labrador Peninsula do not seem to have appreciably large bills. If this form is worthy of recognition at all, which I very much doubt, it will probably prove to be a northern race and perhaps identical with *alascensis* Swarth, about which I have expressed my views elsewhere. But, pending further investigation, we may as well consider all the willow ptarmigan of the Labrador Peninsula as referable to this form.

In this connection it might be well to consider what Lucien M. Turner has to say, in his unpublished notes, concerning the great degree of individual variation that he noted in the hundreds of Ungava ptarmigan that he handled. He found a great "individual variation in the size and shape of the head and beak." He collected a large series of skulls and found them to vary from "1.94 to 2.30 inches in length from occiput to tip of bill." He says that "the large beak is especially noticeable in some of the birds, more especially in the males," but that "some only of the males alone have a noticeably large beak." He concludes by saying that this character should not, in his opinion, warrant the separation of this form.

I found and collected a few specimens of willow ptarmigan near Hopedale and at Ukjuktok Bay, on the northeast coast of Labrador, in 1912. They were breeding there, and I was told that they also breed on some of the islands. They are said to migrate farther south in October and return in April and May. I have a set of eggs taken near Okkak in July. Dr. Harrison F. Lewis (1928) said that this species "bred in the summer of 1928 on the large island at the Bluff Harbor where they had nested every year since 1925. It seems probable that two pairs of willow ptarmigan nested on this island in 1928."

O. J. Murie has sent me the following interesting notes:

Along the Nastopoka River, on the shore of Hudson Bay, lies a region that is apparently an ideal nesting ground for willow ptarmigan. On the morning of May 11, my Eskimo guide led me inland over the granite hills about parallel with the river. The immediate coast in this vicinity is bare and treeless. Ten or 15 miles inland we found scattered patches of small stunted spruces, covering some of the lower hills. Here we found a ptarmigan paradise. The birds were everywhere, apparently all paired, the males thrilled with the energy of the mating season. It was invariably the male that flew out first when we approached small clumps of willows. He went off like a rocket, with a clattering racket of harsh notes and whir of wings. He would generally take a commanding position on a rocky point or top of a hill and, strutting pompously with head high and tail raised and spread, let out his clattering crow.

At this time of the year I was struck by the perfect blending of the plumage with the surroundings, especially in the case of the female. Her plumage at this time is speckled irregularly with various shades of brown and white, in varying stages of molt, which was difficult to distinguish on a ground of mosses, lichens, rocks, and willows, sprinkled with remnants of snow. The females showed a tendency to keep well under cover. The males were not particularly difficult to see, and I spied them readily at a considerable distance on occasion, but sometimes I thought I could detect a principle of protective coloring. The body was pure white, with a brown head and neck, apparently the usual plumage in the season of courtship, judged from this one spring. Usually the white plumage flashed out prominently, but the brown neck was lost on the background of rock and moss. Although the eye was attracted at once, there was a tendency to pass it over without recognizing the headless body as a bird. Similarly, on a snowbank, the brown neck was seemingly detached, and not a part of a bird.

Nesting.—Turner says in his notes:

In the vicinity of Fort Chimo, nesting of this species begins during the latter part of May. The nest is usually placed in a dry spot among the swamps or on the hillsides where straggling bushes grow. The nest is merely a depression in the mosses and contains a few blades and stalks of grass together with a few feathers from the parent bird, which is now in the height of the molt from the winter to the pre-aestival plumage. The first eggs obtained were two on June 1, 1884.

Young.—Referring to the period of incubation, Turner writes:

It is a rather difficult matter to determine, as the female is compelled, during a stress of severe weather, to sit upon her eggs to prevent them being lost by cold or rain. It is not unusual for severe snow and sleet accompanied by cold rains and even a severe freeze to occur during the early half of June at Fort Chimo. Some of the most dismal days of the year occur in early June. The parent bird, during such weather, may be two or three days on the nest after the first young bird has appeared and thus prevented from giving such attention to the young as these tender creatures require. It is not rare to find a nest containing two or three eggs and near by to find one or more young which have perished while the mother has perhaps wandered off with three or four young which were able to follow her.

He says that "the Indians consider the downy young of the ptarmigan a special delicacy. Even taken from the shell the bird serves in lieu of an oyster." He frequently saw them eating the embryos taken from eggs that they were blowing. He says that they make special excursions to collect the small chicks for food. One party that he saw returning from such a hunt had more than 250 of these helpless young.

At Ukjuktok Bay, on August 3, 1912, I surprised a family party of willow ptarmigan in a boggy, grassy hollow. The young, which were about half grown, rose with a startling rush of wings and went whirring off like a flock of quail. The old birds did not flush. The female feigned lameness, in spite of the fact that the young had all flown; I could not make her fly, and she finally walked away. The male walked boldly out into the open marsh, looking at me, too close to shoot, then ran behind some spruces and flew away to join the young. I followed them up and flushed the male first; then three of the young rose singly. I could not find any more of the young, but a little later I found both old birds in the exact spot where I had first seen them.

Voice.—O. J. Murie has sent me the following good description of the ptarmigan's notes:

The call of the Ptarmigan is very striking. It consists of a rattling *krrr-r-r-ruk-uk-uk-uk-uk*, followed by a more deliberate, low-toned, throaty *puk-que'-o, puk-que'-o, puk-que'-o*. I thought it fitted well with the surrounding hills of rough granite and the scant growth of ragged, twisted spruces. Sometimes the female was heard responding with a peculiar whirring sound, a nasal *nyek, nyek*, somewhat similar to some notes of other members of the grouse family. By imitating this note we frequently drew the male to us in a headlong flight. He would drop on a knoll near by and send out his startling call. The Eskimos take advantage of this trait and decoy the birds to be shot.

Winter.—Ptarmigan are great wanderers in winter, but very erratic in their movements, appearing in enormous numbers during some seasons on the eastern and southern coasts of the Labrador Peninsula, and being nearly or quite absent during other seasons. Dr. Harrison F. Lewis writes to me:

This species often appears on the coast in great numbers in autumn, but the numbers so appearing vary from year to year. When these birds come out from the interior to the coast they seem to come as far out as possible, and often go in numbers to the very outermost islands which, of course, are often united to the mainland by ice at that time of year. Large numbers are taken by the residents for food. No matter how plentiful the ptarmigan may be in the fall and the month of December, comparatively few are usually to be seen after the 1st of January, although a few may often be observed from time to time until spring.

But vast numbers remain in northern Ungava, at least during some winters, according to Turner. One day he saw a gyrfalcon flying across the Koksoak River, and writes:

We had seen hundreds of ptarmigans on the left bank among the thickets. The hawk plunged among these birds which began to rise as soon as the hawk was sighted. I am certain that not less than 1,800 ptarmigans rose before that hawk; and, as the latter did not reappear, we suspected there was at least one less ptarmigan. The air fairly trembled as these birds arose. In my notes I find the following for the locality of Fort Chimo and date of December 7, 1882. Hundreds of this species of ptarmigan have made their appearance in this vicinity during the past week where two weeks ago not a dozen birds of the kind could be found during a tramp of an entire day.

The traffic in ptarmigan feathers is very heavy and will give an idea of the enormous number of birds killed. The Indians save only the clean feathers from the breasts and backs of the birds, pack them in bags, and trade them to the Hudson's Bay Co. They are then packed in barrels and shipped to England. Mr. Turner estimated that it required the feathers of 16 ptarmigan to make a pound and says that 31 barrels of feathers were shipped from Fort Chimo during the year ending June 1, 1883. As the average weight of feathers in a barrel was 51 pounds, he calculated that it accounted for 25,296 ptarmigan killed. And to this must be added a very large number of birds killed whose feathers were soiled or for some other reason were not saved. Fortunately the ptarmigan is a prolific breeder; otherwise, the wholesale slaughter of chicks and old birds would soon exterminate the species.

LAGOPUS LAGOPUS ALASCENSIS Swarth

ALASKA PTARMIGAN

HABITS

Harry S. Swarth (1926) gave the name *alascensis* to this supposed subspecies, which he characterized as "slightly larger than *albus*. A large billed race; bill slightly smaller than in *ungavus*, much larger than in *albus*. In summer plumage, generally more reddish-colored than either *ungavus* or *albus*, a difference that is most conspicuous in females in the barred breeding plumage."

He gives as its range "the Alaskan mainland except on the southeastern coast, northern Yukon Territory (specimens from vicinity of Forty-mile), and eastward for an undetermined distance."

At least three authors have attempted to subdivide the willow ptarmigans of North America, with decidedly confusing and somewhat unsatisfactory results. Austin H. Clark (1910) after studying 115 specimens, 20 from Newfoundland, 60 from Labrador, 3 from central arctic North America, 18 from the mainland of Alaska, 2 from Kodiak Island, and 12 from the Shumagin Islands, says:

All those from Labrador and central arctic America, with others from Point Barrow, Kotzebue Sound, Cape Lisbourne, Kowak River, Yukon River, and near St. Michaels, belong to a well-differentiated race, with the beak very large, high and stout, the culmen strongly arched, and usually with a prominent ridge from the inferior corner of the maxilla to in front of the nostril. They are identical among themselves, it being impossible to tell from the examination of any one specimen whether it was taken in Alaska or in Labrador.

J. H. Riley (1911) evidently differed from him, for when he described and named the Ungava bird, from the Labrador Peninsula, he gave as the range of *albus* "from the west side of Hudson Bay, west through northern Alaska to eastern Siberia." Thus two investigators, with practically the same material for study, have arrived at quite different conclusions.

Now Swarth (1926) comes along with a still different theory, based on a study of a large series of western birds in California collections, together with 3 from the west coast of Hudson Bay and 10 from Fort Chimo, Ungava. He writes:

Comparison of these birds with the series in this museum convinced me of the existence of the following recognizable subspecies of the willow ptarmigan on the North American mainland: (1) *Lagopus lagopus ungavus* from the region east of Hudson Bay, as defined by Riley; (2) *Lagopus lagopus albus* from the west shore of Hudson Bay westward to the coast ranges of northern British Columbia, and for an undetermined distance northward; (3) an undescribed subspecies from the Alaskan mainland and extending for an undetermined distance eastward in the extreme north.

Swarth (1926) then goes on to describe the northern bird, *alascensis*, as quoted above and says further:

Conditions in these western races of willow ptarmigan parallel to some extent those found in the rock ptarmigan. In each species the northern Alaskan subspecies is an extremely ruddy-colored bird compared with the others, and in each the British Columbia subspecies seems to reach an extreme of grayness. In each species, too, the Labrador birds are much more grayish than are those from Alaska. Thus the Labrador willow ptarmigan (*ungavus*) and the British Columbia bird (*albus*) are much alike as regards color but differ in size of bill.

The Labrador bird and the northern Alaskan bird (*alascensis*) are both large-billed forms, but differ in coloration.

Perhaps, after studying the three papers referred to above, the reader may get a clear idea of the subject; but I must confess that I can not. In my opinion, all the willow ptarmigan of the North American mainland (excluding *alexandrae* and perhaps *alleni*) are of one subspecies, *Lagopus lagopus albus*. The differences pointed out above are too slight and too variable to be worthy of recognition in nomenclature.

LAGOPUS RUPESTRIS RUPESTRIS (Gmelin)

ROCK PTARMIGAN

HABITS

Much has yet to be learned about the relationships of the various forms of the rock ptarmigan in North America and their distribution. There have been three forms named in Greenland, one in Newfoundland, three on the mainland, and six on the Aleutian Islands. With the possible exception of some of the Aleutian forms, which are quite distinct, all might well be regarded as subspecies of the European *Lagopus mutus*. Of the three mainland forms Harry S. Swarth (1926) has made an extensive study, the results of which are quite enlightening. The three forms are: A gray-colored bird, *rupestris*, a ruddy-colored form, *kelloggae*, and a dark-colored form, *dixoni*. He says of their ranges:

> First, there is a gray-colored bird that extends from Labrador westward to the coast ranges of northern British Columbia. In the east it apparently extends northward into the Arctic regions; it also occurs on islands north of Mackenzie, but elsewhere in the west it is restricted to the southern part of the region covered by the species *Lagopus rupestris*. Second, there is a ruddy-colored form that occupies almost the entire mainland of Alaska and extends eastward along the Arctic coast about to the one hundredth meridian. Third, there is a dark-colored form with a rather limited range in the coastal region of southeastern Alaska.

The latest contribution on the subject is a paper by P. A. Taverner (1929) based on a careful study of 105 adults of both sexes and young. Even with this quantity of material, the results are not wholly satisfactory, because of lack of adequate series of breeding birds in comparable plumages and because of a number of puzzling specimens that seem to have wandered far away from their normal breeding ranges. Taverner, however, has recognized two wide-ranging mainland forms, a southern, generally grayish bird (*rupestris*), ranging from northern British Columbia to Newfoundland, and a northern, generally yellowish bird (*kelloggae*), ranging from the interior of Alaska to northwestern Greenland. I recognize this revision and include *welchi* in the grayish form (*rupestris*).

It is the wide-ranging gray rock ptarmigan that we are now considering. Lucien M. Turner, in his unpublished notes, has given us the best contribution to the life history of this species. Bendire (1892) quoted some of these notes under Reinhardt's ptarmigan, but, for reasons stated under that form, I prefer to use them here. As to the haunts of these birds he says:

They prefer more open ground and rarely straggle even into the skirts of the wooded tracts. The hilltops and "barrens" (hence often called barren ground bird) are their favorite resorts. As these tracts are more extensive in the northern portions of Labrador and Ungava, these birds are there very abundant. During the summer months they are quite scarce in the vicinity of Fort Chimo, retiring to the interior and the hills of George River for that season. In the month of May the nuptial season arrives and is continued until about June, when nesting and laying begin. The birds are by this time scattered, each pair now taking possession of a large tract of stunted vegetation, among which they make their nest and rear their young. I was never able to procure the eggs of this species.

As to the haunts of rock ptarmigan in Newfoundland, all recent observers seem to agree that "Welch's ptarmigan" is confined, at present at least, to the barren tops of the highest hills along the west and south coasts. Dr. G. K. Noble (1919) says that "all of the Welch's Ptarmigan observed were found on the very highest ranges of the Lewis Hills. These are composed mostly of syenite, very much weathered or fragmented." Ludlow Griscom (1926) agrees with him that this species "is found only on the highest diorite and syenite rock barriers"; he found it also "on the summit of Blomidon." J. R. Whitaker, who lived in Newfoundland for many years, says in some notes he sent to me:

These birds are distributed through the high tundras and hills of Newfoundland. In the middle section of the country they are not nearly so numerous as along the hills nearer the coast. They remain on the high barrens far more persistently than Allen's ptarmigan, but it was quite a rare event to see any on the lower level even during severe winters; this statement applies to central Newfoundland. I am told on good authority that along the west coast, Cape Ray especially, they come down to sea level in numbers during the fall and winter months.

From extensive information recently gathered, I should extend the range of the rock ptarmigan, in Newfoundland, to include suitable highlands along the south coast from Fortune Bay to Cape Ray, with several records for Gafftopsail and Kettys Brook on the railway line. Probably the whole of the northern peninsula, north of Bonne Bay and White Bay, should be included also. I am told that 80 per cent of the ptarmigan that come into the market from this southwestern region are "rockers."

F. Seymour Hersey, who spent two months in Newfoundland in the fall of 1913 and collected 13 specimens of "Welch's ptarmigan" on the Lewis Hills, contributed the following notes on its distribution and haunts:

Welch's ptarmigan is much more local in its distribution than Allen's. I did not find it at Cape Ray or Gafftopsail, and I confidently expected to find it there, but notwithstanding repeated hunting over many miles of ground I found *Allen's* only. After I collected in the Lewis Hills I returned to Gafftopsail and showed my birds to the few people who lived at this lonely place, and all agreed that they had never seen a ptarmigan like them. One man who hunted and trapped a great deal did state that they occasionally shot a small ptarmigan in winter which I believe is Welch's.

The only birds I obtained were collected in the Lewis Hills, which, at the place where my collecting was done, run parallel with the beach and only a short way back from it. They have the form of a continuous ridge, rather than of individual hills, and rise abruptly at a steep angle, to a height of several hundred feet. The sides, as well as the strip from the shore to their base, are well wooded with trees of fair size. The top is undulating and stretches away for long distances without descent, which makes hunting comparatively easy. Individual hills are indicated by slightly increased elevations of the general level, but there are no distinct and separate hills.

The whole surface is devoid of trees, but the ground is well covered with low-creeping vegetation, prostrate blueberry bushes, mosses, and grass. In more protected spots are patches of dwarf spruces. These are usually 1½ to 3 feet high and are so twisted and interlaced that it is often possible to walk across a patch of them without breaking through.

My first day in these hills was foggy and I got no birds. Once I heard a ptarmigan call and following the sound come upon three Allen's. The next day was clear and crisp with a good breeze and only scraps of fog. We had been tramping for several hours with little success when we came to an elevation thickly strewn with large granitelike rocks of a very dark gray color. My guide halted and whispered that if we went cautiously he thought we would soon see some birds. It was not long before I made out a ptarmigan perched on a rock, and very soon after seeing the first one I discovered several others. Their dark colors blended with the gray rocks, but they were nevertheless rather easy to see. As there was no concealment we walked straight toward them and had no difficulty in getting within gunshot. At the first shot they arose, flew a few hundred yards, and alighted again among the rocks. Leaving the guide to retrieve the birds I had shot, I followed the flock and soon flushed them again. This was done several times until I had collected a number of specimens, when the remainder flew out of sight.

In the Flatbay Brook country where I later camped were some high rounded hills. My guide stated there were ptarmigan to be found well up toward their summits, but only in winter, when they had formed into large flocks, could they be successfully hunted. I was also told they were to be found in the mountains about Fortune Bay. Judging from my experience this species is restricted to the tops of certain ranges of hills, mainly near the coast, and they are partial to areas covered with dark gray rocks rather than the lighter colored rocks where Allen's ptarmigan is so abundant.

Spring.—Bernhard Hantzsch (1929), referring to the spring migration at Killinek, northern Labrador, writes:

Suddenly in early spring, mostly in April, seldom sooner, at times not until well into May, the wanderers appear from the south. Usually at first rather small advance posts are established. A short time after that the whole throng of birds follows. As I was assured by the missionaries, Messrs. Waldmann and Perrett, who each have passed a year up to 1906 in Killinek, by Mr. J. Kane who lived there six or seven years, as well as the Eskimos of the neighborhood agreeing, countless large flocks of these birds appear at times, usually passing through rather high in the air. For hours they hasten in many thousands through the sky, so that their numbers cause astonishment. Many flights of the kind are observed from the same place. The birds mostly fly directly across Hudson Strait without delaying. This is almost always covered with ice in the spring and little to be distinguished from the land. The flight is so swift and high, that Missionary Perrett was in doubt whether the birds were migrating to Greenland, which can, however, be safely denied, according to the unanimous reports from there. The Canadian *Neptune* Expedition 1903–1904 observed a great migration of these birds at Fullerton, northwest Hudson Bay. Only a small percentage of the ptarmigan make a stop in our region in order to rest up and hunt food. The forerunners and the stragglers stop more frequently than the main swarm, the latter having perhaps not much farther to go to reach their breeding places. The birds which stop, halt mostly in flocks of ten to thirty, occasionally still more together, and they usually do not act particularly shyly. When contrary winds and hunger tire the creatures out, they are so tame that they can be killed with the long dog-whip. The captured birds form a much-preferred article of food for whites and Eskimos, indeed the latter devour even the entrails, especially when these are warm.

Courtship.—Turner describes the courtship of the rock ptarmigan in his notes as follows:

The mating season begins in May, and during this period the male acts in the strangest manner to gain the affection of his chosen mate. He does not launch high in air and croak like the willow ptarmigan, but runs around his prospective bride with tail spread, wings either dragging like those of the common turkey, or else his head and neck stretched out, and breast in contact with the ground, pushing himself in this manner by the feet, which are extended behind. The male at this time ruffles every feather of his body, twists his neck in various positions, and the supraorbital processes are swollen and erect. He utters a most peculiar sound, something like a growling *kurr-kurr;* as the passion of the display increases the bird performs the most astonishing antics, such as leaping in the air without effort of wings, rolling over and over, acting withal as if beside himself with ardor.

The males engage in most desperate battles; the engagement lasts for hours or until one is utterly exhausted, the feathers of head, neck, and breast strewing the ground. A maneuver is for the pursued bird to lead the other off a great distance and suddenly fly back to the female, who sits or feeds as unconcerned as it is possible for a bird to do. She acts thoroughly the most heartless coquette, while he is a most passionately devoted lover. He would rather die than forsake her side, and often places himself between the hunter and her, uttering notes of warning for her to escape, while attention is drawn to him who is the more conspicuous.

Nesting.—The nest of the rock ptarmigan is a very simple affair, a hollow in the ground or moss of the open tundra, lined with grasses, mosses, or other convenient material, and a few feathers of the bird. It may be partially sheltered beside a hummock, tuft of grass, or low bush, but it is usually in plain sight. But Roderick MacFarlane (1908) says:

It proved no easy matter, however, to find the nests of this species, as the plumage of the birds and the color of the eggs both strongly resembled the neighbouring vegetation. At the same time the female sat so very closely that more than one was caught on the nest, and I recollect an instance where the parent, on the very near approach of our party, must have crouched as much as possible in the hope that she might not be noticed, which would have happened had not one of the smartest of our Indian assistants caught a glance of her eye.

Eggs.—The eggs of the different races of rock ptarmigan are all much alike and are very well described by MacFarlane and Brandt under Kellogg's ptarmigan, but the six sets in my collection show some types different from those described by others. The ground colors vary from "clay color" or "pinkish cinnamon" to "pale pinkish buff," "cream-buff," or "cartridge buff." They average more heavily marked than willow ptarmigans' eggs; two sets are nearly covered with great blotches and splashes of very dark browns, "chestnut-brown" to "bone brown," or nearly black; others are marked like willow ptarmigans' eggs with similar colors. I also have one set of 12 eggs. The measurements of 99 eggs in the United States National Museum average 42 by 30 millimeters; the eggs showing the four extremes measure **45** by 31, 44 by **32.5**, **39** by 29, and 41 by **28** millimeters, including eggs of both mainland forms.

Young.—Turner says in his notes:

When the young are with the parents they rely upon their color to hide themselves among the nearly similar vegetation from which they procure their food. I am certain I have walked directly over young birds that were well able to fly. If the parent birds are first shot, the entire number of young may be secured, as they will not fly until nearly trodden upon, and then only for a few yards, where they may easily be seen. I have found on two occasions an adult female with a brood of 13 young. All the flocks were secured without trouble. At other times only three or four young would be found with both parents. The young are very tender when first hatched; no amount of most careful attention will induce them to eat, and after only a few hours' captivity they die. I could never keep them alive more than 12 hours. The changeable weather, sudden squalls of snow or rain, must be the death of scores of these delicate creatures. Their note is a soft piping *pe-pe-pe,* uttered several times, and has the same sound as that of the young bobwhite, *Colinus virginianus.*

Captain MacMillan tells me that ptarmigan occur in flocks about Bowdoin Harbor, Baffin Island, all through the breeding season, and the residents say that these are young birds, which do not breed until their second spring.

Plumages.—The newly hatched chick of the rock ptarmigan is much like that of the willow ptarmigan but is usually somewhat paler and grayer. The crown patch is "chocolate," mixed with and heavily bordered by black; the rest of the head, neck, and breast is "cream-buff" or "chamois," shading off to "colonial buff" on the chin and underparts; a spot on the forehead, a rictal stripe, and a broad auricular stripe are black, the remaining upper parts are heavily blotched and banded with black, "chamois," "honey yellow," and "tawny."

The juvenal plumage begins to appear almost immediately, the wings coming first, in which the two outer primaries on each wing are white; these white, juvenal primaries are retained all through the first year. Young birds reach the flight stage before they are half grown. In the full juvenal plumage the young birds are darker, more heavily barred, and have less rufous than in the same stage of the willow ptarmigan. The sexes are alike. The entire upper parts are variegated with black, brownish black, "ochraceous-buff," "ochraceous-tawny," and white; the feathers are mainly black, tipped, barred, edged, or notched with the buffs; many feathers in the mantle have a terminal white spot; the breast and flanks are from "cinnamon-buff" to "pinkish buff," fading out to whitish on the chin and belly, spotted on the chin, throat, and neck, and barred on the breast and flanks with sepia or dusky. About the end of July the molt into the autumn plumage begins; this is also called the tutelar or preliminary winter plumage; it is common to both young and adult birds and is a transition plumage between the summer and the white winter plumages. The molts overlap and feathers of all three plumages are often seen in the same bird. During August and September the birds are molting almost continuously. The colored body feathers of both young and adult birds in this autumn plumage are much alike and are quite different from the feathers of the juvenal or summer adult plumages, being more finely vermiculated or sprinkled, less heavily barred, and therefore lighter in effect. The black tail feathers are acquired at this postjuvenal molt, which is nearly complete, and, according to Dr. Jonathan Dwight (1900), "the wings (except the median coverts and inner remiges) become white together with the abdominal wedge of the ventral tract and all posterior to it including the flanks, legs and feet; while the head, throat, breast, sides and back become more or less dusky according to the extent of the renewal in different individuals and probably according to the latitude."

From this time on the colored feathers are gradually replaced by white feathers, until the full, first winter plumage is acquired by this supplementary molt. This is wholly white except for the black rectrices, and in some young males there are traces of black lores.

In adult males the black lores are complete and in some old females there are traces of them.

In the far north, where the ground is not wholly free from snow, the prenuptial molt, in both young and old birds, is much delayed and is more or less incomplete, as the short arctic summer gives hardly time for the birds to acquire and shed the summer and autumn plumages. But farther south, where the birds live on a wholly bare ground, the molt of the body plumage is more complete, and the sexes become distinguishable in plumage, as the birds are now practically adult. In this and subsequent nuptial plumages the males assume a coarsely vermiculated and mottled dress of grayish buff and dusky, except that the wings, tail, abdomen, and feet remain as in winter. The females make a more extensive molt, becoming coarsely mottled and barred with buff and black, but retaining the white remiges, white belly, and black rectrices.

Soon after this nuptial, or summer, plumage is acquired, the molt into the autumn plumage begins, which helps to make the complete molt into the white winter plumage. The autumn plumage is much like that of the young bird and is very finely vermiculated in both sexes, instead of coarsely vermiculated or barred, as in summer. The sexes are now much alike again. These two plumages are well illustrated in the colored plates published by A. L. V. Manniche (1910).

Food.—In summer the food of the rock ptarmigan includes a number of insects, but it is made up more largely of leaves, buds, and berries, such as crowberries, whortleberries, bearberries, and the tender leaves and buds of birches and willows. Numerous seeds are eaten, as well as sphagnum and other mosses, and the leaves of Labrador-tea. In winter, when food is scarce, they have to feed on buds and twigs and the seeds of such weeds as they find above the snow. Their long, strong claws, which are highly developed in winter, enable them to dig down through the snow to reach the mosses and such of the above plants as they can find.

Behavior.—Turner says in his notes:

Their flight is rapid, and when flying with a stiff wind they require the quickest shot to stop them. The beat of the wing is so rapid that it is scarcely discernible, and when the bird is sailing the somewhat decurved wings are held almost motionless as it rolls from side to side. The direction of flight is always in a straight line, rising only sufficiently to clear a patch of trees or intervening ridge, the latter at times passed over only by a few inches in height, to the plain or valley beyond. Sometimes they will fly more than a mile before alighting and at other times only a few rods, depending altogether on the character of the weather. If it is a cold, blustering day with much snow drifting or falling, these birds dislike to take flight, and by using a slight amount of discretion the birds of an entire flock may be all secured, but if it is calm and cold, or warm, they take flight and at this time are rarely approached within shooting distance. Warm, damp, weather with gusty wind is best suited for hunting this ptarmigan.

I have often been amused at these birds' actions when descending from a high bluff to the level ground below. If there is a place that gradually slopes to the bottom, they seem to prefer to slide or tumble down rather than take flight. Just back of my house, and only a hundred or so yards, there was a bluff nearly 100 feet in height. This was the side of a level tract of ground above, and to it great numbers of " rockers " came every morning either to sun themselves or to descend to the lower ground nearer the houses and beyond. Their growling and " snoring " could be heard nearly every morning. I often watched them descend. Some individuals would push their feet forward and with outspread tail on the snow slide to the bottom, while others would roll and tumble over and over until they came to the level ground, where they ran as unconcernedly as birds could do.

Voice.—Mr. Whitaker likens the call of this ptarmigan to "the tattoo of a woodpecker on an especially mellow tree." Mr. Hersey describes it as " a low guttural croak reminding me of the spring song of the crow." O. J. Murie has sent me the following notes:

When slightly alarmed or annoyed by too close approach, these birds uttered a strange sound, like a short interrupted purr—*prrt! prrt!* This usually indicated that they were about to fly. Later, in spring, I heard a more prolonged utterance, possibly the " crowing " of this quiet bird. On several occasions in May, when I disturbed a group of these birds, some of them produced this sound, which might be described as a rolling " snore "—*k-r-r-r-a-r-r-r-uk-r-r-r-a*, plainly varied by a change in the middle which resolves it into three parts. Once when I shot a male for a specimen, the female called with a " whining " note.

Doctor Noble (1919) writes:

On August 24th, during a heavy rainstorm, while making my way across one of these fields of grotesquely shaped stones, I came suddenly upon an old male bird. It had just emerged from between two great blocks, and stood looking at me. After a few moments' hesitation, it stretched out its neck and gave a long cackle, unlike any call I had ever heard. It was a crescendo of clucks, somewhat pheasant-like in quality—*kuk, kuk, kuk, kuk*—each syllable stronger and of a higher pitch than the last.

Ludlow Griscom (1926) had a similar experience:

An old grey cock suddenly appeared on the top of a rock in a field of huge boulders, not more than fifty feet from me. It stretched its neck, cackled loudly and long, and exhibited no fear at all as I drew nearer. It then disappeared, but for some seconds I could hear it clucking to itself in great dissatisfaction as it threaded its way through the maze of its chosen home.

Fall.—From the northern portions of its range the rock ptarmigan makes quite extensive migrations. Capt. Moses Bartlett told me, and others have confirmed it, that during the last of September beginning with the first heavy snow squalls and lasting up to about the middle of October, a heavy flight of ptarmigan occurs across Hudson Straits to Cape Chidley, hundreds of birds being in sight at one time. Often they alight on ships and are easily caught. A similar flight occurs from Ellesmere Island to Greenland. In northern Labrador, according to Bernhard Hantzsch (1929), the flight is irregular. He says:

> In the middle of September, 1905, countless flocks are said to have flown southward over Killinek. In 1906 the first ptarmigan were not observed until 28th September. From 4th October they appeared somewhat more numerously in a heavy, driving snow and some cold, but rather large hunting parties did not get many, particularly as they could not yet travel in the light fall of snow. Whether the unusually long and mild fall kept the birds in their northern dwelling places, or induced them to choose another migration route, must remain unknown.

Winter.—Many rock ptarmigan evidently spend the winter in the lowlands of Ungava, mainly near the coasts, for Turner says:

> During the summer months they are quite scarce in the vicinity of Fort Chimo, retiring to the interior and to the hills of Georges River vicinity for that season, to return about the first of November. This month will be considered as a starting point for their wanderings. Then they appear in full winter plumage and in flocks of various sizes, often numbering over 200 individuals. They resort to the open ground and rock ridges where the snow is more apt to drift from, or during severest weather they retreat to the sheltered places amongst the bushes of alder and willow along the streams and gullies. These situations afford food of berries of various kinds, which yet remain on the stems of prostrate shrubs, twigs and buds of dwarf birch, alder, and willow, together with a few blades of grass. These ptarmigan remain until the latter part of March and disappear as suddenly as they came.

O. J. Murie writes to me as follows:

> During winter the rock ptarmigan along the eastern coast of Hudson Bay were very tame and easily approached, much more so than the willow ptarmigan. The latter birds confined themselves generally to willows or alders or even in the edge of the woods. *L. rupestris*, on the other hand, were generally found in the open or rocky slopes, feeding on the willow tips protruding from the snow or the berries of *Empetrum*, in spots where the ground was blown bare of snow. At times the feathers around the beak were stained purple as a result of feeding on this fruit. Even when resting these hardy birds did not always seek shelter. Once I found a group of two or three crouching on the bare top of a little rise; all hunched up in a little ball and facing a persistent, cold north wind. One of them remained in the same position while I approached within about 4 or 5 feet to photograph it, finally walking off quietly when I reached out still nearer. Generally this species was found in the shelter of a little bush, a single stunted tree, or small clump, but not often in heavy cover.

Enemies.—Hantzsch (1929) writes:

> The flights of ptarmigan are accompanied by birds of prey; especially the proud Gyrfalcon, the smaller Duck Hawk, and the beautiful Snowy Owl follow them. If the flight is smaller, as in the autumn of 1906, then these birds are observed only in small numbers. In addition, foxes assemble, particularly *Vulpes lagopus* (Arctic Fox), in places where there are many Ptarmigan, and all the other beasts of prey, in like manner, probably take a share at the appearance of our much desired bird.

Game.—Hantzsch describes the ptarmigan hunt as follows:

The arrival of the first flocks of these birds is greeted as an event of the day, which controls all the conversation. Now everyone cleans his gun, and even the little eight or ten-year-old chap is happy, whenever a gun is occasionally loaned him. If the ptarmigan appear in great numbers, an occurrence that varies much from year to year, then each one who has a gun and ammunition, from the missionary to the youngest Eskimo lad, betakes himself out into the wild mountainous landscapes. And the district is so large for the few people—at the most 15–20 men assemble near the Killinek mission station—that no one is in another's way. They prefer to go alone or in pairs with the dog-sleds, with a young man along for assistance, in order to overtake the birds more quickly and be able to take the bag home more conveniently. In the few days when the birds are present, a good hunter is often able to shoot several hundred. To be sure the hunt is strenuous. They travel across the wide, snowy landscape until they see a flock flying up somewhere. Sitting down they do not see the ptarmigan until rather near, as I convinced myself. The hunter now usually springs from the halting sled and approaches the birds in order to get one or two goods shots at them. The unwounded birds rise at once and fly away, and it is now a matter of paying attention to where they stop again. After the game has been put on the sleds, they journey farther, seeking either the part of the flock which has flown away, or new bands. Now and then they see several at the same time, at other times they have to wait a long time before coming across a single one.

O. J. Murie says in his notes:

The Indians and Eskimos take advantage of a trait of these birds to trap them. Ptarmigan have a tendency to gather on a conspicuous dark spot on the level white expanse of snow, such as an exposed sandbar. The natives stretch some kind of netting over a frame, which is tilted up over an exposed plot of ground or a spot where sand or earth has been spread on the snow. As the birds gather under the net a string is pulled, which allows the net to fall. The trapped birds are then killed by biting them in the neck.

The rock ptarmigan is an important game bird in Newfoundland, as about 20 per cent of the ptarmigan that come into the St. Johns market are "rockers." But evidently the hunters prefer the larger willow ptarmigan, which are, perhaps, more easily obtained. From reliable parties who have hunted in the vicinity of Quarry and Gafftopsail, as well as on certain sections between Fortune Bay and Cape Ray, I have learned that, when they have been unsuccessful in their hunt for willow ptarmigan, they turn their attention to the hilly sections of the same locality to find "rockers." Sometimes they would go to hunt for willow ptarmigan and not bother about "rockers" unless their hunt for willow ptarmigan was unsuccessful; then they would climb to hilly sections of the same locality where they could always be sure of getting rock ptarmigan.

DISTRIBUTION

Range.—Alaska (including the Aleutian Islands), northern Canada, Greenland, and Newfoundland.

The range of the rock ptarmigan extends **north** to Alaska (Cape Prince of Wales, Point Barrow, Humphrey Point, and Demarcation Point); northwestern Mackenzie (Cape Bathurst, Franklin Bay, and Pierce Point); northern Franklin (Winter Harbor, Gaesefiord, Lake Hazen, and Floeburg Beach); and northern Greenland (Thank God Harbor, Newman Bay, and Lockwood Island). **East** to Greenland (Lockwood Island, Sabine Island, Clavering Island, and Ivigtut); and Newfoundland (mountain ranges). **South** to Newfoundland (Fortune Bay to Cape Ray); Anticosti Island (Fox Bay); northern Quebec (Fort Chimo and Sorehead River); southern Franklin (Cape Fullerton and Bernard Harbor); and central British Columbia (Ingenika River and Ninemile Mountain). **West** to British Columbia (Ninemile Mountain); and Alaska (Baranof Island, St. Lazaria Island, Mount Edgecumbe, Port Frederick, Hinchinbrook Island, Montague Island, English Bay, Unalaska Island, Atka Island, Adak Island, Tanaga Island, Amchitka Island, Kiska Island, Attu Island, Askinuk Mountains, Nome, Teller, and Cape Prince of Wales). It is of casual occurrence in summer on Bonaventure Island, Quebec, where one was taken July 8, 1922 (Stoddard).

Although the majority of the rock ptarmigan withdraw from the extreme northern part of their summer range upon the advent of severe winter weather, they do not appear to retreat beyond the southern limits of the breeding range. Instead, a concentration is noticed at some of the more southern points, as at Ivigtut, Greenland. Among late departure dates in the northern part of the range are: Franklin, Floeburg Beach, September 29, and Winter Harbor, October 15. The species has been observed to return to this general area as follows: Franklin, Newman Bay, March 9, Floeburg Beach, March 11, Thank God Harbor, March 24, Discovery Harbor, April 10, and Winter Harbor, May 12.

The range as above described is for the entire species, under which 10 subspecies are recognized. True *rupestris* is found from Newfoundland and the Ungava Peninsula west to northern British Columbia and southern Yukon; Reinhardt's ptarmigan (*Lagopus r. reinhardi*) is found in southwestern Greenland, north to the vicinity of Disco; Nelson's ptarmigan (*Lagopus r. nelsoni*) is found on Unalaska, Akutan, and Unimak Islands, Alaska, and also on some of the other eastern Aleutian Islands; Turner's ptarmigan (*Lagopus r. atkhensis*) is confined to Atka Island, of the Aleutian group; Chamberlain's ptarmigan (*Lagopus r. chamberlaini*) is found on Adak Island of the Aleutians; Sanford's ptarmigan (*Lagopus r. sanfordi*) is confined to Tanaga Island of the Aleutians; Townsend's ptarmigan (*Lagopus r. townsendi*) is found on Kiska Island, of the Aleutian chain; Evermann's ptarmigan (*Lagopus r. evermanni*) is found on

Attu Island, of the Aleutians; Kellogg's ptarmigan (*Lagopus r. kelloggae*) occupies northwestern Greenland, the Arctic Islands (except Baffin Island), northern Yukon, the interior of Alaska, and the west Arctic coast to Coronation Gulf; and Dixon's ptarmigan (*Lagopus r. dixoni*) is found on the alpine summits of Baranof, Chichagof, and Admiralty Islands, and the adjacent Alaskan mainland.

Egg dates.—Northern Alaska: 15 records, May 28 to July 29; 8 records, June 9 to July 1. Arctic Canada: 23 records, June 3 to July 9; 12 records, June 17 to 28. Iceland: 18 records, May 7 to July 21; 9 records, May 20 to June 19. Greenland: 17 records, May 20 to July 6; 9 records, June 16 to 30. Labrador Peninsula: 12 records, June 11 to July 7; 6 records, June 3 to 20. Aleutian Islands: 7 records, June 10 to 26. Newfoundland: 11 records, June 2 to 12.

LAGOPUS RUPESTRIS REINHARDI (Brehm)

REINHARDT'S PTARMIGAN

HABITS

The race *reinhardi* of the rock ptarmigan was described from southern Greenland, where it undoubtedly forms a well-marked subspecies. A. L. V. Manniche (1910) treats of the rock ptarmigan of northeast Greenland under the name *Lagopus mutus*, that of the Old-World species, from which the Greenland birds differ only slightly. I am inclined to think that a careful study may show that *mutus* is a circumpolar species, of which the American forms are only subspecies. The 1910 American Ornithologists' Union Check List includes in the range of *reinhardi* the northern extremity of Ungava and western Cumberland Sound. Unfortunately summer specimens of rock ptarmigan from anywhere on the Labrador Peninsula are too scarce for us to form any satisfactory opinion as to what the birds of that region really are. Until we know more definitely what the birds of Labrador and Ungava really are, it seems more logical to refer them to *rupestris* and confine *reinhardi* to western Greenland south of Disco. The Greenland birds have been subdivided into three races, as fully explained in Dr. R. M. Anderson's footnotes in his translation of Bernhard Hantzsch's (1929) Labrador paper. But for life-history purposes, I prefer to treat the Greenland birds as all of one form.

Nesting.—What little information we have on the nesting habits of Reinhardt's ptarmigan indicates that it does not differ in this respect from other races of the rock ptarmigan. Capt. D. B. MacMillan told me that he once found a nest, on Baffin Island, that was placed on a nubble, not more than 10 feet long, surrounded by water in a pond; it would be interesting to know how the mother transported her young to the shore.

Eggs.—The eggs are indistinguishable from those of the rock ptarmigan. The measurements of 51 eggs average 41.9 by 30 millimeters; the eggs showing the four extremes measure **44.5** by 29.5, 44 by **32**, **39** by 30, and 40 by **29** millimeters.

Food.—A. L. V. Manniche (1910) says:

Their principal food consisted of buds and short bits of stalks of *Salix arctica*. According to Dr. Lindhard's analyses stomachs of Ptarmigans shot at this season also contained leaves of *Dryas octopetala* and crowns of leaves of *Saxifraga oppositifolia*.

In stomachs of shot young ones I found remnants of plants as well as of insects. The old birds in summer also partly feed on insects.

Behavior.—The same observer writes:

In fine weather these hardy birds did not seem inconvenienced by the temperature frequently as low as some 40° below zero. But it was hard for the ptarmigans to support their lives during severe snow-storms and when the earth was covered by thick, evenly lying crusted snow.

At sunset they flew to the rocks and remained there over night. In the heavy snowmasses on the leeside of the rock they digged holes some 20 centimeters deep, just large enough for the body of the birds, and here they spent their nights apparently without ever altering position judging from the manner in which the excrements were deposited. When several ptarmigans had spent a night in company, their holes were always placed within a rather narrow circumference sometimes nearer and sometimes at a longer distance, but never quite close to each other. The ptarmigans would also often spend their nights in narrow ravines in the rocks filled up with snow.

Sometimes I found my old foot-prints taken possession of by the ptarmigans as night-quarters. They were by night not seldom frightened out of their holes of polar foxes and erimines, which could be easily seen on new fallen snow. I found, however, in no case, signs that ptarmigans were caught in this way.

When a female ptarmigan was going to fly up, she would raise the feathers on the back of her head to a pointed crest and lay the tips of her wings on the upper rump uttering a suffocated clucking, that could best be compared with the call of *Fringilla montifringilla*; at the same time she would execute some courtesying movements with her head and the forepart of her body.

Just after a heavy snow-storm, that covered all the earth evenly with snow, the ptarmigans would prove extremely shy. For a few moments at a time they would settle on summits of rocks or stones, that reach over the snow, and then, by a rapid soundless flight, disappear around corners of rocks through deep ravines or out over extensive plains. When the ptarmigans after some hours had found places with food, they would again become tranquil.

Winter.—It seems to be well established that this and other true rock ptarmigan are migratory. Mr. Manniche (1910) says:

In the absolutely dark season ptarmigans or foot-prints of them were nowhere found in spite of numerous researches on different places, and there can be no doubt, that this species for some three months leaves this part of North-East Greenland.

It may be supposed, that the birds only migrate to somewhat more southerly lying parts of East Greenland, as they already begin to return in the beginning of February, when the sun has not yet appeared. (In 1907 the first ptarmigan was seen at the ship's harbour February 4th and the next year 4 days later.) The migration lasted through February, March, and the larger part of April, and the number of ptarmigans within a certain place might differ a good deal in this time.

Captain MacMillan (1918) recorded them as "common at Etah in spring and fall migration"; and as "not seen in July and August." The migration is very early, as he shot some at Etah on February 13 and saw them on the inland highlands of Ellesmere Island and Grant Land in March.

Hagerup (1891), referring to the region about Ivigtut in southern Greenland, writes:

During winter the number is considerably increased by the birds coming from the north, but the abundance is very variable. Thus the first winter I was at Ivigtut, an uncommonly cold season, comparatively few were seen, though about 400 were shot; but the following winter, which was much milder, the birds were much more numerous, and about twice as many were killed. When snow covers the ground they are less frequent in the valleys than on the mountain slopes and in the clefts; but on the high lands they are not so numerous. They usually resort to side hills, where there are large bowlders, and where some herbs are easily accessible. They change their feeding-ground very often, and sometimes in the course of a single night they arrive in such numbers that on the following day the birds or their tracks may be seen everywhere, while at other times one may travel for days without seeing any sign of one.

LAGOPUS RUPESTRIS NELSONI Stejneger

NELSON'S PTARMIGAN

HABITS

Nelson's ptarmigan is perhaps the best known of the half dozen forms of the rock ptarmigan found in the Aleutian Islands. It is a permanent resident on Unalaska Island, particularly on the eastern and more mountainous end, on the Krenitzin Islands, east of Unalaska, and on the Alaska Peninsula, at the base of which it apparently intergrades with the mainland rock ptarmigan. Dr. Wilfred H. Osgood (1904), who collected specimens on Portage Mountain, at the base of the peninsula, says: "With the material at hand I have been unable to satisfactorily distinguish the rock ptarmigan of the Alaska Peninsula from those of Unalaska Island."

About Iliuliuk Village, on Unalaska Island, we found ptarmigan scarce and wild, even on the mountains back of the village, where they are persistently hunted by the natives. Dutch Harbor is on Amaknak Island, which is separated from Unalaska by only a narrow channel. Here, between June 5 and 9, 1911, we found them quite common on a small mountain, locally known as "Ballyhoo," where we collected most of our birds, shooting nine one day and seven another. This mountain is only about 1,800 feet high, but is very steep and is topped by a knife-edged ridge covered with crusted snow. It is a stiff climb up the south side where the steep slope is clothed in soft mosses, with a sparse growth of coarse grasses, cow-parsnip, and other small herbaceous plants and with bare soil or rocks here and there. The north side breaks off suddenly into precipitous, rocky cliffs, straight down to the bay below. One day I started out to hunt ptarmigan on this mountain in a cold driving rain typical of Aleutian Islands weather; when halfway up the rain changed to snow, and when I reached the crest of the ridge the bay below was hidden in a blanket of fog, and across the bay a snow-capped mountain stood out in bright sunlight. Austin H. Clark (1910) found these birds at "the rugged northeastern end" of this island "on the mossy lower slopes, and one or two on the seacoast itself about the mouth of snow-filled ravines." But all of our birds were taken well up on the sides of the mountains.

Courtship.—At the time of our visit the ptarmigan were busy with their courtships. The males were very noisy and conspicuous; usually several could be seen sitting on little hummocks, as we looked up the mountainside; evidently each cock bird has his own special hummock, which he defends against intruders, for it is well decorated with droppings and molted feathers. Here he sits or struts about, clucking and displaying his charms, with the flaming red combs above his eyes fully extended, while his prospective mate, now inconspicuous in her mottled summer dress, walks about in the vicinity quite unconcerned. At frequent intervals he rises into the air 30 or 40 feet above the ground and floats or flutters downward, sometimes scaling on down-curved wings, uttering during the descent his loud clucking, or rattling, call, *wuck, wuck, wuck*, many times repeated. A pair of birds could often be located by seeing this song flight of the male, his white wings being quite conspicuous at a long distance. It is a very pretty performance and makes a striking display. Lucien M. Turner (1886) refers to a performance that is evidently part of the courtship: "In the male the neck is stretched along the ground, the tail spread and thrown over the back, the wings outstretched, while he utters a rattling croak that may be heard for a long distance."

Nesting.—We found no nests of this ptarmigan, and none of the females collected were anywhere nearly ready to lay. We assumed that they would nest later in the season on the lower, more grassy slopes. Turner's (1886) remarks probably refer in part to this ptarmigan, which he says is extremely abundant on some of the islands in the eastern part of the Aleutian Chain. He writes:

The mating season begins in the early part of May, and is continued for about three weeks, by which time a site for the nest is chosen, usually amidst the tall grasses at the mouth of a wide valley, or else on the open *tundra* among the moss and scanty grass.

The nest of this bird is composed of a few stalks of grass and a few feathers that fall from the mother's breast. The nest is a very careless affair, and often near the completion of incubation the eggs will lie on the bare ground surrounded by a slight circle of grass stalks that have apparently been kicked aside by the mother impatient of her task. The number of eggs varies from nine to seventeen, eleven being the usual number.

Dr. Richard C. McGregor (1906) found a nest on a small island in the Krenitzin Group, east of Unalaska, that unquestionably belonged to the Nelson's ptarmigan; it was taken on Egg Island on July 6; "the nest was a mass of grass, leaves, and a few feathers" and contained six eggs.

Eggs.—The eggs referred to above are described by Doctor McGregor (1906) as follows:

In color the eggs are dull creamy brown overlaid with irregular spots of dark reddish brown, almost black. The larger markings tend to form a ring near the large end of each egg, but this ring is rather poorly defined. The eggs measure as follows, in millimeters and tenths: 42.3×30.1; 42.4×30.6; 40.8×30.5; 41.7×31.3; 42.4×31.8; 45.2×31.4. Incubation was begun.

There are three eggs in the National Museum collection that resemble certain types of rock ptarmigan's eggs. They are ovate in shape with very little gloss. The ground colors are creamy white or pale buff. In one the ground color is nearly covered with small spots and fine sprinkles of very dark brown; in another it is nearly covered with large blotches and small spots of "claret brown" and "liver brown." The measurements of nine eggs average 42.5 by 30.5 millimeters; the eggs showing the four extremes measure **44.2** by 30; 42.5 by **31.8**, and **41.6** by **29.9** millimeters.

Food.—Doctor Osgood (1904) reported that "an examination of the crops of 10 birds killed at Cold Bay showed a variety of food, but buds, particularly willow buds, predominated. Tiny buds and twigs of some small species of *Vaccinium* were found in large numbers, which must have been secured by a very tedious process. Some of the craws contained nothing but buds, others had a few leaves of *Dryas* and *Ledum*, and occasionally one contained some broken pieces of large aments of *Alnus viridis.*"

Behavior.—Turner (1886) writes:

The young are able to follow the mother as soon as they are hatched. As this bird never collects into large flocks, I always supposed the flocks seen in winter were the parents with the brood reared the previous summer. The power of flight of this bird is much stronger than its congener. It is sustained for a longer period and much more rapid. The flesh of this species is better than that of the Willow Ptarmigan and is much sought for as food. The best time to hunt this bird is early in the morning when the wind is calm and a moist snow is falling. The birds are then sluggish and dislike to rise to the hill-tops.

There is not much more to be said about the habits of this ptarmigan, which apparently do not differ materially from those of the mainland rock ptarmigan. This form is much closer to the mainland form in appearance and habits than it is to the other Aleutian forms. It is essentially a bird of the mountains and foothills. It is tame enough where it is not molested, but it soon becomes sophisticated where it is hunted persistently. On the mountains back of Iliuliuk Village on Unalaska Island, where we collected a few specimens, we found it a really sporty game bird. It usually flushed at long range, with loud clucking notes, and flew very swiftly for a long distance, often across some deep ravine, where a long hard walk was necessary to flush it again. The man who makes a good bag under such circumstances earns his birds.

LAGOPUS RUPESTRIS ATKHENSIS Turner

TURNER'S PTARMIGAN

HABITS

The four forms of ptarmigan found on four of the more central islands of the Aleutian Chain, *atkhensis* on Atka Island, *townsendi* on Kiska Island, *chamberlaini* on Adak Island, and *sanfordi* on Tanaga Island, seem to me to be sufficiently different from their nearest neighbors, *nelsoni* on the east and *evermanni* on the west, to warrant recognizing them as four subspecies of a species distinct from *rupestris.* They are all somewhat larger than the mainland rock ptarmigan, with larger and heavier bills, and their eggs are decidedly larger. They are birds of the lowlands, living on the low, rolling hills, grassy plains, and sand hills near the coast; whereas the rock ptarmigan, as well as *nelsoni* and *evermanni,* is essentially a bird of the mountains and moss-covered foothills, coming down to the lowlands only on the arctic tundra. The dark colors of the mountain forms match their habitat, as well as the light colors of the lowland birds match theirs. The Aleutian Islands appear to be the summits of a submerged mountain chain, which at one time may have formed a land bridge between Asia and North America. It seems likely that the central portion of the chain may have subsided first, isolating the central islands long before the eastern and western islands were separated from the two continents. This might have given the birds on the central islands a much longer time to differentiate, while the birds on the two ends of the chain have remained more like the mainland birds. This might account for the presence of light-colored birds in the center and dark-colored birds at both ends, as they now exist.

On Atka Island we found this form of ptarmigan very abundant; there seemed to be more ptarmigan here than on any island we visited. There were comparatively few of them on the hillsides, but in the grassy hollows and among the low, rolling hills of the valleys we were constantly flushing them. They were apparently mated and breeding on June 13, the day of our arrival, but we failed to find any nests during the next two or three days. The male usually flushed first with loud clucking notes, and the female was sure to follow soon after him.

Courtship.—They were very tame and always in pairs, so we had plenty of chances to observe their courtships. On his song flight the male rises 30 or 40 feet in the air and floats down again on decurrent wings, giving a few rapid wing strokes before alighting; sometimes, after checking his descent by rapid wing strokes just above the ground, he sails along and upward to repeat the same performance; during his descent, and particularly during the rapid wing strokes, he utters his loud croaking notes, *kruk, kuk, kuk, kuk, kuk*, or *krrru-ru-ru-ru, ru-ruk*, a prolonged, rattling, nasal, clucking sound of great carrying power. Once I saw a male chasing a female in the nuptial pursuit flight; she led him a long chase up and down the valley and over some low hills, until they finally settled near me on the tundra, where they strutted about in plain sight. The male carried his head high, with the bright orange-vermilion comb over the eyes swollen and distended and with the tail erected and spread as he walked about in a slow and stately manner.

Nesting.—All we know about the nesting habits of this ptarmigan is contained in the following brief statement by Lucien M. Turner (1886), who discovered and described the bird:

> The nest is built amongst the rank grasses at the bases of hills and the lowlands near the beach. The nest is carelessly arranged with few dried grass stalks and other trash that may be near. The eggs vary from eleven to seventeen, and are darker in color than those of *rupestris*, and but slightly inferior in size to those of *L. lagopus*. A number of eggs of this species were procured, but broken in transportation; hence, can give no measurements of them.

Eggs.—There are 24 eggs of this ptarmigan in the United States National Museum, which do not differ in appearance from those of the other Aleutian races. The measurements average 44.8 by 32.9 millimeters; the eggs showing the four extremes measure **48.2** by 32.1, 45 by **34.5**, **42.4** by 32.3, and 43.6 by **31.3** millimeters.

Plumages.—Not much is known about the plumage changes of Turner's ptarmigan. Our birds were all in full summer plumage in mid-June. A series of 15 birds collected on April 4 by Hamilton M. Laing (1925) were just beginning to molt into the summer plumage; a few were still in full, white, winter plumage.

Food.—Of the birds collected by Laing (1925), "the cocks had little in their crops; the hens in most cases were stuffed with the foliage of the crowberry."

Behavior.—Laing (1925) first met this ptarmigan on Atka Island on April 4, 1924, of which he writes:

> On a cold, windy morning, with snow squalls between periods of sunshine, a Duck Hawk posting along the shore was seen to rout some of these white chickens of the north almost from the water's edge and send them whirling over the white hilltop. On going ashore it was found that about fifty ptarmigan were in the vicinity. They refused to fly very far and during the hunt seventeen specimens were secured. Rank grass in tussocks and crowberry patches grew on the hills and as there had been a good fall of snow, walking was difficult. Sometimes the birds were wild, again rather stupid. They were first found cuddled in the sun against the sheltered wall of a small canyon enclosing a brawling stream. Afterwards they flew from one hilltop to another. They were very speedy on the wing, usually flew downwind, and were extremely difficult to kill.

We noticed that the flight of the male is particularly strong and vigorous; he seems to delight in sailing against a strong wind, when he can soar for a long distance, rising and falling again and again, or even remaining perfectly stationary in the air like a poised falcon. When walking the head is carried high, and the motions are very deliberate and stately, almost stealthy in appearance, with frequent nervous twitches of the tail.

Voice.—The male utters his loud clucking notes, similar to those mentioned under courtship, while walking on the ground, when starting to fly, or when alighting. The female has a much softer note, like *cook*, which is very seldom heard, as she is usually a silent bird. Laing's (1925) impressions were as follows:

> The strangest thing about them was their purring snort like that of a startled horse. This seemed an alarm call. A sentinel stood on a hill and gave it again and again. It could be heard 300 or 400 yards, but was very elusive and difficult to locate. One bird gave it in flight, with opened beak. A cock that was winged purred again and again when chased and stopped purring only when caught.

LAGOPUS RUPESTRIS TOWNSENDI Elliot

TOWNSEND'S PTARMIGAN

HABITS

In his original description of this subspecies, Dr. D. G. Elliot (1896) gave as its range both Kiska and Adak Islands, although he observed that "there is a slight difference in the appearance of the birds from the two localities, and this can be attributed possibly somewhat to the difference of date in their capture, the Adak birds having been obtained one month later, but more to their geographical distribution, as Adak is several hundred miles east of Kyska, and the birds' environment has produced a different result upon them but one, not yet sufficiently pronounced to establish even a subspecific form."

It is now well known that *townsendi* is confined to Kiska Island and is a well-marked race, being darker and more heavily barred than any of the other races found on the central islands. The Adak bird has since been separated by Austin H. Clark (1907), under the name *chamberlaini.*

We were on Kiska Island from June 17 to 21 and on Adak Island on June 26 and 27. We collected good series on both islands, in which the characters of both races are well marked, showing that the two forms are quite distinct. It is, therefore, clearly a geographical difference and not a seasonal change, as Doctor Elliot (1896) evidently thought it might be.

At Kiska Island, on June 17, the ptarmigan of this form were still in the uplands, were much wilder than the Atka birds, and not nearly so abundant; but during the few days that we were there we succeeded in collecting a good series. No nests were found; probably we were too early for complete sets, and nests would be found later on in the long grass of the lowlands. So far as we could see, the habits and behavior of the Kiska birds are similar to those of the other races on the neighboring islands, as their environment is practically the same. These islands are all so widely separated that it seems very unlikely that the ptarmigan ever fly from one to the other. Hence each island has produced its peculiar form, which is completely isolated and permanently resident. All these ptarmigan proved to be very good to eat, and we found their plump bodies very welcome additions to the ship's stores of canned food.

LAGOPUS RUPESTRIS CHAMBERLAINI Clark

CHAMBERLAIN'S PTARMIGAN

HABITS

The Adak ptarmigan was separated from the Kiska and Atka birds and given the name *chamberlaini* by Austin H. Clark (1907). It differs from the Kiska bird "in its finer vermiculations above, which give the bird a grayer appearance—the whole plumage presenting a much more delicate pattern." He calls it intermediate between the Kiska and the Atka birds, "but in general coloration it is much grayer than either, being the grayest of all the Aleutian ptarmigan."

We spent June 26 and 27, 1911, on Adak Island, where we found this ptarmigan common but not abundant; we obtained all that we needed for specimens. We found them mostly in a broad valley of small, low, rolling hills, with a number of small ponds scattered through it; on either side of the valley were mountainous peaks, with rocky summits and with plenty of snow on them. The valley was mostly dry tundra carpeted with a dwarf species of reindeer moss, which the gray plumage of the ptarmigan matched very well. The birds were also found on the low hills and in the grassy hollows and lowlands but not on the mountains. They were much wilder than we had found them on other islands, which seemed strange, as this island is uninhabited.

Nesting.—On June 26, 1911, I found a nest of seven fresh eggs in a little valley on a hillside; it was a deep hollow in the ground between a tuft of grass and a little cow-parsnip; it was carelessly lined with dry grass and a few feathers. Dr. Alexander Wetmore also collected a similar set of seven fresh eggs on the same day. Evidently all the Aleutian ptarmigan are late breeders.

Eggs.—The eggs we collected are ovate in shape, and the shell is smooth with little or no gloss. The ground colors vary from "pinkish cinnamon" to pale "pinkish buff"; some eggs are washed with "cinnamon" at one end or the other, giving the egg a richly colored effect. In one set the eggs are thickly covered with very small spots and fine dots of dark browns, which are sometimes concentrated near the small end; two of the eggs have a few large, irregular blotches near the small end. The other set is marked like the common types of ptarmigan's eggs. The markings vary in color from "chestnut-brown" to "bone brown." The measurements of these 14 eggs average 46.1 by 32.9 millimeters; the eggs showing the four extremes measure **47.5** by 33.2, 45.6 by **33.6, 45** by 32.8, and 46.5 by **32.4** millimeters.

Plumages.—Five males collected by Hamilton M. Laing (1925) on Adak Island, April 13, 1924, had already begun to molt into the summer plumage.

Food.—All the Aleutian ptarmigan that we collected had been feeding entirely on green food, principally the young, green leaves and buds of the dwarf willows, the tops of ground evergreens and mosses, and the flower buds and blossoms of herbaceous plants.

Behavior.—We noted nothing peculiar in the habits of this ptarmigan, which were similar to those of its neighbors. Laing (1925) says:

Ptarmigan were even more numerous at Kuluk bay, Adak island, than on Atka island. On April 13 ptarmigan were purring everywhere and were all noted in the grass at low levels. There was no time to ascertain whether they were also numerous on the dark, crowberry-covered hills above, but they were scattered over the flats near the lagoon and the nearby lower grass-covered hills. In the distance was seen what, apparently, was a lively fight between two birds. There was only one round. Several times birds when routed whirled aloft 50 feet or so and then settled again slowly, purring loudly and perhaps threateningly, though what this manoeuvre was for was not clear. Several were shot with the .22 rifle, but they were very tenacious of life and a shot through the body with a hollow-point seldom actually killed them. Some so hit flew a hundred yards before falling. Some of the birds were quite wild, others comparatively tame. They were wildest during the cold, blustery snow squalls.

LAGOPUS RUPESTRIS DIXONI (Grinnell)

DIXON'S PTARMIGAN

HABITS

Dixon's ptarmigan is one of the many dark-colored races so characteristic of the humid coast belt of the northwest. It was discovered by Joseph Dixon and was named for him by Dr. Joseph Grinnell (1909), who ascribed to it the following characters: "Resembling *Lagopus rupestris nelsoni* in corresponding plumage, but much darker; in extreme blackness of coloration nearly like *Lagopus evermanni*, but feathers of chest and back more or less finely vermiculated with hazel."

Dixon's account (Grinnell et al., 1909) of securing the first specimens is interesting and gives an idea of the inaccessible haunts of this bird:

I was crawling down a ledge on the north side of the rocky summit of a mountain at 2700 feet altitude. About twenty-five feet below me a sharp rock jutted out, forming the crest of a hundred-foot cliff. I had glanced along the ledge below but saw nothing, when suddenly a gray-backed ptarmigan rose from a bunch of heather on a narrow ledge and trotted out on a jutting rock, bobbing its head and watching me intently the while. I fired a light charge at the bird which dropped over the cliff. At the report two other ptarmigan jumped up and started swiftly away. I dropped one with the remaining barrel. Then I began the descent to retrieve the birds. By going down to one side of the cliff I had almost reached its base when I came to a sheer drop; so I had to dig my fingers into the crevices and work my way back up again. By going a long way around I finally reached a twenty-foot snow drift at the foot of the cliff and there I found my two birds dead. Both had their crops stuffed with heather buds.

George Willett (1914) says of its haunts:

During the summer months these birds keep well up toward the summits of the mountain ranges, above timber line, where they feed on heather buds and berries. Owing to the difficulties in ascending these mountains, specimens are hard to secure at this season. They apparently move in bodies from one section of the mountains to another, and locating them is largely a matter of luck. I have been in sections of the mountains where sign less than a week old was abundant, but the most diligent search failed to locate a single bird. Whether these changes of location are due to the weather or food supply I am unable to state.

Nothing seems to have been published about the nesting habits of this ptarmigan. There is a set of six eggs in P. B. Phillipp's collection, of which the average measurements are 42.2 by 29.7 millimeters; the eggs showing the four extremes measure **43.9** by 29.9, 42.3 by **30.4**, and **40.8** by **29.4** millimeters.

Doctor Grinnell (1909) quotes from Mr. Dixon's notes, as follows:

Their flight was very swift, more like that of a falcon than a quail. The males would fly out over the mountain side, hover for a moment and then swoop down, and alight on a rock, uttering their loud, rasping call, which sounds similar to the noise produced by running a lead pencil over a stiff rubber comb.

Alfred M. Bailey (1927) gives us two good pictures of Dixon's ptarmigan in its autumn habitat:

During October there appeared a wealth of small birds, and many Ptarmigan were seen and collected. The vegetation was in the height of its "autumn glory," and a peculiar "lily pad," which flourishes abundantly, colored the hills an intense yellow above timber line, while still higher, among the piled boulders, there was a small ever-green growth upon which the Ptarmigan were feeding. A few were found in such a site, and some of them were extremely wild. The Ptarmigan in Granite Creek were taken among the boulders and slide rock on the summit of the highest mountains surrounding the valley, at an altitude of over 4,000 feet. There was absolutely no vegetation. That they are well named "Rock Ptarmigan" there can be no doubt after noting their habit of sunning themselves upon the tops of large boulders; one rested upon a little overhanging ledge which left a sheer drop to the valley floor far below. Several small flocks were seen flying about like so many Doves. They raised from the mountain on which we were hunting and sailed across the valley to the foot of a hanging glacier. One band flew over me and I tried to drop a bird on our narrow ridge, but the tumbling Ptarmigan sailed on into space and dropped at least 1,500 feet to the valley floor.

Again, on November 11, he wrote in his notes:

When just above timber line I saw a Three-toed Woodpecker on a dwarfed hemlock, and, on the snow fields above, about thirty Ptarmigan. The tops were icy, making creepers a necessity. The birds were in full winter plumage, wonderfully handsome fellows, the white of the males being relieved by the black eye patch. Their call notes could be heard from all sides of the snow covered mountains, and here and there cream-colored birds, gleaming in the sun-light, could be seen. Overhead an eagle circled, and soon the air was filled with flying Ptarmigan, although I did not see the Eagle make a swoop toward them. Of all the birds seen, only five were in one band, while the others were scattered in singles or pairs, and I wondered if they spread to feed among the little patches of grass sticking through the snow, or for the protection which isolation sometimes brings. The call note of the males was constantly heard. This note has an individuality about it which can be mistaken for no other bird. The Ptarmigan were tame, and often allowed us within good photographing distance, especially if we tried to imitate their note. They rise from the ground with great speed, and usually their flight is direct, although when flying out over a valley, they often slant down as though to attain greater speed. A few specimens were taken, and we found it difficult to secure our birds, for immediately they were hit, they started sliding down the slippery mountain side, and did not stop until they reached the brush line far below. This particular habitat was picturesque, to say the least, and on this day was remarkably beautiful; for the cloudless sky was a deep blue, the horizon was the serrated white line of the mountain tops, and the winding glacier—from its colorless snow fields at the summit to the seamed and rugged ice field below—with its characteristic shadows and high-lights of blue and white, made a wonderful panorama.

Bailey (1927) says of its winter habits:

After the winter snows have covered the mountains to a considerable depth, these birds drop to the valley floors where they feed among the alders and willows. They were often encountered during the following winter days, sometimes in large flocks, and many specimens were secured. The species may be considered a rather common bird in its proper habitat, near Juneau; it is simply a matter of looking in the proper place—and often involves some rather strenuous work.

LAGOPUS RUPESTRIS SANFORDI Bent

SANFORD'S PTARMIGAN

HABITS

Although I described and named this race myself (1912), in honor of my friend Dr. Leonard C. Sanford, who cooperated with me in organizing our expedition to the Aleutian Islands, I must confess that it is only slightly differentiated from the Adak ptarmigan. We all noticed a difference when our birds were collected, and when we laid our series of about 40 specimens of *sanfordi* beside nearly as many of *chamberlaini*, it was easy to see that the Tanaga birds were appreciably paler in color than the Adak birds. The Tanaga birds are therefore the lightest in color of any of the Aleutian ptarmigan and have the finest vermiculations.

We landed on Tanaga Island on June 25, 1911, and spent only half a day on shore; so far as I know, no one had ever collected birds on this interesting island before; we found it very rich in bird life, and it is a great pity that we were not able to spend more time there. Back of the sandy beach on which we landed was a series of sand hills or dunes, covered with long grass, and beyond these was a flat, alluvial plain or tundra, with one large and several small streams flowing through it from the mountains farther inland, and dotted with a number of small ponds and wet meadows. Northern phalaropes were breeding commonly among the ponds and meadows, and Aleutian sandpipers were abundant, indulging in their flight songs and nesting on the little knolls and hummocks on the tundra, where a brood of downy young was found. At the base of a steep hillside a colony of fork-tailed petrels was beginning to breed. The ptarmigan were tamer and more abundant here than on any of the other islands that we visited; we shot more than 40 in one afternoon. They were commonest on the rolling, grassy hillocks and grassy hills on the tundra. They flushed at short range, did not fly far, and were easily shot.

Nesting.—Five nests were found, but only three sets of eggs were collected; the other two were left to be photographed the next day, but we were forced to go away and leave them, as well as some nests of Aleutian sandpiper and an eagle's nest found by some of the crew. One nest containing nine fresh eggs was on the side of the steep overhanging bank of a stream; it was in a hollow between two large tufts of grass and well hidden under one of them; the hollow measured 7 by 8 inches and was lined with coarse grass and feathers. Another nest, containing eight fresh eggs, was a hollow in the ground, measuring 7 by 6 inches and 3 inches deep, between two little mossy hummocks and under a scraggly cow-parsnip; it was on a little grassy hillock near the beach and was lined with coarse grass and feathers. Other nests were well hidden in the long grass and were found by flushing the birds.

Eggs.—Judged from the three sets of eggs that we collected, consisting of eight or nine eggs each, the eggs of this ptarmigan are very handsome, in fact the prettiest ptarmigan's eggs I have ever seen. They are ovate in shape with a smooth and slightly glossy surface. The ground colors vary from "ochraceous-tawny" or "cinnamon" to "cream-buff" or "cartridge buff"; some of the eggs are washed at the large end or at the small end with "tawny," giving them a very rich appearance. They are boldly and heavily marked with large irregular blothes and small spots of the colors usually seen on other ptarmigan's eggs, dark browns, "chestnut-brown" to "bone brown," or nearly black. The measurements of 25 eggs average 46.5 by 33.9 millimeters; the eggs showing the four extremes measure **48.2** by 34, 46.6 by **34.6**, **44.2** by 34.6, and 45.5 by **32.8** millimeters.

Plumages.—We know nothing about the plumage changes of Sanford's ptarmigan, except that a series of two adult males and six adult females, collected by Donald H. Stevenson on September 18, 1921, now in the Biological Survey collection, show a postbreeding plumage quite different from the June breeding plumage. In the male this is darker, browner, or redder than the breeding plumage; the prevailing color of the breast, head, neck, and flanks is from "tawny" to "ochraceous-tawny," instead of "cinnamon-buff" mixed with pale grayish buffs, as in the June birds; on the upper parts the tawny shades are much more heavily peppered, variegated, or barred with black, entirely unlike June birds. The differences are similar in the female; the "cinnamon-buff" feathers, barred with black, of the June plumage are being replaced by white feathers on the belly, and by "tawny" or "ochraceous-tawny" feathers on the breast and flanks, more finely barred or peppered with black or dusky; the upper parts are also more tawny with finer barring or peppering and with more black than in June birds. Breeding females have no peppered feathers and no white on the belly. A juvenal bird, collected at the same time and place, is like the adult female, but the colors are duller.

LAGOPUS RUPESTRIS KELLOGGAE Grinnell

KELLOGG'S PTARMIGAN

HABITS

The name *kelloggae*, "Montague rock ptarmigan," was first applied by Dr. Joseph Grinnell (1910) to a bird that he described as a new subspecies from Montague Island, Prince William Sound, Alaska. But a later and more extensive study by Harry S. Swarth (1926) has shown that the specimens on which Doctor Grinnell's name is based are merely variants toward *dixoni* of a race, distinct from *rupestris*, that inhabits the whole of northern and western Alaska. He says of the characters and distribution of the race he now calls *kelloggae:*

> The notable feature of this bird is its bright ruddy tone of coloration, a character that is evident in both sexes and in all stages of the summer plumages. As compared with *rupestris*, the general tone of color throughout is brighter and more reddish, and there is notable restriction of the dark areas on individual feathers.
>
> The extreme manifestation of this race is reached on the northwestern and northern coast of Alaska, it occupies practically the whole of the Alaskan mainland, and it extends eastward of Alaska along the Arctic coast for some distance. In the latter region the duller color of specimens from Baillie Island, Coronation Gulf, and Bathurst Inlet, is to be interpreted, to my mind, as indicative of intergradation with *rupestris*.
>
> Southeastward there is intergradation again with *rupestris* as occurring in British Columbia, about at the Alaska-Yukon boundary line. A series of seventeen skins from the vicinity of Eagle (U. S. Biol. Surv. coll.), in the upper Yukon region, demonstrates such intergradation satisfactorily. Certain selected skins from this series and from the British Columbia aggregation are hardly to be distinguished, and none of the Eagle specimens shows the extreme of ruddiness that is seen in Alaskan birds from more northern points. The Eagle series as a whole, however, certainly belongs with the northern Alaska subspecies rather than with *rupestris*. On the southern coast there is apparent intergradation with *dixoni*, as shown by skins from Kodiak Island, Seward, and Prince William Sound.

According to the latest revision of this species by P. A. Taverner (1929), as recognized in the new American Ornithologists' Union Check List, the range of this form is extended eastward along the arctic coast and islands to western Greenland, north of Disco.

As the habits of this ptarmigan do not differ materially from those of other rock ptarmigan, I shall not attempt to duplicate what I have already written on the species.

According to Roderick MacFarlane (1908) the summer home of the rock ptarmigan in northern Canada

consists of vast plains or *steppes* of a flat or undulating character, diversified by some small lakes and gently sloping eminences, not dissimilar in appearance to portions of the North-West prairies.

The greater part of the Barren Grounds is every season covered with short grasses, mosses, and small flowering plants, while patches of sedgy or peaty soil occur at longer or shorter distances. On these, as well as along the smaller rivulets, river and lake banks, Labrador tea, crow-berries, and a few other kinds of berries, dwarf birch, willows, etc., grow.

Referring to his trip to Hooper Bay, Alaska, Herbert W. Brandt writes to me:

Our first acquaintance with the rock ptarmigan was made in the upper solitudes of the Beaver Mountains high above timber line on April 6. On these bald snow-beaten hills we found a number of straggling flocks; one that numbered about 20 birds contained only males, while the others were evidently mated couples, banded together. On the wind-swept slopes were numerous mossy hummocks, and in the leeward side of this scant protection, the bird scoops out a snug little snowy igloo. This is its only retreat and roosting place during the long cold winter, for Mr. Twitchell advises me that this hardy species seldom descends to the larger willows and spruces, which line the streams below. The rock ptarmigan is a rather common summer resident in the Askinuk Mountains, where it seems to prefer the sterile open ridges in the vicinity of 1,000 feet in altitude.

Nesting.—Brandt says on this subject:

The contents of the nest of this species range from 6 to 11 eggs, but the usual number found is 8. The nesting site is so chosen that protection is afforded by a hummock, a small tree, or even a growth of frost-dried grass, but occasionally no concealment whatever is present. The lining of the nest consists solely of surrounding materials, such as grasses, lichens, and moss, together with a few feathers.

Eggs.—The rock ptarmigan, according to MacFarlane (1908), lays fewer eggs than the willow ptarmigan, the usual number being six or seven and rarely more than nine. He describes the eggs, based on the very large series collected by him, as follows:

The eggs are ovate or short ovate in form, resembling the eggs of *Lagopus lagopus* considerably, both in colour and markings, but they average smaller. The majority are readily distinguished from those of the latter, the markings, as a rule, being smaller and better defined, and seldom running into indistinct and irregular blotches, as is frequently the case in the eggs of that species. The ground colour ranges from a pale cream to a decided yellowish-buff, and in many specimens this is entirely hidden by a vinaceous rufous suffusion. The spots and blotches range from a dark clove-brown to a dark claret-red, with paler coloured edgings; they are of various sizes, from the size of a buckshot to that of No. 10 shot, and are irregularly distributed over the egg.

Brandt gives a very good description of the eggs as follows:

The egg of the rock ptarmigan is ovate to elongate ovate in shape and has slight to considerable luster, which apparently increases as the egg incubates. The surface is smooth and greasy, and the sturdy shell strongly resists the drill. The vivid markings on this beautiful egg are so numerous that they often all but envelope the pale ground color and produce rich-mottled decoration that gives the egg a noteworthy appearance. These spots are distributed evenly over the surface, except that often a confluent cap intensifies the color at the larger end. The eggs laid by one bird are similar in shape, plan of markings, and coloration, but seldom are there two sets from different parents exactly alike. The inconspicuous ground color follows the paler buffs and creams, often with a reddish suggestion; shell pink, pinkish buff, or cream color is often observed, while many eggs are still lighter than these pale colors. The markings range in size from the smallest spots to those approaching thumb-nail in size and are more or less confluent. As a rule, the larger the spots are in size, the more the ground is shown. These markings when dry are blackish brown, but where the pigment has been sufficiently thinned, the color ranges from walnut brown and maroon to blackish brown. Underlying spots apparently are not present.

Behavior.—Doctor Grinnell (1900) says of his experience with this ptarmigan in the Kotzebue Sound region:

I first met with this species on September 17, 1898, about the summit of the Jade Mountains on the north side of the Kowak Valley. On that day I saw three flocks of 6, 7, and 20 birds, respectively. In each case they were flushed from ridges at some distance, and were probably feeding on heath and blue-berries, which fairly covered the ground on favorable slopes. At a distance the birds appeared to be entirely white, at this date, though no specimens were obtained. I rather think the summer plumage of the Rock Ptarmigan is of much shorter duration than that of the Willow Ptarmigan in the lowlands. The Rock Ptarmigan, according to my experience, are confined exclusively to the higher hill-tops and mountains in summer, and at such elevations the snow remains later in the spring and comes much earlier in the fall than in the valley, leaving a very brief summer. No Rock Ptarmigan were detected in the Kowak Valley until February 11th. On account of the light snow-fall in the early part of the winter, they probably found sufficient forage on the mountain sides up to this date. However, during March and April flocks of from a dozen to a hundred were often met with in the lowlands. These flocks could be traced up by following their tracks, especially if the snow was freshly fallen or laid by the wind. Then tracks of a large flock of Rock Ptarmigan would form a broad swath and extend across the tundra for miles, the individual lines of tracks zigzagging back and forth so as to take in every willow twig or bunch of grass sticking up through the snow, but all tending in the same general direction. The birds, when on these feeding marches, apparently seldom take flight unless disturbed, and I have followed these roads from one set of "forms" in the snow, where the birds had passed the preceding night to the second set of "forms" of the succeeding night, and then finally found them, doubtless on their second day's walk without taking flight; except occasional individuals left behind. The tracks of the Rock Ptarmigan are easily distinguishable from those of the Willow Ptarmigan by their much smaller size and the shorter strides; and they seem not to be in the habit of dragging their middle toes over the ground at each step, as evidenced by the tracks in the case of the Willow Ptarmigan.

W. Sprague Brooks (1915) says:

The males are quite pugnacious, when in flocks, often pursuing each other and going through antics suggesting the young males of domestic fowls.

Rock Ptarmigan exhibit considerable curiosity at times, a trait I have not noticed in the Willow Ptarmigan. When one of its kind is dead or wounded the rest frequently show great concern and interest in the unfortunate one.

Many times while walking over the tundra I would be startled by the rattling call of a male Rock Ptarmigan, and turning about see him alight within a few yards of me with tail spread and eye wattles erect. After strutting about and "showing off" a moment he would busy himself searching for food as though no man were in the country. In the winter plumage the males are very beautiful.

Voice.—Dr. E. W. Nelson (1887) came across a pair of these birds near St. Michael, of which he writes:

They allowed me to approach within 20 feet, and paid no attention beyond looking curiously at me as I walked slowly along. The suspicion of the male being slightly excited, he uttered a low, rolling or whirring sound, like that produced by rolling the end of the tongue. The female answered with a low, clear *yop-yop*, with a peculiar intonation, strikingly like that of the female hen-turkey, except it was much lower. When we were about 15 feet from the birds, they stood looking at us for a moment with a pretty air of innocent curiosity, and then, without showing the slightest signs of alarm, arose and flew off to the hill-side, a hundred yards or more away.

Migrations.—Nelson says of their migratory movements:

During the entire year these birds are resident north at least to Bering Straits, as I obtained specimens from that vicinity on one of my winter expeditions. In summer it extends still beyond this, to all portions of the country crossed by mountain chains and hills. In autumn, toward the last of October and first of November, this bird unites with the common Ptarmigan in great flocks, on the northern shore of Norton Sound, and migrates thence across the sound to Stuart's Island, thence reaching the mainland. The birds are frequently seen by the natives while they are passing Egg Island, on their way to the island just mentioned. They are said to commence their flight just before dark in the evening, and at this season, as mentioned under the preceding species, many are snared at the head of Norton Bay. In April the birds return to the north, always traveling in the evening or night, as they do during their autumnal migrations.

LAGOPUS RUPESTRIS EVERMANNI Elliot

EVERMANN'S PTARMIGAN

HABITS

The well-marked subspecies *evermanni*, the darkest colored of all the ptarmigans, the males being almost black, is confined to Attu Island, the westernmost of the Aleutian Islands, 1,400 miles west of

Unalaska. Dr. D. G. Elliot (1896) in his original description of it observes:

The males of *L. evermanni* bear a certain resemblance to specimens of *L. mutus*, of the Eastern Hemisphere, where these have much black in their plumage; but between Attu and the continent of Asia is found *L. ridgwayi*, a very distinct form from Bering Island, about 300 miles west of Attu. This would seem to bar any possible relationship between *L. evermanni* and any continental species, though it is a surprising fact, and one that can only be theorized upon and not thoroughly explained, that species which are closely allied can be separated by many miles of sea and land, and yet retain their specific characteristics, though distinctly different species may be found occupying interlying territory. This is one of the curiosities of geographical distribution, the solution of which is probably beyond the power of man to fathom.

Everyone who has visited Attu Island has remarked on the scarcity of this ptarmigan. Perhaps it may live so far up on the mountains that its favorite haunts have not been visited; this would fit my theory that the dark-colored birds are mountain birds and that the light-colored birds are lowland birds. But perhaps the scarcity may be due to persistent hunting by the inhabitants of Chichagof Village or to the presence of blue foxes, which are very common here. Doctor Elliot had only seven specimens on which to base his description. Austin H. Clark (1910) writes:

I did not find this bird at all common on Attu, doubtless because I did not succeed in locating its favorite haunts. During an entire day's trip over the mountains on the right of the harbor, behind the town, and about the large lake at the summer encampment, only three were seen, one in the mountains above the lake and two in the lowlands between the town and the lake. All three were shot. On arriving at the ship one of the men told me he had never seen ptarmigan so common as about the summit of the mountains at the left of the harbor entrance. As he had had considerable experience with ptarmigan in seldom visited portions of Alaska, and was a reliable man, I arranged to visit the locality the next day with him as a guide in order to obtain a series of this little known species. We started early and reached the place a little before noon, but, although the droppings of the birds were extremely abundant everywhere, we saw none of the birds themselves. Just as we were preparing to leave, after searching the whole district thoroughly, a fine cock came flying over from one of the neighboring peaks and was promptly secured. On our way back to the shore we saw one other which was chased for over a mile but without success.

Hamilton M. Laing (1925) says:

During our three days' stay at Attu Island only three ptarmigan were found. A single bird on April 20 bounded up from the shore and flew wildly away. Next day, which was warm and sunny, two single birds were seen sitting on the brow of the bluff above the shore and both were secured. Even in life the difference between these birds and the previous forms was evident, the new blackish feathering giving them a decided speckled appearance. A climb to one of the hilltops disclosed no evidence of the birds at higher elevations.

We spent only a day and a half at Attu Island and collected eight ptarmigan, one pair that I shot in the valley and six that Rollo H. Beck shot on the mountains; the bare and moss-covered rocky sides of the mountains seem to be their favorite haunts; the eight birds secured were all that we saw.

Nothing seems to be known about the nesting habits or eggs of Evermann's ptarmigan. All we know about molts and plumages is that the male described by Doctor Elliot (1896) was just completing the molt into the summer plumage on June 4, and that the males collected by Laing (1925) were just beginning this molt on April 21. Our birds, seven males and one female, were in full summer plumage on June 22 and 23, and the female had a bare patch on the belly, showing that she was incubating. We learned nothing further about their habits.

LAGOPUS LEUCURUS LEUCURUS (Richardson)

NORTHERN WHITE-TAILED PTARMIGAN

HABITS

The name "northern" white-tailed ptarmigan, adopted in the new American Ornithologists' Union Check List, seems hardly suitable for this, the type race of the species, for the Kenai race, *peninsularis*, ranges entirely north of it. Typical *leucurus* is the bird of the western Canadian mountains in British Columbia and Alberta. This form is darker, with more black in the plumage and has shorter wings and tail than the southern bird, *altipetens*. The southern bird being the best known race, the reader is referred to *altipetens* for the principal life history of the species.

Dr. Frank M. Chapman (1908), in his attractive account of this bird in the Canadian Rockies, says:

> They are said not to descend below timberline during the summer, but we noted a striking exception to this rule at Lake Louise, where numbers of them came regularly to feed about the forest-surrounded stable. They were evidently attracted by the fallen grain and may have learned of this supply of food during the winter when the heavy snowfall drives them to lower levels.

Referring to their behavior, he writes:

> The first evidence they gave of being aware of my presence, was to remain perfectly motionless, then, as I made no further advance, they attempted to combine action with rigidity of pose and were almost successful in achieving this impossible feat. With painful slowness, one foot was placed in advance of the other, at the rate of about three steps to the minute. If I drew so near that the birds seemed convinced that they were seen, the male assumed a more alert, bantamlike attitude, ducking his upraised head and flirting his tail as though inviting me to conflict. The pose of the female was more henlike, and less aggressive. She showed virtually no concern when I was within three feet of her, feeding about the rocks, and even stopping to scratch her head. After an

hour or two, the male became more accustomed to me, and seemed as much at ease as his mate, uttering a low, crooning note suggesting that of a comfortable chicken on a sunny day.

DISTRIBUTION

Range.—Alpine sections of southern Alaska, western Canada, and the United States.

The range of the white-tailed ptarmigan extends **north** to Alaska (Lake Clark, Savage River, and Robertson River); Yukon (probably La Pierre House); and southwestern Mackenzie (Nahanni Mountains). **East** to southwestern Mackenzie (Nahanni Mountains); western Alberta (Henry House, Laggan, and Sulphur Mountain); western Montana (St. Marys Lake, Piegan Pass, and Beartooth Mountain); Wyoming (Medicine Bow Mountains); Colorado (Mount Zirkel, Arapahoe Mountain, Longs Peak, Bald Mountain, James Peak, Breckenridge, St. Elmo, Cochetope Pass, and Summit Peak); and New Mexico (Costilla Peak, Taos Mountains, and Mora Pass). **South** to northern New Mexico (Mora Pass); southwestern Colorado (Dolores Mountain); and northwestern Oregon (Mount Jefferson). **West** to northwestern Oregon (Mount Jefferson and Mount Hood); Washington (Mount St. Helena, Mount Rainier, Pyramid Peak, Cloudy Pass, Mount Sahale, and Mount Baker); British Columbia (Mount Arrowsmith, Della Lake, Ninemile Mountain, Groundhog Mountain, head of the Iskut River, and Dochda-on Creek); and Alaska (Admiralty Island, Hooniah, Glacier Bay, Valdez, Kenai Mountains, and Lake Clark).

The white-tailed ptarmigan is confined entirely to mountainous regions, and it does not perform a migration comparable to that of the willow and rock ptarmigans. A slight vertical movement usually takes place in winter when the birds descend from the peaks and ridges to sheltered valleys in search of food.

The range as described is for the entire species, which has been divided into four subspecies. The "northern" white-tailed ptarmigan (*Lagopus l. leucurus*) is found from northern British Columbia and central Alberta south to Vancouver Island; the Kenai white-tailed ptarmigan (*Lagopus l. peninsularis*) occurs from central Alaska, northern Yukon, and Mackenzie south to the Cook Inlet region, Kenai Peninsula, and central Yukon; the Washington white-tailed ptarmigan (*Lagopus l. rainierensis*) is found in the Cascade Mountains of Washington; and the southern white-tailed ptarmigan (*Lagopus l. altipetens*) occupies the Rocky Mountain region of the United States, from Montana to northern New Mexico.

Egg dates.—Colorado: 14 records, June 19 to July 15; 7 records, June 26 to July 6.

LAGOPUS LEUCURUS PENINSULARIS Chapman

KENAI WHITE-TAILED PTARMIGAN

HABITS

The type of the local race of the white-tailed ptarmigan named *peninsularis*, one of a series of 26 specimens, was taken by J. D. Figgins in the Kenai Mountains, on the Kenai Peninsula, Alaska, on August 11, 1901. It is in the gray fall, or transition, plumage. Dr. Frank M. Chapman (1902) in his original description gives it the following subspecific characters: "In nuptial plumage differs from corresponding phase of plumage of *Lagopus leucurus* in having the black areas of great extent, the buff areas much paler. In fall, transition or 'preliminary' plumage differs from similarly plumaged specimens of *Lagopus leucurus* in being decidedly grayer."

Chapman quotes from Mr. Figgins's notes, as follows:

> Reared far above all timber, these interesting birds must depend upon their color for protection at all times. Found only on the bleak barren grounds, not even a blade of grass rises to offer them a retreat. Their color is an exact imitation of their rocky surroundings, and if the bird remains at rest it is impossible to detect it though only a few feet distant. When approached they crouch as closely to the ground as possible, usually near some small boulder, and remain thus while you are in motion, but if a stop is made they try to steal away and in that way reveal themselves. As soon as a movement is made they resume their former position. They are hard to flush, depending rather upon their color for safety than their wings. A low cackling when their young are disturbed are the only notes I have heard. The food of this ptarmigan is berries and the leaves of small plants. The principal berry resembles our blueberry in appearance and remains fresh the year round, falling from the plant only when a new crop is grown.

LAGOPUS LEUCURUS ALTIPETENS Osgood

SOUTHERN WHITE-TAILED PTARMIGAN

HABITS

Dr. Wilfred H. Osgood (1901) discovered that the white-tailed ptarmigan of the Colorado mountains is subspecifically distinct from the bird of the mountains north of the United States and gave it the name *altipetens*. He gives as its characters: "Adult in fall plumage similar to *Lagopus leucurus*, but general color of upperparts buff instead of gray; adult in summer plumage indistinguishable in color from *leucurus;* wings and tail decidedly longer than in *leucurus*."

The white-tailed ptarmigan is an alpine species, a permanent resident in the high mountains, above timber line during most of the year at least. In the southern portion of its habitat it ranges from 10,000 to 14,000 feet altitude and somewhat lower farther north. It is the only ptarmigan known to breed within the limits of the United

States. Mrs. Florence M. Bailey (1918) gives the following attractive description of its haunts in the Glacier National Park:

Skirting an acre of snow, I zigzagged back and forth over the face of the "ideal ptarmigan slope," open to swift-winged enemies, but by its broken surface and variety of colors affording a safe background for ptarmigan in the mixed summer plumage. Even the wide expanse of slide rock was broken by occasional dwarf evergreens and streaks of grass, and many of its red shales were patterned with lemon-yellow or curly-brown lichen covering deep ripple marks. Above the main mass of slide was a wide grassy slope of soft yellowish brown tones that would soon match the brown of the ptarmigan. Above this the narrow outcropping ledges and stony slopes made a terraced Alpine flower garden, one of the gardens that are among the choicest of all nature's lavish gifts to man; this one, with its maturing seed harvest, providing veritable grain fields for hungary bird and beast. Some of these Alpine terraces were fairly white with the lovely low, wide-smiling *Dryas octopetala.* In other places the beds of white were spotted with the pink mossy cushions of *Silene acaulis,* while in still others there were clumps of dwarf sedum, whose dark-red flowers and seed pods contrasting strikingly with their pale green leaves might well attract the attention of furry vegetarians locating granaries, and make good feeding grounds for the Arctic grouse.

Nesting.—Bendire (1892) quotes A. W. Anthony as writing:

In southern Colorado, where I have met with this species, nesting must begin some time from the first to the middle of June, as I have found young birds but an hour or so from the egg, from July 1 to the 18th. The nests I have seen were located in the loose rocky débris of steep hillsides, a simple depression in the short fine grass which grows in small patches between the rocks above the timber line. Although utterly devoid of protection from bush or shrub, so nearly does the sitting bird resemble the gray bowlders which surround her on every side that the discovery of the nest is due largely to accident. When incubating it is nearly impossible to flush the bird, according to my experience. Twice have I escaped stepping upon a sitting Ptarmigan by only an inch or so, and once I reined in my horse at a time when another step would have crushed out the life of a brood of nine chicks but an hour or so from the egg. In this case the parent crouched at the horse's feet, and, though in momentary danger of being stepped on, made no attempt to escape until I had dismounted and put out my hand to catch her. She then fluttered to the top of a rock a few feet distant, and watched me as I handled the young, constantly uttering low anxious protests. The chicks were still too young to escape, mere little awkward bunches of down that stumbled and fell over one another when they attempted to run.

He quotes Doctor Coues's description of a nest as follows:

The nest in its present state measures scarcely 5 inches in diameter by about an inch in depth. It thus seems rather small for the size of the bird, but is probably somewhat compressed in transportation. The shape is saucer-like, but with very little concavity of surface. The bottom is decidedly and regularly convex in all directions, apparently fitting a considerable depression in the ground. The outline is to all intents circular. The nest is rather closely matted, the material interlacing it in all directions, and retains considerable consistency. The material is chiefly fine dried-grass stems; with these are mixed, however, a few small leaves and weed tops and quite a number of feathers. The latter, evidently those of the parent birds, are imbedded throughout the substance of the nest, though more numerous upon its surface, where a dozen or so are deposited; there may have been some loose ones lost in handling.

Illustrating the perfectly concealing coloration of the bird on its open nest, Evan Lewis (1904) writes:

On reaching timber-line a Junco was seen building, and a search was made for a loose stone to mark the spot for a photograph when the set was complete. In the search I was just about to put my hand on a Ptarmigan when I saw what it was. I then made two exposures with the small camera and left the camera on top of a large rock to mark the spot, the nest being three steps and one foot due south from the mark. I went to the cabin at the lake and got the large camera and tripod. When I returned I took three rather shorter steps, as I supposed, and looked for the bird or its nest. For ten minutes I looked over the ground foot by foot. I could not believe my own eyes that the bird was not there, yet I could not see her. At last I was about to return to the mark and step the ground over again, when a reflection from the bird's eye showed her to me just one foot from where I was standing.

W. C. Bradbury (1915) had some similar experiences. Nests, which he had previously located and marked, he had difficulty in finding again even when standing within a few feet of them. One nest was right beside a stone that his foot was on; but it had been lightly covered with grass when the bird left it. In one case he found that only three eggs had been deposited in a period of five days.

Eggs.—The white-tailed ptarmigan has been credited with laying as many as 15 eggs and as few as 4; probably the usual numbers run from 6 to 8. In shape they vary from ovate to elongate-ovate, and they have little or no gloss. They are quite unlike other ptarmigans' eggs and are colored more like small eggs of the dusky grouse. The ground color is usually "cartridge buff" or "pinkish buff" and rarely "cinnamon-buff." The lightest-colored eggs are sometimes nearly immaculate; most of the eggs are more or less evenly covered with small spots or fine dots; some are more heavily marked with larger spots or small blotches; but the ground is never largely concealed. The markings are in various shades of brown, usually very dark brown, "Vandyke brown" to "bone brown," but sometimes as light as "snuff brown" or "tawny-olive." The markings are rarely concentrated into a ring. The measurements of 31 eggs average 42.9 by 29.5 millimeters; the eggs showing the four extremes measure **49.3** by 29.6, 44.7 by **32.3**, **39** by 28.3, and 42.9 by **28.2** millimeters.

Young.—Mrs. Bailey's (1918) account of the behavior of mother and young is worth quoting, as follows:

Listening, I caught it again—the softest possible call of a mother ptarmigan! There she stood, only a few feet from me, hard to see except when in motion, so well was she disguised by her buffy ground color finely streaked with gray. A round-bodied little grouse with a small head, she was surrounded by a brood of downy chicks, evidently just hatched, as their bills still held the sharp projection for pipping the shell. Preoccupied with the task of looking after

her little family, as I talked reassuringly to her, she ignored my presence. Nothing must hurry the unaccustomed little feet, nothing must interfere with their needed rest. Talking softly she gradually drew the brood in under her motherly wings and sat there only a few yards from me, half closing one eye in the sun and acting oblivious to all the world. Once the downy head of a chick appeared between the fluffed-out feathers of her breast, and once she preened her wing so she showed the white quills remaining from the white plumage of winter.

Her bill opened and her throat palpitated as if she were thirsty, as she sat brooding the young, and I imagined that the last hours of hatching high above water had been long and trying to the faithful mother. But though water—clear cold mountain brooks—were below, no need of her own could make her careless of her little ones. Keeping up a motherly rhythmic *cluck-uk-uk, cluck-uk-uk,* interlarded with a variety of tender mother notes, she led them down by almost imperceptible stages, slowly, gently, carefully, raising a furry foot and sliding it along a little at a time, creeping low over the ground with even tread, picking about as she went, while the little toddlers gradually learned the use of their feet. Like a brood of downy chickens, some were more yellowish, some browner than others, but they all had dark lines on head and body giving them a well-defined color pattern. Peeping like little chickens, while their mother waited patiently for them they toddled around, trying to hop over tiny stones and saving themselves from going on their bills by stretching out wee finny wings. As chickens just out of the shell instinctively pick up food from the ground, they gave little jabs at the fuzzy anthers of the dryas, little knowing that pollen was the best food they could find, a rich protein food from which the bees make bee bread to feed their larvae.

Plumages.—In the small, downy chick of the white-tailed ptarmigan the crown, shoulders, central back, and central rump are "tawny," bordered and sprinkled with black; the rest of the head shades from buffy white on the forehead to dull white on the chin and throat, with black spots and bands on the front and sides of the head in somewhat different patterns; the rest of the upper parts are variegated with pale buffs, grays, and black; the underparts are grayish white, with a slight buffy tinge on the breast. As with all other grouse, the juvenal plumage appears very early, the wings first when the chick is very small; and the neck and head are the last to be feathered.

By the time that the young bird is half grown it is fully feathered in juvenal plumage. In this the crown is barred or mottled with black, white, and pale buff; the back and rump are mainly vermiculated or peppered with black on a grayish white to "cinnamon-buff" ground color, but some feathers show larger black areas; the scapulars, wing coverts, and tail coverts are similar, with more black on the scapulars; the chin and throat are grayish white, barred with dusky; the breast and flanks are pale buff, barred with black, and the belly is buffy white; the two outer primaries are white and the others dusky, the tail feathers are dusky, banded and mottled with "cinnamon-buff." This plumage is hardly complete in August,

before the postjuvenal molt into the late summer plumage begins. This is similar to the preliminary winter, or tutelar, plumage of adults. About a month later, in September, the molt into the pure-white winter plumage begins. These two molts effect a complete change of plumage between August and October, except that the two outer primaries are retained for a year, which serve to distinguish young birds from adults.

What little material is available seems to indicate that the molts and plumages of adults are similar to those of other ptarmigan. A partial prenuptial molt of the contour plumage, head, neck, and back takes places from March to June. The late summer, or tutelar, plumage is assumed by a partial postnuptial molt beginning late in July; this is much grayer, more finely vermiculated, and with less black than the nuptial plumage, which is more heavily vermiculated, with more black spots, on a more "ochraceous-buff" ground color. Females are more buffy or ochraceous than males in both plumages. A supplementary molt produces, in September and October, the complete change into the pure-white winter plumage.

Food.—Dr. Sylvester D. Judd (1905a) summarizes the food of this ptarmigan, as follows:

> During winter in Colorado, according to Professor Cooke, they subsist, like other ptarmigan, largely on willow buds. The stomachs of two birds collected at Summitville, Colo., in January, 1891, at an altitude of 13,000 feet, were found to contain bud twigs from one-third to one-half inch long, but the kind of bush from which they came could not be determined. Doctor Coues, quoting T. M. Trippe, states that the food of this bird is insects, leguminous flowers, and the buds and leaves of pines and firs. According to Major Bendire, the flowers and leaves of marsh marigold (*Caltha leptosepala*) and the leaf buds and catkins of the dwarf birch (*Betula glandulosa*) are eaten. Dr. A. K. Fisher examined the stomachs of two downy chicks collected on Mount Rainier, Washington, and found beetles and flowers of heather (*Cassiope mertensiana*) and those of a small blueberry.

Mrs. Bailey (1928) adds:

> The crop of one New Mexico specimen was filled mainly with leaves of the dwarf willow, and fruiting spikes of *Polygonum viviparum*, with one flower of *Geum rossii*, while the gizzard held mainly *Polygonum* seeds, a few other small seeds, a few small grasshoppers, and other small insects.

Behavior.—Quoting Trippe, Coues (1874) writes:

> In localities where it is seldom molested it is very tame, and I have been informed by persons whose word is worthy of belief, that they have frequently killed it with sticks. But when persistently persecuted, it soon becomes wild, and leaves the range of a shot-gun with surprising quickness. After hunting several large flocks for three or four days, they grew so shy that it was difficult to approach within gunshot, although at first they had been comparatively tame. Nimble of foot, the Ptarmigan frequently prefers to run away on the approach of danger, rather than take wing, running over the rocks and leaping from point to point with great agility, stopping every

little while to look at the object of alarm. I have sometimes chased them half a mile or more, over the rocky, craggy ridges of the main range, without being able to get within gunshot, or force them to take wing. The flight of the Ptarmigan is strong, rapid, and at times sustained for a considerable distance, though usually they fly but a few hundred yards before alighting again. It resembles that of the Prairie Hen, consisting of rapid flappings of the wings, alternating with the sailing flight of the latter bird. The note is a loud cackle, somewhat like the Prairie Hen's, yet quite different; and when uttered by a large flock together, reminds one of the confused murmur and gabble of a flock of shore-birds about to take wing. It is a gregarious bird, associating in flocks throughout the year, except in the breeding season. The different broods gather together as soon as they are nearly grown, forming large flocks, sometimes of a hundred or more.

Dr. D. G. Elliot (1897) says:

They were not what may be called tame, unlike the Willow Grouse in this respect, but were always very uneasy at my presence, and ran about with uplifted tail as if uncertain which way to fly, but when they once got started there seemed to be no farther difficulty in their minds as to the proper direction, which I noticed never led near where I stood. Sometimes I have seen them light on the bare limbs of a stunted tree or large bush at the edge of the timber line, where they stood perfectly motionless for quite a length of time, observing every movement I made, and then suddenly burst away with great speed, uttering a low cackle as they flew. They are very skillful in concealing themselves, either squatting in the snow with only the head exposed to view, or else crouching behind some stone or large bowlder. In summer their peculiar gray plumage assimilates so well to the hue of the ground and the moss-covered stones lying about in all directions that it is next to impossible to perceive them, and at this period, especially during the breeding season, they rarely move when approached, perhaps only going a few feet on one side to avoid being stepped upon.

Denis Gale wrote to Major Bendire (1892):

Irrespective of season, as a general rule, a single bird will not flush unless urged to it. During the summer months this is especially noticeable; they will only move out of your way when directly in your path, and close upon them, by short tacks right and left, sidling off from you, at each tack changing sides, moving quickest on the short run just before slowing up for the turn. Two or more together are much more likely to flush, and if alarmed while flying will utter a quick repeated *köck, köck*, very like the note uttered by ***Pediocaetes phasianellus campestris*** under similar circumstances.

Game.—The white-tailed ptarmigan is a fine game bird for those who are hardy enough to stand the hard tramping necessary for its pursuit in the high mountains. Edwyn Sandys (1904) writes:

Unlike many of its kin, this bird is not troubled with overconfidence in man, but is apt to fly smartly and present none too easy a mark. It is also quite a runner, and taken altogether, the "snow quail," as the miners call it, is a fit quarry for an expert, especially if he be a "tenderfoot," unused to Alpine work and the pure, thin air of the heights; for this ptarmigan is a lover of high altitudes, seldom, if ever, being seen lower than five or six thousand feet.

Fall.—Dr. George Bird Grinnell wrote to Major Bendire (1892):

In the autumn the birds are generally rather wild, and if nearly approached become quite uneasy and run about, holding the tail elevated and looking very much like a white Fan-tail Pigeon. At this season the only cry that I have heard is a sharp cackle like that of a frightened hen. This the bird begins to utter a short time before it takes wing, and continues it for quite a little while after having begun to fly.

On the high plateaus where this bird is found the wind often blows with a tremendous sweep and is almost strong enough to throw down a man. When such a wind is blowing the Ptarmigan dig out for themselves little nests or hollows in the snow banks, in which they lie with their heads toward the wind and quite protected from it. Often on the rocky slopes where there is no snow they may be seen lying crouched on the ground behind rocks or small stones, with their heads directed to the quarter from which the wind blows. If startled from such a place they all take wing at once, looking like a flock of white Pigeons, and fly for a short distance, but as soon as they touch the ground again they throw themselves flat on it behind the most convenient shelter.

Winter.—Sandys (1904) says:

At the approach of winter the broods of a district frequently join forces in a packlike formation. I have seen 40 or 50 together, and heard the miners speak of packs of several hundreds; this, however, is hearsay, and perhaps 100 birds together would be a large pack. During rough weather the birds will go under the snow; in fact, they will hide in snow whenever it is available.

During severe winters and when the snow is so deep that their food supply is covered, these ptarmigan desert their normal home above timber line and descend into the edges of the spruce timber on the hillsides or into the creek bottoms among the willows, where they can find food and shelter.

LAGOPUS LEUCURUS RAINIERENSIS Taylor

RAINIER WHITE-TAILED PTARMIGAN

HABITS

Dr. Walter P. Taylor (1920) described and named this dark race from a series of eight adults and four young birds collected on Mount Rainier, Wash. He says that adults in nuptial plumage are

similar to *Lagopus leucurus leucurus*, but dark areas more blackish; buffy wash over light areas not so consistently present, and when present paler.

Comparison with specimens of *Lagopus leucurus leucurus* from Moose Pass, British Columbia, Moose Pass, Alberta, and Moose Branch of Smoky River, Alberta (one specimen from Henry House, Alberta), practically topotypes of *leucurus*, all in nuptial plumage, indicates that the dark areas in *rainierensis* average more blackish than in *leucurus*. In the latter the shade is close to mummy brown (Ridgway, Color Standards, 1912), while in *rainierensis* they approximate one of the darker shades of blackish brown. The buffy portions of the feathers in *rainierensis* are paler than in *leucurus*, being, in the former, near light ochraceous-buff, while closer in the latter to ochraceous or ochraceous-tawny.

There is no evidence that this race differs materially in habits from other white-tailed ptarmigan. Doctor Taylor (1927) says that it ranges in altitude between 6,000 and 8,000 feet

in the Arctic-Alpine Zone all around the mountain, rarely dropping down into upper Hudsonian, except in winter.

The ptarmigan finds congenial surroundings on the pumice slopes at and above timber line on Mount Rainier. Here the combination of bright light, freezing boreal blasts, dwarfed and wind-blown vegetation, and extensive snow and ice fields provide Arctic conditions in fact.

Nesting.—Taylor describes a nest as follows:

Consciously or unconsciously the ptarmigan had here selected a nest site which for grandeur of outlook would be hard to equal. The nest was on the ground on the south side of a rock on a southwest slope of Pyramid Peak, at an altitude of about 6,100 feet, where the hardy conifers, dwarfed and matted in their unequal struggles with the elements, had at length given up completely. At first glance the nest did not appear to have been specially constructed; but it was later found that a hollow had been excavated and filled with dried vegetation. The nest itself was comfortably dry, though the soil below was damp, and doubtless usually frozen solid. A few feathers were scattered about the nest. Plants in the immediate vicinity were the red and white heathers and the Siberian juniper. There were five eggs, one infertile, one addled, and three in various stages up to approximately 10 days' development.

Voice.—Taylor gives the best description of notes of this species that I have seen. Referring to the female, he writes:

While on the nest she several times uttered a *hoot, hoot, hoot, hoot, hoot,* a low, almost inaudible, soothing series of grouse-like notes. Another note uttered by the ptarmigan as she turned the eggs was a *cluck! cluck!* much resembling the call of a barnyard fowl to her little chickens. When away from the nest she stalked about rather slowly, occasionally jerking back her head in a characteristic manner, and regarded with evident anxiety the nest site about which we were grouped. If we approached the nest too closely the gentle bird was not a little perturbed and warned us *perrt! perrt!* or sometimes *pit-prrrrt! prrrrt!*

Of the male he says:

One of the birds, a cock, remained in the vicinity for upward of an hour, watching the observer and calling for his mate. His principal call was somethink like *Su-squeek! cluck-luck-a-luck, cluck-luck-a-luck!* or sometimes *Squeek! cluck! cluck! cluck! cluck! Cluck lucka-lucka-lucka-cluck!* Occasionally the call is blurred at the end, *Cluckrrrrrrrr!* The squeaking note, which is of staccato quality, high pitched and conspicuous, may be twice repeated, as follows: *Squeek! chuck chuck chuck chuck chuck chuck chuck Squeek! chuck chuck chuck!* A call somewhat resembling that of the red-shafted flicker was heard *yip! yip! yip! yip!* Another combination *Yip, yip squeech! yip! yip!* A warning note may be represented by the syllable *chirrr chirr chirr chirr chirr chirr.*

TYMPANUCHUS CUPIDO AMERICANUS (Reichenbach)

GREATER PRAIRIE CHICKEN

HABITS

CONTRIBUTED BY ALFRED OTTO GROSS

The prairie chicken ranks first among the game birds of the prairies of our Middle West. It is to the prairie what the ruffed grouse is to the wooded sections of the country. As intensive agriculture pushed to all sections of the range of the prairie chicken and as interest in hunting increased, this fine game bird at one time seemed in grave danger of following the course taken by the heath hen, to extinction as a game bird. In fact, it is gone from much of its former range, and its original numbers have been greatly reduced in practically the entire area of its distribution.

Because market hunting has been made a thing of the past since the beginning of the twentieth century and also because of the increasing restrictions on hunting by State departments, as well as various effective conservation programs, the prairie chicken is now holding its own and is increasing its numbers in many sections of its present range. Another hopeful sign is the fact that it has been extending its range to the northwest, and to-day the species is well represented on the prairies of Manitoba and is gradually spreading westward through Saskatchewan and Alberta, where formerly it did not exist.

The State Department of Conservation of Wisconsin has undertaken a comprehensive investigation of the prairie chicken to ascertain all the facts that affect its life, with the expectation that the department will be able to carry on a more effective program of conservation. Until the fundamental facts in the biology of our game birds are clearly known, conservation commissions will be handicapped in handling questions of game legislation and game management.

Prairie chickens, in common with other grouse, go through definite cycles of numbers. The problem of fluctuations in numbers of various species of wild life is not yet definitely solved, but work on it in relation to the ruffed grouse is being undertaken by many institutions and individuals in different parts of the country; hence there are excellent prospects of this work being brought to a successful conclusion.

The weather condition during the nesting season, especially during the height of the hatching period, is so important that it is frequently the determining factor in the number of young birds available for the next hunting season. A series of torrential cloud-bursts followed by long, cold, rainy spells during the first two weeks of June will cause hundreds of broods to perish.

During severe winters, especially when deep snows cover the ground, the birds are severely pressed to obtain enough food. A successful attempt has been made in Wisconsin to relieve this condition by the establishment of winter feeding stations. Crops of buckwheat and other grains are planted and left in the field to provide food to tide over the birds during these severe times.

One of the major problems involved in the conservation of the prairie chicken is the menacing fires that have swept the prairie regions during the nesting season of the birds. A fire at this time will destroy hundreds of nesting birds and their nests and eggs and in the course of a few hours undo the work of years of conservation work. Fires in fall destroy quantities of prairie-chicken food and the much-needed cover, without which the birds are left exposed to predators. The encroachment upon the breeding and feeding area by agriculture has long been recognized as a factor that has affected the status of the prairie chicken in the Middle West. This unfavorable situation is being relieved somewhat by the establishment of large State game preserves, on which the birds are given absolute protection and where conditions are systematically improved for the birds. The maintenance of winter feeding stations has been especially helpful in tiding the birds over the times when deep snows cover most of the normal food supply.

Intensive hunting has done much toward decimating the numbers of prairie chickens. The automobile and the fine modern roads have all been in favor of the hunter and against the birds.

Predators in their relation to game birds are important, but the value of vermin control is frequently overestimated. The wholesale killing of all hawks and owls, for example, should be rigidly avoided, for in the past this practice has actually acted as a boomerang to the objective of conservation of game birds.

Diseases and parasites of birds have not been well known in the past, but they are now becoming to be recognized as important factors in the life of our game birds. Under ordinary conditions, diseases and parasites may be of minor importance, but just as soon as the vitality and normal resistance of the birds are lowered by a series of adverse conditions, such as severe weather and scarcity of food, diseases and parasites manifest themselves and become of prime importance. It is the exceptional bird that is not parasitized, and hence this menace is ever present. There is also danger of infectious diseases, such as blackhead, which has been found to affect the prairie chicken and which figured in the decline of the heath hen. It is highly probable that the cycles in the grouse population are primarily dependent on some disease, either in itself or in combination with other factors. The evidence points to the conclusion that the vast majority of the parasites and diseases of our game birds have been introduced through poultry and exotic game birds. It is apparent that the adaptation of the prairie chicken to the conditions imposed by civilization is not a simple matter. In this adjustment, birds such as the prairie chicken will require much organized assistance on the part of conservation commissions, sportsmen, and bird lovers.

Courtship.—The courtship of the prairie chicken generally begins during the first warm days that lay bare the open fields of the winter's accumulation of snow. Though an early beginning may be made, the courtship does not reach its maximum until the latter part of April or the first of May, when companies of prairie chickens may be seen collected in favorable, often traditional, spots of the open fields throughout the prairie-chicken country.

O. M. Bryens, of Luce County, Mich., reports the first booming, or "crowing," as it is generally termed in the Middle West, as March 22, 1925; March 13, 1926; April 17, 1928; and March 27, 1929. According to Prof. W. W. Cooke, the booming of the prairie chicken was from March 7, at Caddo, Okla., to March 24, at Barton, N. Dak.

The courtship season continues through the month of May, but the vigor of its execution diminishes and the number of individuals that take part decreases as the sets of eggs hidden in the grasses of the prairie are completed and the domestic duties of incubation on the part of the female begin.

A few birds were still booming on the prairies near Hancock, Wis., when I arrived there the first week of June, 1929, and birds were also booming the second week of June, 1930, in various parts of Wood and Waushara Counties, central Wisconsin. I heard no booming and obtained no authentic accounts of birds booming after the second week of June.

The courtship of the prairie chicken is similar to that described for the heath hen, but since these performances are such an important part of the behavior of this bird a number of interpretations as made by other observers are of great interest.

Dr. Frank M. Chapman (1908) has given us a very vivid account of the prairie chicken as he observed it in the sandhills of Nebraska:

At short range the bird's note suggested the mellow resonant tone of a kettledrum, and when bird after bird, all still unseen, uttered its truly startling call, the very earth echoed with a continuous roar. As a rule, each bird had its own stand separated by about ten yards from that of its neighbor. The boom is apparently a challenge. It is preceded by a little dance in which the bird's feet pat the ground so rapidly as to produce a rolling sound. This cannot be heard for a greater distance than 30 yards. It is immediately followed by the inflation of the great orange air sacks at the side of the neck, which puff out as quickly as a child's toy balloon whistle; the tail is erect and widely spread, the wings drooped, the neck tufts are raised straight upward, giving the bird a singularly devilish look, then with a convulsive movement of the lowered head, the boom is jerked out and at its conclusion the air sacks have become deflated.

One might imagine after so violent a performance the bird would feel a certain sense of exhaustion or at least quiescent relief, but his excess of vitality seeks still other outlets; uttering hen-like calls and cacks he suddenly springs a foot or more straight into the air, whirling about as though he were suffering from a combined attack of epilepsy and St. Vitus dance. But all this activity is only a prelude to the grand finale of actual combat. Like a strutting turkey cock, the neighboring birds go towards each other by short little runs, head down, the orange eye-brow expanded and evident pouch inflated, neck tufts, and tail straight up, and looking like headless birds with two tails. Their mating is followed by no make-believe duel but an actual clash of wings. Uttering a low, whining note they fight as viciously as game cocks; and the number of feathers left on the ground testifies to effective use of bill and claws.

First bird called at 4.40 and by seven o'clock the performance was practically over.

A prairie cock when in the lists is a strikingly conspicuous creature; he wears no adornment which cannot be concealed at a moment's notice. The sight of a passing hawk changes the grotesque beplumed, be-oranged bird into an almost invisible squatting brownish lump, so quickly can the feathers be dropped and air sack deflated. With woodland birds so great a change is unnecessary, but the prairie hen can hide only under its own feathers.

H. L. Stoddard (1922), in notes from southern Wisconsin, says that "the 'cooing ground' at the sandy west end of Sauk Prairie has been used each spring for over 30 years, the birds always using the same knoll whether in rye, stubble, or grown to grass." Cooing started early in March and continued well into June. The birds arrived early; some were on the grounds before daylight, but on other occasions the bulk came shortly after daylight. The cooing is a "resonant C-A-O-O-O-O-O, H-O-O, H-O-O, rising and in the same tones as do re mi of the musical scale." This note carries a long distance. "I have heard it over water when the nearest land was nearly 2 miles away." Two cackling calls were like that of roosters, "one a loud Ka-Ka-Ka-Ka-a-a-a-a and the other a long-drawn q-u-a-h."

Alexander Sprunt, jr., says in his notes:

I witnessed a dance one afternoon of five pairs of the birds which came to a sudden end in a strange manner, and one which would thrill the heart of an ornithologist. Lying ensconced behind a log, I was reveling in the eaves-dropping act of witnessing the ludicrous antics of five males, who, with air sacs inflated, tails spread, and wings drooping, were bobbing up and down like corn in a hopper, about an admiring group of hens. The booming was intense and incessant, all having something to say at once. Suddenly, without a moment's warning, a huge snowy owl appeared from behind a low ridge at the far edge of the dancing ground and on widespread wings shot low over the dancers at a height of about 3 feet. Like so many feathered bombs, the chickens scattered to right and left, and in an instant the dancing ground was deserted. No attempt was made by the disturber to follow any of the revelers.

Nesting.—The nest of the prairie chicken is invariably on the ground, but the character of the vegetation in which it is built reveals considerable individual variation. Generally the nesting site is among grasses and weeds or low shrubbery in very open situations, but sometimes it may be adjacent to trees and woodlands and in rare instances may be surrounded by trees of considerable size. The vegetation about the nest is usually very thick and effectively conceals the eggs and the incubating bird from view. It also serves as a protection from extremes of temperature. There are sometimes killing frosts during the nesting season, in May, and there are many days in June when the heat is great enough to kill the embryos if left exposed to the direct rays of the sun for any great length of time.

The nest is placed in a natural hollow of the ground, or a slight excavation may be made by the bird by scratching out the loose earth and then molding the cavity to conform to the size and shape of the body. In this cavity the bird places a scant quantity of nesting material, in some instances the nest lining being little more than the bent-over blades and weeds growing about the structure.

The following descriptions serve to represent the character of the nesting site as well as the nature and construction of the nests built in three different types and situations located in central Wisconsin: A prairie-chicken nest containing 17 eggs was found 4 miles southeast of Bancroft, Portage County, on June 4, 1929, in a small clearing of a jack-pine grove, the trees of which ranged from 35 to 50 feet in height. The trees of the clearing had been cut the year before and piles of brush left in place. Some of the brush was more or less hidden by the rank growth of grass and weeds which had sprung up around it. The nest, built in a very shallow depression, 4 centimeters deep and 18 by 20 centimeters in diameter, was near one of the piles of brush. Some of the smaller branches were arched over the nest when found. It was protected by the brush on one side, but on the other it was well exposed to view, a condition very favorable for observations and photography from the blind that was later placed in position. Several of the pine trees were so near that they provided shade for the nest during certain hours of the day. Although this nest was not built in the usual surroundings, it is interesting to note that there were extensive marshlands all about the site.

In the drainage area of Wisconsin there are isolated areas of high ground locally called "islands," a name originating from the days before the drainage ditches, when they were in reality "islands" during the rainy period of the year. One such island of about 100 acres near Bancroft, Wis., is called Prairie Chicken Island, because these birds have always lived and nested there in unusually large numbers.

Since agriculture has been encroaching on the original habitat of the prairie chicken, many of the birds have adapted themselves with more or less success to conditions created by farming activities. A nest of the prairie chicken was found on June 24, 1929, on a farm 10 miles northeast of Friendship, Adams County, Wis. The nest was on rather high meadowland and was completely surrounded with a luxuriant growth of clover, timothy, and other grasses. The eggs were well concealed by a beautiful canopy of red-clover blossoms. The bowl of the nest measured 15 by 19 centimeters in diameter but only 4 centimeters in depth. The lining of the nest consisted of grasses and weed stems, all apparently picked up from the vicinity of the nest. The clover field was bordered on one side by a low, wet marsh, in which various sedges and rushes prevailed; on the other side was a cleared area being used as a potato field.

Other individuals cling to the more remote prairie districts away from farms, often in situations occupied in common with the sharp-tailed grouse. A nest containing 11 eggs was found in such a situation in Portage County on June 17, 1929. It was placed among tufts of sedge (*Carex stricta*). The nesting cavity was 18 by 21 centimeters in diameter and 8 centimeters deep. The lining of the nest was made up entirely of sedge, among which were a few feathers of the incubating bird. Near the nesting site was a low, wet marshy area, and on the other side there were thickets of small willows and poplars. No farmland or farm buildings were within 3 miles of this location.

Eggs.—The background color of the eggs varies from a "dark olive-buff" to a "grayish olive" tint. Most of the eggs are dotted with many fine and a few larger spots of "sepia." The spotting varies considerably in different eggs from those with scarcely any marks to those with many fine dots and 20 or more well-defined spots ranging from 1 to 2 millimeters in diameter. Sets of eggs found in open prairie regions seemed to have less spotting and the color of the spots was a "vinaceous-buff" rather than the dark markings of "sepia" present on eggs found in nests located in the wooded sections of the State. This difference in coloration, however, is probably a mere coincidence and is not to be correlated with a consistent difference in habitat.

Nine sets comprising 100 eggs, all found in central Wisconsin during the summer of 1929, were weighed and measured. The average long diameter was 44.86 millimeters; average short diameter,

33.59 millimeters. The eggs showing the four extremes measured **49.1** by 34.1, 45 by **35.3, 40.5** by 32.9, and 40.8 by **30.2** millimeters.

The laying of the eggs as determined by studies of captive birds extends over a period of days equal to nearly twice the number of eggs in the set. Apparently approximately the same ratio holds for birds living under natural conditions. In a nest in Adams County, Wis., the first egg was found on May 5, 1929; on May 12 there were 5 eggs; and the set of 11 eggs was completed on May 22. The time required to complete a set of eggs depends on a number of factors, such as the condition of the weather, the health of the bird, and the available food supply. The laying of the eggs is not necessarily on alternate days, but more apt to be very irregular. Certain eggs of the set are laid on successive days to be followed by a lapse of two days before the next egg is deposited.

During the laying period of one individual under observation the bird covered the rather exposed eggs with nesting material before she left them. This instinctive habit may be for the purpose of concealment or for protection from extremes of temperature or for both.

The number of eggs layed by the prairie chicken based on studies made of 40 nests, in which the number of eggs was presumably complete, varied from 7 to 17. The average number was 11.5 eggs per set. A nest containing 21 eggs has been reported, but such unusual sets probably represent the eggs of two females using the same nest.

The dates when the nests were found are of some interest to those who may desire to know the probable date when a nest of the prairie chicken may be found. Of 41 nests in which the date of finding was recorded, 20 were found in May, 18 in June, and 3 in July. The earliest date was of a nest found on May 5, when it contained a single egg, and the latest was a nest found July 10 that still contained eggs on July 15, the last time it was visited. The average date when the 41 nests were found is June 3.

Judging from these results we may say the last week of May and first week of June are the times when one may expect to find the largest numbers of nests of the prairie chicken in central Wisconsin.

Young.—The incubation of the eggs is performed by the female. Soon after the courtship season the male goes into retirement and undergoes the ordeal of molting. I have never seen the male bird near the nest, nor have I ever observed him participating in the care of the young.

Incubation begins soon after the set of eggs is completed, but in certain cases where we have found live embryos in eggs left in the nest after the rest of the brood had departed it was evident that incubation had started before these eggs were laid. Sometimes an unusual disturbance about the nest may delay the start of incubation for several days, thus making the determination of the incubation period under normal conditions in the field a difficult task. This accounts for discrepancies in the determinations made by different observers, which vary from 21 to 28 days as the incubation period for the prairie chicken. In the case of a nest under continuous observation, incubation began on May 22 and the eggs hatched on June 14, establishing an incubation record of 23 days. The incubation period for the eggs of the closely related heath hen, as determined by Dr. George W. Field, is 24 days.

After incubation is started the prairie chicken, under normal conditions, remains faithful to her duties through the vicissitudes of weather, storms, and dangers of attacks from enemies. Unless flushed from the nest she leaves only for very short intervals to feed, usually at dawn or late in the afternoon about sunset, times when the eggs will not be exposed to extreme heat. Excessive heat is very destructive to the embryos, and great care must be exercised in flushing birds away from nests in which the protecting vegetation has been removed for purposes of photography. The birds that left the nest normally slipped off it quietly and made no attempt to cover the eggs, as they did during the laying period when the nest was abandoned for a longer time. One bird, after nervously surveying her surroundings, sneaked off the nest and walked briskly in a crouched position until she had gone several yards from the nest. There she hesitated, elevated her head, and looked about as if to determine whether her movements had attracted any attention. She then casually nipped at the grasses as she walked along, and finally when about 25 yards away from the nest she arose with a loud whir of wings and disappeared in the scrub pines, where she probably found at least a part of her meal. After an absence of half an hour she flew into view, circled the nest, and alighted in the tall grass 15 to 30 yards away from the nest. At first she crouched in the grass completely concealed from view, but after being assured all was well she walked along stealthily though not directly toward the nest. It was not possible to keep her in view at all times, but now and then she would come to an open place and from this vantage point more carefully scrutinize the surroundings. Sometimes she completely encircled the nest and blind with a wide radius and frequently retraced her steps to make a careful inspection in order to satisfy herself that no spying enemy was near. One could not be sure whether this behavior was prompted by apprehension of harm to herself, fear of revealing the presence of the nest, or to both. After these maneuvers, which usually took about 20 minutes, were completed, she apparently was assured, and then without hesitation she walked quickly and directly to the nest. The position she assumed on the nest varied, and one could never be sure whether she would be facing toward the blind or away from it. The ruffed grouse, which builds its nest at the base of a stump, a log, or tree, invariably faces away from the side of the nest thus protected from the sneaking approach of some prowling enemy. Prairie chickens, which nest in thick vegetation, will usually face toward the side that is opened up for purposes of photography, a very desirable position for the photographer.

Many of the prairie chickens studied exhibited extreme restlessness and made much more of a task of incubation than does the ruffed grouse. One continuously shifted her position during the course of the day; at other times she would pick aimlessly at the nesting materials, and not infrequently she would reach far out of the nest at some unsuspecting grasshopper or other insect that chanced to alight on the tall grass. Certain females revealed a sensitive, nervous temperament and quickly responded to any stimulus, whether it was the *caw* of a crow or the hum of a distant tractor; and even the shadow of a passing cloud was sometimes sufficient to make her respond. When startled she would frequently elevate her head to command a wider view of her surroundings. If she caught sight of the source of her alarm, such as the passing of a dog, she would retract her head, become perfectly motionless, and retain a "frozen" position until the source of danger passed. If the animal ventured too near she would fly off with a violent "whirrrr" of her wings, which was sure to attract, if not startle, the intruder. She would drop into the grass a short distance away, utter a sharp distressing cry, and feign a wounded bird. After several repetitions of the performance, until she had attracted the enemy away to a safe distance from the nest, she would sail gracefully away, leaving the bewildered creature behind. This behavior is common to many birds, but the deception is remarkably well executed by the prairie chicken. I have seldom seen a prairie chicken try these tactics with a human being; apparently they have learned that it is best in such cases to get out and away without the least delay. That the leaving of the nest when flushed by a human being is a quick performance is revealed by moving pictures. The usual speed is 16 frames per second in the ordinary moving pictures. It requires only three or four frames of the picture to show the bird until she is away from the scene. This means that it requires only three-sixteenth to one-fourth of a second for the bird to leave the nest.

As it is not possible in most cases definitely to establish the cause of nest destruction, it is necessary to depend on circumstantial evidence. In 1929, out of 12 nests studied in Wisconsin only three reached the period of hatching. The adult birds of two of the nests were killed and the eggs destroyed, presumably by coyotes, as the tracks of the animals were found in the sandy soil around the nests, and the mass of feathers left behind seemed to indicate the work of such animals. One incubating prairie chicken was killed by a horned owl; the eggs of another located in an open situation were destroyed by crows, and in two other cases the eggs disappeared without a trace of the intruder. One nest was accidently destroyed by a farmer while plowing, and the eggs of two others were deserted where it was problematical whether the incubating birds were taken by some predacious bird or mammal or were merely frightened away in some manner.

During the summer of 1930, 28 prairie chicken nests were found, of which 17 reached the period of hatching and the others failed for one reason or another. Indirect evidence indicates that crows were responsible for the destruction of three nests of eggs. One was probably broken up by a dog, four were drowned out by floods, in one case the bird was killed, presumably by a mink, within a few feet of the nest, in another case the embryos in the eggs were killed by exposure to the heat of the sun, and one nest was deserted. Combining the records of 1929 and 1930, we have 40 nests of which 3 hatched in 1929 and 17 in 1930, 20 in all, or an average of 50 per cent for all the nests observed.

Every egg hatched of the three nests that reached the stage of hatching in 1929. The 17 nests that succeeded in reaching the hatching stage in 1930 contained 208 eggs. Twenty-nine eggs, or approximately 14 per cent, failed to hatch. Only 8 of the 17 birds succeeded in hatching every egg, and in 9 nests there were one or more sterile eggs, eggs with dead embryos, or both. Of 29 eggs that failed to hatch, 6 were sterile and 23 contained dead embryos. The latter were killed by excessive heat of the sun or by failure of the eggs to hatch in time before the old bird left the nest with her young.

During the summers of 1929 and 1930, the date of hatching was noted for 23 sets of eggs in the field or by special incubation. Of these, 3 hatched in May, 17 in June, and 3 in July. The earliest date of which we have a record was May 29, 1930, and the latest July 7, 1929. Nest No. 12 contained eggs on July 15, 1929, but it was not possible to record its hatching. The average date of hatching of the 23 nests was June 10. These records indicate that the majority of the nests hatch during the first two weeks of June. The condition of the weather at this time is a most important factor in the determination of the number of birds to be expected the following season. A long continuous cold rainy spell with cloud-bursts, such as is sometimes experienced in the Middle West during the first part of June, is certain to have a disastrous effect on the broods of young birds.

The first eggs are pipped on about the twenty-second day after incubation starts. The shell is slightly raised at the point on the circumference between the blunt and pointed ends, but slightly nearer the larger end. The embryo inside the egg can be heard to "peep" at this time, an event that greatly excites the mother and at once becomes a great stimulus in her behavior. It causes her to reach under her breast feathers to turn the eggs every few minutes and to exhibit a great deal of nervousness in her response to various other stimuli.

On the morning following the day on which the eggs are slightly pipped, all the eggs destined to hatch are pipped. In those eggs where there was but a slight elevation in the shell the day before, there are now well-defined openings through which the tip of the bill of the embryo can be clearly seen. At this stage the mother bird may frequently roll the eggs with her body in addition to turning them with her bill. I have even observed the old bird pick out bits of shell from a pipped egg, as if attempting to facilitate the process of hatching. After a few hours the calcareous shell is cracked for its entire circumference, but the shell membrane may remain intact for a longer time. The struggles of the young, however, soon fling the cap open with a part of the shell membrane on one side serving as a kind of hinge. After the embryo has kicked itself out of the prison shell, the tension of the drying shell membrane pulls the cap back in place and thus prevents the youngster from being cupped by its own shell.

The time of hatching of the various eggs of a set is remarkably uniform, and in some instances the time elapsed from the time of hatching of the first to the last egg is less than an hour. In cases where incubation started before the last one or two eggs were laid, the latter may be delayed, and in several cases under observation, during the summer of 1930, they failed to hatch in time, and the contained young were left behind when the brood left the nest.

The precocious young of the prairie chicken are ready to leave the nest as soon as their down is dry, and the mother bird often has difficulty in preventing the first young from leaving the nest before the last to hatch are prepared to go. The brood may leave in a few hours after hatching, but if the hatching takes place late in the afternoon the old bird, unless disturbed, will brood them on the nest during the night, but leaves the next morning just as soon as the temperature and weather conditions are suitable for the chicks to move. The eggshells are never removed from the nest by the old bird.

The chicks, as well as embryos in the eggs, are very sensitive to extremes of temperature. Young left exposed to the cool damp night air will quickly perish, and brooding at such times is absolutely necessary. Though the young quickly perish if cold, especially if wet, they will make rapid and remarkable recoveries from nearly paralyzed conditions. Young that seem almost lifeless can be quickly restored to an active condition by merely holding them in your hand and blowing your breath over them for 10 or 15 minutes. In one instance a mother bird was frightened away from the nest at night and failed to return. The young, still damp from the fluids of the eggs, seemed destined to die. They were taken inside the blind, revived, and then kept alive and contented all night by placing them against my body inside of a flannel shirt. At daybreak the old bird appeared on the scene and claimed the youngsters one by one as they were released under the burlap of the blind.

The language of the prairie chicken is readily understood by the young, even when first hatched. If the young are taken inside of the blind at the time of hatching they are indifferent to various sounds and notes of other birds calling outside of the blind, but as soon as the adult prairie chicken appears and begins calling they respond at once. If she gives her *brirrrb–brirrrb* call, they struggle all over one another trying to get out, but if the old bird becomes alarmed and suspicious she gives a sharp shrill call of caution and immediately each little chick cows down and "freezes" to a perfectly motionless pose. Chicks set free at the edge of the blind made no effort to go to the mother bird unless called. As soon as the call note is given there seems to be an irresistible impulse on the part of the young to follow that call, although they can not see the old bird. When the chicks wander from the nest at hatching time that same call brings them back, and hence this response is important in their preservation and doubtless is a matter of evolutionary development. Though the character of the call has a distinct meaning to the birds, one can with little effort imitate it and completely deceive the adult or young. I have often made use of this fact by inducing the bird to come near enough to the blind to obtain large portrait pictures of her. One can so excite the bird that is nearing the hatching time that she will exhibit an unusual behavior, such as turning the eggs over and over or twisting and squirming about the nest. She will, if the call is well imitated and continued, leave the eggs to search for the young. The bird at one nest under observation circled the blind again and again and even attempted to get under the burlap to reach what she apparently supposed was a young in distress. The old bird seems to have but little resourcefulness in aiding a young in unusual situations. In one case where the bird was taking her young away from the nest, two of the young accidentally fell into a deep horse track. The young called desperately, but the mother seemed helpless. She raced around the opening several times and then settled down in the grass to call for them. These two young would have perished without my intervention. An experiment of placing young in a hat near the nest also proved the bird's lack of resourcefulness to cope with an unusual situation.

Few things in nature have a greater human appeal than a family of gallinaceous birds. The whole scene from the hatching of the first young to the departure of the brood is one brimming with thrilling incidents. The motherly interest of the old bird when the first youngster pokes its head through the breast feathers and gives a contented peep, as it picks at its mother's bill or her eye, is an event never to be forgotten. Then the unexpected poking of a downy head through the plumage, first at the side, then through a rear window, and perhaps two youngsters surprising each other as they appear simultaneously, all are experiences that make a long vigil in the blind well worth the effort. As more of the young hatch they become more daring and may vigorously compete for a position on the mother's back. They make repeated attempts to scale the slippery feathered dome, and finally when one does succeed he has an unmistakable look of triumph. All these things seem to have a truly human aspect, and surely the most skeptical can not help but take an anthropomorphic attitude toward their behavior.

Ordinarily a few hours after the young are hatched the old bird leaves the nest, allowing the young to follow after her. She generally goes a few yards, settles in the grass and then continues calling until all the young are gone. This procedure is repeated until she is well away from the nest. During this time the young are brooded a great deal, but before many hours, especially if the day is warm, they become active in searching for insects and other food.

Plumages.—In a prairie chicken a few hours old the chin and lores are "primuline yellow," sides of head, including down on eyelids, "naphthalene yellow," throat and breast "wax yellow," remainder of underparts "barium yellow," and down on tarsus "straw yellow."

The lighter areas above are yellowish, strongly tinged with "cinnamon," which approaches "Mikado brown" on the rump, and the entire upper parts are marked with numerous irregular black spots and patches. There is a small black patch back of the eye and three irregular shaped black spots in the auricular region. The iris is "dark Quaker drab," the base of the upper mandible is "pecan brown" and tipped with a lighter color, and the upper surface of the mandible is black, which extends for a distance equal to two-thirds the length of the bill. The lower mandible is pale "flesh color," tipped with "straw yellow." The posterior part of the tarsus, not covered with down, is "yellowish citrine." The upper surface of the toes is "honey yellow," the undersurface "mustard yellow," and the nails "flesh color."

The young of the sharp-tailed grouse is similar to that of the prairie chicken, but with the following minor differences: The yellow of the underparts, not so deep or so bright as in the prairie-chicken young. Upper parts with much less black, especially in the region of the back and rump. The brown of the rump is a paler shade. The black on the upper mandible extends down only one-half the distance of the length of the bill. When the young of the prairie chicken and sharp-tailed grouse are compared with young of the ruffed grouse, they are seen to be much yellower and much less reddish brown, so characteristic of the day-old ruffed grouse.

The average weight of 17 young prairie chickens that hatched in a nest near Bancroft, Portage County, Wis., was 15.9 grams. The average weight of the eggs 2 days before hatching was 19.4 grams.

[AUTHOR'S NOTE: As with all young grouse, the wings and scapulars begin to grow soon after the chick is hatched, and juvenal plumage is rapidly acquired, long before the young bird attains its growth. In this plumage the crown is "hazel," spotted with black; the feathers of the back, scapulars, tertials, and wing coverts are boldly patterned with "ochraceous-tawny," black, and "snuff brown," many feathers, especially the scapulars, having broad white shaft streaks spreading out into a white tip; the primaries are spotted with pale buff on the outer web; the pointed tail feathers are barred or patterned with the colors of the back; the chin is white, and the underparts are dull white, washed with buff and spotted with dusky on the breast and flanks.

Before the young bird is fully grown, in July, the postjuvenal molt begins with the primaries. This is a complete molt, except that the outer pairs of juvenal primaries are retained for a whole year. Otherwise the first winter plumage is practically adult. Adults have a partial prenuptial molt, about the head, in March and April, and a complete postnuptial molt in August and September.]

In addition to the normal plumages of the prairie chicken there are unusual types that have attracted the attention of sportsmen and ornithologists. In certain individuals of the prairie chicken there is a prevalence of rufous or reddish brown, which is due to an excess of red pigment in the feathers, a condition known as erythrism. The red phase of the plumage is a common occurrence among ruffed grouse, but as yet it has been noted in comparatively few cases of pinnated grouse. The following cases, which have come to my attention, are of interest:

George N. Lawrence (1889) described a specimen in which all of the light markings were tinged with light, bright rufous and the entire underparts, throat, and neck tufts were deep rufous (reddish brown).

William Brewster (1882 and 1895) called attention to four specimens exhibiting the red phase in which there was but little variation with respect to the depth and extent of the reddish brown or chestnut coloring. The upper parts of the birds were strongly suffused with reddish brown, while most of the underparts were clear reddish or rusty chestnut and the usual blackish chestnut bars were nearly or quite wanting on the sides. All four specimens examined by Brewster were males.

Cases of albinism, in which there is a lack of pigment, resulting in a white plumage, have been frequently noted. Some of the albinos are not pure, but may have a little pigment in certain of the feathers, giving them a dusky appearance.

An albinistic specimen collected March 6, 1893, near the Missouri River, Iowa, is in the collection of the Museum of Comparative Zoölogy, Cambridge, Mass. This bird is white, with the exception of minor, pale, rusty brown crossbars and markings.

J. A. Spurrell (1917) records a very interesting case of an albino in Sac County, Iowa, which attained quite a local reputation because the bird was so clever that it eluded all attempts to trap it. The hunters in the vicinity made a point to spare the "white chicken" in the hope that it might be captured alive.

Hybrids between the prairie chicken and the sharp-tailed grouse have been noted by many observers. In Wisconsin, where the ranges of the two species overlap, it is a common experience to see them associated together at all seasons of the year, and it is not at all surprising that they frequently interbreed.

J. H. Gurney (1884) described a male hybrid between the sharp-tailed grouse and the prairie chicken in which the pinnae were present, but only one-fourth of an inch long. The tail was a hybrid gray between the brown of the sharp-tailed grouse and the white of the prairie chicken. The sides of the toes were only slightly feathered and the general coloration was intermediate between the two species.

A hybrid between the prairie chicken and the sharp-tailed grouse in which the elements of the prairie chicken predominated has been reported by F. C. Lincoln (1918). The author has examined a hybrid specimen in the possession of Mrs. H. M. Hales, of Hancock, Waushara County, Wis., which also resembles the prairie chicken in most of the characters of its plumage.

William Rowan (1926) figures and describes two female hybrids of the sharp-tailed grouse and the prairie chicken collected in Alberta, Canada. One individual collected near Edmonton resembles the pinnated grouse more closely than the sharp-tailed grouse, while the other, shot at Gough Lake in the southern part of the Province, resembles more nearly the sharp-tailed grouse. The ovary of the Edmonton bird, according to Mr. Rowan, was normal. These are the only Alberta hybrids known to him, but he states that hybrids between the prairie chicken and the sharp-tailed grouse are frequent in Manitoba, where the pinnated grouse is more numerous than it is in Alberta.

Glenn Berner, of Jamestown, N. Dak., writes that he killed a hybrid grouse in 1923 in which the back, head, and tail resembled the prairie chicken, whereas the breast, legs, feet, and under tail parts were like those of the sharp-tailed grouse. The breast was not barred as in the prairie chicken but spotted as in the sharp-tailed grouse. The bird when flying had the characteristic cackle of the sharp-tailed grouse.

According to O. A. Stevens, Fargo, N. Dak., there is a hybrid in the collection at the North Dakota State Agricultural College.

I have examined a female and three male hybrid specimens of the prairie chicken and sharp-tailed grouse in the Museum of Comparative Zoölogy, at Cambridge, Mass., which were obtained in the Boston markets, March 24, 1873, February, 1887, December 29, 1899, and January 24, 1898, respectively. So far as I know, these hybrids do not reproduce themselves, and in most cases this is probably due to the sterility of the individuals.

Food.—The prairie chicken, like other grouse, is adaptable in its food eating habits, varying its diet from season to season and sustaining its life on the food that is most abundant and easily obtained.

Dr. Sylvester D. Judd (1905a) reported on the examination of 71 stomachs of prairie chickens collected in the Middle West and representing all months of the year except July. The food consisted of 14.11 per cent animal matter, chiefly grasshoppers, and 85.89 per cent vegetable matter, made up of seeds, fruit, grain, leaves, flowers, and bud twigs.

According to Judd's report the prairie hen is highly insectivorous from May to October inclusive, insects constituting one-third of the food of the specimens shot during this period. The species is particularly valuable as an enemy of the Rocky Mountain locust. During an invasion by this pest in Nebraska, 16 out of 20 grouse killed by Prof. Samuel Aughey (1878) from May to October inclusive, had eaten 866 locusts. Beetles and miscellaneous insects were eaten in smaller numbers.

From October to April, inclusive, according to the Biological Survey report (Judd, 1905a), the prairie hen takes little but vegetable food, consisting of fruit, leaves, flowers, shoots, seeds, grain, and miscellaneous vegetable material. It is especially fond of rose hips, which comprised 11.01 per cent of the food. When the deep snow causes scarcity of other supplies, the sumac affords the prairie hen with abundant food. Seeds make up 14.87 per cent of the annual diet, of which grass seeds form 1.03 per cent, seeds of various polygonums 8.49 per cent, and miscellaneous weed seeds 5.35 per cent.

The prairie chicken eats more grain than any of the other native gallinaceous birds; the food examined by the Biological Survey was 31.06 per cent grain. The stomach of one bird shot in June in Nebraska contained 100 kernels of corn and 500 grains of wheat. Buckwheat, barley, oats, and millet are relished, but corn appears to be the favorite cereal, amounting to 19.45 per cent of the food. Wheat was next in order represented by 11.61 per cent. Like other gallinaceous birds, it is fond of mast such as hazelnuts and acorns, though it obtains much less than the ruffed grouse. A bird shot in Minnesota in March had bolted 28 scarlet-oak acorns.

An analysis of organic material in the food of 17 prairie chickens collected in Wisconsin for the Wisconsin Conservation Commission, chiefly in fall, revealed that about 28 per cent of the food was animal and 72 per cent vegetable matter. Gravel constituted 6 per cent of the combined organic and inorganic material of the crop and stomach contents. The average weight of the crop contents was 25.7 grams; the maximum 83 grams. The average weight of the stomach contents was 14.3 grams, and the largest quantity found in any one stomach weighed 23.1 grams.

There were 84 kinds of vegetable matter and 82 kinds of animal matter represented in crops and stomachs of the prairie chickens collected in Wisconsin. Arranged in order of the percentages of the entire food eaten by the birds, the 25 more important foods follow: Short-horned grasshoppers, 26.7; ragweed, 11; oats, 10.8; clover, 7.7; black bindweed, 6.2; acorns, 4.5; greenbrier, 3.6; dogwood, 3.5; crickets, 3.3; buckwheat, 3.1; bramble, 3.1; blueberries, 2.4; rose, 1.7; hawkweed, 1.4; chokeberry, 1; galls, 0.94; ants and wasps, 0.88; poison ivy, 0.80; birch, 0.80; pin cherry, 0.64; woody débris, 0.64; bunchberry, 0.53; wild black cherry, 0.53; smartweed, 0.47; pigeon grass, 0.47.

It will be seen that the chief difference between the above list of foods and the results published by the Biological Survey in 1905 is the absence of corn in the recent list. In Doctor Judd's list corn made up 19.45 per cent of the entire contents of all the birds examined. This may be accounted for by the change in methods of farming. In the past, corn was husked in the field and much grain was accidently left behind by the harvesters. To-day, in Wisconsin, a dairy State, all the corn that is raised is cut and made into ensilage for the cattle. Practically no corn is allowed to ripen in the field; hence it does not appear as a food for the prairie chicken. Both lists agree in the large percentage of grasshoppers comprising the food. The prairie chicken and sharp-tailed grouse are notable grasshopper consumers, which fact, together with their fondness for weed seeds, makes their presence a great asset to the farmers.

Except in the northern part of their range, where very severe weather and deep snows prevail, there is sufficient natural food for the prairie chickens at all seasons of the year. The prairie chicken is there hard pressed for an existence, since it does not seem to be able to subsist on buds and other foods above the snow to the extent that it is done by the ruffed grouse. In Wisconsin, experiments conducted with winter feeding stations by the conservation commission have proved a great success. Plots of ground ranging from a half acre to two acres in extent are planted chiefly to buckwheat, with sorghum, sunflowers, broomcorn, and corn planted as accessory foods on most of the plots. Half of the crop is left standing and the other part is cut and placed in covered shocks, which are opened up after the deep snows arrive. According to the reports of the wardens in charge, as many as 200 to 300 birds visited a single station at one time, a strong testimonial for the practicability of such stations in game management.

Migration.—The prairie chicken is a permanent resident in much of its range, but in the Northern States there is a regular annual movement of the birds southward at the approach of winter weather. There are counties in Wisconsin where prairie chickens do not breed, or are present in very small numbers during the summer season, whereas they are represented by large numbers of individuals during the winter months, especially when deep snows and extremely cold weather exist in the more northerly sections of the range from which the birds apparently come. Observers in Door County, Wis., have reported seeing flights of prairie chickens approaching the land from Green Bay. The birds supposedly came from the opposite shore, a distance of 12 to 15 miles, which, if true, means that they sometimes take flights exceedingly long and continuous for a bird of the type of the prairie chicken. A. E. Doolittle, superintendent of Peninsular Park, Door County, saw a flock of 300 prairie chickens headed northeast, up the shore of Green Bay, which he thought were en route for the Michigan side of the bay. William Fairchild, former keeper of Chambers Island (near the middle of Green Bay, a distance of about 7 miles from the mainland), saw two prairie chickens arrive from Marinette in April, 1927. They remained several weeks, then flew eastward to Door County proper.

R. M. Anderson (1907), writing of the prairie chicken in Iowa, states: "While a certain number remain throughout the winter, large flocks pass southward early in the winter, returning in March."

The migration was even more marked in the past when the birds were abundant. J. A. Spurrell (1917) states that there was a marked migration of birds away from Sac County, Iowa, until about 1875–1880. After that date, he says, corn became a common crop and birds wintered as well as nested abundantly in that section of the State.

Prof. W. W. Cooke (1888), in writing of migrations of prairie chickens in Iowa, stated:

In November and December large flocks of prairie chickens come from northern Iowa and southern Minnesota, to settle in northern Missouri and southern Iowa. This migration varies in bulk with the severity of the winter.

During an early cold snap immense flocks come from the northern prairies to southern Iowa, while in mild, open winters the migration is much less pronounced. During a cold, wet spring the northward movement in March and April is largely arrested on the arrival of the flocks in northern Iowa; but an early spring, with fair weather, finds them abundant in the southern tier of counties in Minnesota, and many flocks pass still farther north. The most remarkable feature of this movement is found in the *sex* of the migrants. It is the females that migrate, leaving the males to brave the winter's cold. Mr. Miller, of Heron Lake, Minn., fairly states the case when he says: "The females in this latitude migrate south in the fall and come back in the spring about one or two days after the first ducks, and they keep coming in flocks of from ten to thirty for about three days, all flying north. The grouse that stay all winter are males."

In the spring of 1884, at Iowa City, Iowa, the first flocks passed over March 10, and the bulk March 22; at Newton, Iowa, the bulk was noted March 23.

Glenn Berner, of Jamestown, N. Dak., says in his notes:

During the spring of 1924 I witnessed a decided northward movement of prairie chicken flocks numbering 10 to 100, some quite high in the air—late in the afternoon—possibly 30 flocks being seen from one location in two hours and very few of them alighting.

O. A. Stevens, of Fargo, N. Dak., writes that he saw a flock of prairie chickens fly high overhead on October 25, 1930, a time that coincided with a marked migratory movement of other birds. Mr. Stevens has noted for several years this annual movement taking place during the latter part of October. He considers these flights as a distinct migration, as the birds were always moving in the same direction.

Information on the migration of the prairie chicken supplied by F. C. Lincoln, of the Biological Survey, is as follows:

Although not a true migrant in the strict sense of the word, the prairie chicken has been known to make more or less regular flights north and south. Curiously, these movements appear to be confined chiefly to the females, the males remaining in the breeding areas during winter. The flights apparently were not infrequent a generation ago when the birds were much more abundant than at present. The extent of the movement and the number of individuals participating were dependent upon the severity of the winter in the northern part of the range. The exodus usually took place in November or December (Wisconsin, Madison, November 25; Iowa, Ogden, December 14, and Osage, December 30), while the return trip was made in February, March, or April (Iowa, Sioux City, February 4, Marshalltown, February 7, and Osage, February 13; Wisconsin, Unity, February 15, Elkhorn, March 2, and Whitewater, April 1; and Minnesota, Fort Snelling, March 21, and Minneapolis, April 8).

Thus it is clear that the prairie chicken, at least in the Northern States, makes flights of considerable length, which we can consider of a migratory character. In addition to these movements there are shifts and concentrations of the birds that are very local and mainly concerned with the food supply. At the feeding stations established in Wisconsin it was not unusual for 200 to 300 birds to feed at a single station at one time, a total far in excess of the numbers breeding in the vicinity. Some of the birds may have migrated from the north, but it is probable that the mass of these flocks are merely aggregations from a limited region of a few counties.

Winter.—In regions where deep snows prevail, the prairie chickens often dig themselves into the deep drifts to avoid the excessive cold. One observer at Green Bay, Wis., relates observing five prairie chickens alight on the surface of the snow, which was about 2½ feet deep. The birds walked up to some weed stalks that projected through the snow and then dug themselves in at places about 10 to 12 yards apart. A day later the same observer flushed the birds from the snow bank and found well-molded places on the ground among the weeds. There was an accumulation of droppings in each burrow, indicating that the birds had remained in the same spot during the night. The practice of digging into the snow has proved disastrous at times when it becomes covered over with a resistant layer of ice.

Gale W. Monson, of Argusville, N. Dak., says in his notes:

In winter the prairie chickens are our most conspicuous birds. They spend the nights in the tall grass of marshy meadows making small pockets for themselves in the snow. At sunrise they leave their beds and fly to the nearest cornfield, there to eat their fill of that grain. In the afternoon they return again to their sleeping quarters. Their chief enemy at this time is the snowy owl, which sometimes depletes their numbers to a noticeable extent.

Several observers in Wisconsin report that the tall marsh grass is frequented by the prairie chickens as soon as the water of the swamps is solidly frozen over.

John Worden, of Plainfield, Wis., states that during times of deep snows the prairie chickens are often in a semistarved condition. At such times the birds showed little fear of man and often allowed him to approach within a few yards before attempting to fly. A farmer living near Babcock, Wis., stated that in collecting shocks of corn following a heavy snowstorm, he had virtually to drive the birds away, and when flushed they flew but a few yards to the next shock of corn. Such behavior is probably very unusual except under very extraordinary circumstances when the birds are suffering with extreme hunger. According to Mr. Worden, the birds frequently alight in trees during winter, but he says they invariably roost on the ground at night.

F. Hall, of Babcock, Wis., states that during the winter of 1928–29 a flock of about a dozen prairie chickens came regularly to the poplar trees of his back yard, but he was not certain to what extent they fed upon the buds.

DISTRIBUTION

Range.—South-central Canada and the United States east of the Rocky Mountains, except the Southeastern States.

The full range of the prairie chicken extended **north** to southern Saskatchewan (Quill Lake and Indian Head); southern Manitoba (Oak Lake, Carberry, Westbourne, Ossono, and Shoal Lake); northern Minnesota (Crooked Lake); central Wisconsin (Unity, Wild Rose, and West Depere); Michigan, (Chatham, McMillan, Sault Ste. Marie, and Fourmile Lake); southern Ontario (Wallaceburg and Chatham); and Massachusetts (Springfield, Newton, and Cape Ann). **East** to Massachusetts (Cape Ann and Marthas Vineyard); Long Island, N. Y. (Miller Place and Hempstead); New Jersey (Barnegat); and southern Maryland (Marshall Hall). **South** to southern Maryland (Marshall Hall); District of Columbia (near Washington); probably Virginia and perhaps North Carolina; southwestern Pennsylvania (Blairsville); central Ohio (near Columbus); southern Indiana (Bloomington, Marco, and Bickwell); northwestern Kentucky (Henderson); southern Louisiana (Iowa Station and Calcasieu Pass); and Texas (Beaumont, Richmond, Edna, Port Lavaca, St. Charles Bay, Austin, and Tascosa). **West** to northwestern Texas (Tascosa); Colorado (Barton and Barr); southeastern Wyoming (Chugwater); northwestern Nebraska (Chadron); South Dakota (Pine Ridge Reservation, Kadoka, and Short Pine Hills); North Dakota (Bismarck, Charlson, and Crosby); and southern Saskatchewan (Johnston Lake and Quill Lake).

A specimen taken in the fall of 1917 near Huntley, Mont., is at present the only record for that State. Prairie chickens have been noted as rare in winter near Fayetteville, Ark., and are said to occur in that season at De Witt, Ark.

The prairie chicken and its eastern relative, the heath hen, have been extirpated over great areas in their former range. The heath hen is, in fact, extinct except for a single bird, which at the time of writing (November, 1930) was still living on Marthas Vineyard, Mass. Western Indiana marks the present eastern boundary of the species.

Many attempts have been made to transplant the prairie chicken into other parts of the country. These have all met with failure, except for an apparent introduction in northern Michigan. They are reported as thriving in the vicinity of Sault Ste. Marie and McMillan.

The range as described is for the entire species, which has, however, been separated into three subspecies. True *americanus*, the greater prairie chicken, formerly occurred from the southern parts of the prairie Provinces of Canada and eastern Colorado east to southwestern Ontario, northwestern Ohio, and western Pennsylvania. The range of the heath hen (*T. c. cupido*) included New England (to southern New Hampshire), New York, and other States of the Atlantic seaboard, probably south to and including Maryland. Attwater's prairie chicken (*T. c. attwateri*) is found in the coastal region of Texas and southwestern Louisiana.

Egg dates.—Manitoba: 6 records, May 24 to June 20. Minnesota and Dakotas: 30 records, May 1 to June 18; 15 records, May 18 to 29. Wisconsin: 37 records, May 5 to July 10; 19 records, May 28 to June 20. Illinois and Iowa: 32 records, April 20 to June 6; 16 records, May 6 to 25. Marthas Vineyard: 3 records, June 2 and 5 and July 24. Texas: 4 records, April 3 to May 16.

TYMPANUCHUS CUPIDO ATTWATERI (Bendire)

ATTWATER'S PRAIRIE CHICKEN

HABITS

The small dark race of the prairie chicken named *attwateri* is confined to southwestern Louisiana and eastern Texas, mainly in low prairies in the coastal counties. It was described by Maj. Charles E. Bendire (1894) and named by him in honor of Prof. H. P. Attwater. He gives as the subspecific characters:

> Smaller than *T. americanus*, darker in color, more tawny above, usually with more pronounced chestnut on the neck; smaller and more tawny light colored spots on wing coverts, and much more scantily feathered tarsus, the latter never feathered down to base of toes, even in front; a broad posterior strip of bare skin being always exposed, even in winter, while in summer much the greater part of the tarsus is naked.

George Finlay Simmons (1925) describes its haunts as "rolling open, grassy, fertile upland prairies, where the grass is from 1 to 3 feet tall, old and thick and mixed with weeds; wheat and corn fields; takes to timber only during snow and sleety storms."

Referring to its history, Simmons writes:

Formerly abundant on the open prairies, these wonderful game-fowl became extirpated in the Austin Region through two agencies; civilization and hunting. They disappeared rapidly as the country was settled up and as cultivated fields took the place of the extensive, wild, unfenced prairies; and hunters quickly killed the few remaining birds.

Its courtship performances, nesting habits, eggs, plumages, and molts are all similar to those of the common prairie chicken.

Eggs.—The eggs are indistinguishable from those of the common prairie chicken. The measurements of 27 eggs average 42.3 by 31.5 millimeters; the eggs showing the four extremes measure **44.9** by 32, 42.4 by **33.5**, **38.8** by 28.9, and 39.8 by **28.6** millimeters.

Food.—Simmons (1925) says that it

sometimes flies to treetops to inspect corn fields before alighting in them to feed; frequently feeds in the open in plain sight of observers several hundred yards away. During early breeding season, feeds largely on insects, such as grasshoppers, crickets, potato bugs, and other beetles; in fall and winter, tops and seeds of leguminous plants, tender buds and green leaves of late winter, fruits, berries, and waste grain of stubble and corn fields.

Behavior.—Simmons says that it is

observed singly and in pairs in spring; in fall and winter, roamed about in flocks of from 10 to 12 up to a 100 or more, moving about over the prairies and grain fields, generally keeping among bushes and tall grass inland, the open prairies and grassy knolls along the coast. Stately of bearing, but otherwise very much like the domestic fowl in its actions. In spring, a "scratching ground" or smooth, open courtship ground is selected, where pairing takes place.

Voice.—The same observer describes the notes as "nondescript calls; strange cackles, with a muffled booming love-call, *uck-ah-umb-boo-oo-oo-oo-oo-oo;* and a loud beating *boom-ah-boom*, perhaps produced by a beating of the wings; when alarmed, a rapidly repeated *cluk-cluk-cluk-cluk;* female, when flushed, utters a low *kuk-kuk-kuk-kuk-kuk-kuk-kuk-kuk.*"

TYMPANUCHUS CUPIDO CUPIDO (Linnaeus)

HEATH HEN

CONTRIBUTED BY ALFRED OTTO GROSS

HABITS

The heath hen and the prairie chicken are so closely related that they are now considered as geographical races and not as distinct species. In 1885 William Brewster (1885a) called our attention to differences between the pinnated grouse of Marthas Vineyard (heath hen) and the western pinnated grouse, or prairie chicken. He named the western form as the new species, because *Tetrao cupido* of Linnaeus was the eastern form from the fact that its habitat is given as Virginia. The differences between the eastern and western birds are so slight and the variations of the individuals so great that ornithologists now concede that Brewster was not well justified in the establishment of a new species.

In prehistoric times the common ancestors of the heath hen and the prairie chicken probably ranged in an uninterrupted distribution from the Atlantic seaboard to the plains east of the Rocky Mountains. Later the birds of the East became separated from those of the West, and as a result of this isolation and differences in environment certain modifications arose that have resulted in the establishment of the two geographical races—*Tympanuchus cupido cupido*, the heath hen, and *Tympanuchus cupido americanus*, the prairie chicken.

The prairie chicken is still flourishing and in recent years has been rapidly regaining its numbers in favorable sections of the Middle West, but the heath hen has been unable to cope with the changing conditions of its restricted environment, and to-day is represented, so far as we can ascertain, by a single male individual, which is living out its normal life on the scrub-oak plains of Marthas Vineyard Island, Mass.

The following account is made up primarily of modified excerpts of the contributor's monograph on the heath hen (Gross, 1928) and from subsequent annual census reports:

Historical.—The heath hen is among the first of the American birds to be mentioned in the writings of the early colonists who came to our shores. There is, however, such a dearth of material concerning the heath hen during these early times that we know but little concerning the conditions under which it existed, and the records are so incomplete that we are unable to determine with any degree of accuracy its relative abundance and distribution prior to the nineteenth century. Some of the earlier American writers designated the heath hen by the name "heathcocke," "pheysant," or "grous," but their notes and descriptions are such that they can be clearly referred to this species. William Wood (1635) in his New England Prospect writes as follows: "Heathcockes and Partridges be common: he that is husband, and will be stirring betime, may kill halfe dozen in a morning. The Partridges be bigger than they be in England, the flesh of the Heathecocks is red, and the flesh of the Partridge white, their price is four pence a piece." Wood resided at what is now the city of Lynn in Massachusetts. His map included Cape Ann and the Merrimac River; hence it is evident that the heath hen existed in northeastern Massachusetts in his day. Thomas

Morton (1637), writing concerning the heath hens, which he called "pheysants," stated that these birds were like the pheasant hen of England in size but were rough footed and had "stareing" feathers about the neck. The birds, according to Morton, were so common that they seldom wasted a shot upon them. The writings of many others who followed indicated that the birds were distributed along the Atlantic seaboard from Maine and Massachusetts southward to Virginia and possibly the Carolinas. They were by no means evenly distributed over this region, but were restricted to certain areas whose features and productions were suitable for their existence. There were large heavily timbered areas that probably were never visited by the heath hen. In favorable localities, such as the brushy plains of eastern Massachusetts, they were abundant. Thomas Nuttall (1832) wrote as follows: "According to information I have received from Governor Winthrop, they were so common on the ancient brushy site of Boston, that laboring people or servants stipulated with their employers not to have Heath Hen brought to table oftener than a few times a week." No published statement has ever been found that more impressively reveals to us the abundance of these birds in early colonial times. It was chiefly on the sandy scrub-oak plains of Massachusetts, Connecticut, Long Island, New York, New Jersey, and Pennsylvania that they existed in large numbers when the white man first came to America. The birds served as a valuable source of food, and because they were easily tricked and killed they were exterminated at an early date in the more accessible areas, and soon after 1840 were entirely gone from the mainland of Massachusetts and the State of Connecticut. The birds persisted for a longer time on Long Island, and a few continued to battle for existence on the plains of New Jersey and favorable places among the pines and scrub oaks of the Pocono Mountains in Northampton County, Pa. Since 1870 the surviving members of this interesting race have been restricted to Marthas Vineyard Island, off the southeastern coast of Massachusetts.

Because of conflicting reports and uncertain statements we can not be positive whether the heath hen was native to Marthas Vineyard or was introduced there from the mainland by man. In either case it is truly remarkable that the heath hen, after being so greatly depleted in numbers, has persisted for more than half a century in this very restricted area where excessive interbreeding has occurred and where the birds have been subjected to all the vicissitudes of diseases, enemies, and other adverse conditions.

In 1890 William Brewster (1890) made a careful census and at that time estimated that there were 200 birds on the entire island. Kentwood (1896) stated that there were less than 100 birds in 1896.

By the beginning of the twentieth century the birds had reached a very low ebb in their existence. The year 1908 witnessed one of the most notable steps taken in the history of the heath hen in an effort to preserve it from extinction, in the establishment of a reservation in the midst of the breeding range where the birds could be protected from poachers and predators by competent wardens. Six hundred acres were purchased by private subscription, and an additional tract of 1,000 acres was leased by the Commonwealth for a reservation, which was systematically improved to make it attractive for the birds. There is no doubt that the prolongation of the life of the heath hen on Marthas Vineyard Island has been due to the interest taken in it by the State of Massachusetts, conservation organizations, bird clubs, sportsmen, and bird lovers. The State Department of Conservation expended $70,000, and thousands more were contributed by individuals, in the unprecedented efforts to prevent the bird from becoming extinct. Many attempts were made when the birds were abundant to transplant them to other favorable places on the mainland and to other islands such as Long Island, one of their former strongholds. Furthermore, the most experienced sportsmen and game breeders were unable to breed the birds in captivity, a fact indicating that the heath hen was very sensitive to radical changes in its environment and that it would not yield to such methods of conservation. All the many experiments of introducing the western prairie chicken to the East have likewise proved unsuccessful.

When the reservation was established in 1908 there were only about 50 heath hens, but as a direct result of the efficient protection the birds increased very rapidly and by 1915–16 they were to be found in all parts of Marthas Vineyard with the exception of Gay Head, the extreme western end of the island. It was then possible to flush a flock of 300 or more birds almost any day from the corn and clover plots planted on the reservation for the birds. An estimate made by William Day, then superintendent of the reservation, indicates there were probably 2,000 birds on the island. This was a great triumph for those who had encouraged and fostered the reservation, but unfortunately success was not long-lived.

In spite of the unusual precautions taken to prevent the spread of fire, a terrific conflagration broke out during a gale on May 12, 1916, which swept the greater part of the interior of the island, destroying brooding birds and their nests and eggs, as well as the food and cover of the birds on more than 20 square miles, right in the heart of the breeding area of the heath hen. This fire undid in a few hours the accomplishment of many years of work. A hard winter followed the fire, and in the midst of this came an unprecedented flight of goshawks, which further decimated the number of birds. The net result of this catastrophe was an amazing decrease in the number of heath hens, which according to official estimates was reduced to less than 150 birds, most of which were males.

There was a slight rally in numbers during the following few years, but the birds were too far gone to overcome the surmounting uncontrollable conditions of extensive interbreeding, declining sexual vigor, the condition of excess males, and, worst of all, disease.

In 1920 many birds were found dead or in a weak and helpless condition, indicating that disease was then exacting its toll. The heath hen is very susceptible to poultry diseases, and when domestic turkeys were introduced to the island in large numbers the dreaded disease blackhead came with them. The turkeys and heath hens fed on the same fields, and thus the disease was readily transmitted through droppings to the native birds.

The heath hen continued to decrease in numbers, and by 1925 it was apparent that they had reached their lowest ebb in history. The Federation of the Bird Clubs of New England then came to the front and offered to raise $2,000 annually to support additional warden service. In spite of this splendid cooperation the birds, after two years of effort on the part of all concerned, continued to decrease. The 1927 spring census revealed but 13 birds, only 2 of which were females. During the fall of 1928 only two birds were seen and after December 8 but one was reported. This bird was photographed from a blind on April 2, 1929, at the farm of James Green located on the State highway between Edgartown and West Tisbury. At that time it was the common expectation that the bird would step out of existence before the end of another year. (See pl. 1, frontispiece.) It was seen regularly until May 11, 1929, but after that date it disappeared among the scrub oaks to live a life of seclusion, as was customary for the heath hen to do in the past, during the summer months. After the molting season it again appeared at the Green farm to announce to the world that it was still alive. It was seen at irregular intervals during the winter, and after the first warm days of March it appeared daily at the traditional booming field at the Green farm. The bird was studied and photographs were taken again at the time of the annual census in March-April, 1930. The lone bird continued to appear at the Green farm during April and May, where it was observed by many ornithologists and bird lovers who journeyed to the island to get a glimpse of the famous last bird. The bird again disappeared during the summer, and no reports were received until it almost met a tragic death on September 15, when it was nearly run over by an automobile traveling one of the little-used roads leading across the scrub-oak plains. In October it resumed its daily visits to the open field on the Green farm and at the time of this writing

(November 15, 1930) was still alive. It is the first time in the history of ornithology that a bird has been studied in its normal environment down to the very last individual. How long this bird will live no one can safely predict; its going is inevitable, and the death of this individual will mean the death of its race, and then another bird will have taken its place among the endless array of extinct forms. Ornithologists, bird lovers, and sportsmen the world over, however, will have the satisfaction of knowing that all that could be done has been done to save this bird from extinction. The State department has assured us that the last bird will be allowed to live, and when death comes, whether it is due to old age, disease, or violence, we shall know that the life of the last heath hen was not wilfully snuffed out by man. [During the fall of 1931, this lone survivor disappeared.]

Courtship.—There was no part of the behavior of the heath hen more unique, more interesting, and more specialized than the extraordinary performances during the courtship season. I vividly remember the thrill of hearing and seeing the heath hen's boom for the first time on Marthas Vineyard Island during April, 1923. At that time the birds came regularly to a definite part of the meadow west of the reservation house. A wooden blind, 4 by 6 by 6 feet, had been erected several years before for the convenience of the large numbers of ornithologists who journeyed to the island each year to get a glimpse of the heath hen. The blind had become a part of the environment of the drumming field around which the birds came to enact their fantastic dances without any fear of being harmed. At times one or two of the birds would even alight on the flat top of the structure, offering unexcelled opportunities for study of the intimate details of their behavior.

The following observations, as recorded in my notebook for April 11, 1923, are typical of the many mornings spent inside the blind:

"I left the reservation house at 3.30 a. m. It was very dark and only the faint light of the stars illuminated the way. The cold, enervating air quickened my step, and as I walked along, the frosted grass crunched under my feet with a metallic resonance. At this early hour all was quiet so far as voices of birds were concerned. I entered the blind, closed its door on creaky hinges, and prepared myself to wait patiently for the first note of the heath hen. A slight fog rolled in from the sea and for a time hid the stars. At 3.55 a. m. with the first dim light of dawn I heard the clear whistled notes of a bobwhite perched somewhere among the scrub oaks. At 4.05 the first robin was heard chirping, and five minutes later a vesper sparrow was singing its awakening song. In a short while a host of other birds were adding their notes to the morning chorus. At 4.21 the first "toot" of the heath hen was heard, a note that has often been mistaken for a muffled blast of a tug boat or a fog horn. Though I was well prepared for this deception, I must admit that I did not at first associate this curious note with the heath hen, for the light was yet dim and the fog obscured the view of the bird. At 4.27 a heath hen appeared from the scrub oaks on the south side of the meadow at a point relatively near the blind. The calls of this individual at once stimulated the birds on the western end of the field to greater activity in tooting. After 20 minutes a second and then a third bird came out of the scrub oaks on the south, and for a time all were busily engaged in feeding. One of the males flew to the roof of the blind, where he commanded a splendid view of the field and his companions. Later it was found that the resonant wooden roof proved to be an admirable place to conduct their stamping and courtship performances. At 4.45 the birds without any warning interrupted their feeding and began "booming or tooting" at a point only a few yards from the blind. The male on the roof joined his fellows on the ground. At this close range the call resembled *whhoo–doo–doooh*. The note varied somewhat in subsequent renditions and was variously interpreted as *whoo–oodul–doo–o–o–o–o*, *whoodle–dooh*, or *whoo–dooh–dooh*. The sound was accented on the second syllable or the first part of the second and then gradually diminished in intensity. It required from 1½ to 3 seconds to render the different versions given above. The number of calls a minute varied greatly, according to weather conditions, temperature, time of day, and the season. The booming was interspersed with henlike calls resembling *cac*, *cac*, *cac*, or *oc*, *oc*, *oc*, *oc*, *goc*, *goc*, *goc*, *goc*, occasionally ending with a queer call that sounded like *auk–ae–e–e–e–e–ek*. The males frequently leaped into the air to a height of 3 or 4 feet and so doing uttered a piercing rolling *wrrrrrrrrb*, followed by a curious indescribable laughterlike sound. In this wild demonstration the bird completely reversed its orientation in the air and landed on the ground, usually facing in the opposite direction. This leaping and screaming seemed to be augmented by similar performances of the birds on the other side of the field, and it was an evident challenge to their fellow antagonists.

"At 5 a. m. the sun pierced the screen of fog and appeared like a giant fiery ball above the eastern horizon. The morning chorus of birds then rapidly diminished in volume, but the heath hen now prepared for real action. One male from each group ran rapidly toward each other in a defiant warlike attitude. When near together they hesitated, lowered and waved their heads, leaped at each other, and struck their wings vigorously as they lunged forward. A few feathers flew, but no real harm was done, and they settled back in a comfortable position and occasionally uttered a long-drawn-out but shrill cry, which fluctuated greatly in tone and intensity. One bird arose after a few minutes, circled, paced a few steps, and went through his repertoire of toots and calls without any interference from his antagonist. Later one ambitious male insisted on chasing one opponent after another, following after them rapidly on foot until they took wing. He flew after them for a distance of 30 to 50 yards, then returned to repeat the performance with another weaker member of the group of a dozen birds. It reminded one of boys at play after being pent up in school all day. These thrilling spectacles continued until 6.50 a. m., when with one common impulse all the birds left the field to the seclusion of the scrub oaks to remain quiet until the afternoon, when they again appeared on the drumming field during the few hours preceding sunset. But when the last glow of twilight faded into darkness the fantastic dance ceased to be resumed at the coming of dawn the next day."

The first "tooting" calls of the year were usually heard the last week of February or the first week of March, the date varying from year to year and depending largely on the nature of the weather. In 1927 a series of warm days started the birds booming as early as February 12. Though an early beginning was sometimes made it was not until the latter part of April or the first week of May that the courtship reached its maximum intensity. It then gradually diminished, and by the end of May the performance was generally over, but a few more persistent males often continued a few weeks longer. In 1923 the last "boom" for the year was recorded on June 11; and in 1920 a small group of males was still performing as late as June 20. After the month of June the birds ceased their nuptial displays until the mating season of the next year.

The following details of the courtship performance were obtained by repeated observations from blinds of the birds at close range and are supplemented by a study of captive birds and detailed laboratory dissections: The tooting is usually prefaced by a short run, followed by a very rapid stamping of the feet, a part of the performance that is not readily detected unless the observer is very near to the birds. The stamping is vigorous enough, however, to be distinctly heard at a distance of 25 or 30 feet, and certain males, which did their stamping on the resonant roof of the blind, produced a noise second only to the tooting that followed. In preparation for tooting the neck is outstretched forward; the pinnae (neck tufts) are usually directed upward or forward; the primaries are spread and held firmly against the sides of the body and legs; and the tail is thrown upright at right angles to the axis of the body, thus displaying the white under tail coverts when viewed from the rear.

During this procedure the whole musculature of the body seems to be in a strained state of contraction, as if it required great effort on the part of the bird. As the inflation of the orange-colored sacs (it is really one large sac with two lateral areas devoid of feathers) begins, the tooting sound is heard. Sometimes there is a slight inflation before any sound is given. The inflation seems uniform and does not fluctuate with the inflections and accents of the tooting call. At the end of the tooting the sacs collapse suddenly by the release of air through the nares or more rarely through the opened mandibles. The sacs do not produce the notes, as was thought by some of the earlier ornithologists, but have much to do with modifying the sounds produced by the syrinx (the vocal mechanism at the junction of the bronchial tubes). The sounds are produced by the air forced from the lungs, which vibrates specialized membranes of the syrinx under control of a complex set of muscles. The sound waves then issue through the trachea and glottis to the pharynx. In the production of such notes as the ordinary cackle the mandibles are opened and the air accompanied by the sound waves issues out of the mouth. In the tooting performance the mandibles are tightly closed, the throat patch is elevated, and the tongue is forced against the roof of the mouth (palate) by the mylohyoides muscles, which close off the exit through the internal nares. The tongue is bent in such a way that it causes the glottis at the base of the tongue to open directly in front of the esophagus. The air now coming from the respiratory system is forced to fill the modified anterior end of the esophagus, or gullet, which becomes distended like a balloon.[1] While the air sac is filling, the sound waves produced by the syrinx beat against these tense drumlike membranes, which serve as resonators for the sounds and give them their great carrying power. The ordinary cackles and screams of these birds seem louder than the tooting or booming calls when one is near to the birds, but at a distance of 200 yards or more you can scarcely hear these calls, whereas the booming carries for long distances, often 2 miles or more under favorable conditions.

The female's part in the courtship is a passive one. She minds her own business, and I have never heard her utter any calls or notes or show any concern in response to the ardent attentions of the males. When a male or pair of males came strutting and circling about a female she kept on with her feeding. If the males came too near she merely stepped to one side and continued with the serious business of procuring food.

---

[1] A number of detailed experiments were performed with both dead and living birds, which clearly demonstrate the nature of the vocal mechanism as described above. See Gross (1928).

The females gave frequent calls and notes when attending their young. If the mother bird was suddenly surprised she gave a characteristic sharp call signal for her young to scatter and hide. If the members of the brood were very young and unable to fly, she feigned a wounded bird and cried out as if in great distress as she fluttered along the ground. If the young were older she usually sailed out over the scrub oaks and uttered a loud cackling call, which apparently was also for the purpose of attracting attention away from the young.

Nesting.—The nest of the heath hen was built upon the ground and was usually composed of leaves, grasses, and twigs already in place, to which were added materials found near the nesting site. The nests were concealed by the low dense vegetation of the scrub-oak plains. Indeed they were so well hidden from view and the eggs so well covered when the bird was away that few of the nests were ever found, in spite of the great efforts various observers have expended to locate them.

William Brewster (1890) states: "Only one person of the many I have questioned on the subject has ever found a heath hen's nest. It was in oak woods among sprouts at the base of a large stump and contained either 12 or 13 eggs." There was a set of six eggs in the Brewster collection that were found in a nest in the woods near Gay Head on July 24, 1885. This set was described and one egg was figured by Capen (1886). One of the eggs given to the United States National Museum is figured by Bendire (1892, pl. 3, fig. 2). Bendire stated that the six eggs referred to above were the only eggs in any collection known to him.

In 1906, E. B. McCarta found a nest and nine eggs in a low but dense growth of scrub oak near the central part of Marthas Vineyard Island. Dr. George W. Field photographed the nest on June 2, and two days later the eggs were placed under a bantam hen. One of the eggs hatched on June 20, but unfortunately the chick was killed by the hen, and the other eggs failed to hatch. This set of eggs with the chick is now a part of a display group in the American Museum of Natural History, New York. On June 5, 1912, Deputy Warden Leonard, after a most prolonged and diligent search, found a nest and four eggs covered with leaves in a slight hollow surrounded by a dense mass of sweet ferns growing among the scrub oaks. The oaks in the vicinity were 2 or 3 feet in height. When the nest was visited on June 12 the bird was incubating. On June 21, Doctor Field was able to determine that there were eight eggs, and a week later he took an excellent series of photographs of the bird on the nest. The bird sat so closely on the eggs that it was dislodged only by active effort. Deputy Leonard had no difficulty in approaching the bird, and she fought the approach of the hand in the same manner as would a sitting hen, ruffling her feathers, opening her beak, and striking viciously. The incubation period of the heath hen, according to Doctor Field, is 24 days.

Eggs.—In addition to the eggs mentioned above in connection with the account of the nests, there are the following: An egg in the Brewer collection of the Museum of Comparative Zoölogy marked *Tympanuchus cupido*, Holmes Hole, Mass. (Holmes Hole is the old name for Vineyard Haven). There is an egg in the John E. Thayer Museum and another in the collection of the Boston Society of Natural History. The latter was picked up on the plains of Marthas Vineyard after the destructive fire of 1916. Bendire (1892) describes the eggs as "creamy-buff in color with a slight greenish tint, ovate in form and unspotted." They are regularly oval in form, all specimens being quite uniform in this respect. The color is yellowish green of a peculiar shade. I have compared the colors of this set of eggs with Ridgway (1912) and find all of them to be a "deep olive buff." One egg has a small spot of drab. All other heath-hen eggs that I have examined are of this same deep olive-buff color.

The measurement of the five eggs in the Brewster collection are as follows, 43.5 by 32.5, 43.5 by 32.1, 43.8 by 32.5, 43.9 by 32.4, and 46.2 by 32.9 millimeters, and the sixth egg, now in the United States National Museum, measures, according to Bendire, 44 by 33 millimeters. The egg in the Brewer collection collected at Holmes Hole measures 44.2 by 32.6, and the egg at the Boston Society of Natural History collected at Marthas Vineyard in 1916 measures only 40.3 by 30.4, the smallest egg of the species I have examined.

Young.—The first young of the heath hen for the season usually made their appearance during June. The earliest record is of a brood of 8 or 10 young seen near Edgartown on June 14, 1913, by Dr. Charles W. Townsend. During the season of 1913, nine broods, with an average of four chicks to the brood, were seen. On July 15, 1914, a dead chick was found that was estimated to be about five days old. In 1915, 14 broods were reported. The first brood, seen on June 19, consisted of six chicks about five days old. Allan Keniston saw seven broods with an average of five chicks each during the summer of 1918, and in 1919 he reported broods on June 29, July 1, and July 11, the members of which were able to fly well at that time. For 1920 the following were noted: June 20, 10 young; June 24, 6 young; July 4, 2 small broods, the numbers were not recorded; July 9, 5 or 6 young. During the summer of 1921 a brood of seven was seen on July 3, and a few days later a brood of eight was recorded. On July 31 a brood of six, about two-thirds grown, was seen. In 1922, between June 15 and June 30, there were five broods reported containing four to eight chicks; the exact dates were not recorded. In 1923 the writer made a continuous search for nests and young throughout the summer, but the birds by that time were so greatly reduced in numbers that only one brood of two chicks was seen, and that on July 3 during a downpour of rain, as we suddenly surprised the brooding mother in the middle of one of the little-used crossroads. Since 1923 there have been no authentic records of broods of young birds, and I very much doubt if any young have been reared since 1925.

Plumages.—The following description is based on a downy young about four days old: Underparts "cream-buff," the throat and middle of the belly approaching "colonial buff." Sides of the head "Marguerite yellow" with three small black spots back of the eye. Upper parts "tawny-olive" or "Isabella color," the region of the rump "snuff brown" and "russet," variously marked with black. There is a conspicuous mark on the forehead. The remiges and coverts marked with various tones of brownish drab and black, the feathers tipped with dingy white. The measurements of this specimen are as follows: Bill, 8; tarsus plus third toe, 39.8; wing, 42.5; length, 85; third toe, 23 millimeters. The natal down of a 2-day-old specimen hatched in captivity by Dr. John C. Phillips is similar to the above description but with the following differences: The underparts brighter yellow, the throat and sides of the head "amber yellow." The bright yellows fade rapidly when exposed to air and light, and in chicks two weeks old the bright yellows of the underparts are faded to a uniform "cream-buff." The measurements of the 2-day-old chick are: Bill, 7.5; tarsus plus third toe, 37; wing, 28; length, 79; third toe, 19 millimeters.

I have been unable to obtain a specimen of the heath hen in the completed juvenal plumage, but it is reasonable to infer that the sequence of the molts and plumages are similar to those described for the prairie chicken.

The first winter plumage of the heath hen is acquired by a complete postjuvenal molt except on the two distal primaries. The first winter plumage is similar to the adult plumage, but the younger birds are readily distinguished from the adults by their smaller size, by the more rufescent color of the upper parts, and by the coloration of the throat, which is "cinnamon-buff" in contrast to the "warm buff" or "cream color" of the adults and the white throat of the juvenals. The first plumages of the heath hen and the prairie chicken are so nearly alike that one can not readily distinguish them from each other. This ontogenetic resemblance indicates the close relationship of the two races.

The first nuptial plumage is acquired by a partial prenuptial molt, which does not involve the wings and tail and body plumage but is restricted to the head region. The nuptial plumage was usually completed by April. The second or adult winter plumage is acquired by a complete postnuptial molt during August and September. Specimens collected in October and November had the adult winter plumage completed. From the adult winter plumage onward the plumages are repetitions of the nuptial and winter phases.

Food.—The heath hen in its feeding habits was similar to other gallinaceous birds, such as the ruffed grouse, in being very adaptable to the changing food supply throughout the seasons of the year. It was not dependent on any particular item but subsisted on what was most abundant and most easily procured.

During the spring months the heath hen congregated in the open fields and meadows of the farms to feed upon the tender shoots of grasses, sorrel, and other plants, but when these became hardened and less palatable with the approach of summer the birds changed their diet to fruits and insects. In fall, berries and insects, such as grasshoppers, were freely eaten, and in the winter months, acorns, seeds, and certain berries found throughout the range of the heath hen on Marthas Vineyard provided the birds with a livelihood. Comparatively little snow falls on the island, and hence it was an exceptional winter when the birds were not able to obtain sufficient food from native plants. Even at times when the ground was covered with snow the scrub oaks held out an exhaustless supply of food in the form of acorns.

Our chief knowledge of the food of the heath hen is based upon the meager notes of food included in the data with birds collected and preserved as skins in various museums, upon field notes, and upon studies of birds kept in captivity. The following comprise the principal foods:

The crops of three heath hens collected by C. E. Hoyle on January 10, 1891, and two on December 28, 1895, contained bayberries (*Myrica carolinensis*). A specimen of a heath hen killed by a snowy owl was reported by Allan Keniston to have had its crop gorged with bayberries. Birds kept in captivity at the reservation in 1915 ate very freely of bayberries. During the winter of 1923–24 a flock of 15 birds frequented a large bayberry thicket near the south shore where they subsisted chiefly on these berries.

The bearberry (*Arctostaphylos uva-ursi*), sometimes wrongly named mountain cranberry or cranberry, is extremely abundant throughout the central portion of the island and was freely eaten by the heath hen during the winter months when the trailing plants were not covered with snow.

Audubon (1840) stated that the barberry (*Berberis vulgaris*) was the chief food of the heath hen. This is also included in the lists of the food of the heath hen by other earlier writers, but there is no evidence that the birds in recent years depended very largely on the barberry for a source of their food.

The fleshy wild-rose berries, more frequently called rose apples or rose hips, were eaten by a bird collected March 7, 1896.

The birds fed very freely on wild strawberries, which were abundant in the meadows and open areas of the reservation. They were also fond of the cultivated varieties, as evidenced by their frequent depredations in gardens grown near the haunts of the heath hen.

The partridgeberry (*Mitchella repens*) was so frequently eaten by the heath hen that the earlier settlers called this berry the heath-hen plum. Not only the berries but the leaves of this plant were often eaten by the birds during the early fall and winter months.

The dryland blueberry (*Vaccinium vacillans*), the low-bush blueberry (*V. pennsylvanicum*), and the black huckleberry (*Gaylussacia baccata*) were eaten during the berry season. On August 24, 1913, William Day saw a flock of 51 heath hens feasting on blueberries, and he also states that captive birds ate freely of blueberries provided for them.

The acorns of the scrub oak (*Quercus ilicifolia*) have been called the "bread" of the heath hen. No natural food is more abundant on Marthas Vineyard, and no food is more dependable during the winter months. The scrub-oak acorns are small and were swallowed whole by the birds. The scrub oaks provide one of the reasons why these birds have persisted in the scrubby plains.

Leaves were found in the crops of many specimens examined by Mr. Hoyle. In the spring months I found that the birds showed a great preference for the leaves of sheep sorrel (*Rumex acetosella*) over other plants such as clover and tender blades of grasses that were equally abundant in the same field. The distribution of the birds on the field usually corresponded roughly with the distribution of this plant. One male specimen trapped on April 5, 1924, had its crop gorged with leaves of sorrel. There were in this specimen, 1,846 leaves and parts weighing 32.2 grams. Another bird trapped on May 20, 1924, had its crop completely filled with 38.6 grams of seeds of the sheep sorrel. Although other leaves, such as those of clover, alfalfa, and other herbaceous plants, were eaten, the birds exhibited a decided preference for the leaves of the sorrel.

Buds, including those of the scrub pine (*Pinus virginiana*), were eaten by certain birds during the winter months.

The heath hen was especially fond of the grains of cultivated crops, such as corn, buckwheat, millet, and sunflowers, and in late years these were planted on the reservation, especially for the use of the heath hen.

The animal food of the heath hen consisted primarily of insects, chiefly grasshoppers, which were sometimes excessively abundant on the island late in summer and in autumn.

There are very few data concerning the food of the young, but, judged from the food of the young prairie chickens, it probably consisted mainly of insects, especially in the case of the younger birds. The crop contents of one 5-day-old heath-hen chick accidentally killed on July 15, 1914, contained more than 80 insects representing 10 species. The vegetable matter was merely incidental and negligible in this specimen.

Game.—The heath hen even when in its prime was never given a high place among game birds by the sportsmen. It was easily shot, because of its direct and laborious flight, and the habit of massing in flocks in the open fields made it too easy a victim for the pot hunter and the market gunner. The ease with which the heath hen was tricked and killed readily accounts for the rapidity of its early disappearance from the mainland, whereas the clever ruffed grouse of the woodlands still holds its own and is ever ready to challenge the wits of the most skilled sportsman.

The attitude of the sportsman in the past toward the pinnated grouse is well illustrated by the following excerpt from an article by Elisha J. Lewis in his book The American Sportsman, 1885.

> So numerous were they a short time since in the barrens of Kentucky, and so contemptible were they as game birds, that few huntsmen would deign to waste powder and shot on them. In fact they were held in pretty much the same estimation, or, rather abhorrence, that the crows are now, as they perpetrated quite as much mischief upon the tender buds of the orchards, as well as the grain of the fields, and were so destructive to the crops, that it was absolutely necessary for the farmers to employ their young negroes to drive them away by shooting of guns and springing loud rattles all around the plantation from morning till night. As for eating them, such a thing was hardly dreamed of, the negroes themselves preferring the coarsest food to this now much admired bird.

It is apparent that the heath hen was not considered an ideal bird from the point of view of the sportsman, and our efforts to save the heath hen were not made on the plea of its economic importance as a game bird. It is interesting to note, however, that it was the sportsman who took the initiative and who provided a large part of the funds to assist the State in the vain attempt to preserve this interesting race of birds.

Enemies.—Man directly or indirectly is in part responsible for the disappearance of the heath hen from most of its former range. Comparatively soon after the coming of the white man, it was driven from one locality to the next, until it was forced to entrench on the scrub-oak plains of Marthas Vineyard Island. It is a striking example of a bird that has not been able to adapt itself to the changing conditions brought about by civilization.

Audubon (1840) wrote: "We frequently meet with the remains of such [heath hen] as has been destroyed by the domestic cat which prowls in the woods in a wild state." What was true in Audubon's day was true in the more recent years of the heath hen, when cats ranked high among the enemies of the bird. In addition to the cats reared on the island, large numbers were introduced by people who dumped them in the interior of the island when they left their summer homes in autumn. A large part of the effort of the State Department and the special wardens in the control of vermin was directed toward the semiwild house cat.

Large numbers of hawks are attracted to the island because of the abundance of mice and shrews, which live among the scrub oaks. Unfortunately, many of the hawks, for example the marsh hawk, which have a good reputation elsewhere, are frequently tempted, on Marthas Vineyard, to prey upon birds; and when the heath hen was common, these birds were also numbered among their victims. The goshawks, notorious for their killing of game birds, played their part in the history of the heath hen. The most notable instance of their wholesale depredations was in the winter of 1916–17, following the destructive fire that swept over the island during the preceding spring. Other hawks, such as the red-shouldered, the rough-legged, the pigeon hawk, and others, as well as the different species of owls, were killed on sight by the wardens in charge of heath-hen protection. It is probable that the wholesale killing of hawks and owls so upset the balance of nature that it acted as a boomerang to the heath hen.

Disease was one of the most important factors in the recent decline of the heath hen and was one that man was unable to control. Blackhead is a disease common to poultry, but, so far as we know at present, it is unusual in birds living in a free, wild state. The heath hen, however, had the peculiar habit of congregating in the open fields near farmhouses where poultry was kept. In most instances, chickens and turkeys had access to the fields visited by the heath hen, and thus the dreaded disease was readily transmitted to the native birds. Blackhead was found in the adult heath hen, and this is presumptive evidence that the disease was very destructive to the young.

Internal and external parasites were found on the few heath hens examined, but these were all of minor importance as compared to the disease blackhead.

In the recent history of the heath hen it was well known that there was a great excess of male birds. This abnormal ratio may have been brought about in part by some hereditary influence, but it is certain that this condition was aggravated by the fires that ravaged the island during the breeding season. At such times the females were destroyed on the nests, whereas the males escaped the conflagration. Furthermore, a female with young was subject to more danger of being killed than the male, which never cared for the young.

Other factors that played their part were the excessive interbreeding, which was destined to occur after the heath hen was restricted in range and to exceedingly small numbers of individuals. It was also found upon examination of dissected specimens that many of the birds were sterile.

TYMPANUCHUS PALLIDICINCTUS (Ridgway)

LESSER PRAIRIE CHICKEN

HABITS

Comparatively little seems to be known and still less has been published on the habits and distribution of the small, light-colored, lesser prairie chicken, which is found in the Upper Sonoran Zone of the Great Plains from Kansas and Colorado to central Texas and eastern New Mexico. It has disappeared from many sections where it was once abundant; too much grazing on, and extensive cultivation of, the grassy plains have driven it out. But it is still to be found in fair numbers in its restricted range, where it is protected, or not disturbed.

We are greatly indebted to Walter Colvin for the information that follows regarding this fine bird, which I have gathered from his published article (1914) and from the full notes and photographs he has sent to me. Writing of its distribution and haunts in 1914, he says:

The natural habitat of this beautiful grouse is far remote from the habitat of its allied cousin, the heath hen, and still less remote from its nearer cousin, the common prairie hen of the Middle States. Its present confine is the southwestern counties in Kansas, extending west from Meade, through Seward, Stevens, and Morton and north into Stanton, Grant, and Haskell counties, crossing the line into Colorado some fifty miles, extending south through Beaver, Texas, and Cimarron counties of Oklahoma, into the panhandle of Texas, but how far south and east I haven't sufficient data at hand to determine, although I believe it is safe to assert that they do not extend farther south into Texas than two degrees by air line. In northwestern Oklahoma I have seen the chickens within a few miles of the New Mexico line.

Formerly this variety of chickens was common in Woodward County, Oklahoma, and Captain Bendire, in his Life Histories, mentions securing their eggs near Fort Cobb, Indian Territory, in 1870. At that time reliable information goes to show that they were far more plentiful south of the great Indian highway than north. The pan-handle is a typical bunch-grass country, and during the early eighties a great prairie fire broke out in its southern extremity, sweeping north to the narrow strip of short grass land in "No Man's Land," where it died. The chickens that were driven north found an ideal home in the rolling, sandy bunch-grass country that abounded just across the line.

Their range in its entirety would probably cover no greater area than a fourth of the State of Kansas, and the most abundant nucleus is in Stevens and Morton counties. Here they are quite plentiful in its sandhill and bunch-grass fastness, where, in the fall of the year they sometimes gather in flocks of several hundred birds, roaming where they will, a typical bird of the long-grass country.

Courtship.—Concerning the courtship of this species, Colvin says:

The nuptial performances of the cocks are similar to those of the common variety, but the ventriloquial drumming sound does not appear to be quite so rolling or voluminous. In May, 1907, I put up in the heart of the nesting-ground, where I had an excellent opportunity to study their habits. The cocks generally select for drumming-ground a slight rise covered with buffalo grass, where they gather each spring for the nuptial performance. They are very partial to their drumming-grounds, and even though disturbed will return to their old haunts year after year. I saw one drumming-ground that had been used for many years.

Here the cocks would gather sometimes as high as fifty birds to perform their antics. The drumming of so many cocks would be of such volume as to sound like distant thunder. Hens attracted by the drumming would cause disturbance. Cock fights and a general all-around rumpus would begin. A great deal of strutting and clucking would be done by the males. Finally, when, with lowered head and wings and air-sacks full, a successful cock would drive his hen from the bunch, peace would reign again, and the drumming would be resumed.

Nesting.—A nest that Colvin found near Liberal, Kans., on May 28, 1920, was "located in a bunch of sage, growing in a swale just below a brow of a hill;" it consisted of "a hollow scooped out in the sand and lined with grasses. So well concealed was the nest that one could observe only a small portion of the female as she sat upon the nest. Disturbing the sage brush she left the nest, disappearing over the hill. The nest contained 12 eggs on the point of hatching."

Another set of eggs, which he kindly presented to me, was taken in the same general region on June 2, 1920; these eggs were only slightly incubated. He says in his notes:

On John Napier's farm I was shown a nest of the lesser prairie hen. It was placed under the south side of a bunch of sage, and was a mere hollow in the sand lined with grasses. When I saw the nest it had been exposed by a corn lister. The nest originally contained 12 eggs, but one had been broken by the lister, leaving 11. Mr. Napier informed me that the team had passed over the nest and sitting hen twice before she was finally raked off by the doubletree.

Of a third nest he writes:

Through the efforts of an old-time trapper, Ed. Ward, I was successful in securing a set of 13 straw-buff-colored eggs. The nest, a mere hollow in the sand, was lined with a few grasses, and was situated under a tumbleweed, which had lodged between two tufts of grass on the north side of a sloping hillock. The sitting hen allowed us to approach quite close before taking wing.

Mr. Ward informed me that the nests were almost invariably placed on top of a rise, or on its sloping sides. The nests, though usually placed in open situations, are extremely difficult to find, owing to the dichromatic arrangement of the feathers, which so harmoniously blends with the surroundings of the sitting bird. A far greater protection to the sitting hens is their non-scent-giving powers during the nesting season, which was fully demonstrated the following spring, when I again visited that vicinity in order to secure a series of photographs.

In company with one of the best-known chicken dogs, I thrashed over several sections of bunch grass land where chickens were common and known to nest each year, but without success. I found no hens off the nests during the heat of the day, but quite frequently saw them flying to the feeding grounds after twilight. Several times while hunting their nests I felt sure that I was within a few feet of the sitting birds, but was compelled to give up the search. The hens are close, hard sitters, and very few nests are found. Prairie fires expose many nests and are the nesting hen's worst enemy.

Eggs.—The full set seems to consist of 11 to 13 eggs, so far as we know. The eggs are ovate in shape, smooth, and rather glossy. The colors vary from "cream color" to "ivory yellow" in my set; most of the eggs are sprinkled with very fine dots of pale brown or olive; but some are nearly or quite immaculate. Mr. Colvin calls them straw color or straw buff. Bendire (1892) says: "The ground color varies from pale creamy white to buff. The markings, which are all very fine, not larger than pin-points, are lavender colored. More than two-thirds of the eggs are unspotted, and all look so till closely examined."

The measurements of 47 eggs average 41.9 by 32 millimeters; the eggs showing the four extremes measure 43.5 by 33.5, 40.5 by 33, and 40.7 by 30.4 millimeters.

Plumages.—I have never seen a downy young of the lesser prairie chicken, but probably it is much like the chick of its northern relative. The sequence of molts and plumages is doubtless similar to those of the prairie chicken. A young bird, about one-third grown but in full juvenal plumage, has the crown and occiput mottled with "tawny" and black; the chin and throat are white; the feathers of the back and scapulars are variously patterned with transverse bars of "ochraceous-tawny," "tawny-olive," "cinnamon-buff," and black, with median white stripes or tips; the central tail feathers are similarly barred and tipped, the pattern diminishing on lateral rectrices; the underparts are dull whitish, heavily spotted or barred on the breast and flanks with black, sepia, and pale dusky, darkest on the chest, and more or less heavily tinged on the flanks and breast with "ochraceous-buff."

Food.—Colvin (1914) says that "during the summer months they feed largely on grasshoppers, but in the fall and winter they feed almost entirely on kaffir corn and maize, cane seed, and other varieties of semi-arid cereals. As to the palatableness of the meat I much prefer duck."

Behavior.—The same observer says:

In general characteristics and makeup the lesser prairie hen is of a sturdy, robust nature, being some two-thirds the size of the common prairie hen. They are veritable dynamos of "git up" and energy. Such vivacity and activity I have never seen displayed in any other game bird. On a cold, snappy day they have the life and energy of half a dozen quails, and for speed they put their first cousin to shame. A full-grown cock, well fatted, will weigh from a pound eleven ounces to a pound fourteen ounces.

Game.—Judged from Mr. Colvin's published accounts (1914 and 1927) the lesser prairie chicken must be a fine game bird, sufficiently wild and swift of wing to make sporty shooting, and large and plump enough to make a desirable table bird. These birds were wonderfully abundant in earlier days, as a few quotations from Mr. Colvin's writings will show. "In a cane field near the State Line," he says, "we saw a flock of 500 or more, and when they arose it seemed that a hole had been rent in the earth." He wanted to stop and shoot a few, but his companion urged him on, saying, "Those are only rovers. I'll show you some chickens when we get up in the State." Evidently he made good, for Mr. Colvin (1914) writes:

Two miles farther along we came to Ed Ward's. He informed us that there were a "few" chickens in a cane and kaffir corn field a quarter of a mile east. We flushed several birds from the tall bunch grass just before we reached the field, which were promptly despatched; however, in the field things became more lively. Such a sight I have never seen before nor since. Chickens were flushing everywhere, and droves of fifty to a hundred would take down the corn rows, sounding like a moving avalanche as they touched the blades of corn. Still birds were quite wary, and the only good shots were to be had over the dog.

As we thrashed back and forth across the grain field, the chickens arose in flocks of fifty to five hundred, and generally sixty to eighty yards distant, making shooting difficult. The majority of the birds, after being flushed, would fly back into the field, while some would go to the bunch-grass covered hills half a mile away. Mr. Ward and I estimated that there were from thirty-five hundred to four thousand chickens in this one field, a sight never to be forgotten.

A few years later, hunting over the same ground, he found the chickens much diminished in numbers; his thoughts are expressed as follows:

Gathering our duds together, we started for our long journey home. A few clouds, fringed with gold, freckled the western sky, and over all a red mantle was cast while the sun slowly lowered to the horizon. My mind went back to the events of the day and to the time when the chickens were more plentiful, and I realized with a shudder that we were nearing the sunset life of the king of upland game birds. But the decrease in their numbers is not due so much to the gunners, as gunners are few per capita in those parts, but is due largely to the cutting up of this vast wilderness into small farms. The bunch-grass land can not be mowed for hay; therefore, in such land the chickens have found an ideal home in which to rear their young and harbor themselves during the winter. Such land is soon destroyed by cultivation and small pastures. With the advancement of civilization the flocks scatter and become depleted.

He wrote to me in December, 1927:

The saving of the lesser prairie hen for future time is assured, as some recent laws were enacted in Kansas that gave the game commission the power that enables them to close and open seasons without any special legislation from the State. After visiting the nesting grounds this summer and noticing that many of the birds had been destroyed, I took it up with the State game warden, J. B. Doze, and Lee Larabee of the commission, and they established the close season this year. We are therefore assured of a good crop of birds next year.

Winter.—Of their winter habits, Colvin says:

Though naturally lovers of the free range, during the winter they rely largely upon the farmer and rancher for their food. A large amount of grain is consumed by the flocks as they roam from one grain field to another. In the eighties a man by the name of Hatch settled in the sandhills just inside the Kansas line in Seward County. Here he planted a grove of black locust trees and spread out his broad fields of maize and kaffir corn. The Texas bobwhites, mountain quail, and lesser prairie hens soon learned that this man was a friend of the birds, and straightway made it their rendezvous. Here, each fall, the chickens gathered by the thousands, and each spring spread out over the vast prairies, nesting and rearing their young. In the fall of 1904 my brother estimated that he saw in a single day, 15,000 to 20,000 chickens in and around this one grain field. Though timid if persecuted, if unmolested they become quite tame, coming to the barn lots to feed, and will put as much confidence in man as quails when protected.

DISTRIBUTION

Range.—The Great Plains region, from southeastern Colorado and Kansas south to west-central Texas and probably southeastern New Mexico.

Breeding range.—The breeding range of the lesser prairie chicken extends **north** to southeastern Colorado (Gaumes Ranch and Holly) and southwestern Kansas (Cimarron). **East** to southwestern Kansas (Cimarron) and Oklahoma (Ivanhoe Lake, Fort Reno, and Fort Cobb). **South** to southwestern Oklahoma (Fort Cobb); northern Texas (Mobeetie and Alanreed); and east-central New Mexico (Portales). **West** to east-central New Mexico (Portales); and southeastern Colorado (Cimarron River and Gaumes Ranch).

Winter range.—Confined chiefly to central Texas. **North** to Monahans, Midland, and Colorado City. **East** to Colorado City, Middle Concho River, and Bandera. **South** to Bandera, Fort Clark, and the Davis Mountains. **West** to the Davis Mountains and Monahans. Casual in winter at Lipscomb in the Panhandle area, and it appears probable that some winter in southeastern New Mexico (vicinity of Carlsbad).

No information is available relative to the movements of this species between breeding and wintering areas.

Casual records.—Widmann (1907) reports a specimen in the Hurter collection said to have come from southwestern Missouri, and that in January, 1877, large numbers were shipped to Fulton market, New York City, from Pierce County. Neff (1923) states the species was noted in Lawrence County, Mo., in 1887. A specimen in the collection of the Academy of Natural Sciences of Philadelphia was taken between January 24 and 28, 1894, near Garneth, Kans., while another was collected at Oakley, Kans., January 1, 1921. The exploring party of Capt. John Pope collected two specimens (later made the types of the species) on the Staked Plains, N. Mex., on March 3 and March 11, 1854.

Lesser prairie chickens also have been reported from Nebraska, but in the lack of specimen evidence it is thought that the records refer to *T. c. americanus.* At the present time the Arkansas River appears to be a very definite northern boundary to their range.

Egg dates.—Colorado to Texas: 12 records, May 5 to June 12; 6 records, May 20 to June 1.

PEDIOECETES PHASIANELLUS PHASIANELLUS (Linnaeus)

NORTHERN SHARP-TAILED GROUSE

HABITS

As the specimen on which Linnaeus bestowed the type name of the species came from the Hudson Bay region, the name *P. p. phasianellus* is now restricted to the dark-colored race, which ranges through the forested regions of northern Canada to central Alaska. Its center of abundance seems to be in the vicinity of Great Slave Lake, Mackenzie. Swainson and Richardson (1831) say that " it is found throughout the woody districts of the fur-countries, haunting open glades or low thickets on the borders of lakes, particularly in the neighbourhood of the trading-posts, where the forests have been partially cleared." According to Major Bendire (1892) it was found breeding at Fort Rae, in latitude 63° N., and at Fort Good Hope, in the Mackenzie River Basin. MacFarlane (1908) found it breeding in the valley of the Lockhart and Anderson Rivers, where two nests were found, but the eggs were afterwards lost. Herbert W. Brandt says in his Alaska notes:

The sharp-tailed grouse proved to be the most common gallinaceous bird we encountered during the early stages of our dog-sled trip to Hooper Bay. We first collected it on March 22, when two handsome males were taken, but small flocks were seen from time to time during the previous day. After we reached the Koskokwim River the birds became scarcer, and we did not see it at all along the Yukon River. The center of abundance of this grand grouse appears to be in the vicinity of Lake Minchumina, with its sheltered, scenic, birch-clad hills, that transcends in exquisite beauty any region I saw in glorious Alaska.

Nesting.—The nesting habits of this grouse are apparently similar to those of its more southern relatives. The only published account of it I can find is by Baird, Brewer, and Ridgway (1905), as follows:

> Mr. Kennicott found the nest of this bird at Fort Yukon, at the foot of a clump of dwarf willows. It was in dry ground, and in a region in which these willows abounded and were quite thickly interspersed with other trees, especially small spruces, but no large growth. The nest is said to have been similar to that of *Cupidonia cupido*. Mr. Lockhart also found it breeding in the same region. The nests seen by him were likewise built on a rising ground under a few small willows.

Eggs.—Major Bendire (1892) describes the eggs very well, as follows:

> The number of eggs to a set varies from seven to fourteen and their ground color from a fawn color with a vinaceous rufous bloom to chocolate, tawny, and olive brown in different specimens. The majority of the eggs are finely marked with small, well-defined spots of reddish brown and lavender, resembling the markings found on the eggs of *Tympanuchus americanus*, only they are much more distinct. Compared with the eggs of the two southern subspecies, *P. phasianellus columbianus* and *P. phasianellus campestris*, they usually are very much darker colored, even the palest specimens being darker than the heaviest marked eggs of either of the two subspecies. These markings are entirely superficial, and when removed leave the shell a creamy white in some cases and a very pale green in others. In shape they are usually ovate.

The measurements of 27 eggs average 43.1 by 32.3 millimeters; the eggs showing the four extremes measure **48** by 33, 44.5 by **33.5**, **40.5** by 32, and 42 by **30** millimeters.

Food.—Swainson and Richardson (1831) say: "They feed on the buds and sprouts of *Betula glandulosa*, of various willows, and of the aspen and larch, and in autumn on berries." Macoun (1909) says: "These birds keep in pairs or small flocks and frequent the juniper plains all the year. The buds of these shrubs are their principal food in winter, as their berries are in summer."

DISTRIBUTION

Range.—Alaska, western Canada, and the Western United States.

The full range of the sharp-tailed grouse extends **north** to Alaska (Allakakat and Fort Yukon); Yukon (Ramparts); Mackenzie (Fort Good Hope, Fort Norman, Grandin River, Fort Rae, Fort Resolution, and Fort Smith); northeastern Alberta (Fort Chipewyan); east-central Saskatchewan (Cumberland House); Manitoba (Grand Rapids, Norway House, Oxford House, and York Factory); northern Ontario (Severn House and Fort Albany); and Quebec (Great Whale River). **East** to Quebec (Great Whale River and Fort George); eastern Ontario (Lake Abitibi, Lake Temiskaming, Parry Sound, and Beaumaris); formerly northeastern Illinois (Waukegan); formerly east-central Iowa (Grinnell); and formerly central Kansas (Fort Hays). **South** to formerly Kansas (Fort Hays, Ellis, and Banner); New Mexico (Raton); Colorado (Pagosa Springs, Fort Lewis, and Cortez); southern Utah (Parawan Mountains); and formerly northern California (Canoe Creek and Fort Crook). **West** to formerly northern California (Fort Crook and Camp Bidwell); Oregon (Fort Klamath, Caleb, and The Dalles); Washington (Toppenish and Dosewallops River); British Columbia (Nicola, Kamloops, Cariboo Road, 158-mile House, Quesnelle, and Hudsons Hope); southwestern Yukon (Tagish Lake, Alsek River, and Lake Kluane); and Alaska (Kolmakof, Holy Cross, Tacotna, Lake Minchumina, and Allakakat).

Sharptails are not now known to breed east of Ontario, Minnesota, South Dakota, and Colorado, and apparently they have been entirely extirpated from Iowa, Kansas, and California. The species occurs only as a summer straggler in western Alaska (Allakakat, Holy Cross, Lake Minchumina, and Kolmakof), and as a winter straggler in southeastern Ontario (Lake Abitibi and Temiskaming) and Quebec (Great Whale River). It has, however, occurred in summer at Fort George, Quebec. In some winters it is abundant at Vermilion and Watertown, S. Dak.

Migration.—In common with some of the ptarmigans, there appears to be a definite migration from the northern part of the range, governed by the severity of winter conditions and the available food supply. This exodus, however, does not extend south of the breeding range. It has been observed, apparently, only by E. A. Preble, who noted flocks moving southward at Fort Norman, Mackenzie, on October 1, 1903, and who detected early spring arrivals at Fort Simpson, Mackenzie, on March 12, 1904.

The range as outlined is for the entire species, which has been separated into four subspecies. The northern sharp-tailed grouse (*P. p. phasianellus*) occupies the northern part of the range south to Lake Superior, the Parry Sound district (casually) of Ontario, and the Saguenay River, Quebec. The Columbian sharp-tailed grouse (*P. p. columbianus*) is found from the interior lowlands of British Columbia south (formerly) to northeastern California, Utah, Colorado, and northern New Mexico. The range of the prairie sharp-tailed grouse (*P. p. campestris*) extends from eastern Colorado, Kansas, northern Illinois, and Wisconsin north to southern Manitoba, Saskatchewan, and Alberta.

Egg dates.—Northern Canada: 12 records, May 1 to June 3; 6 records, May 16 to 21.

Washington and Oregon: 9 records, April 6 to June 18. Montana: 5 records, May 19 to June 5. Alberta to Manitoba: 52 records, May 2 to June 22; 26 records, May 30 to June 11. Dakotas and Minnesota: 27 records, May 2 to June 26; 14 records, May 19 to June 8. Colorado: 2 records, April 1 and May 8.

PEDIOECETES PHASIANELLUS COLUMBIANUS (Ord)

COLUMBIAN SHARP-TAILED GROUSE

HABITS

This is the grayest of the three races of sharp-tailed grouse; the northern race, typical *phasianellus*, is much darker, and the eastern race, *campestris*, is more buffy or rufous. This western race inhabits the lowlands of the Great Basin from the Rocky Mountains to the Cascades and Sierra Nevadas. It was discovered by Lewis and Clark on the plains of the Columbia River in 1805 and was named by Ord in 1815. As it is not so widely distributed or so well known as *campestris*, the reader is referred to the prairie sharp-tailed grouse for the life history of the species.

Major Bendire (1892) says of this grouse:

> It is one of the most abundant and best known game birds of the Northwest, inhabiting the prairie country to be found along the foothills of the numerous mountain chains intersecting its range; seldom venturing into the wooded portions for any distance, and then only during the winter months, when it is partially migratory in certain sections. According to my own experience the Columbian Sharp-tail breeds more frequently on the sheltered and sunny slopes of the grass-covered foothills of the mountains than in the lower valleys and creek bottoms.

As to its past status in California, Grinnell, Bryant, and Storer (1918) say:

> When Newberry, Cooper, Henshaw, and other early naturalists were making observations upon the fauna of California, previous to 1880, they found this species numerous in the plateau region northeast of the crest of the Sierra-Cascade range. Since then, man's occupancy of that territory, and uncontrolled levy upon its birds for food or sport, has resulted in the apparently complete disappearance of this species. They say: Coming north from San Francisco, we first found it on a beautiful prairie near Canoe creek (near Cassel, Shasta County), about fifty miles northeast of Fort Reading; subsequently, after passing the mountain chain which forms the upper canon of Pit River, we came into a level, grass-covered plain, through which the willow-bordered river flows in a sinuous course like a brook through a meadow (probably near Lookout, Modoc County). On this plain were great numbers of birds of various kinds, and so many of the sharp-tailed grouse, that, for two

or three days, they afforded us fine sport and an abundance of excellent food. We found them again about the Klamath lakes.

Grinnell, Bryant, and Storer (1918) say that it is "now almost or entirely extirpated" as a California bird, and that "the disappearance of this bird can be attributed to no other cause than to its incessant pursuit by man. As long as a single bird remained hunting persisted. Moreover the fact that this grouse prefers grassy localities, just such as are selected for ranch sites, indicates another of the factors that led to its extermination."

The southern limit of its present range seems to be in northern New Mexico, where Mrs. Florence M. Bailey (1928) says of it:

Though naturally a bird of more northern country with abundant rank grass for breeding places, the high altitude of the grassy, broken-rimmed mesas northeast of Raton, some 8,000–9,000 feet in elevation, "appears to create a little world suitable to it in New Mexico," which, the oldest settlers attest, has long been inhabited by it.

Nesting.—Major Bendire (1892) says of its nesting habits:

Nidification began usually from about April 15 to May 1, according to the season. I found a set of fifteen eggs, which had been sat upon about a week or ten days, on April 22, 1871. Some birds must have laid earlier still, as it was no uncommon sight to find fully grown birds by July 10. All the nests of this species which I examined were invariably well concealed and rather difficult to find. You might search daily for a couple of weeks and be unsuccessful in finding a nest, and again you might stumble on two or three on the same day. A bunch-grass covered hillside, with a southerly exposure, seemed to be a favorite nesting site with this Grouse at Fort Lapwai, while at Camp Harney, Oregon, they confined themselves during the breeding season to the sage brush covered plains of the Harney Valley, interspersed here and there with a low grassy swale, nesting along the borders of these, where the grass attained a heavier growth. The nest, like that of all the Grouse, is always placed on the ground, usually close alongside some tall bunch of coarse grass, which hides it completely from view. Even if it did not, the female harmonizes in color so thoroughly with her surroundings that she is not apt to be noticed, unless she should leave her nest, which she does not do very readily, as she is a very close sitter. A slight hollow, usually scratched out on the upper side of a bunch of grass, if the nest is placed on a hillside, is fairly lined with dry grass, of which there is ordinarily an abundance to be found in the vicinity, and this constitutes the nest. A few feathers from the lower parts of the bird are usually mixed in among the eggs, each one of which is often imbedded about two-thirds in its own mould and does not touch the others. Once only did I find the eggs placed on top of each other, eight in the lower and five in the upper layer.

Eggs.—The eggs of the Columbian sharp-tailed grouse are indistinguishable from those of its prairie relative. The measurements of 58 eggs average 43.3 by 32.1 millimeters; the eggs showing the four extremes measure **46.5** by **34.5**, **39** by 31, and 42.5 by **30.5** millimeters.

Young.—Major Bendire (1892) writes:

The young are active, handsome little creatures, and able to use their legs at once on leaving the shell. They are at first fed mostly on insects, young grasshoppers and crickets forming the principal portion of their bill of fare. The former are always abundant and easily obtained; later, when the young are able to fly, the mother leads them to the creek bottoms, where they find an abundance of berries and browse. They are especially fond of the seeds of the wild sunflower, which grows very abundantly in some places, and when these are ripe, many of these birds can be found in the vicinity where these plants grow.

Coues (1874) says:

The young, as usual among gallinaceous birds, run about almost as soon as they are hatched; and it is interesting to witness the watchful solicitude with which they are cherished by the parent when she first leads them from the nest in quest of food, glancing in every direction, in her intense anxiety, lest harm befall them. She clucks matronly to bring them to brood under her wings or to call them together to scramble for a choice morsel of food she has found. Should danger threaten, a different note alarms them; they scatter in every direction, running, like little mice, through the grass till each finds a hiding place; meanwhile, she exposes herself to attract attention, till, satisfied of the safety of the brood, she whirrs away and awaits the time when she may reassemble her family. In the region where I observed the birds in June and July, they almost invariably betook themselves to the dense, resistant underbrush, which extends for some distance outward from the wooded streams, seeking safety in this all but impenetrable cover, where it was nearly impossible to catch the young ones, or even to see them, until they began to top the bushes in their early short flights. The wing and tail-feathers sprout in a few days and are quite well grown before feathers appear among the down of the body. The first coveys seen able to rise on wing were noticed early in July; but by the middle of this month most of them fly smartly for short distances, being about as large as Quails. Others, however, may be observed through August, little, if any, larger than this, showing a wide range of time of hatching, though scarcely warranting the inference of two broods in a season.

Behavior.—The habits of this grouse are essentially the same as those of the prairie sharp-tailed grouse, but Bendire (1892) observes:

The habits of the Columbian Sharp-tailed Grouse vary very materially in different portions of the country where I have met with them. At Fort Klamath, Oregon, where they are rather rare, I have found them inhabiting decidedly marshy and swampy country, and keeping close to, if not in the edges of, the pine timber throughout the year. At Fort Custer, Montana, this Grouse, during the winter, was much more arboreal than terrestrial in its habits, moving around on the limbs of the large cottonwood trees as unconcernedly as on the ground; spending in this way almost all their time, except when feeding. At Harney, Oregon, and Lapwai, Idaho, they might be frequently seen in small trees and bushes which grow along the creeks, but scarcely ever in large trees, of which there was an abundance. Here, they uttered very few notes at any time, while at Fort Custer I have frequently heard them cackling in the tall cottonwoods which grew along the Big Horn River bottom, before I had approached within several hundred yards of them, evidently giving notice to other birds in the vicinity of my coming. This fine game bird is decreasing very rapidly throughout its range. It does not seem to prosper in the vicinity of man, and as the country is becoming more and more settled, it recedes before civilization. As it is not a particularly shy bird, it falls an easy victim to the gunner.

PEDIOECETES PHASIANELLUS CAMPESTRIS Ridgway

PRAIRIE SHARP-TAILED GROUSE

HABITS

On my various trips to North Dakota, Manitoba, and Saskatchewan, I became quite familiar with the sharp-tailed grouse of the eastern plains. It is not so much a bird of the open prairie as is the prairie chicken; but we found it very common in the sandhills, among the willow thickets, and on the low, rolling hills overgrown with shrubbery. Its range is becoming more and more restricted as the Central West becomes more thickly settled and more land comes under cultivation. In some places it is decreasing in numbers where prairie chickens are increasing. Edwyn Sandys (1904) writes:

It has been claimed by more than one well-known expert that the sharptail and pinnated grouse are bitter foes, but this I am inclined to doubt. I am well aware of the belief among western sportsmen that the one species drives the other from its haunts, but believe that the true reason for the supplanting of one species by the other is nothing more than the closer settlement of what a few years ago were wild regions. In other words, one bird follows the farmer, while the other retreats before him.

Courtship.—The courtship performance of the sharp-tailed grouse is no less interesting than that of other grouse. It is quite similar to that of the prairie chicken and the heath hen; perhaps not quite so grotesque but more animated. These birds have favorite spots, generally small knolls, to which they resort for this purpose every spring; these are known as "dancing hills." Frank L. Farley writes to me from Alberta:

On my farm at Dried Meat Lake, an average of a dozen pairs nest every year. There are two dancing hills on this farm that the Indians told me had been used as long as they could remember; one is a little knoll right overlooking the lake, and the other is half a mile away. I have seen as many as 50 birds dancing on each hill at a time; that is, waiting until the ground was vacated by the previous dancers. They would wait patiently for their turn. These dances take place every April and May, and often the grain, when up, is tramped entirely away. I can generally get up to within 25 feet of the dancers with my car to watch them.

Dr. D. G. Elliot (1897) gives a very good account of the "dancing," as follows:

In the early spring, in the month of April, when perhaps in many parts of their habitat in the northern regions the snow still remains upon the ground, the birds, both males and females, assemble at some favorite place just as day is breaking, to go through a performance as curious as it is eccentric. The males, with ruffled feathers, spread tails, expanded air sacs on the neck, heads drawn toward the back, and drooping wings (in fact, the whole body puffed out as nearly as possible into the shape of a ball on two stunted supports), strut about in circles, not all going the same way, but passing and crossing each other in various angles. As the "dance" proceeds the excitement of the birds increases, they stoop toward the ground, twist and turn, make sudden rushes forward, stamping the ground with short quick beats of the feet, leaping over each other in their frenzy, then lowering their heads, exhaust the air in the sacs, producing a hollow sound that goes reverberating through the still air of the breaking day. Suddenly they become quiet, and walk about like creatures whose sanity is unquestioned, when some male again becomes possessed and starts off on a rampage, and the "attack" from which he suffers becomes infectious and all the other birds at once give evidences of having taken the same disease, which then proceeds with a regular development to the usual conclusion. As the sun gets well above the horizon, and night's shadows have all been hurried away, the antics of the birds cease, the booming no longer resounds over the prairie, and the Grouse scatter in search of food, and in pursuit of their daily avocation. While this performance is always to be seen in the spring, it is not unusually indulged in for a brief turn in the autumn, and while it may be considered as essentially a custom of the breeding season, yet like the drumming of the Ruffed Grouse, it may be regarded also as an exhibition of the birds' vigor and vitality, indulged in at periods of the year even when the breeding season has long passed.

Ernest Thompson Seton (Thompson, 1890) says:

The whole performance reminds one so strongly of a Cree dance as to suggest the possibility of its being the prototype of the Indian exercise. The space occupied by the dancers is from 50 to 100 feet across, and as it is returned to year after year, the grass is usually worn off and the ground trampled down hard and smooth.

Hamilton M. Laing (1913) noticed that the birds danced in pairs and that each pair usually kept to a certain section of the hill. He describes a lively fight he saw as follows:

The fun had reached its frenzy pitch when suddenly I noted that something other than dancing was taking place. It very much resembled a fight; and soon I realized that such it really was, though it had a most absurdly comic side to it. The fray was a three-cornered affair. The first fellow fled in circles; the second followed him; and the third brought up the rear. I decided that it was two cocks fighting, and that the cause at issue, and root of the trouble, was merely following the contestants. They whirled about the hill at lightning speed, running on legs that fairly spun, or dashing short snatches on the wing, through the set or over the dancers. The second fellow had blood in his eye, and the first—evidently an interloper, who was not wanted—lacked the courage or fiber to turn and fight it out. Yet, when the pursuer caught him, they bit and held on with a grip like bull-dogs, and rolled over, and beat each other with their wings, and shed each other's feathers. The interloper always got thumped, but not until he was properly mauled would he retreat.

E. S. Cameron (1907) writes:

At this date (April 18) the ball is opened by a single cock making a run across the open space as fast as he can use his legs, the tail being inclined stiffly over the back, while the wings are dragged, so that a large white area is exposed behind. The vivid yellow supraciliary fringe is erected, and, all the feathers of the neck standing on end, a pink inflated sac is disclosed. At the same time the head is carried so low as almost to touch the ground, so that the bird is transformed in appearance and, as observed through binoculars at some distance, looks to be running backwards. He then returns at full speed, when another cock comes forward toward him, both advancing slowly, with vibrating tails, to meet finally and stand drumming their quills in a trance with tightly closed eyes. After perhaps a minute one bird peeps at the other, and seeing him still enraptured, resumes an upright graceful carriage, anon stealing gently away. His companion is thus left foolishly posing at nothing, but presently he too awakes, and departs from the arena in a normal manner. Meanwhile the remaining cocks, one after another, take up the running till all have participated, but the end of each figure seems to be the same. Two birds squat flat on the ground with their beaks almost touching for about twenty minutes, and when they do this they are out of the dances for that day. The dance appears to terminate by some bird, either a late starter or one more vigorous than the rest, being unable to find a partner to respond to his run. Having assured himself of this, he utters a disgusted clucking, and all the grouse fly away at intervals as they complete their term of squatting.

Nesting.—The sharp-tailed grouse is not very particular as to a nesting site and is not much of a nest builder. The nest is a hollow in the ground, scantily lined with whatever loose material is available. It is usually partially concealed under a thick tuft of tall grass on the prairie or under bushes or thick herbage in a bushy tract or near a stream. We found these grouse very common around Crane Lake in 1905. On one day, June 5, we discovered three nests on the bushy prairies near Bear Creek. One of the nests that I photographed was under a little rosebush; the hollow was 7 inches across and 2½ inches deep; it was lined with fine twigs, straws, and feathers. Another similar nest was under a "grease bush," or "silver willow." A nest found June 2, 1913, near Lake Winnipegosis, Manitoba, was a hollow in the ground 8 inches across and 4 inches deep; it was concealed in long grass on a grassy and bushy dry place on a low, wet prairie near the lake. We found that the birds were usually close sitters, flushing almost underfoot. In one case, where we wanted to photograph the bird on a previously located nest, we had difficulty in finding it again, although we had marked the spot with a tuft of cotton. After scanning the ground very carefully, foot by foot, for a long time, we finally discovered the bird sitting on her nest within less than 10 feet. She shows up plainly enough in the photograph, but in life her color pattern matched her surroundings so well that she was nearly invisible, though in plain sight.

Dr. Alfred O. Gross has sent me his notes on five nests found by him in Wood County, Wis., June 1 to 4, 1930. Following are his notes on three of them, containing 13, 11, and 12 eggs, respectively:

Nest built among the grass and moss of a rounded knoll about 4 feet in diameter, located in a wet, swampy region remote from any farmed land. Growing throughout the swamp were small willows, poplars, and swamp grass.

It was lined chiefly with grass, much of which was the bent-over tips of the grass growing about the nest. The nest and eggs were partially concealed by the overhanging grasses.

Nest lined with oak leaves, dry ferns, grass, and a few feathers of the nesting bird. About three-fourths of the lining consisted of oak leaves, which had fallen from a scrub oak near by. About the nest was a thick growth of moss and checkerberry (*Gaultheria*). Very near to the nest was a thicket of pin cherries, poplars, and willows, and here and there were masses of blueberry bushes about a foot in height. The country about the nest was a sand plain, with a few marshy sloughs, but most of the land was brushy, chiefly willows, cherry, hazel, black and scrub oaks, and poplars.

This nest was located near the margin of a scrub-oak area that bordered an extensive grassy marsh. The edge of the marsh was only a few feet distant. The nest was very shallow and lined with sticks, grass, and a few leaves and grouse feathers. Growing about the nest were vines and grass, and the whole was well concealed by a mass of blueberry bushes.

Gross says further:

Judged from the situations in which these five nests were found, the sharp-tailed grouse chooses a nesting site remote from farms and in places where there is considerable brushy growth. The prairie-chicken nests, however, are nearly all near farms or on the farmed fields, such as the meadows and alfalfa and clover fields. Though the nesting sites vary considerably, the birds are frequently seen together during the hunting season, and Mr. Cole, local game warden at Wisconsin Rapids, states that on one occasion he killed a prairie chicken and a sharp-tailed grouse with one shot. Several times in traversing the prairies I have seen both species together, this summer (1930).

Eggs.—From 10 to 13 eggs generally constitute the full set for the sharp-tailed grouse, but as many as 14 or 15 are sometimes found in a nest. They are ovate in shape, and the shell is smooth, with a slight gloss. They are quite dark colored when first laid and have a purplish bloom, but the bloom disappears very quickly and the color gradually fades. The colors vary from "buckthorn brown" or "old gold" to "dark olive-buff" or "olive-buff." Many eggs are almost, or quite, immaculate, but more often they are speckled with very small spots or minute dots of dark brown. The measurements of 58 eggs average 42.6 by 32 millimeters; the eggs showing the four extremes measure **45.2** by 33.8, 44.2 by **34.1**, **40** by 31, and 41 by **30** millimeters.

Young.—The period of incubation is about 21 days, and this duty is probably performed by the female alone. Only one brood is raised in a season. Seton (Thompson, 1890) writes:

A partial history of the young in a wild state is briefly as follows: At the age of 6 weeks they are fully feathered and at 2 months fully grown, although still under guidance of the mother at this time. There is usually not more than six or seven young ones left out of the original average brood of fifteen, which statement shows the number of chicks which fall a prey to their natural enemies, while many sets of eggs also are destroyed by the fires which annually devastate the prairies. As the fall advances they gather more and more into flocks and become regular visitors to the stubble fields, and, in consequence, regular articles of diet with the farmers until the first fall of snow buries their foraging grounds and drives them en masse to the woods.

Doctor Gross says in his notes:

So far as my observation goes only one bird incubates and cares for the young. I did not see the male about the nest at any time. The female is a nervous, excitable creature and continually flits her tail when walking about or approaching the nest. The tail is usually widespread and displaying the patches, and the long tail feathers are held up in an upright position, making a striking picture.

Plumages.—Downy young sharptails are decidedly yellowish; the general color varies from "mustard yellow" above to "straw yellow" below, washed on the crown and back with "ochraceous-tawny"; they are spotted on the crown and blotched or streaked on the back with black; there is a black spot at the base of the culmen and a black spot on the auriculars.

As with all grouse, small chicks show the growth of juvenal plumage, beginning with the wings; before the young birds are half grown they are fully feathered, and even before that they are able to fly. In full juvenal plumage, in July, the crown and occiput are "hazel," centrally black; the feathers of the mantle and wing coverts are brownish black, broadly tipped, barred, or notched with "ochraceous-tawny" on the back and with "ochraceous-buff" on the coverts; the feathers of the mantle have a broad median white stripe; the underparts are dull white, spotted or barred on the breast and flanks with "sepia" and "cinnamon-buff"; the four central tail feathers (sometimes nearly all of them) are patterned with "ochraceous-buff" and black, with a broad median white stripe, and they are all decidedly pointed; the chin and throat are "colonial buff."

Within a month or so, during August, the postjuvenal molt takes place. This is a complete molt, except that the outer primaries on each wing are retained for a full year; it produces a first winter plumage, which is practically adult in September. Adults may have a partial prenuptial molt, about the head and neck. They have a complete postnuptial molt, during July, August, and September. The sharp-tailed grouse has several times been known to hybridize with the prairie chicken.

Food.—Dr. Sylvester D. Judd (1905a), by his examination of 43 stomachs of sharp-tailed grouse, showed that

animal matter (insects) formed only 10.19 percent of the food, while vegetable matter (seeds, fruit, and 'Browse') made 89.81 percent. The insect matter consists of bugs, 0.50 percent; grasshoppers, 4.62 percent; beetles, 2.86 percent, and miscellaneous insects, 2.21 percent in a total of 10.19 percent of the food. Vernon Bailey, of the Biological Survey, found that three birds shot by him in Idaho August 29 had eaten chiefly insects, including grasshoppers, small bugs, and small caterpillars. The young of the sharp-tailed grouse, like those of other gallinaceous species, are highly insectivorous. A downy chick from 1 to 3 days old, collected on June 27, in Manitoba, by Ernest Thompson Seton, had eaten 95 percent of insects and 5 percent of wild strawberries. The sharp-tailed grouse is fond of grasshoppers. Vernon Bailey shot 3 birds at Elk River, Minn., September 17, 1894, which had eaten, respectively, 7, 23, and 31 grasshoppers. The species is a destroyer also of the Rocky Mountain locust. Of 9 birds collected by Professor Aughey from May to October, inclusive, 6 had eaten 174 of these pests. The bird eats also a few crickets and, like other gallinaceous game birds, devours the Colorado potato beetle (*Leptinotarsa decemlineata*).

The vegetable food of the sharp-tailed grouse, so far as ascertained in the laboratory, comprises weed seeds, 7.39 percent; grain, 20.50 percent; fruit. 27.68 percent; leaves, buds, and flowers, 31.07 percent; and miscellaneous vegetable food, 3.06 percent; making a total of 89.81 percent. Like many other game birds, the species feeds on mast (largely acorns), including acorns of the scarlet oak (*Quercus coccinea*). Corn is eaten, but wheat is the favorite grain. It formed 17.21 percent of the food. A thousand kernels of wheat were sometimes found in one stomach. The sharp-tailed grouse is a great browser. It makes 31.07 percent of its food of leaves, buds, and flowers. Ernest Thompson Seton found it eating the buds of willow and birch. It feeds on the leaves of cottonwood, alder, blueberry, juniper, and larch; also leaves of quillwort (*Isoetes*), vetch, dandelion, grass, and rush (*Juncus*). Hearne says that in winter it eats the tops of the dwarf birch and the buds of poplars. Flowers form 19.90 percent of its diet, the species leading all other birds in this respect. A half pint of the showy, bluish blossoms of the pasque flower (*Pulsatilla hirsutissima*) which brightens the western prairie are often taken at a meal, and those of the dandelion also are eaten. Inflorescence of grasses, alder, willow, maple, and canoe birch are plucked along with leaf buds. Like the prairie hen and the ruffed grouse, the sharp-tailed grouse is frugivorous, and fruit forms 27.68 percent of its diet. Hips of wild rose alone form 17.38 percent. Ernest Thompson Seton, who examined hundreds of stomachs of the sharp-tailed grouse, says that he can not recollect an instance in which they did not contain the stony seeds of the wild rose. Mr. Seton states that in places in Manitoba, where he has collected during the winter, gravel to pulverize the food is not to be had, and the stony rose seeds act in its stead. Rose hips appear difficult to digest, and, furthermore, are sometimes thickly set with bristles that would irritate the human stomach, but appear to cause no inconvenience to the grouse. The persistent bright-colored hips are readily seen above the snow, and they are a boon to the birds in wintry northern regions where the struggle for existence is bitter. It feeds on blueberries and cranberries and on the snowberry (*Symphoricarpus racemosus*), various species of manzanita, bearberry (*Arctostaphylos uva-ursi*), buffalo berry (*Lepargyrea argentea*), juniper berries, huckleberries, and arbutus berries. It takes also the partridge berry (*Mitchella repens*), a favorite with the ruffed grouse, Like many other species, it eats with relish the fruit of cornel (*Cornus stolonifera*) and poison ivy (both *Rhus radicans* and *Rhus diversiloba*).

Behavior.—The sharp-tailed grouse rises with a loud whir of wings and flies away with considerable speed, usually in a straight line in the open, with rapid beats of the wings alternated with short periods of sailing on down-curved wings. While hunting for nests, where these grouse were common, we kept flushing them from their roosting and dusting places among the bushes. As they rose, and for some distance afterwards, they uttered a peculiar clucking note sounding like *whucker*, *whucker*, *whucker*. They flew rapidly for some distance and then set their wings and scaled downward into some good cover or behind some little hill. Seton (Thompson, 1890) says:

Their mode of flight is to flap and sail by turns every 40 or 50 yards, and so rapid and strong are they on the wing that I have seen a chicken save itself by its swiftness from the first swoop of a peregrine Falcon, while another was seen to escape by flight from a Snowy Owl.

Aretas A. Saunders says in his notes:

Though usually found on the ground, these birds sometimes perch in trees, either cottonwoods along a river valley or in yellow pines where hills and prairie come together. In southeastern Montana and northwestern South Dakota they come into the pine hills when the snow gets deep and spend the nights there. In that region I saw them in larger flocks than elsewhere, often 50 or 60 birds in a flock. In the pine hills I have seen as many as 30 birds perched in a single yellow pine and have watched them fly from tree to tree, half a dozen birds leaving at a time, and then another group following till the whole flock had reached the new tree.

Voice.—Besides the hollow booming sound made by the birds in the courtship dance and the guttural clucking notes so commonly heard when they are flushed, "the birds have several cackling notes, and the males a peculiar crowing or low call, that in tone sounds somewhat like the call of the turkey" (Goss, 1891).

Enemies.—These grouse are subject to the attacks of the numerous enemies that beset all ground-nesting birds. Their eggs and young are preyed upon by the smaller and slower predatory mammals and birds, while the larger beasts and birds of prey attack the adults. While absorbed in their courtship dancing they are easy prey to the crafty coyote. Laing (1913) tells of a family of coyotes that established their den, skilfully hidden, on one of the dancing hills; here he found ample evidence that the wise old coyote and her pups had levied regular toll on the unsuspecting grouse. The larger falcons and the goshawk take their share of the grouse, though the latter are fast enough in flight to give even these swift-winged hawks a lively chase. C. L. Broley has sent me the following notes:

One cold day this fall, while bird observing with Mrs. Broley on the banks of the Red River, north of Selkirk, we saw three sharp-tailed grouse fly out of the woods on the opposite side of the river as if they had urgent business elsewhere, and a moment later a fourth one burst out with two goshawks in hot pursuit. The grouse flew directly out over the water, where one of the hawks, reluctant to follow in the open, dropped out of the chase. The grouse fled upstream, keeping some 50 feet above the water and an equal distance from each shore, followed, some 30 feet behind, by the other hawk, which to our surprise kept below the altitude of the thoroughly frightened chicken, which was taking advantage of a strong north wind and making wonderful time, using his wings almost continually and sailing very little.

Fortunately we had a clear view up the river for nearly 2 miles and while the birds were covering this distance their respective positions remained unchanged, but when the bend of the river was reached the grouse changed its course, going over the land. The goshawk immediately rose from below to well above its quarry, with the full intention, we thought, of making its swoop, but unfortunately intervening trees close to us hid the rest of the incident.

Norman Criddle (1930), in an interesting study of the fluctuations in numbers of grouse in Manitoba, from 1895 to 1929, has shown on a graph "the remarkable regularity with which sharp-tailed grouse fluctuate in company with grasshoppers." His theory is that young grouse are fed to a large extent on grasshoppers and that during periods of abundance of these insects the birds rear large broods. He says:

Referring again to our graph it will be observed that all the high points of Sharp-tailed Grouse abundance were preceded and accompanied by grasshopper outbreaks. Of these outbreaks the one of 1900–1903 was much the worst that we ever experienced in the Aweme district, and it was not finally subdued until 1905. The peak of grouse abundance was also attained at this time, and we have no records of Sharp-tailed Grouse ever being in greater numbers.

Game.—My experience with sharp-tailed grouse was confined to the breeding season, when they are very tame, flush near at hand, fly rather straight and evenly, and give one a good, open, easy shot. We considered them easy marks and had no difficulty in shooting all we needed for specimens. But in fall, after they have been shot at, they are evidently more gamy and seem to be popular as game birds. Sandys (1904) writes:

The sport afforded by this grouse is of a very high order. At the opening of the season it lies well to the dog, and springs with the usual whirr of wings, at the same time uttering a vigorous clucking, which is repeated again and again as the birds speed away, alternately flapping and sailing. When driven to brush, they very frequently behave not unlike quail, flushing close at hand, and offering the prettiest of single chances. The flesh is excellent, light-colored in young birds, and darkening with age, but always worthy of a place on the board. Not seldom, as one nears the pointing dog, he will see the birds squatted in the grass, and perhaps, have one after another turn and run a few yards before taking wing. When thus seen they are very handsome, the crest is raised, and the white hinder feathers show like the flag of a deer, or the scut of a cottontail rabbit. Almost invariably the flush is straggling, giving a quick man a fine opportunity for scoring again and again. At the proper season, i. e., just before the broods begin to pack and become wary, this bird affords sport to be long remembered. I have enjoyed it to the full, and know of nothing better for a business-harassed man than a day on the sunny open with the sharptails behaving well. Like all prairie-grouse, this bird, rising close, is an easy mark for whoever has learned not to be hurried by the sound of wings. A good twelve-gauge, properly held, should stop its buzzing and clucking fully three-fourths of all reasonable chances.

Winter.—Seton (Thompson, 1890) writes:

During the summer the habits of the chickens are eminently terrestrial; they live, feed, and sleep almost exclusively on the ground; but the first snow makes a radical change. They now act more like a properly adapted perching bird, for they spend a large part of their time in the highest trees, flying from one to another and perching, browsing, or walking about among the branches with perfect ease, and evidently at this time preferring an arboreal to a terrestrial life. When thus aloft they are not at all possessed of that feeling of security which makes the similarly situated Ruffed Grouse so easy a prey to the pothunter. On the contrary, their perfect grasp of the situation usually renders them shy and induces them to fly long ere yet the sportsman has come near enough to be dangerous. Like most of the members of its family, the Prairie Chicken spends the winter nights in the snow, which is always soft and penetrable in the woods although out on the plains it is beaten by the wind into drifts of ice-like hardness. As the evening closes in the birds fly down from the trees and either dive headlong into a drift or run about a little and select a place before going under. The bed is generally about 6 inches from the surface and a foot long from the entrance. Each individual prepares his own place, so that a flock of a dozen chickens may be scattered over a space of 50 yards square. By the morning each bird's breath has formed a solid wall of ice in front of it, so that it invariably goes out at one side. The great disadvantage of the snow bed is, that when there the birds are more likely to become the prey of foxes and other predaceous animals, whose sagacious nostrils betray the very spots beneath which the unsuspecting bird is soundly slumbering. I am inclined to think this is the only chance a fox has of securing one of the old birds, so wary are they at all other times.

Laing (1913) says:

The question of winter food is never a big problem with these grouse. There is always an abundance of hawthorn hips, rose fruit, snowberries, or other winter-cured fruits; in addition to these, edible buds of many kinds, are in abundance. Best of all, they seem to relish the sweetish, frost-ripened berries of the dwarfish snowberry that peep above the snow just far enough to invite picking. Of the tree-buds, poplar, willow, and dwarf birch are winter staples, and these are consumed in great quantities. The quest of wheat in winter frequently leads these grouse to come right into the towns. They first took this bold step after finding wheat dropped along the roads and railway. As the clue led in the direction of the big red elevators, they followed it in at first, but soon needed no grain trail by way of invitation. They were quick to learn that they were not molested in the winter, and indeed received a ready welcome. The daily itinerary of this flock in mid-winter is about as follows: With the first peep of dawn they leave their snow beds and mount to the tops of the willows and poplars, close at hand. At sunrise or a little before it, they whizz off into town and scatter around the various feeding grounds mentioned above. About ten o'clock they usually take a run out along the railway track, evidently to get a supply of gravel; then they return to spend the warm part of the day in the scrubby sandhills. Here they sit about in the sun, and pick a few buds; or if the day is very cold and the snow light and deep, they burrow for their noon-day nap. At three o'clock they return to their feeding-place of the morning, and then shortly before sundown go back to the scrub to make their beds for the night. Here while the cruel wind sweeps across the plains and the thermometer ofttimes registers in the minus forties, they remain cuddled snugly away from the bitter night-world.

CENTROCERCUS UROPHASIANUS (Bonaparte)

SAGE HEN

HABITS

The recent American Ornithologists' Union check lists, both old and new, call this the sage *hen*, but I prefer to call it a *grouse*, as it really is and as it was called in earlier editions. I see no reason for calling it a hen, except that the cackling notes of the female remind one of that familiar domestic fowl. It is a true grouse and a grand one, by far the largest of our American species and the largest in the world except the European capercaillie, which far exceeds it in size. A fully grown sage-grouse cock is said to weigh as much as 8 pounds, but the hen will not weigh more than 5 pounds, probably both usually weigh much less.

It was discovered by Lewis and Clark about the headwaters of the Missouri River and on the plains of the Columbia; they named it "cock of the plains" and gave the first account of it. The technical description of it and the scientific name, *urophasianus*, were supplied by Bonaparte in 1827.

The range of the sage grouse is limited to the arid plains of the Northwestern States and the southwestern Provinces, where the sagebrush (*Artemisia tridentata* and other species) grows; hence it is well named sage grouse or cock of the plains. Its range stops where the sagebrush is replaced by greasewood in the more southern deserts. Like the prong-horned antelope, another child of the arid plains, it has disappeared from much of its former range, as the country became more thickly settled and these large birds were easily shot. It has been said that the sage was made for this grouse and this grouse for the sage, where it is thoroughly at home and where its colors match its surroundings so well that it is nearly invisible while squatting among the lights and shades of the desert vegetation. It seldom wanders far from the sagebrush, but may be found occasionally in the shade of the narrow line of trees that marks the course of some small stream. Dwight W. Huntington (1897) describes its haunts very well, as follows:

> I found the Sage Grouse most abundant in the vicinity of Fort Bridger and south to the Uintah Mountains. Here the tufted fields of the gray-green sage sweep up to the sides and walls of the adjacent "bad lands," or buttes, devoid of vegetation but beautiful in color and fantastic in form. The buttes are strangely fashioned by erosion, and are full of the fossil remains of animals and fishes. Numerous domes, spires, and pinnacles surmount the buttes and the conglomerate layers running about them have been compared to Egyptian carv-

ing. Towards the southwest are the blue Uintah Mountains, with snow flashing on their crests all summer, and towards the east the vast plain of sage extends as far as the eye can reach, blending at the horizon into an azure sky. The trout streams which issue from the mountain side become the small rivers of the plains, flowing at long intervals and nourishing a narrow line of verdure or a yellow screen of cottonwood, which marks their course. It is along such streams that the sage grouse hunter must pitch his camp.

Courtship.—Much has been written about the courtship of the sage grouse, which is the most spectacular performance indulged in by any of the grouse. It has been variously described by different observers. Frank Bond (1900) was one of the first to describe and illustrate this with a drawing; he writes:

During the months of April and May the Sage Cocks are usually found in small flocks of a half dozen or more, stalking about with tails erect and spread after the manner of the strutting turkey cock, but I have never seen the Grouse dragging their wings upon the ground, turkey fashion, and in the manner described by Dr. Newberry in the quotation from this author found on page 406 of Dr. Coues's "Birds of the Northwest," nor have I ever found a wing of a Sage Cock in this or any other season, which exhibited the slightest wearing away of the primaries. Instead of dragging its wings upon the ground the Sage Cock will enormously inflate the air sacks of the neck until the whole neck and breast is balloon-like in appearance, then stooping forward, almost the entire weight of the body is thrown upon the distended portion and the bird slides along on the bare ground or short grass for some distance, the performance being concluded by the expulsion of the air from the sacks with a variety of chuckling, cackling or rumbling sounds. This performance is continued probably daily, during the pairing and nesting season, and of course the feathers are worn away by the constant friction.

William L. Finley's account of it differs somewhat. He has sent me the following notes on the subject:

On May 13 we rose at 3.15 and were in the blind a little after 4, still very dark. The birds at this time were already strutting. We could hear them and occasionally see a flash of white from the breasts. The birds were active between 4.30 and 6.30 a. m. and had left the strutting ground by 8 a. m. When the sage cock starts to strut, his tail spreads and the long pointed tail feathers radiate out in a half arc. The air sacs are filled and extend nearly to the ground, hiding the black breast feathers. This is the first movement. Then the bird takes one or two steps forward and throws up the pouch, apparently by drawing back the head and neck. The next movement is a repetition of throwing the air sacs up and down and getting under headway for the last toss of the pouch, which is brought down with a jerk, as one would crack a whip, making a "plop" that on a quiet morning we easily heard for a distance of 200 or 300 yards. The whole movement gives one the idea that the bird inflates the air sacs and then, by the rigid position of the body and throwing the head and neck back, gives these air sacs a very vigorous shaking. In the movement when the pouch spreads, the bare yellow skin on the lower part of the pouch or chest shows clearly. As the pouch is thrown up and down, the wings are held rigid, the tips of the wing feathers sometimes touching the ground. The white feathers that cover the chest are exceedingly stiff; these grate against the wing feathers, giving out a wheezy sound that at first I thought came from the inhaling and exhaling of air. I soon discovered that this rasping noise was made by the stiff feathers rubbing together. This rubbing of the breast feathers against the rigid wing feathers seems to account for the very worn appearance of the breast later in the season. If there were any gurgling or chuckling noises I failed to catch them. As the strutting ends, the air sacs are deflated and each time the bird goes through the motion as if gulping or swallowing something.

There were 56 cocks scattered around in an area of 4 or 5 acres, each bird having a space for himself. Occasionally when one bird came too near another it resulted in a fight. Once I saw two fight with lowered heads. Occasionally they would jump in the air, striking very much as an ordinary rooster strikes. Two or three times I saw a female feeding near by, where the males were strutting. One passed by several cocks, but they paid no attention to her, or she to them. This gave me the impression that the strutting was not so much a courting performance or even a nuptial dance; it seemed to be a gathering place where the males came together and " showed off " among themselves.

E. S. Cameron (1907) observed:

By ruffling up all their feathers, spreading their tails, and dragging their wings along the ground they looked much larger than they really were, while they produced a rattling sound with their quills after the manner of turkey-cocks and peafowl.

Some additional information is given by L. E. Burnett (1905), who writes:

I have heard them drum as early as December. This performance is most often observed where hundreds of males and females have congregated together, a custom which they have in the fall of the year. By February, the males are all drumming, but this is not continued during bad weather which closes the session until fair weather returns. By the latter part of the month the males are in full dress. Their protracted meetings last until the first days of May. After the violets and buttercups have come and the song of the sage thrush begins, their drumming is heard but occasionally. When drumming they stand very erect, holding the wings away from the sides and nearly perpendicularly, while the large loose skin of the neck is worked up, and the head drawn in and out until the white feathers are brought to the chin. At the same time the galls are filled with air until the birds look as if they were carrying snowballs on their shoulders. Then the skin which lies between the galls is drawn in with a sucking movement, thus bringing the galls together or nearly so. With this action the air is expelled from the throat producing the noise, which is hard to mimic and which resembles that of an old pump just within hearing distance. After the bird has accomplished this feat he walks away a few paces either in a straight line or a circle, with wings down, hanging loosely, but not grating on the ground. At times they do drag the wings as they strut along with tail spread and erect, though not so perpendicular as that of a turkey. Again they will dance about with all the pomp of a male pigeon.

Maj. Allan Brooks (1930) says that " the feathers of the breast and neck of the male sage grouse are specialized feathers only " and are not worn away by rubbing on the ground during the display.

Nesting.—Major Bendire (1892) says:

The nest is always placed on the ground, in a slight depression, usually under the shelter of a small sage bush. I have found several, however, some little distance from sage brush flats, alongside and sheltered by a bunch of tall rye grass (*Elymus condensatus?*), near the borders of small creeks. The nest is usually very poorly lined, and in fact the eggs frequently lay on the bare ground without any lining whatever, and are often found in quite exposed situations. I found such a one on May 11, 1875. My notes read as follows: "I stumbled accidentally on this nest. It was placed within a yard of a much-used Indian trail, in a very exposed position, so much so that I saw the eggs while still 5 yards off. There really was no nest, simply a mere depression scratched out by the bird on the south side of a very small sage bush, which afforded no concealment or protection from rain whatever. The bush itself was not over a foot and a half high, growing on a rocky plateau about 3 miles east of Camp Harney. A few feathers were scattered among the eggs which laid on the bare ground, and were separated from each other by bits of grass and dry leaves of the sage. One of the eggs was nearly covered with dirt and almost buried out of sight. The set contained eight eggs, and these were nearly hatched. They were cold when found, and the nest had evidently been abandoned for some days.

Illustrating the concealing colorations of the close-sitting bird, Bendire quotes Capt. William L. Carpenter as follows:

I found a nest at Fort Bridger, Wyoming, where this species is numerous, June 1, with nine fresh eggs. I was standing alongside a sage bush watching butterflies; several times looking down carelessly without seeing any thing unusual, when happening again to glance at the foot of the bush, in the very place before observed, I saw the winking of an eye. Looking more intently a grayish mass was discerned blending perfectly with the color of the bush, which outlined itself into the form of a Sage Hen not 2 feet from my foot. She certainly would have been overlooked had not the movement of her eyelids attracted my attention. I stood there fully five minutes admiring the beautiful bird, which could have been caught in my butterfly net, then walked back and forth, and finally passed around the bush to observe it from behind. Not until then did it become frightened and fly away with a loud cackling. The nest was a depression at the foot of a sage bush, lined with dead grass and sage leaves. The spot was marked and visited several times, always passing within a few feet without alarming the bird.

D. E. Brown tells me of a nest found by a sheep herder. The bird did not flush from the nest until the sheep were all around her; she then flushed with a great noise, scattering the sheep in all directions. This habit may often prove very useful in preventing cattle from trampling on the eggs.

Eggs.—The number of eggs laid by the sage grouse usually varies from 7 to 9, in some localities from 10 to 13; as many as 15, or even 17, have been found in a nest. Bendire (1892) found but one set of 10, and found more sets of 8 than any other number. They vary in shape from ovate to elongate ovate, and the shell is smooth with little or no gloss. The ground colors vary from pale "ecru-olive" or "deep olive-buff" to "yellowish glaucous," "olive-buff," or "greenish white." They are generally quite evenly marked with small spots and fine dots of dark brown, "bister," or "brownish olive"; in very light-colored eggs the spots are in very pale shades of brown or olive. The markings are very easily washed off when the eggs are fresh. The eggs in a set are seldom, if ever, uniform in type; there are usually two or more conspicuously different types in each set. The measurements of 110 eggs average 55 by 38 millimeters; the eggs showing the four extremes measure **59.5** by 39.5, 58.5 by **40.5**, **51** by 37, and 56.5 by **35.5** millimeters.

Young.—Bendire (1892) gives the period of incubation as 22 days. This duty is performed by the female alone, as the polygamous males desert the females as soon as the eggs are laid and associate in flocks by themselves. Consequently the full care of the young rests on the devoted mother. Mrs. Irene G. Wheelock (1904) writes:

Like all grouse nestlings, they run about as soon as the down is dry, which is about fifteen minutes after the shell breaks. They pick up food at her scratching all day, and at night they nestle on the ground under her wings, only a row of little heads being visible. As soon as their own feathers are developed, they sleep every night in a circle about her, each one with head pointed to the outside as before, and always on the ground; for the Sage-Grouse never trees. It is not difficult to come upon a brood sleeping this way on a moonlight night; but the only satisfaction will be to hear the sharp alarm of the mother, a whirr as she runs by you, and a knowledge that though the young are hiding on the dust at your feet, you could not find them were your eyes tenfold sharper. I have groped carefully on hands and knees among them, and actually touched one before I saw it at all. For the desert hides its secrets well, and the little grouse have learned to trust to it for safety.

Grinnell, Bryant, and Storer (1918) say:

After the young birds have learned to fly, they descend from the uplands down along the larger canons, often invading the meadow lands, where small tender weeds are added to their diet. At such places the young birds may gather into large flocks. When approached they crane their necks and make a weak attempt at cackling. When closely pressed they run rather than fly. By the last of August or early September the young birds are joined by the old male birds, which come off the higher slopes and ridges where they have stayed during the summer, and large flocks become the rule.

Plumages.—The sage-grouse chick is well colored to escape detection when crouching on the ground in the gray shadows of the desert. The crown, back, and rump are mottled and marbled with black, dull browns, pale buff, and dull white; the sides of the head and neck are boldly spotted and striped with black; there are two large spots of "sayal brown" bordered with black, on the fore neck or chest; underparts grayish white, suffused with buff on the chest.

The juvenal plumage comes in first on the wings, while the chick is very small, then on the scapulars, back, tail, sides of the breast, and flanks, lastly on the rump, head, neck, and belly. The juvenal plumage is much like that of the adult female; but the breast is more buffy and more spotted than barred; the feathers of the black breast patch are tipped with white; and the feathers of the mantle are conspicuously marked with a broad shaft streak of white. Young males seem to be much darker than young females.

A nearly complete postjuvenal molt, including all but the outer two primaries on each wing, produces the first winter plumage, which is practically adult and in which the sexes are fully differentiated. There may be a partial prenuptial molt, but I have seen no evidence of it in either young or old birds. Adults have a complete postnuptial molt, mainly in August.

Food.—Dr. Sylvester D. Judd (1905a) says:

The feeding habits of the sage grouse are peculiar, and its organs of digestion are unlike those of other grouse. The stomach is not differentiated into a powerful grinding gizzard, but is a thin, weak, membranous bag, resembling the stomach of a raptorial bird. Such an organ is evidently designed for the digestion of soft food, and we find that the bulk of the sage grouse's diet consists of leaves and tender shoots. A stomach collected September 7, 1890, in Idaho, by Dr. C. Hart Merriam, contained leaves of sage and other plants, seeds, and a ladybird beetle (Coccinellidae). Four birds shot in Wyoming during May and September by Vernon Bailey had gorged themselves with the leaves of sagebrush (*Artemisia tridentata*). This and other sages, including *A. cana* and *A. frigida*, furnish the bulk of the food of the sage grouse. Other food is taken, but it is comparatively insignificant. B. H. Dutcher, formerly of the Biological Survey, examined a stomach which, besides sagebrush leaves, contained seeds, flowers, buds of *Rhus trilobata*, and ants and grasshoppers. Three birds collected by Vernon Bailey on September 5, in Wyoming, had varied their sagebrush fare with ladybird beetles, ground beetles (Carabidae), fly larvae, ants, moths, grasshoppers (*Melanoplus* sp.) and the leaves of asters and yarrow. Of two birds killed in May, one had fed wholly on the leaves of sagebrush (*Artemisia tridentata*), while the other in addition had taken insect galls from sagebrush and the flowers and flower buds of a phlox (*Phlox douglasii*), together with some undetermined seed capsules, pieces of moss, and several ants. A third bird, killed in July, had eaten a few plant stems and numerous grasshoppers.

During the winter the sage grouse feeds almost entirely on the leaves of the sage, but "in summer," according to Bendire (1892), "its principal food (in Wyoming and Colorado) is the leaves, blossoms, and pods of the different species of plants belonging to the genus *Astragalus*, and *Vicia*, commonly called wild pease, which are always eagerly sought for and consumed in great quantities."

Robert S. Williams reported to Bendire from Montana that he scared up a flock among tall grass in a mountain meadow; one of these birds had its crop full of the blossoms of a species of goldenrod. Bendire also quotes George H. Wyman as stating that a sage grouse will go a long way for food in a wheat field; some that he examined had traveled at least 8 miles to fill their crops with ripe wheat.

Behavior.—Dr. D. G. Elliot (1897) says:

It is not always easy to flush these birds, as they will run long distances before taking wing, and skulk and hide at every opportunity. But when forced to rise, they flush with a great fluttering of the wings and utter a loud *kek-kek-kek*, which kind of cackle is kept up for quite a considerable time. They seem to have difficulty in getting well on the wing, and rise heavily, wabbling from side to side as if trying to gain an equilibrium, but once started they go far and fast enough, with intermittent quick beats of the pinions and easy sailing on motionless wings.

Mr. Finley refers in his notes to their morning flights for water, as follows:

September 22, Horsfall and I were up at 4.40 a. m., left the cabin at 5.40, just as it was getting light. As we walked down the long draw, through which Warner Creek winds back and forth, we saw sage grouse coming in to water. They came from the rimrocks and higher plateaus, perhaps from several miles away, as some of them sailed down from high over the rimrock, showing that they had come from a second high rimrock about a mile back. They came in singles, in small flocks of from 8 to 10, occasionally larger flocks numbering 30 or 40. They lit out in the open at 100 or 200 yards back from the creek, and then walked down to water. In a short distance I counted 150 birds. From the cabin down to a small reservoir, where the water was backed up covering several acres of ground, there must have been 1,500 to 2,000 birds. Around the reservoir site there was also a larger number. It was the same along the creek above the cabin, where it winds through meadow and sagebrush. The grouse do not frequent the water at all during the middle of the day, and we found very few toward evening. They seem to come in before daybreak, and a little after many have departed for the high plateau, by 6 o'clock or before sunrise. By 8 o'clock all had departed.

Bendire (1892) quotes Dr. George Bird Grinnell as stating:

On a very few occasions I have seen the Sage Grouse standing on the branches of a sage bush, sometimes 2 or 3 feet from the ground, but I imagine that this is quite an unusual position for the bird. This species, commonly, I think, goes to water twice a day, flying down to the springs and creek bottoms to drink in the evening, then feeding away a short distance, but roosting near at hand. In the morning they drink again and spend the middle of the day on the upland. The young birds, when feeding together, constantly call to one another with a low peeping cry, which is audible only for a short distance. This habit I have noticed in several other species of our Grouse, notably in the Dusky Grouse and the Sharp-tail.

Coues (1874) quotes from Doctor Newberry's account, as follows:

A very fine male which I killed there was passed by nearly the whole party, within thirty feet, in open ground. I noticed him as soon, perhaps, as he saw us, and waited to watch his movements. As the train approached he sank down on the ground, depressing his head, and lying as motionless as a stick or root, which he greatly resembled. After the party had passed I moved toward him, when he depressed his head till it rested on the ground, and evidently made himself as small as possible. He did not move till I had approached within 15 feet of him, when he arose and I shot him.

And from Mr. Holden, Coues quotes:

They roost in circles on the ground. I have seen a patch of ground fifteen feet in diameter completely covered with their excrement. I think they resort to the same place many nights in succession, unless disturbed.

Voice.—The vocal efforts of the sage grouse seem to be limited to a deep guttural clucking note, *kuk, kuk, kuk*, slowly repeated as the bird flushes, a rapidly repeated scolding note, *tuk-a-tuk*, and a cackling note of the female, like the cackle of a domestic hen. The chicks call to one another with faint peeping notes.

Enemies.—Sage grouse have been found to be infested with tapeworms and probably they are also infected with some of the other parasites and diseases to which other grouse are subject. The eggs and young are preyed upon by various predatory animals and birds, mainly crows and magpies, but their worst enemy is man. Natural enemies of all wild creatures have merely checked their increase, but when man comes on the scene it means extermination. So it is with this fine large grouse, an easy mark for the gun. It has been extirpated from much of its former range and is disappearing very steadily in many other places.

Sandys (1904) relates the following incident:

One day I was watching an old male which had taken up a position upon an almost bare knoll. It was before the open season, a very idle period on the plains; so, partly to pass away time, and partly in the hope of discovering something, the field-glass was brought into play. Before the bird had been thoroughly scrutinized, some falcon, which looked like a male peregrine, shot into the field of vision, and made a vicious stoop at the huge quarry. Whether or no the grouse had been watching the hawk is impossible to say, but in any event he was ready. As the hawk was almost upon him, up went the long tail, down went the head, and the wings were a trifle raised. Most readers, probably, have seen a man hump his back and get his shoulders about his ears when he expected to be struck from behind by a snow-ball. The action and attitude of the grouse were comically suggestive of that very thing. The hawk appeared to be only fooling, for certainly it made no determined strike, but presently rose and curved away. An instant later the grouse took wing.

Game.—Sandys says of the game qualities of this grouse:

As an object of the sportsman's pursuit, the sage-grouse is greatly inferior to most of its relatives. The young, the only ones worth shooting, are great runners, and only take wing when compelled to, and once in the air their size is against them, although they fly fairly fast. Another objectionable feature is their ability to carry off shot, which sometimes borders on the marvelous. A light gun, deadly on other grouse, will hardly serve for these big fellows, the use of it surely meaning a lot of wounded birds. The coveys usually are small, as the young have many enemies, among which the chief are fierce storms, wet, wolves, foxes, and rapacious birds, while man plays no unimportant part in the work of destruction.

Dwight W. Huntington (1903), however, speaks very highly of it as a game bird, and says of its value as a table bird "that these birds, like others, often receive a flavor from their food, and when the wild sage is their exclusive diet they have a more or less bitter taste. When, however, the birds are young and have been feeding on grasshoppers, their flesh is as good as that of the sharp-tails or prairie-grouse."

As to his method of hunting sage grouse, Huntington writes:

My shooting of these birds was mostly done from the saddle while on the march. When we flushed a covey of birds I took a shot at them, and marking those that flew away to the particular bush where they settled, rode at once to the spot and sometimes dismounted to shoot at the scattered birds. Upon several occasions I went out with a friend especially to shoot them, riding here and there (we had no dog) until the horse flushed a covey, and following them so long as we could make them take wing. Birds often escaped by hiding in the sage and refusing to fly. The most likely places seemed to be depressions where the water evidently flowed in wet seasons and little knolls adjacent, but we stumbled upon the birds almost anywhere in the sage, and often made very good bags. It was next to impossible to miss one, since the shots were always in the open and the marks large. The birds required hard hitting, however, to bring them down, and I would not advise the use of shot smaller than number 5 or 6. A wounded bird is difficult to recover without a dog where the sage grows thickly, and I always tried to kill the birds outright. The side shots, or those at quartering birds, are more likely to be fatal than those at birds going straight away, since the shot then penetrates the lighter feathers beneath the wings.

Burnett (1905) says:

The counties of Albany, Converse, Natrona, and Carbon are the places where grouse are most abundant in Wyoming. A single hunter has been known to kill a hundred birds in a day without a dog. The best hunting is found over lands adjacent to springs, down green draws and the bottoms along streams, and the best time to find coveys is in the morning or evening when the birds are feeding. After feeding they hide either on the feeding ground or at some distance from it where the sage is large enough to screen them from enemies and the rays of the sun.

Its large size, the ease with which it can be killed, and the accessibility of its haunts combine to make this grouse a popular game bird during summer and fall when its flesh is most palatable. Consequently it is disappearing very fast, notably in California, Oregon, and Washington, where the extension of good roads and the increase in automobiles have made the sagebrush plains more accessible. To save this fine bird from extinction, as civilization spreads, the open season for shooting it must be shortened and the bag limits reduced. Even then, it probably can not be saved except on protected reservations.

Fall.—These grouse are usually resident throughout the year wherever they are found; but on some of the elevated plateaus, in the more northern portions of their range, the sagebrush, on which they feed in winter, becomes buried under the snow; they are therefore obliged to migrate in search of a food supply.

In the days of their abundance they used to gather in immense packs in fall. Bendire (1892) quotes the following from notes sent to him by Dr. George Bird Grinnell:

In western Wyoming the Sage Grouse packs in September and October. In October, 1886, when camped just below a high bluff on the border of Bates Hole, in Wyoming, I saw great numbers of these birds, just after sunrise, flying over my camp to the little spring which oozed out of the bluff 200 yards away. Looking up from the tent at the edge of the bluff above us, we could see projecting over it the heads of hundreds of the birds, and, as those standing there took flight, others stepped forward to occupy their places. The number of Grouse which flew over the camp reminded me of the oldtime flights of Passenger Pigeons that I used to see when I was a boy. Before long the narrow valley where the water was, was a moving mass of gray. I have no means whatever of estimating the number of birds which I saw, but there must have been thousands of them.

Winter.—Unless the snow is too deep the sage grouse seek shelter from the winter storms and blizzards in the denser clumps of sagebrush on their favorite plains or find protection in the brushy valleys of the streams, in coulees, or in sheltered hollows. Sandys (1904) writes:

As winter tightens its grip upon the sage lands, the birds of many broods unite into packs of from fifty to one hundred and odd. The flush of one of these large packs is something to be remembered, for great is the tumult of wings, and piercing the cackling, as the heavy fowl beat the air in frantic efforts to get squared away upon their chosen course. At this season the only way to get any sport out of them is by using the rifle.

DISTRIBUTION

Range.—Western United States and casually in the interior of southwestern Canada.

The range of the sage grouse has been greatly restricted through the development of the West and through grazing activities, particularly of sheep, which do much to extirpate the birds over wide areas. The full range apparently extended **north** to (casually) the interior of southern British Columbia (Osoyoos Lake); southern Saskatchewan (casually Skull Creek, Val Marie, and casually Pinto Creek); North Dakota (Marmarth, Deep Creek, and formerly Fort Berthold); and formerly northeastern South Dakota (Grand River Agency and Fort Sisseton). **East** to South Dakota (formerly Fort Sisseton, Indian Creek, formerly Rapid City, and formerly Sage Creek); northwestern Nebraska (Antelope Creek); southeastern Wyoming (Marshall, Arlington, and Cheyenne); Colorado (Walden, Kremmling, Dillon, Lone Cone, and Dolores); and formerly northern New Mexico (Tres Piedras). **South** to formerly New Mexico (Tres Piedras and Tierra Amarilla); southern Utah (Grass Valley and Hamblin); Nevada (Belmont and Queen); and eastern California (Big Pine and the headwaters of the Owens River). **West** to eastern California (the headwaters of the Owens River, Long Valley, Ravendale, Madeline Plains, Eagleville, and Tule Lake); Oregon (Klamath Falls, Silver Lake, Fort Rock, Silvies River, Turtle Cove, and Haines); Washington (Rattlesnake Mountains, Yakima, and Ellensburg); and (casually) the interior of southern British Columbia (Osoyoos Lake).

Egg dates.—Washington and Oregon: 16 records, March 11 to May 28; 8 records, April 11 to May 6. Idaho, Montana, and Wyoming: 25 records, April 25 to June 15; 13 records, May 16 to 29. Colorado and Utah: 11 records, May 10 to June 3; 6 records, May 19 to 28.

Family PHASIANIDAE, Pheasants, Peacocks

PHASIANUS COLCHICUS TORQUATUS Gmelin

RING-NECKED PHEASANT

HABITS

CONTRIBUTED BY CHARLES WENDELL TOWNSEND

Although some of the earlier English settlers in North America called the ruffed grouse the pheasant, a name that is still retained in the southern parts of its range, no true pheasants are native, nor were they successfully introduced into America until 1881, when Judge O. N. Denny, then American consul general at Shanghai, China, after a previous unsuccessful attempt, sent 30 ring-necked pheasants to Oregon. Of these 26 survived and were liberated in the Willamette Valley. Two years later more were sent (Shaw, 1908). Although several early attempts at introduction were made, the first successful introduction of pheasants into the East was in 1887 by Rutherford Stuyvesant, who brought over a number of birds from England and liberated them on his estate at Allamuchy, N. J. In the nineties, pheasants were brought from England and liberated in various places in Massachusetts and elsewhere.

The bird proved to be remarkably hardy and prolific and spread rapidly, partly by natural increase and partly by artificial breeding in private and State farms, and by the shipment of eggs and birds to new sections of the country. The bird thrives in the North, but south of Baltimore and Washington, according to Dr. J. C. Phillips (1928), although there have been many attempts at introduction, "the stock does not hold out long if thrown on its own resources."

W. L. McAtee (1929) says of this bird that it

now has an almost continuous distribution over the Northern States from coast to coast. It has proved hardy as to climatic conditions, wary as to enemies, and without doubt is more numerous than any native game bird in the area occupied. The success of the introduction of pheasants in the Northwestern States is well known, but how amazingly the birds have thrived in certain other sections is not generally appreciated. In South Dakota, according to the Director of the State Department of Game and Fish, pheasants increased steadily from the first, a fact justifying almost steady lengthening of the open season and increase in the daily bag limit. The total bag in 1926 was estimated at a million birds, and in 1927 from one and a half to two millions, a record that has scarcely been approached in all our history by a single species of game bird in a single State.

William L. Finley contributes interesting details on the status of the bird in Oregon, which, somewhat condensed, are as follows:

For more than 20 years the success of the ring-necked pheasant seemed to be complete in Oregon. The numbers of the birds increased and they spread to all parts of the Willamette Valley and over into other valleys, although thousands of the birds were killed each hunting season. Then there came a period when the bird seemed to be just holding its own, and for the past 20 years the population has been on the decrease. Many reasons have been given for this decrease, as, for example, too much shooting or an increase in enemies.

One reason for the decrease in the number of pheasants in the Willamette Valley may be a change in the life of the bird itself. It survives best where it nests, roosts, and lives out in the open fields, where its watchfulness is always a check upon its enemies. In the Willamette Valley during the past 20 years the pheasant has come to be a bird of the woods, often lays its eggs in other birds' nests, and is not so good a mother to its chicks as formerly, so that fewer survive. It was the common thing late in summer or early in fall to flush a covey of 8 to 15 birds, while nowadays it is very rare to see more than 4 to 6 young pheasants in a covey.

Dr. J. C. Phillips (1928) says:

The extraordinary vitality of the first birds set out by the writer at North Beverly, Mass., in 1897 and 1898 was a most interesting feature. The broods were at first large and the species did not appear to meet any natural checks to its spread for a number of years. This initial "vigor," however, seems to have been lost here as well as in other places where the pheasant has been planted for 25 to 30 years.

The following information regarding the pheasant from W. H. Hudson (1902) is of interest in this connection:

In Britain, where it has been permitted to run free in the woods for the last sixteen to seventeen centuries, it is still scarcely able to maintain its existence without the strictest protection and a great deal of attention on the part of man. It is known that when the birds are left to shift for themselves they soon decrease in numbers, and eventually die out, except in a few rare cases where the conditions are extremely favorable. How heavy the cost is of keeping pheasants in numbers sufficient for the purpose of sport is well known to all those who have preserves.

The many thousands killed annually may well be the cause of the failure of these birds to hold their own without artificial aid. The extent to which this aid was given in Massachusetts in the year 1929 is shown by William C. Adams, director of the Division of Fisheries and Game, who writes: "From the four game farms we have shipped out 13,200 pheasant eggs, 1,554 adult pheasants, and 12,728 pheasants at least three months old." Incubators and brooders are used, and the eggs and birds are distributed to individuals and local game clubs throughout the State for liberation of the birds in the wild.

The pheasant is generally believed to be of Asiatic origin and at a very early date *Phasianus colchicus* was introduced into Europe, although Newton (1893–1896) thinks it not impossible that it may have been indigenous to Europe. This species introduced into England was called the English pheasant. Later the so-called Chinese, or ring-necked, pheasant (*Phasianus colchicus torquatus*) was introduced and hybridized with the pure *colchicus*, so that for the past 50 years nearly all English pheasants, according to F. C. R. Jourdain, have had some trace at least of a white collar. Those first introduced into Oregon were *P. colchicus torquatus*, the Chinese pheasant, or Denny pheasant, as they were sometimes called, after Judge Denny who introduced them; but those introduced in the eastern part of the United States were the mixed *colchicus* and *torquatus* race then existing in England, and birds and eggs have been widely distributed over the country. The resulting mixture, however, is generally called the ring-necked pheasant, *Phasianus colchicus torquatus*.

Courtship.—The cock ring-necked pheasant is a magnificent bird, and the display of his brilliant plumage and beautiful form to the best advantage before the demurely plumaged hens may well arouse their admiration and passion. On English game preserves, where only the males are killed, or in many of our own States where only the cocks are allowed to be shot, each cock may have a large harem, yet where the hens are not spared polygamy doubtless is still the rule. J. G. Millais (1909) notes, however, that in a wild state one rarely meets with more than three hens to a cock. Leffingwell (1928) says: "My opinion is that pheasants are naturally polygamous in the wild, but that some males may have monogamous tendencies. In the spring it is most usual to see one male bird with several females." He thus describes the display:

> The male runs around the female with short steps, usually with the tip of the partly outstretched wing describing an arc on the ground, and stops in front of her. The feathers of the upper back, lower back, rump, and tail are shifted over to the side on which the female is, and the tail partly spread. The neck is bent and the head kept low. Apparently the air sacs are partly inflated, for after the pose is held for several seconds, the plumage is allowed to fall back to its natural position as the bird gives out a hissing sound.

Millais (1909) described the courtship display as follows:

At this season the usual walk is seen to be more reserved and dignified, and the whole of the feathers are held out so as to give the bird a puffed appearance; the brilliant scarlet patch of skin round the eye is inflated and lowered beyond the angle of the jaw and the purple ears erected and inclined outwards. The bone-coloured bill is of a brighter hue and the eye, especially in the Mongolian subspecies, very brilliant. Thus he proceeds until the moment of show. The wing nearest the female is then lowered and extended, the scapulars dropped a little, the tail is also spread and turned over towards her, so that she may see its full beauty. The feathers of the rump are also opened as far as possible, the neck is lowered and curved and the head slightly turned to display the extended eye ornaments. If the female walks coquettishly away, or picks about with apparent indifference, he is not annoyed but walks ahead to stop her and displays the other side of his person.

Courtship displays of captive pheasants are easily studied, but under favorable circumstances these may be seen in the wild. I have thus described my own observations (Townsend, 1920):

In courtship the ear-tufts of the cock are erected and the bare skin about the eyes is prominent and very red. He struts before the hens turning in all directions to display his gorgeous plumage, or walks, with an exaggerated bobbing motion. Every now and again he flaps his wings almost inaudibly, crows and flaps again with a loud clapping sound.

The rivalry among the cocks leads to more or less fighting, after which the victor, according to Beebe (1918), crows and flies off.

Nesting.—The ring-necked pheasant, as a rule, nests on the ground in the open in fields of grass or grain and in bushy pastures, by hedgerows or roadsides or haystacks, very rarely in the woods and rarely at any great distance from water. The female is so protectively colored that one may pass within a few feet of the incubating bird and not see her, and she trusts so much to this protective coloration that she rarely leaves the nest until almost stepped upon. When she does leave she generally skulks quietly away and rarely flushes. She is much more difficult to see than the clutch of exposed eggs. A tuft of grass or a group of luxuriant weeds helps in the concealment of the nest, but in a field of grass or grain the eggs lie bare to the sky when the crop is harvested if they are not destroyed in the process. F. J. Rice writes of such a nesting situation found on August 10:

The oats had, of course, been harvested at this date and the nest lay in the stubble, with no concealment beyond some weeds, which grew up 6 or 8 inches from the ground. There was but slight pretense of a nest, the nine eggs lying on the earth among the weeds. In an hour [after his dog had flushed her] the bird was incubating again. She seemed very fearless and did not leave the nest again until a team and hayrack were driven within 10 feet of her.

The dog, in hunting about the field, may have happened upon the bird, for the sitting hen is protected not only by her coloration but by her lack of scent. Leffingwell (1928) reports tests on prize-winning dogs conducted by M. C. Ware, secretary of the Southern Idaho Field Trail Club. Two of these dogs, "in ranging over a large area, pointed twelve birds, all males. As pheasants are very abundant here, many females were undoubtedly present but were not found because of their scentlessness." Newly hatched young, like the incubating female, appear also to be scentless, for "three newly hatched pheasants found by Mr. Ware could not be located by the dogs until the birds were seen."

Pheasants have been known to lay their eggs in other birds' nests. Leffingwell (1928) reports that A. A. Allen found a nest of a ruffed grouse containing eight eggs of the owner and four of a pheasant, and that Finley has found pheasant eggs in the nests of domestic fowl, bobwhites, ruffed grouse, and sooty grouse. In England pheasants have been observed to use the old nest of another bird in a tree (Van de Weyer, 1919, and Tegetmeier, 1911). The nest itself consists generally of a slight natural hollow in the ground, or one made by the female, lined, sometimes very scantily, with leaves, grasses, or weed stalks.

Eggs.—[Author's note: The ring-necked pheasant lays 6 to 14 or 15 eggs, usually 10 to 12. They vary in shape from ovate to short ovate. The shell is smooth with a very slight gloss. The usual color is a rich, brownish olive, varying from "wood brown" or "avellaneous" to "dark olive-buff" or "olive-buff." Some of the darkest eggs are "buffy brown" and very rarely some eggs are pale blue. The measurements of 29 eggs in the author's collection average 41.85 by 33.5 millimeters; the eggs showing the four extremes measure **44.5** by 33.2, 42.6 by **35.1**, **39.3** by 31.3, and 40 by **31.1** millimeters. Witherby's handbook (1920) gives the average of 35 British eggs, a mixture of both races, as 45.93 by 36.04, the maxima 49.2 by 37 and 48 by 39, and the minima, 39 by 36.5 and 46.8 by 34.5 millimeters.]

Young.—The length of incubation is stated to vary between 23 and 25 days. Leffingwell (1928) says of 656 birds constituting the first hatch at a game club that 81.7 per cent hatched on the twenty-third day, 15.5 per cent hatched on the twenty-fourth day, and 2.7 per cent hatched on the twenty-fifth day. In the second hatch 94.5 per cent hatched on the twenty-third day, the rest on the twenty-fourth day. As a rule, the female alone incubates, but Tegetmeier (1911) records an instance of a cock pheasant not only building the nest but incubating and hatching the eggs, and he states that the incubation by the male has been observed in several cases of wild birds.

As soon as the down on the chicks has dried after they have escaped from the eggshell and have fluffed out their feathers, they show great activity, and the mother bird leads them in search of food. If enemies, including the chief one, man, should disturb the family at this time, the mother resorts to the "broken-wing" tactics and flops over the ground, endeavoring to lead the enemy away. At other times the mother springs into the air with a loud whirring of wings, while the young scatter and hide. Leffingwell (1928) says that "normally all the care of the young is undertaken by the female until the young birds are from six to seven weeks old. After this time the cock bird occasionally wanders with the flock. I have seen cock pheasants with a flock of six or seven week old birds on several occasions and as no females were observed it is not unlikely that they were setting or raising another brood."

The mother bird leads her brood about exactly like the domestic hen, helps them to find food, and broods them in cold or storm and at night.

The gain in weight of the young pheasants is at first rather slow. The weight is doubled soon after the second week, but is increased nearly sevenfold over the initial weight at 3 weeks, and at 5 weeks the male chick weighs, according to Leffingwell's figure, about 15 times his birth weight and is fully feathered. Leffingwell writes me:

It is somewhat difficult to define the age at which young pheasants can fly. The birds are able to clear a 1-foot obstruction when they are 4 or 5 days old, but whether or not you can call this flying I do not know. Certainly when they are a week old they can fly distances of 4 or 5 feet.

I copy the following from my notes of September 1, 1913:

From my window this morning at 7 I saw a hen pheasant with a number of half-grown young scale down the hill, alight on the edge of the "forest," and disappear within. A few minutes later I heard a loud, insistent calling, *kee kee kee*, which suggested somewhat a flicker. Then I saw the hen pheasant standing erect, with head and neck stretched up, in the field, close to the trees. Soon she stopped calling and disappeared in the grass, and three young pheasants came sailing down the field to her. She had been drumming up the laggards.

Plumages.—[Author's note: The chick in down is thus described in Witherby's handbook (1920):

Fore-head and sides of crown buff to yellow-buff with blackish line or spots down sides, centre of crown dark red-brown to blackish-brown; nape rufous; back of neck buff to yellow-buff with short blackish line in centre; rest of upper-parts rufous-buff with three wide black lines and wings with black blotches; sides of head pale yellow-buff to pale buff with a brownish streak from base of upper mandible and a black spot on ear-coverts; under-parts buff white to pale buffish-yellow, sometimes with a tawny tinge at base of throat.

The same writer describes the juvenal plumage, as follows:

Crown and back of neck dark brown, feathers with subterminal pale buff marks giving a spotted appearance; mantle and scapulars brown-black, feathers with margins and central streaks of buff to pale buff, those of upper-mantle and base of back of neck more or less tinged rusty-rufous; back, rump, and upper tail-coverts browner (less black) and with rather more pale buff margins and centres and often bars; sides of head pale buff minutely speckled black; chin pale buff; sides and base of throat same but streaked and with concentric lines of dark brown; flanks same but with thicker concentric lines or broken bars of brown to black-brown; tail feathers small and narrow, buff and rufous closely barred black; wing feathers much as adult female, but with pale buff barring; wing-coverts as scapulars, but with larger centres, and greater often with bars of pale buff.

The postjuvenal molt begins when the young bird is about half grown; it is a complete molt, except that the two outer primaries on each wing are retained for a full year. By November or December young birds have completed the molt into the first winter plumage, which is practically adult.

I can find no evidence of a prenuptial molt in adults, but they have a complete postnuptial molt late in summer and early in fall, beginning in July or August and ending in September or October. Old females, whose ovaries have become inactive through old age or disease, sometimes assume more or less of the male plumage. I have seen such a bird in a local collection in Taunton and another in the collection of W. F. Peacock in Marysville, Calif. This bird, Mr. Peacock told me, had been raised in captivity and had laid some 280 eggs. When 6 years old she quit laying, and a post-mortem examination, 2 years later, showed that the ovaries were completely atrophied. From the time she stopped laying the male plumage gradually developed, until finally she could easily be mistaken for a first winter male, having acquired the head and body plumage and the long tail but not the comb and wattles of the male.]

On the development of the male genital organs, Paul D. Dalke has sent me the following notes:

The size and development of the testes in the pheasant have been noted in all the cock birds taken since January 22, 1930. A specimen taken on the above date had testes that measured 4 by 10 millimeters. The next male was collected on February 28, 1930. This was a young cock, but the testes had already grown considerably and measured 12 by 18 millimeters. Another young cock collected on March 5, 1930, had testes that measured 13 by 21 millimeters. A specimen of March 28 showed an increase in size of the testes to 15 by 25 millimeters. On May 1, 1930, two cock pheasants were collected, one an old cock and the other of the previous year's crop. The latter had testes measuring 16 by 30 millimeters and the former 22 by 35 millimeters.

Food.—The food of the ring-necked pheasant consists of insects, weed seeds, wild fruits and berries, and cultivated crops. Occasionally small rodents are eaten. The economic status of the bird, whether harmful or beneficial, is dependent on local conditions and seasons of the year, and on the proportion between weed seeds and injurious insects eaten on the one hand and cultivated crops on the other.

Pheasants are at times very destructive to sprouting corn and even to corn in the ear. They also are known to eat tomatoes, beets, peas, beans, and other farm crops, including grain of all sorts. The stomachs of those shot doing damage in a garden often contain, however, a large number of injurious insect pests as well as many weed seeds. Much of the grain—wheat, oats, rye, barley—eaten by pheasants is waste grain that has fallen to the ground during harvesting.

In summer, according to an unpublished report kindly sent me by Leffingwell, 20.4 per cent of the contents of 11 pheasant stomachs consisted of injurious insects, and insects are consumed almost exclusively by the broods of young. Stomach examinations made by various observers show that grasshoppers, crickets, potato beetles, squash bugs, curculio beetles, and larvae of all kinds, including those of the gypsy and brown-tail moths and including also the tent caterpillar, are among the items in the food. With the injurious insects, a certain number of beneficial ones are eaten.

McAtee (1912) records that the stomach of one pheasant contained 360 larvae of March flies, *Bibio*, and another 432. "Twenty-three acorns and 200 pine seeds were taken by the bird who ate the largest amount of mast, and about 800 capsules of chickweed, containing more than 8,000 seeds were in the stomach of the weed-seed eater." McAtee sums up his report as follows: "What is most evident is that pheasants are gross feeders, their capabilities for good or for harm are great. If a number of them attack a crop they are likely to make short work of it, or if they devote themselves to weed seeds or insect pests they do a great deal of good."

Leffingwell (1928) says that "in Minnesota, F. D. Blair, the superintendent of game laws, believes that pheasants destroy more mice per bird than do most of the hawks and owls." At the seashore, especially when the uplands are covered with snow, pheasants visit the salt marshes, where they probably consume small crustaceans and mollusks.

Clarence Cottam (1929) says: "Except during the coldest winter those stomachs collected at mid-day contained little food material, while those taken in the morning and evening were full or being filled. This suggests that pheasants usually feed during the early morning or late afternoon."

Paul D. Dalke says in his notes:

On January 22, 1930, I made a trip to collect pheasants 5 miles south of Ann Arbor, Mich., where the birds are fairly abundant. I flushed one cock pheasant from a field overgrown with wild sweetclover; the cock had traveled only 100 feet from the time he left the roost until 2 p. m. While traveling this 100 feet he had rested three times, as indicated by the hollowed-out forms in the snow and by the droppings. Although corn was available in an adjacent field this pheasant had nothing in his crop and only grass in his gizzard.

Behavior.—The introduction of the ring-necked pheasant has added greatly to the charms of the countryside. To watch a cock pheasant striding along the ground or launching into the air, showing his metallic-blue head, his snow-white neck ring, his golden-brown back, and his long magnificent tail is always worth while. The males, as if aware of their conspicuous shape and colors, generally run away or flush at some distance, while the modest-hued hens depend more on their protective coloration and allow a much closer approach. Frequenting for the most part cultivated fields and bushy pastures, they are also to be found in swamps and moist thickets and among sand dunes. In the sand their tracks are as easily seen and recognized as in the snow—three toe marks widely diverging, with a dot or a line behind and on the inner side made by the hind toe if the sand or snow is soft. Occasionally the tracks show that they drag the middle toe, but the birds usually step clear. When the bird is walking, the distance between tracks is generally 4 or 5 inches, but a stride of 7 or 8 inches is not uncommon, and I have measured several tracks where 18 inches was cleared at each running step. In running the tail is cocked up at an angle of 45°, but when the bird is feeding it is generally held horizontally. Pheasants have been seen swimming across bodies of water (Tegetmeier, 1911).

When suddenly startled, pheasants flush with a loud metallic whir of the wings, but not so thunderous as that of the ruffed grouse. Occasionally a bird flies off almost silently. They are able to shoot up nearly vertically if they are hemmed in by trees or a building, and they make off at a great speed, which has been estimated to reach 38 miles an hour (Tegetmeier, 1911). When not much frightened they soon set their wings and scale, each primary standing out like fingers on a hand, while the long tail is spread to its utmost and curved downward for a brake, and their wings are fluttered rapidly as they approach a landing. In flight the feet are at first drawn up in front but are quickly jerked back and are held extended behind under the tail. In very short distance flights this backward extension may not take place. It is generally believed that a pheasant's flight is limited to somewhat less than a mile, although, when helped by a strong wind, longer distances are accomplished. In a very strong wind a cock pheasant in attempting to fly over my house one January day collided with a chimney and met his instant death.

As to the distance to which pheasants wander from their coverts, Leffingwell (1928) states that of 16 banded birds recovered within three years " but two exceeded two miles, while the average distance covered was but one and one-fifth miles. One bird, however, went six miles and another three miles." Merriam (1889) records that a pair in Oregon traveled in two months 50 miles from the point where they were released.

Pheasants, like domestic fowls, are fond of dusting themselves to get rid of lice, and dusting places are common in pastures where these birds are found. Pheasants spend the night on the ground and also in trees, the latter especially where foxes or other ground enemies abound. Several times early in the morning I have seen pheasants roosting in trees, singly, and once as many as five. As the sun rose, their breasts glistened like burnished copper in its rays.

Under complete protection, where shooting is not permitted, as in public parks, pheasants become very tame, but when persecuted they quickly develop great wariness, and they seem to be able to distinguish the harmless farm laborer from the man with a gun. Any unusual noise, such as blasting, makes pheasants crow, and they are usually sensitive to any shock, whether from an explosion or an earthquake, and respond by crowing. This response to earthquakes or distant explosions is apparently due to the sense of feeling rather than of hearing. In Japan pheasants are believed to give warnings of earthquakes. It is found that they respond to earthquake shocks so slight that they are unnoticed by human beings, and the birds may in this way foretell a more severe earthquake shock that follows. Hartley (1922) states that "during the World War the pheasants in England developed into fairly responsible sentinels against Zeppelin attacks. The birds seemed particularly sensitive to far-off explosions and a raid generally was heralded by a concerted crowing of cocks."

William Beebe (1918) writes:

In the spring the cock pheasant invites his mate or mates to share or appropriate some especially delectable morsel of food. The accompanying movement is a picking up and dropping of the food, thus calling it to visual attention, while at the same time a low chuckle or crowing sound is uttered.

Voice.—The courtship "song" of the cock pheasant is his crow, which suggests a juvenile bantam rather than the noble pheasant. It is a challenge call by which the cocks announce their territorial holdings. This crow, which consists of a long followed by a short note, can be heard from a considerable distance, but when the observer is near at hand he hears also a hurried clapping of the wings, which is heard loudest following the crow and not preceding it, as in the case of the domestic cock. If, however, the bird is seen at this time it will be observed that he flaps his wings two or three times almost inaudibly before the crow, and follows the crow with a rapid succession of five or six flaps, which are audible for some distance. Leffingwell (1928) states that

the wings are held rather stiffly and the force of the beat is directed upwards and inwards, somewhat after the manner of a drumming partridge. The two preliminary wing beats are given at intervals of about one twenty-fifth of a second, while those given after the call begin very rapidly but soon diminish in vigor. The force of the latter strokes seem to push the pheasant backwards against its tail, which is partly flattened on the ground and acts as a brace.

Crowing is, of course, commonest in the spring of the year, and I have heard one crowing on a pleasant day as early as mid-January. The birds are generally silent in the summer months, but they are often heard again in October. The young are said to attempt to crow when 7 or 8 weeks old.

Besides the crow, the alarm notes, emitted when the birds are startled, are most commonly heard. These are loud and hoarse croakings, which they emit as they fly away, sometimes in two or three syllables, written *cuck-et* or *tuck-ee-tuck*, sometimes a prolonged and scolding repetition of croaking notes. These croaking notes, suggestive at times of an old domestic hen, may also be given when the bird is on the ground. I have heard a cock pheasant in a small thicket croaking continuously, owing apparently to irritation caused by crows who were scolding him from the trees above. I have also heard a querulous *queep*, *queep*, *queep* given by a hen pheasant. I once disturbed a hen pheasant in feeding, and she looked up in time to see a fox creeping toward her from the other side. The pheasant at once flew off uttering a whistling shriek, and the fox, interrupted in his turn, also departed. Millais (1909) says that "the cock pheasant when he has paired or gathered his wives, makes use of a gentle note or chuckle."

The young have a variety of notes. Leffingwell (1928) lists five distinct calls in birds up to 7 weeks of age—calls expressive of contentment, caution, alarm and fright, and the flock call.

Field marks.—The cock pheasant, with his resplendent plumage, white neck ring, and long tail, is easily identified. The hen and young might be mistaken for ruffed grouse except for the much longer tails.

Game.—The cock pheasant is a prize well calculated to delight the heart of the sportsman. Splendid in plumage—a magnificent trophy—large and heavy, and delicious eating, it tests his skill to the utmost. In game preserves in England, numerous beaters drive the pheasants to the quiescent sportsmen and force the birds to fly high and at great speed over them. In this country, the sportsman seeks the bird, going on foot over the fields and shooting the pheasant as it flushes and makes off. Dogs are generally used, and although the pheasant often lies close to the pointing dogs and allows himself to be flushed by the sportsman, he may sometimes hide, or, worse still from the point of view of the owner of a well-trained dog, he may run long distances and entice the dog to follow. Only by continuous replacement with birds raised on private or State farms can the drain on pheasants by hunters be checked, for in these days of automobiles hunters may visit numerous favorable localities many miles apart in the course of one day.

A very important asset of the ring-necked pheasant, both from the sportsman's and the bird lover's point of view, is that it diverts gunfire from our fast-diminishing grouse and other game birds.

As the natural habitat of the ruffed grouse is in woods, while that of the pheasant is usually in open fields and pastures, there is no direct conflict between these two birds, and from this point of view there seems to be no harm resulting to the ruffed grouse from pheasant introduction. On the other hand, bobwhites, having much the same habitat as the pheasant, might be displaced by the latter bird if there were not enough food for both. Dr. George W. Field (1914) does not think there is any evidence that the pheasant interferes with our native game and says:

More conclusive is the testimony from records of the great shooting estates in England, Scotland, and Germany, where for at least two generations accurate records have been kept of the number of birds produced annually on each estate. These records show conclusively that there has been no diminution in the number of native grouse and quail, and no displacement of these birds by the introduced and naturalized pheasant, which is here produced in larger quantities than on any similar area in the world.

In this connection the remarks of W. L. McAtee (1929) are pertinent:

There is little fear, however, that any of the large and highly edible species classed as game birds will continue for any period as pests. Should they exhibit destructive tendencies their numbers can easily be cut down by the extension of the open season and increase in the bag limits. No bird that is widely prized for food is ever likely to become destructively abundant in the United States.

There is, however, a subtle danger from the introduced pheasant, that of carrying to our native game blackhead, the heterakis worm, and other diseases and parasites. Prevalence of the heterakis worm in wild pheasants has been shown by Dr. E. E. Tyzzer. McAtee (1929) does not think that this danger should be given much weight "in view of the fact that domestic poultry is constantly being introduced, abounds in all parts of the country, and constitutes a source and reservoir of most of the diseases to which our game birds are susceptible. In other words, the disease hazard is scarcely likely to be notably increased by further introductions of game birds. There should, of course, be proper inspections of imported birds and exclusion or quarantine when found necessary." Dr. J. C. Phillips (1928), however, says "the interesting point is that the pheasant may now easily infect territory at a distance from farmyards."

Enemies.—Man, of course, is the pheasants' worst enemy in destroying them, but their best friend in conserving the stock that might otherwise become extinct. All mammals that prey on eggs, and, if they are powerful enough, on the sitting or roosting birds, like rats, skunks, weasels, foxes, and coyotes, are more or less destructive to pheasants. A few hawks and owls also take toll of pheasants, but the damage done by these birds is probably insignificant, and their influence may be of value in eliminating the feebler fliers and the diseased birds. Crows, however, undoubtedly eat many pheasant eggs and the far-wandering domestic cat is a very serious menace to these ground-nesting birds.

Winter.—When the ground is covered with a blanket of snow and ice, pheasants are obliged to wander far for food. At this time, if weed seeds are largely covered, they are driven to eat the buds of bushes and trees and to seek manure and garbage heaps, and even to partake with the poultry on a farm. At such times, if the salt marshes are still open, they frequent these for small mollusks and crustaceans. I have found the thinner snow under a tree scratched away by pheasants in order to get at the seeds and dormant insects on the ground. Sometimes the long tails of the males are frozen to the snow and ice during the night and the birds held captive.

DISTRIBUTION

Range.—Introduced and now fairly well established in approximately the northern half of the United States and in extreme southern Canada. **North** to southern British Columbia (Vancouver Island and Fraser Valley); southern Alberta; southern Manitoba; southern Ontario (Kent County and north shore of Lake Erie); central New York and Vermont (Lake Champlain region); central New Hampshire (Concord, Hanover, and Plymouth); and southwestern Maine. **East** to the Atlantic coast. **South** to Maryland (north of Baltimore); Pennsylvania; Ohio; Kentucky; Missouri; Kansas; Colorado; and California (Inyo, Tulare, and Kern Counties and Santa Clara Valley). **West** to the Pacific coast.

Introductions into the eastern Provinces of Canada and into the States south of those named above have not been successful.

Egg dates.—Washington and Oregon: 5 records, April 13 to June 17. California: 2 records, May 3 and June 10. Michigan, April 17. Massachusetts, May 16. Pennsylvania: 2 records, May 12 and June 4.

Family MELEAGRIDIDAE, Turkeys

MELEAGRIS GALLOPAVO MERRIAMI Nelson

MERRIAM'S TURKEY

HABITS

The wild turkey of the mountain regions of the Southwestern United States and extreme northwestern Mexico was described by Dr. E. W. Nelson (1900) and named in honor of Dr. C. Hart Merriam. He has characterized it as follows: "Distinguished from *M. g. fera* by the whitish tips to feathers of lower rump, tail-coverts, and tail; from *M. g. mexicana* by its velvety black rump and the greater amount of rusty rufous succeeding the white tips on tail-coverts and tail, and the distinct black and chestnut barring of middle tail feathers."

Nelson showed in the same paper that the ancestors of our domestic turkeys were neither of the forms that we now call *merriami* and *intermedia* but the more southern, strictly Mexican form, *M. gallopavo gallopavo.*

That this wild turkey is not nearly so abundant as it was 50 years ago is shown by the following quotation from Henry W. Henshaw (1874):

> The wild turkey is found abundantly from Apache throughout the mountainous portion of Southeastern Arizona. In New Mexico it was met with further to the north, in the mountains, and I was informed by Colonel Alexander that he had found them in large numbers in the Raton Mountains, in extreme Northern New Mexico. It breeds abundantly through the White Mountains, Arizona, and about the middle of August several broods of the young, about two-thirds grown, were met with. Toward the head of the Gila, in New Mexico, the canons, in November, were found literally swarming with these magnificent birds; in many places the ground being completely tracked up where they had been running. As many as eleven were killed by the members of a party during a day's march.

Nesting.—Two brief notes by Major Bendire (1892) are all that he gives us on the nesting habits of this turkey, which are probably not very different from those of other wild turkeys. He quotes William Lloyd as saying that " near a river their nests would be made on small inlets surrounded by reeds; on the hills in shin-oak clumps." He says that Frank Stephens found a nest on the east slope of the Santa Rita Mountains in Arizona, " in the oak timber, just where the first scattering pines commenced, at an altitude of perhaps 500 feet. It was placed close to the trunk of an oak tree on a hillside, near which a good-sized yucca grew, covering, apparently, a part of the nest; the hollow in which the eggs were placed was about 12 inches across and 3 inches deep."

Mrs. Florence M. Bailey (1928) says the nest is "on the ground in tall thick weeds or briers, lined with grass, weeds, and leaves."

O. W. Howard (1900) found a nest in the Huachuca Mountains in Arizona, which he describes as follows:

The nest was in the bed of the canon at the base of the hill, in a natural depression in the soft earth at the side of a rock, and just under a large white oak tree. The nest had a lining of leaves and small twigs, with a few feathers from the old bird scattered about. The nest was about a mile above the place where I had seen the first bird and at about 7,000 feet elevation. Strange to say, the nest was within a stone's throw and in plain sight from a well-traveled trail.

Eggs.—The eggs are indistinguishable from those of other wild turkeys. The measurements of 16 eggs average 65.8 by 47.3 millimeters; the eggs showing the four extremes measure **70.5** by **49**, **61.7** by 46.7, and 64.5 by **46** millimeters.

Food.—Mrs. Bailey (1928) lists the food of this turkey as follows:

In winter pinyon nuts, acorns, and juniper berries; in summer flower buds, grass and other seeds, wild oats, wild strawberries, manzanita berries, rose haws, fruit of wild mulberry and prickly pear, grasshoppers, crickets, beetles, caterpillars, ants, and other insects. In New Mexico "the crop of a Merriam Turkey killed February 10 on Haut Creek contained 76 juniper berries, 25 pinyon nuts, 6 acorns, 30 soft worms an inch long, grass blades and some rock. The crop of a gobbler, weighing about 30 pounds and shot March 25 out of a flock of 50 in the Black Mountains, contained 30 pinyon nuts and 215 juniper berries" (Ligon). The stomach of a specimen collected near the southern end of the range contained fully a half pint of the fruiting panicles of grass (*Muhlenbergia*), a few seeds of *Bromus*, and some grass blades comprising 55 per cent; pinyon pine and other pine seeds, 45 per cent. In some localities considerate ranchmen plant small patches of oats for turkey food (Ligon).

Mrs. Bailey quotes Charles Springer as saying:

At times, and particularly in years when there are few or no nuts, the principal food of the Merriam Turkey is wild rye, which is plentiful in the canyons and draws in our mountains and foothills. On the Suree I have often seen wild Turkeys eating the short blades of Kentucky blue grass which grows wild along the canyon near the stream and remains green all winter. One of the most important winter foods of the Merriam Turkey is the red kinickinick berry which grows on the high ridges and plateaus in our mountains. When acorns, pinyon, and pine nuts, and other foods may be buried deep under snow, the Turkeys may find kinickinick berries on the high ridges and high places from which the snow blows off. Mason Chase tells about the wild Turkeys hunting out, or at least finding and appropriating, caches of nuts made by rodents. He says this occurred during a time when deep snow covered up all the Turkey's food except the buds of shrubs.

Dr. C. Hart Merriam (1890) found it on San Francisco Mountain, Ariz., feeding on wild gooseberries in the balsam belt in August and on pinyon nuts in the cedar belt in September. Major Bendire (1892) says that it also eats the fruits of the giant cactus, "which is alike a favorite article of food with man, bird and beast."

Behavior.—Henshaw (1874) writes:

They roost at night in the large cotton-woods by the streams, and soon after daylight, having visited the stream, they usually betake themselves to the dry hills, where they feed, in the fall, at least, almost exclusively upon the seeds of grasses and grasshoppers. I think they return once or twice during the day to drink, the dry nature of their food rendering a copious supply of water necessary. In these wilds, they appear to be wholly unsuspicious, and without knowledge of danger from man, and, if not shot at, will allow one to get within a few yards without manifesting any distrust. They rarely fly, except when very hard pressed, but, when alarmed, run with such rapidity as to quickly outstrip the fleetest foot, betaking themselves to the steep sides of the ravines, which they easily scale, and soon elude pursuit. Apparently, the only dangers they have to fear in these regions are from birds of prey, which attack the young, but more especially from the panthers. In certain portions of the Gila Canyon the tracks of these animals were very numerous, and always these sections appeared to have been entirely depopulated of Turkeys, an occasional pile of feathers marking the spot where one had fallen a victim to one of these animals.

DISTRIBUTION

Range (entire species).—Southern Ontario; the Eastern, Central, and Southern United States, including the southern Rocky Mountain region; and Mexico, except the extreme western and southern parts.

The range of the wild turkey has been greatly restricted since the advent of civilization, so that the species is now extirpated throughout New England and the Great Plains. It is still common locally in Pennsylvania (largely through introductions) and in some of the Southern States, as South Carolina, Georgia, Florida, Louisiana, Texas, New Mexico, and Arizona.

The complete range of the species extended **north** to Arizona (Bill Williams Mountain, and San Francisco Mountain); formerly Colorado (Coventry, Salida, and Buckhorn); formerly Nebraska (Valentine); formerly southeastern South Dakota (Cedar Island, Fort Randall, Yankton, and Vermilion); formerly Iowa (Grant City, Ames, and Fort Atkinson); formerly southern Wisconsin (Newark, Lake Koshkonong, Waukesha, and Racine); formerly southern Michigan (Grand Rapids, Locke, and Reece); formerly southern Ontario (Mitchells Bay, Plover Mills, and Dundas); formerly northern New York (Niagara County); and probably formerly southern Maine (Mount Desert Island). **East** to probably formerly southern Maine (Mount Desert Island); formerly Massachusetts (Ipswich, Montague, and Mount Holyoke); formerly Connecticut (Northford); formerly New Jersey (Sussex County, Raccoon, and Cape May County); Virginia (Neabsco Creek and Ashland); North Carolina (Walke and the Cape Fear River); South Carolina (Georgetown, Santee, Mount Pleasant, and Charleston); Georgia (Riceboro, MacIntosh, Cumberland Island, and St. Marys); eastern Florida (Port Orange, Oak Hill, Malabar, Fort Kissimmee, and Everglade); Tamaulipas (Soto la Marina and Forlon); and Vera Cruz (Zacuapan, Mirador, and Soledad). **South** to Vera Cruz (Soledad and Paso del Macho); Hidalgo (Real del Monte); and Michoacan (La Salada). **West** to Michoacan (La Salada); Durango (Durango and El Salto); Chihuahua (Colonia Garcia and Cajon Bonito); and Arizona (formerly Huachuca Mountains, formerly Santa Catalina Mountains, Salt River Bird Reservation, Sierra Ancha, Baker Butte, Apache Maid Mountain, and Bill Williams Mountain).

The range as above described is for the entire species, which has, however, been divided into four subspecies. The eastern wild turkey (*Meleagris g. silvestris*) ranged over the entire eastern part of the country north of central Florida and west to eastern Texas, central Kansas, Nebraska, and Oklahoma; the Florida turkey (*M. g. osceola*) ranges through southern Florida; the Rio Grande turkey (*M. g. intermedia*) is found in central and southern Texas, chiefly between the Brazos and Pecos Rivers, north to the Staked Plains, and in northeastern Mexico; and Merriam's turkey (*M. g. merriami*) is found in mountainous regions in New Mexico, Arizona, and Colorado, ranging also into western Texas and northern Mexico.

Wild turkeys have been restored to parts of the range from which they had been exterminated, notably in Pennsylvania. They also have been carried north successfully as far as Minnesota (State game farm at Minneapolis).

Egg dates.—Michigan: 3 records, February 10 and May 5 and 6. Pennsylvania: 5 records, May 5 to June 30. South Carolina and Georgia: 15 records, March 30 to May 25; 8 records, April 25 to May 22. Missouri and Arkansas: 8 records, April 3 to June 2. Louisiana, Oklahoma, and Texas (*silvestris*): 20 records, April 9 to July 25; 10 records, May 3 to 16. Arizona, Mexico, and New Mexico (*merriami*): 7 records, April 8 to May 8. Texas and Mexico (*intermedia*): 23 records, March 4 to June 28; 12 records, May 1 to June 3. Florida (*osceola*): 15 records, March 25 to May 22; 8 records, April 10 to May 3.

MELEAGRIS GALLOPAVO SILVESTRIS Vieillot

EASTERN TURKEY

HABITS

When the noble red man roamed and hunted unrestrained throughout the virgin forests of eastern North America, this magnificent bird, the wild turkey, another noble native of America, clad in a feathered armor of glistening bronze, also enjoyed the freedom of the forests from Maine and Ontario, southward and westward. But the coming of the white man to our shores spelled the beginning of the end for both of these picturesque Americans. The forests disappeared before the white man's ax, his crude firearms waged warfare on the native game, and the red man was gradually eliminated before advancing civilization. In the days of the Pilgrims and Puritans the Thanksgiving turkey was easily obtained almost anywhere in the surrounding forest; the delicious meat of the wild turkey was an important and an abundant food supply for both Indians and settlers; and the feathers of the turkey held a prominent place in the red man's adornment.

Thomas Morton (1637), one of the earliest writers, says:

> Turkies there are, which divers times in great flocks have sallied by our doores; and then a gunne, being commonly in a redinesse, salutes them with such a courtesie, as makes them take a turne in the Cooke roome. They daunce by the doore so well.

They soon began to disappear, however, for John Josselyn (1672) writes:

> I have also seen threescore broods of young Turkies on the side of a marsh, sunning of themselves in a morning betimes, but this was thirty years since, the English and the Indians having now destroyed the breed, so that 'tis very rare to meet with a wild Turkie in the Woods.

Edward H. Forbush (1912) says:

> In Massachusetts Turkeys were most numerous in the oak and chestnut woods, for there they found most food. They were so plentiful in the hills bordering the Connecticut valley that in 1711 they were sold in Hartford at one shilling four pence each, and in 1717 they were sold in Northampton, Mass., at the same price. From 1730 to 1735 the price of those dressed was in Northampton about one and one-half penny per pound. After 1766 the price was two and one-half pence, and in 1788, three pence. A few years after 1800 it was four pence to six pence a pound, and about 1820, when the birds had greatly decreased, the price per pound was from ten to twelve and one-half cents.

Wild turkeys made their last stand in Massachusetts in the Holyoke range, where the last one was killed in 1851. According to Dr. D. D. Slade (1888)

> these birds had the range of a large tract of wild mountainous country, in some parts almost inaccessible and impassable, lying at the base of and comprising Mount Holyoke, and to the Southwest also including Mount Tom and its surroundings. I am unable to state the exact period at which this flock became exterminated but should say it must have been in 1840 or thereabouts.

The last turkey in Connecticut was seen in 1813, a few remained hidden in the Vermont Hills until 1842, and they were said to be numerous along the southern border of Ontario as late as 1856. Albert H. Wright (1914 and 1915) has written a very complete account of the early history of the wild turkey to which the reader is referred. Dr. Glover M. Allen (1921) also has given us a very full history of this bird in New England. Both of these exhaustive papers give far too much information to be included here.

Audubon (1840) wrote, as to its status in his time:

The unsettled parts of the States of Ohio, Kentucky, Illinois, and Indiana, an immense extent of country to the north-west of these districts, upon the Mississippi and Missouri, and the vast regions drained by these rivers from their confluence to Louisiana, including the wooded parts of Arkansas, Tennessee, and Alabama, are the most abundantly supplied with this magnificent bird. It is less plentiful in Georgia and the Carolinas, becomes still scarcer in Virginia and Pennsylvania, and is now very rarely seen to the eastward of the last-mentioned States. In the course of my rambles through Long Island, the State of New York, and the country around the Lakes, I did not meet with a single individual, although I was informed that some exist in those parts. At the time when I removed to Kentucky, rather more than a fourth of a century ago, Turkeys were so abundant, that the price of one in the market was not equal to that of a common barn fowl now. I have seen them offered for the sum of three pence each, the birds weighing from ten to twelve pounds. A first-rate Turkey, weighing from twenty-five to thirty pounds avoirdupois, was considered well sold when it brought a quarter of a dollar.

In the mountains of central Pennsylvania turkeys have always existed up to the present time. C. J. Pennock tells me that they "have multiplied greatly within the last 15 years," but that most of the "stock has been intermixed with domestic birds." This is largely due to the efforts of the game commission in controlling the hunting season and bag limits and by importing birds from other States or transferring them from a section where they are plentiful to one where they are scarce. A report of this has been published in some detail by Bayard H. Christy and George M. Sutton (1929).

M. P. Skinner wrote to me in 1928:

The wild turkey is still found over most of North Carolina wherever there are undisturbed forests of the kind preferred by the turkey. In the sand hills there are still two or three groups living mostly in the swamps and river bottoms, and totaling perhaps 30 birds in all. They are resident and nonmigratory.

In the sand hills, wild turkeys have largely retired to the deep swamps, for they prefer to roost only in trees standing in water; but quite often they feed out on the drier upland.

James G. Suthard says in his notes, sent to me in 1930:

This noble game bird formerly bred in Kentucky, at large, but at the present time has a very restricted range. It is found in the areas bordering Virginia and Tennessee, in Taylor, Larue, and Hart Counties in central Kentucky, and also in the game preserve in Lyon and Trigg Counties. I have some records for Fulton and Hickman Counties. It is possibly found straggling in other counties, but, because of its retiring habits, I have never seen it, nor do I have any authentic records other than those already mentioned. It breeds during April and May, sometimes late in June.

Wild turkeys are essentially woodland birds. When the Eastern States were largely covered with virgin forests, they ranged widely over the whole of these districts. As the land became cleared they often resorted to clearings, open fields, savannas, or meadows in search of grasshoppers, other insects, berries, and other foods. As their numbers were reduced by persistent hunting, they became very shy and were forced to retire to the wooded hills and mountains, where in many places they made their last stand. There are many hills and creeks named for this bird because turkeys were once common there. Turkeys are now found, in the Northern and Eastern States, only in the more remote and heavily wooded mountains, the wildest and least frequented forests, or the most inaccessible swamps, far from the haunts of man. In the Southern States they are much more abundant and more widely distributed. M. L. Alexander (1921) says of their haunts in Louisiana, which are typical:

The determining factor in the distribution of turkeys is the occurrence of oaks, wild pecans, beech and other nut-bearing trees. It is chiefly the oaks that attract them to the flatwoods type of river lands, while the beech, chinquapin, and certain species of oaks furnish the mast on the slopes of creeks, ravines and small rivers in pine regions. Dogwood, holly, black gum and huckleberry are among other trees and shrubs, growing chiefly on slopes and ridges, that furnish food for turkeys. Such food is not generally available, however, unless there is sufficient undergrowth to protect the birds while they are feeding. Late in the winter, after the best of the berries and mast in the bottoms of the hill sections have been picked up, or washed out by the rains, the turkeys frequent southerly slopes, with a good cover of brush, scratching in the fallen leaves and other woodland débris for such seeds and insects as may be concealed there.

Courtship.—The courtship display of the turkey gobbler is too well known to need any description here. The wild turkey's display is similar, with the same expansion of body plumage, erection and spreading of the fan-shaped tail, swelling of the naked head ornaments, and the drooping and rattling of the wing quills, accompanied by gobbling and strutting.

Audubon (1840) mentions a peculiar feature of the gobbler at this season, the "breast sponge," which fills the upper part of the breast and crop cavity. This is a thick mass of cellular tissue, which serves as a reservoir of sweet, rich oil and fat, on which the gobbler draws to supply the loss of flesh and energy during the mating season.

The object of the display and the gobbling notes is, of course, to attract the females. Turkeys are polygamous, the gobbler having many mates and serving them all every day during the laying season until his vigor is exhausted. The females separate from the males before the mating season, and each hen comes to her favorite cock once each day, for a short time, during the laying season. She keeps the nest concealed from him and shuns him after the eggs are laid, lest he might break the eggs to prolong his sexual enjoyment. The gobbler often begins to display and gobble before he leaves his roosting tree. He gobbles, watches, and waits until he sees the hen approaching, or hears her responsive yelp or cluck. He flies down to the ground, struts and gobbles again, and waits for the hen to come to him. He probably knows how many hens he has in his harem and keeps on strutting and gobbling until he has served them all. He roosts in the vicinity and repeats the performance every day until the laying season is over or until he becomes emaciated and takes no further interest in the hens.

Audubon (1840), who had far better opportunities for observing the wild turkey than can ever be had again, writes:

I have often been much diverted, while watching two males in fierce conflict, by seeing them move alternately backwards and forwards, as either had obtained a better hold, their wings drooping, their tails partly raised, their body-feathers ruffled, and their heads covered with blood. If, as they thus struggle, and gasp for breath, one of them should lose his hold, his chance is over, for the other, still holding fast, hits him violently with spurs and wings, and in a few minutes brings him to the ground. The moment he is dead, the conqueror treads him under foot, but, what is strange, not with hatred, but with all the motions which he employs in caressing the female.

When the male has discovered and made up to the female (whether such a combat has previously taken place or not), if she be more than one year old, she also struts and gobbles, turns round him as he continues strutting, suddenly opens her wings, throws herself towards him, as if to put a stop to his idle delay, lays herself down, and receives his dilatory caresses. If the cock meet a young hen, he alters his mode of procedure. He struts in a different manner, less pompously and more energetically, moves with rapidity, sometimes rises from the ground, taking a short flight around the hen, as is the manner of some Pigeons, the Red-breasted Thrush, and many other birds, and on alighting, runs with all his might, at the same time rubbing his tail and wings along the ground, for the space of perhaps ten yards. He then draws near the timorous female, allays her fears by purring, and when she at length assents, caresses her.

Nesting.—Audubon says on this subject, referring to the Southern States:

About the middle of April, when the season is dry, the hens begin to look out for a place in which to deposit their eggs. This place requires to be as much as possible concealed from the eye of the Crow, as that bird often watches the Turkey when going to her nest, and, waiting in the neighbourhood until she has left it, removes and eats the eggs. The nest, which consists of a few withered leaves, is placed on the ground, in a hollow scooped out, by the side of a log, or in the fallen top of a dry leafy tree, under a thicket of sumach or briars, or a few feet within the edge of a canebrake, but always in a dry place. The eggs, which are of a dull cream colour, sprinkled with red dots, sometimes amount to twenty, although the more usual number is from ten to fifteen. When depositing her eggs, the female always approaches the nest with extreme caution, scarcely ever taking the same course twice; and when about to leave them, covers them carefully with leaves, so that it is very difficult for a person who may have seen the bird to discover the nest. Indeed, few Turkeys' nests are found.

When an enemy passes within sight of a female, while laying or sitting she never moves, unless she knows that she has been discovered, but crouches lower until he has passed. I have frequently approached within five or six paces of a nest, of which I was previously aware, on assuming an air of carelessness, and whistling or talking to myself, the female remaining undisturbed; whereas if I went cautiously towards it, she would never suffer me to approach within twenty paces, but would run off, with her tail spread on one side, to a distance of twenty or thirty yards, when assuming a stately gait, she would walk about deliberately, uttering every now and then a cluck. They seldom abandon their nest, when it has been discovered by men; but, I believe, never go near it again when a snake or other animal has sucked any of the eggs. If the eggs have been destroyed or carried off, the female soon yelps again for a male; but, in general, she rears only a single brood each season. Several hens sometimes associate together, I believe for their mutual safety, deposit their eggs in the same nest, and rear their broods together. I once found three sitting on forty-two eggs. In such cases, the common nest is always watched by one of the females, so that no Crow, Raven, or perhaps even Pole-cat, dares approach it.

Bendire (1892) refers to nests found in Nebraska and Texas in more open situations. One is described as "a simple affair, on a grassy hillside, in an exposed position, and lined with dead grass."

George M. Sutton (1929) describes a nest in Pennsylvania, as follows:

On June 6, on a rocky mountainside about twelve miles from Lock Haven, Clinton County, I examined a nest which held seventeen well-incubated eggs. On the day before there had been eighteen eggs in it; it is thought that a skunk or fox had disturbed the nest, though the female bird evidently had been sitting closely most of the time. This nest was built among small, angular rocks, and, while not very well hidden from above, it was screened on all sides by thick laurel, which made photography difficult. The female bird was either very unsuspicious or remarkably brave, for she did not leave her nest while we were near. Her broad back, with its squamate pattern and dull greenish lights, was difficult to discern among the foliage and the intricate interlacing of shadows. When first seen her neck was stretched out at full length in front of her, and her plumage was spread and flattened out noticeably. When she realized she was being observed she drew her head back and moved it slowly about in a snakelike manner, while she gave forth strange hissing and grunting sounds. When she had become accustomed to us she again stretched her neck out in front of her. Occasionally, when disturbed, she gave a characteristic *quit, quit.*

Eggs.—The normal set for the wild turkey numbers from 8 to 15 eggs. The smaller sets are laid by young birds. As many as 18 or 20 eggs have been found in a nest, which were probably laid by one bird. Occasionally two, or even three, birds lay in the same nest, taking turns at incubating or guarding the nest; in such cases the nest may contain many more eggs.

The eggs are usually ovate in shape, but sometimes they are short ovate, or elongate ovate and quite pointed. The shell is smooth, with little or no gloss. The ground colors vary from "pale ochraceous-buff" or "pale pinkish buff" to "cartridge buff" or buffy white. They are more or less evenly marked with small spots and fine dots of "light vinaceous-drab," "pale purple-drab," "clay color," or "pinkish buff." The measurements of 56 eggs average 62.6 by 44.6 millimeters; the eggs showing the four extremes measure **68.5** by 46, 64.5 by **48.5**, **59** by 45, and 64.7 by **42.4** millimeters.

Young.—The period of incubation is 28 days and this duty is performed by the female alone in seclusion. The male does not even know the location of the nest. The following is from Audubon's (1840) matchless account:

The mother will not leave her eggs, when near hatching, under any circumstances, while life remains. She will even allow an enclosure to be made around her, and thus suffer imprisonment, rather than abandon them. I once witnessed the hatching of a brood of Turkeys, which I watched for the purpose of securing them together with the parent. I concealed myself on the ground within a very few feet, and saw her raise herself half the length of her legs, look anxiously upon the eggs, cluck with a sound peculiar to the mother on such occasions, carefully remove each half-empty shell, and with her bill caress and dry the young birds, that already stood tottering and attempting to make their way out of the nest. Yes; I have seen this, and have left mother and young to better care than mine could have proved, to the care of their Creator and mine. I have seen them all emerge from the shell, and, in a few moments after, tumble, roll, and push each other forward with astonishing and inscrutable instinct.

Before leaving the nest with her young brood, the mother shakes herself in a violent manner, picks and adjusts the feathers about her belly, and assumes quite a different aspect. She alternately inclines her eyes obliquely upwards and sideways, stretching out her neck, to discover hawks or other enemies, spreads her wings a little as she walks, and softly clucks to keep her innocent offspring close to her. They move slowly along and, as the hatching generally takes place in the afternoon, they frequently return to the nest to spend the first night there. After this they remove to some distance, keeping on the highest undulated grounds, the mother dreading rainy weather, which is extremely dangerous to the young in this tender state, when they are only covered by a kind of soft hairy down of surprising delicacy. In very rainy seasons, Turkeys are scarce, for if once completely wetted the young seldom recover. To prevent the disastrous effects of rainy weather the mother, like a skilful physician, plucks the buds of the spice-wood bush and gives them to her young.

In about a fortnight the young birds, which had previously rested on the ground, leave it and fly at night to some very large low branch, where they place themselves under the deeply curved wings of their kind and careful parent, dividing themselves for that purpose into two nearly equal parties. After this they leave the woods during the day and approach the natural glades or prairies in search of strawberries and subsequently of dewberries, blackberries, and grasshoppers, thus obtaining abundant food and enjoying the beneficial influence of the sun's rays. They roll themselves in deserted ants' nests to clear their growing feathers of the loose scales and prevent ticks and other vermin from attacking them, these insects being unable to bear the odor of the earth in which ants have been. The young Turkeys now advance rapidly in growth and in the month of August are able to secure themselves from unexpected attacks of Wolves, Foxes, Lynxes, and even Cougars by rising quickly from the ground by the help of their powerful legs, and reaching with ease the highest branches of the tallest trees. The young cocks show the tuft on the breast about this time and begin to gobble and strut, while the young hens pur and leap in the manner which I have already described.

C. J. Pennock writes to me that in northern Florida, where the turkeys are somewhat intermediate but rather nearer the northern form, a cold, wet spell late in April or early in May produces considerable mortality among the young and that after such an unfavorable season turkeys are much scarcer for one or more years. The weather also has much to do with the time of laying. He has seen young able to fly as early as May 26 and a brood of very young as late as July 9. At times he has seen two hens together with their combined broods of 20 or more young. The young are able to fly up into the trees when about one-third grown. The broods of young remain with their mothers all through the winter and until the spring mating time comes.

Plumages.—In the wild-turkey chick the crown is "pinkish cinnamon" and the back a somewhat lighter shade of the same, fading off to still lighter shades on the breast and flanks; the crown and upper parts are heavily spotted or blotched with dark, rich browns, "bister" to "Vandyke brown"; the sides of the head and underparts are "pale pinkish buff" to "ivory yellow," nearly white on the chin and throat and almost "straw yellow" on the belly.

As with the quail and grouse, the young turkey starts to grow its wings when a small chick; these are soon followed by the plumage of the back, breast, and flanks; the tail comes later, followed finally by the head and belly. The juvenal feathers of the back are "walnut brown," edged with "russet," with a broad median "russet" stripe, a whitish tip, and large black areas near the tip; the wing coverts are similar, but in duller colors and with less black; the scapulars are "sayal brown," peppered with black and spotted or barred with black along the outer edge and at the tip; the tertials and secondaries are "hair brown," marked like the scapulars on the outer edge; the primaries are "hair brown," mottled and peppered with buffy white; the underparts are "fuscous," with whitish tips and shaft streaks; the tail is barred with dusky and "pinkish cinnamon."

Before the young bird is fully grown, in September, a postjuvenal molt takes place; this is a complete molt, except that the two outer primaries on each wing are retained for a year. In this first winter plumage the sexes begin to differentiate, the males becoming much larger than the females, but the plumages of the two sexes are very much alike, and they resemble the adult female. Wilson (1832) says:

On the approach of the first winter the young males show a rudiment of the beard or fascicle of hairs on the breast, consisting of a mere tubercle, and attempt to strut and gobble; the second year the hairy tuft is about three inches long; in the third the turkey attains its full stature, although it certainly increases in size and beauty for several years longer.

Audubon's (1840) statement is similar.

Wilson (1832) says of the female:

Females four years old have their full size and colouring; they then possess the pectoral fascicle, four or five inches long (which, according to Mr. Audubon, they exhibit a little in the second year, if not barren), but this fascicle is much thinner than that of the male. The barren hens do not obtain this distinction until a very advanced age; and, being preferable for the table, the hunters single them from the flock and kill them in preference to the others. The female wild turkey is more frequently furnished with the hairy tuft than the tame one, and this appendage is gained earlier in life. The great number of young hens without it has no doubt given rise to the incorrect assertion of a few writers that the female is always destitute of it.

Adults apparently have only one complete postnuptial molt in August and September. A fully grown gobbler seldom weighs more than 20 or 25 pounds, even when in good condition; there are some, apparently authentic, records of birds weighing between 30 and 40 pounds, but such cases must be very rare; reported records of 50 pounds are unreliable.

Food.—Dr. Sylvester D. Judd (1905a) found that the stomachs and crops of 16 wild turkeys examined by the Biological Survey contained 15.57 percent of animal matter and 84.43 percent of vegetable matter. The animal food consisted of insects—15.15 percent—and miscellaneous invertebrates, such as spiders, snails, and myriapods—0.42 percent. Grasshoppers furnished 13.92 percent, and beetles, flies, caterpillars, and other insects 1.23 percent. The 84.43 percent of the bird's vegetable food was distributed as follows: "Browse," 24.80 percent; fruit, 32.98 percent; mast, 4.60 percent; other seeds, 20.12 percent; miscellaneous vegetable matter, 1.93 percent.

Judd says that they are very fond of grasshoppers and crickets, and that

during the Nebraska invasion of Rocky Mountain locusts, Professor Aughey examined the contents of six wild turkey stomachs and crops collected during August and September. Every bird had eaten locusts, in all amounting to 259. The wild turkey has been known also to feed on the cotton worm (*Alabama argillacea*), the leaf hoppers, and the leaf-eating beetles (*Chrysomela suturalis*). The grasshopper (*Arnilia* sp.) and the thousand-legs (*Julus*) form part of the turkey's bill of fare. Tadpoles and small lizards also are included.

Of a bird shot in Virginia, he says:

Ten percent of its food was animal matter and 90 percent vegetable. The animal part consisted of 1 harvest spider (Phalangidae), 1 centipede, 1 thousand-legs (*Julus*), 1 ichneumon fly (*Ichneumon unifasiculata*), 2 yellow-jackets (*Vespa germanica*), 1 grasshopper, and 3 katydids (*Cyrtophyllus perspiculatus*). The vegetable food was wild black cherries, grapes, berries of flowering dogwood and sour gum, 2 chestnuts, 25 whole acorns (*Quercus palustris* and *Q. velutina*), a few alder catkins, seeds of jewel weed, and 500 seeds of tick-trefoil (*Meibomia nudiflora*). Another turkey, also shot in December, had eaten a ground beetle, an ichneumon fly, 2 wheel bugs, 10 yellow-jackets, a meadow grasshopper, 75 red-legged grasshoppers, a few sour-gum berries, some pine seeds (with a few pine needles, probably taken accidentally), several acorns, a quarter of a cupful of wheat, and a little corn.

Various other kinds of berries, fruits, and insects are doubtless eaten when available, as turkeys will eat almost anything they can find in these lines.

Behavior.—The turkey's ordinary method of locomotion is walking or running; the long powerful legs enable these birds to travel long distances and very rapidly on foot. But they are also strong fliers when hard pressed or when necessity requires it, and can fly for a considerable distance or even across wide rivers. What few turkeys I have seen in flight looked to me like huge ruffed grouse, with long tails spread and heavy wings beating rapidly, though the speed of these large heavy birds is proportionately much less. Audubon (1840) says:

Their usual mode of progression is what is termed walking, during which they frequently open each wing partially and successively, replacing them again by folding them over each other, as if their weight were too great. Then, as if to amuse themselves, they will run a few steps, open both wings and fan their sides, in the manner of the common fowl, and often take two or three leaps in the air and shake themselves. During melting snowfalls, they will travel to an extraordinary distance and are then followed in vain, it being impossible for hunters of any description to keep up with them. They have then a dangling and straggling way of running, which, awkward as it may seem, enables them to outstrip any other animal. I have often, when on a good horse, been obliged to abandon the attempt to put them up, after following them for several hours.

While traveling about during fall and winter the sexes gather into separate flocks, the females forming the largest flocks; young males also flock by themselves and, for the most part, keep away from the old gobblers. When flocks of old and young males happen to meet they do not ordinarily quarrel; but they seem to have different interests.

What few turkeys still survive, in regions where they are much hunted, have developed a high degree of shrewdness and cunning. An instance of cunning is given by Dr. J. M. Wheaton (1882) as follows:

As if aware that their safety depended on their preserving an incognito when observed, they effect the unconcern of their tame relatives so long as a threatened danger is passive or unavoidable. I have known them to remain quietly perched upon a fence while a team passed by; and one occasion knew a couple of hunters to be so confused by the actions of a flock of five, which deliberately walked in front of them, mounted a fence, and disappeared leisurely over a low hill before they were able to decide them to be wild. No sooner were they out of sight, than they took to their legs and then to their wings, soon placing a wide valley between them and their now amazed and mortified pursuers.

Wild turkeys have a preference for roosting over water, and they will often go a long way in order to obtain such a roost. The backwater from the overflowing streams when it spreads out widely through the standing timber of the river bottoms, affords them great comfort; also the cypress ponds to be found in our southern river districts. They evidently fancy that there is greater safety in such places.

Voice.—The wild turkey has quite a vocabulary, according to E. A. McIlhenny (1914), a language with various meanings. If the strutting gobbler thinks he has heard the cluck and yelp of a calling hen, *cluck*, *cluck*, *keow*, *keow*, *keow*, he drops his broad wings, partly spreads his tail, and listens; then *vut-v-r-r-o-o-o-m-m-i* comes the booming strut, and *gil-obble-obble-obble*. Then let the hen give her low quavering yelp, *keow-keow*, *keow*, and he will yell out in a fierce and prolonged rattle. More calls from the hen, *keow*, *keow*, *kee*, *kee*, or *cluck*, *keow*, *ku-ku*, one interspersed with loud gobblings, until the siren call of the hen, *cut-o-r-r-r*, *cut*, *cut*, *keow*, *keow*, *keow*, indicates that she has gone to him and all is quiet. Should any threatening danger intrude on this pretty love scene, a warning note is given, *cluck*, *put*, *put*, or *put*, *o-r-r-r-r*, or perhaps the turkeys walk quietly away saying, *quit*, *quit*, in irritated alarm.

Enemies.—Although the eggs and young are preyed upon by many predatory animals and birds, it is only the larger species that are strong enough to attack an adult turkey. Audubon (1840) writes:

Of the numerous enemies of the Wild Turkey, the most formidable, excepting man, are the Lynx, the Snowy Owl, and the Virginian Owl. The Lynx sucks their eggs and is extremely expert at seizing both young and old, which he effects in the following manner. When he has discovered a flock of turkeys he follows them at a distance for some time, until he ascertains the direction in which they are proceeding. He then makes a rapid circular movement, gets in advance of the flock, and lays himself down in ambush until the birds come up, when he springs upon one of them by a single bound and secures it. While once sitting in the woods on the banks of the Wabash, I observed two large Turkey-cocks on a log by the river, pluming and picking themselves. I watched their movements for awhile, when of a sudden one of them flew across the river, while I perceived the other struggling under the grasp of a Lynx.

Game.—It is probably safe to say that the wild turkey is the largest and grandest game bird in the world, certainly in North America. It is not so well known and not so popular as the quail or ruffed grouse, because comparatively few sportsmen have had an opportunity to hunt it, on account of its growing scarcity and the remoteness of its haunts. What few turkeys remain within easy reach of civilization have become so highly educated that it requires considerable experience and skill to outwit them. Their eyes can not easily recognize a stationary object, but they are very quick to detect the slightest movement. Their sense of hearing is very acute, and they are always on the alert for approaching enemies, especially human beings. As a food bird the turkey is unsurpassed both in quantity and quality.

The methods employed in hunting turkeys are, or have been, many and varied. An interesting method of capturing turkeys, in the days when they were plentiful and unsuspicious, was thus described by John Hunter in 1824, as quoted by Albert H. Wright (1914) in his excellent history of this bird:

> The turkey is not valued, though when fat, the Indians frequently take them alive in the following manner. Having prepared from the skin an apt resemblance of the living bird, they follow the turkey trails or haunts till they discover a flock, when they secrete themselves behind a log in such a manner as to elude discovery, partially displaying their decoy, and imitate the gobbling noise of the cock. This management generally succeeds to draw off first one and then another from their companions which, from their social and unsuspecting habits, thus successively place themselves literally in the hands of the hunters, who quickly despatch them and await for the arrival of more. This species of hunting, with fishing, is more practiced by the boys than the older Indians, who seldom, in fact, undertake them unless closely pressed by hunger.

A common method of capture, referred to by many writers, was to trap them in an inclosure, or pen, made of logs. The top was covered with logs, leaving narrow open spaces between them. A trench was dug, sloping gradually down, under the log wall and up into the pen. Corn or other grain was sprinkled along this trench and plenty of it spread on the inside of the pen to tempt the turkeys to enter. When, after eating all they wanted, they attempted to escape, they constantly looked upward for an opening but seldom, if ever, had sense enough to crawl out the way they had come in. Large numbers were caught in this way.

Audubon (1840) says that as many as 18 turkeys have been caught in a pen at one time, and as many as 76 within a period of two months.

One of the most popular methods, which is still widely practiced, is calling the gobbler by imitating the call of the hen during the mating season. This requires the utmost skill, experience, practice, and thorough knowledge of the habits and haunts of the birds. Much has been written in various books and numerous articles in sporting magazines on how to succeed in this. The instruments used in calling may be simply the leaf of a tree held between the lips, the box or trough call, the splinter and slate, or a new clay pipe; but the commonest and most effective call is made from the wing bone of a hen turkey. The hunter must know how to use these perfectly, for a false note will drive the turkey away, perhaps never to return. He must also be able to keep perfectly still for a long time, with his gun, or rifle, trained on the spot where he expects the turkey to appear, for the slightest visible movement would spoil his chance. He would better be well concealed, but success may be had, even if he is in plain sight, if seated against a stump or tree large enough to conceal the outline of his body. As to the use of the calls, he had better study the various seductive notes of the hen, the turkey language, or, better still, learn them from an experienced hunter.

Tracking turkeys in the snow on a clear cold winter day is splendid sport. It has been well described by Edwyn Sandys (1904). In following a flock of turkeys a single track may turn off to one side; this means a tired bird, which will soon crouch to rest. If he carries a shotgun, the hunter should follow this bird, for he will soon flush it and get a flying shot. But, if carrying a rifle, he should follow the main flock; sooner or later he will get a long shot at some of them, though it may be a long chase unless the snow is soft and deep. Should the birds take wing they will fly in a straight line, indicated by the direction of the long steps taken in rising, and the trail can be taken up again.

Coursing turkeys with greyhounds, as practiced in the more open western country is exciting sport. It is also vividly described by Sandys (1904). The hunter on horseback, accompanied by a good greyhound, finds his turkeys feeding out on an open plain and tries to flush one headed for the open. The turkey's first flight is his longest, hotly pursued by dog, horse, and man. If the bird comes down and tries to run, he is soon overtaken. His flights and runs gradually grow shorter and shorter, until he becomes exhausted and is caught.

Well-trained turkey dogs are useful in chasing winged birds, which a man could never catch. Audubon (1840) says:

Good dogs scent the turkeys when in large flocks at a great distance; I may venture to say half a mile away, if the wind is right. Should the dog be well trained to the sport, he will set off at full speed on getting the scent and in silence until he sees the birds, when he instantly barks, and, running among them, forces the whole flock to take to the trees in different directions. This is of great advantage to the hunter, for, should all the turkeys go one way, they would soon leave the perches and run again; but when they are separated by the dog, a person accustomed to the sport finds the birds easily and shoots them at pleasure.

Fall.—Turkeys are not migratory, in the strict sense of the word, but they are much given to extensive wanderings, mainly in the fall and winter, in search of food, which varies in abundance from one season to another. Audubon (1840) writes:

About the beginning of October, when scarcely any of the seeds and fruits have yet fallen from the trees, these birds assemble in flocks, and gradually move towards the rich bottom lands of the Ohio and Mississippi. The males, or, as they are more commonly called, the gobblers, associate in parties from 10 to 100, and search for food apart from the females; while the latter are seen either advancing singly, each with its brood of young, then about two-thirds grown, or in connexion with other families, forming parties often amounting to 70 or 80 individuals, all intent on shunning the old cocks, which, even when the young birds have attained this size, will fight with and often destroy them by repeated blows on the head. Old and young, however, all move in the same course, and on foot, unless their progress be interrupted by a river, or the hunter's dog force them to take wing. When they come upon a river, they betake themselves to the highest eminences, and there often remain a whole day, or sometimes two, as if for the purpose of consultation. During this time the males are heard gobbling, calling, and making much ado, and are seen strutting about, as if to raise their courage to a pitch befitting the emergency. Even the females and young assume something of the same pompous demeanour, spread out their tails, and run around each other, purring loudly, and performing extravagant leaps. At length, when the weather appears settled, and all around is quiet, the whole party mounts to the tops of the highest trees, whence, at a signal, consisting of a single cluck given by a leader, the flock takes flight for the opposite shore. The old and fat birds easily get over, even should the river be a mile in breadth; but the young and less robust frequently fall into the water, not to be drowned, however, as might be imagined. They bring their wings close to their body, spread out their tail as a support, stretch forward their neck, and, striking out their legs with great vigour, proceed rapidly towards the shore; on approaching which, should they find it too steep for landing, they cease their exertions for a few moments, float down the stream until they come to an accessible part, and by a violent effort generally extricate themselves from the water. It is remarkable, that immediately after thus crossing a large stream, they ramble about for some time, as if bewildered. In this state, they fall an easy prey to the hunter.

Winter.—During winter, when the snow is too deep or soft to travel on the ground, turkeys often remain in the trees for long periods, subsisting on buds and what fruits, nuts, and berries they can find above the snow. They are great travelers, however, in light or on hard snow. When hard pressed for food they sometimes venture into farmyards or grain fields, or along roadsides or railroad tracks where grain has been spilled. At such times they can be easily baited by scattering corn in such places.

MELEAGRIS GALLOPAVO OSCEOLA Scott

FLORIDA TURKEY

HABITS

The Florida wild turkey, which is resident in the southern half of Florida, was described by W. E. D. Scott (1890) and named for Osceola, a famous chief of the Seminole Indians. Scott says that it is similar to the northern wild turkey,

but perceptibly darker in general tone. *Coloring of tail* and *upper tail-coverts similar* in *both forms*. The white on the primary and outer secondary quills restricted, and the dark color (brownish black) predominating, the white being present only as detached, narrow, broken bars *not reaching* the *shaft* of the feather. The inner secondaries of a generally dirty grayish brown *without* apparent bars, but with brownish vermiculations on the inner web.

Referring later to the Caloosahatchie region, he (1892) writes:

This is still a very abundant bird in this part of Florida, though said to be diminishing in numbers every year and to be not nearly so plentiful as it was ten or fifteen years ago. During my stay at Fort Myers from November till March, the open season, the birds were constantly offered for sale in the markets, the price being on the average ten cents a pound for dressed birds. A hen turkey could generally be bought for from seventy-five cents to one dollar and a gobbler for from one dollar to a dollar and a half. Only a few years back the regular price paid to the hunters was twenty-five cents each. This I was told by many reliable people who had lived there a dozen years or more.

It would seem that these birds, living as they do at this point in cypress swamps and "bay heads," have a natural protection that will not allow of their absolute extermination, but, unless the exceedingly good laws passed by the last legislature of the State are carefully enforced, the Wild Turkey, still very abundant in this region, is doomed to become in a few years as rare as it has already become in the northern part of Florida.

Dr. William L. Ralph, in a letter to Major Bendire (1892), states:

Fifteen years ago I found the Wild Turkey abundant in most parts of Florida, north of Lake Okeechobee, with perhaps the exception of the Indian River region, but they have gradually decreased in numbers since then, and though still common in places where the country is wild and unsettled, they are rapidly disappearing from those parts, in the vicinity of villages and navigable waters.

One can hardly believe that the Wild Turkeys of to-day are of the same species as those of fifteen or twenty years ago. Then they were rather stupid birds, which it did not require much skill to shoot, but now I do not know of a game bird or mammal more alert or more difficult to approach. Formerly, I have often, as they were sitting in trees on the banks of some stream, passed very near them, both in rowboats and in steamers, without causing them to fly, and I once, with a party of friends, ran a small steamer within 20 yards of a flock, which did not take wing until several shots had been fired at them.

Turkeys are still fairly common in the more remote regions of Florida, or where no hunting is allowed, especially around the edges of the larger cypress swamps, such as the "big cypress" in Collier

County. There, in 1930, I saw a small band of them on the outskirts of a protected citrus plantation; and one day I saw one cross the Tamiami Trail from one tract of pine woods to another; shooting is prohibited for a mile on each side of this road. We often saw their tracks around the borders of the pine woods and open savannas, near the cypress swamps; they feed in such places and roost at night in the large cypresses.

Courtship.—Doctor Ralph says further:

These birds are polygamous, and the female takes all the cares and duties of incubation upon herself. The gobblers are very pugnacious, and will often fight fiercely for the favors of the hens. The love season begins in Florida about the middle of February and lasts for about three months, and during this period the gobblers frequently utter their call and are then easily decoyed within gunshot. Native hunters have informed me that the hens roost by themselves at this season of the year.

Nesting.—On this subject Ralph writes:

The nest is a slight depression in the ground, either at the foot of a tree or under a thick bush or saw palmetto. It is lined sparingly with dead leaves and grass, etc., but I could never find out whether this material was placed there by the birds or was there originally. I think these birds raise but one brood a season, though I have found fresh eggs as early as the middle of March and as late as the 1st of May. I have never found more than thirteen eggs in one nest, nor less than eight, unless they were fresh, the usual number being ten. The chicks of this species are very tender, and as they follow their mothers as soon as hatched I have often wondered how the latter could raise so many as they do. The natives of Florida say that a hen Turkey will desert her nest if the eggs are handled. Whether this be true or not I do not know, for I never tried to find out but once, and then, though the bird was gone on my second visit to the nest, I always had a strong suspicion that she was shot, for its whereabouts was known to several persons besides myself.

I have a set of 10 eggs in my collection that was taken on March 28, 1908, near Everglade; the nest was a hollow in the ground under a saw palmetto, near the Big Cypress. We found a nest, from which the young had hatched, on April 19, 1902, on the border of Jane Green Swamps, Brevard County. It was a mere hollow in the sand, lined with strips of palmetto leaves, under a small cabbage palmetto; it was well shaded but not particularly well hidden, and contained the broken shells of nine eggs.

Eggs.—The eggs are similar to those of other wild turkeys. The measurements of 56 eggs average 61 by 46.3 millimeters; the eggs showing the four extremes measure **66** by 46.7, 62.5 by **48.8**, **56.3** by 46.4, and 65.2 by **41** millimeters.

Young.—F. M. Phelps (1914) relates the following experience:

Late on the afternoon of April 18th, as we were working along an open glade bordering a cypress swamp, the dog began to nose excitedly in the grass. Suddenly up popped half a dozen little brown cannon-balls, quail I thought, but when they alighted in some cypress saplings I saw at once they were young

Turkeys. The old hen, hard pressed, soon rose from the grass and sailed away across the tops of the cypress trees. More youngsters kept popping up until there were eleven sitting about in the saplings some twelve or fifteen feet up. Soon one gave a peculiar little *quit*, and then to my utter astonishment flew straight away over the tops of the cypress trees after the old hen, and one by one the rest followed. My guide pronounced them to be about two weeks old and that seemed to me about correct.

The ability of very small young to fly is also attested by Donald J. Nicholson (1928), who writes:

On May 3, my brother Wray and I were going thru the pine and cypress country just east of Turner's River about one mile, and just about noon we came upon a turkey hen with five or six little ones not quite as large as a full-grown Bobwhite. We stood and watched her for a few seconds, and she ran slowly thru the scattered low palmettos with the young scampering along. We conceived the idea that it might be possible to catch one of the youngsters, and began to give chase, but immediately they all rose and flew with strong flight, alighting in the lower limbs of a thirty-foot pine tree. We managed to find two of the young perched on dead branches not far apart, peering down upon us; they did not offer to fly or show any restlessness.

Behavior.—Again Nicholson says:

One day we were driving along the Tamiami Trail not far from where we encountered the young turkeys, and saw five large gobblers feeding right out in the open, 400 yards from the road. It was a sort of prairie or savanna, among the stands of cypress, and had been burned over; short grass had grown up. The birds paid not the slightest heed to us and we sat and watched them for ten or twenty minutes. However, one bird would stand with head and neck erect; as if on guard while the others fed; then another would take its place.

MELEAGRIS GALLOPAVO INTERMEDIA Sennett

RIO GRANDE TURKEY

HABITS

When George B. Sennett (1879) first called attention to the characters in which the Rio Grande turkey differs from the other races of wild turkeys, he evidently thought it was an intermediate and should not be named, for he said, at that time: "All Lower Rio Grande specimens, therefore, must be held as *gallopavo* (the Mexican form), or a var. *intermedia* established—an alternative not to be desired." Later on, however, he (1892) described and named it *ellioti*, in honor of Dr. Daniel G. Elliot. But his earlier name, *intermedia*, must stand under the law of priority. He says that it

* * * can be distinguished from the other forms by its dark buff edgings on tail and upper and lower tail-coverts, in contrast with the white color on the same parts of *mexicana*, and the deep, dark, reddish chestnut of the same parts in *M. gallopavo*, the eastern United States bird. The lower back is a deep blue-black and is wanting in those brilliant metallic tints so prevalent in the eastern bird and in the type of *mexicana*. The primaries of the wing are

black with white bars in contrast with *M. gallopavo* the primaries of which are white with black bars.

The range or habitat of this race, so far as known at the present time, is restricted to the lowlands of eastern Mexico and southern Texas. It will probably not be found south of Vera Cruz, nor is it likely to be met with to the north beyond the Brazos River of Texas, its range being thus restricted within about ten degrees of latitude. Wherever timber and food are in abundance we find this new form common to the coast and lowlands, and we could not expect to find it at an altitude exceeding 2,000 feet above sea-level; while the variety *mexicana* is found only at the higher altitudes from 3,000 to 10,000 feet above the sea.

The Rio Grande turkey is now known to have been quite widely distributed in Texas from the central-northern part southward and westward into northern Mexico, though it has become much scarcer except in the wilder portions of the State. Writing of this turkey in Kerr County, Tex., Howard Lacey (1911) says:

Formerly very common, but getting rather scarce now that the shotgun is becoming almost as common a piece of furniture as the rifle in the ranchman's house. These birds are so foolishly tame when about half grown as they are wild and able to take care of themselves when fully mature; if they were not shot at until fully grown and allowed to roost in peace at night, there is no reason why we should not have them always with us. Armadillos and skunks sometimes roll the eggs out of the nests, and they have plenty of enemies besides the boy with the shotgun.

Austin P. Smith (1916), referring to the same general region, writes:

There can be little doubt that, at the present time, Wild Turkeys exist in greater numbers in Kerr and adjoining counties than in any other part of Texas. Their abundance may be accounted for, as the result of the encroachment of the Cedar and various species of scrubby oaks upon lands formerly under cultivation or in pasture; to the decrease in numbers of the Armadillo (*Tatu novemcinctum texanum*) which of late years have been much hunted for commercial purposes; and to the enactment of a law limiting the open season and the number that may be killed. During the winter spent in the region several heavy snowfalls occurred. These caused many turkeys to seek open spots in the valleys and along fence rows, often in the vicinity of human habitations, and I recall one flock of seven hunting for several hours within a hundred feet of the building I lived in.

George F. Simmons (1925) says its haunts in the Austin region are " wild, rough, brushy, country; dry, big-timbered arroyos running back from watered creeks; hill and valley country; shin oak clumps on hillsides; creek bottoms and lower slopes; wild, less-frequented, thinly settled country, particularly in the mountains and notches in the hills."

Nesting.—Simmons describes the nest as a " slight hollow, scraped out by the bird, lined with grasses and leaves, among low bushes, in dense woods along streams, in tangles of briar vines, and in thick weedy places. Very difficult to find, particularly when placed in growths of underbrush."

Eggs.—The eggs are similar to those of the other wild turkeys. The measurements of 49 eggs average 62.4 by 46.5 millimeters; the eggs showing the four extremes measure **64.8** by 43.2, 64 by **48.6, 57.2** by 43.6, and 61.7 by **43.2** millimeters.

Food.—Simmons (1925) says of its feeding habits:

When pecans are ripe, the birds feed under the pecan trees along the valleys. At other seasons, they wander about wooded slopes in the daytime, feeding among the cedars and scrub oaks which cover the hillsides and ridges of the land that was once prairie; at night, they return to the valleys to roost. Feed on nuts, acorns, seeds, grain, berries, plant tops, insects, crickets, and grasshoppers.

Sennett (1879) says that in April "their principal food was the wild tomato, which attains about the size of a cranberry, and which they devoured whole, together with insects and larvae."

Behavior.—Simmons (1925) writes:

Observed singly or in pairs during the breeding season, at other times in flocks of from 12 to 15 or more; flocks are usually practically all of one sex or the other. Very wary; when in danger, it usually sneaks away or runs through the 'underbrush and into thickets, preferring to trust to its stout legs rather than taking wing. Usually roost each night in the same locality, birds returning singly and by twos and threes at dusk, until all the birds have assembled in their favorite places in the tops of larger, taller trees, generally over water and frequently in partially submerged trees, possibly for protection against prowling coyotes and bob-cats; big pecan trees along wooded creek valleys, in washes, and in bottomlands are generally selected. Birds make their way back to the higher ridges before daylight has half arrived. Males flock together during the period in which females are kept busy with eggs or care of young.

Vernon Bailey, in Mrs. Bailey's book (1902), gives the following account of the habits of the Rio Grande turkey:

Over most of the country where the wild turkeys were once plenty they have now become scarce or extinct, but in a few places may still be found in something like their original abundance, living much as their ancestors lived, breeding unmolested, strolling through the woods in flocks, and gathering at night in goodly numbers in their favorite roosting places. Perhaps the best of these undevastated regions are on the big stock ranches of southern Texas, where the birds are protected not by loosely formed and unenforced game laws, but by the care of owners of large ranches, who would as soon think of exterminating their herds of cattle as of shooting more than the normal increase of game under their control. Here, at least through the breeding season, the turkeys are not more wary than many of the other large birds, and as we surprised them in the half open mesquite woods along the Nueces River, would rarely fly, merely sneaking into the thickets, or at most running from us. The ranchmen say that the turkeys always select trees over water to roost in when possible, and no doubt they do it for protection in this region where foxes, coyotes, and wildcats abound. On the edge of the flooded bottoms of the Nueces River they roosted in the partially submerged huisache trees.

A loud gobble just at dusk led us to their cover, and crouching low to get the sky for a background we could see the big forms coming in singly or in twos or threes, and hear the strong wing beats as they passed on to alight in the huisaches out in the water. When the noise of their wings and the rattling of branches had subsided, with a few gobbles from different quarters they settled down for the night. The next morning, as the darkness began to thin and a light streak appeared in the east, a long loud gobble broke the stillness, followed by gobble after gobble from awakening birds in different parts of the bottoms, and before it was half daylight the heavy *whish whish* of big wings passed overhead, as the turkeys with strong, rapid flight took their way back to the higher ridges.

Family CRACIDAE, Curassows, Guans, Chachalacas

ORTALIS VETULA VETULA (Wagler)

CHACHALACA

HABITS

This curious and exceedingly interesting bird, the chachalaca, brings a touch of Central American bird life into extreme southern Texas in the lower valley of the Rio Grande, where so many other Mexican species reach the northern limits of their ranges and where the fauna and flora are more nearly Mexican than North American.

On May 27, 1923, I spent a good long day, from before sunrise until after sunset, in the haunts of the chachalaca, with R. D. Camp, George F. Simmons, and E. W. Farmer, the last named a chachalaca hunter of many years' experience, who knows more about this curious bird than any man I have ever met. The locality to which he guided us was the famous Resaca de la Palma, where so many other observers have made the acquaintance of the chachalaca, only a few miles outside of the city of Brownsville, Tex. This and other resacas in the vicinity are the remains of old river beds of the Rio Grande, which from time to time in the past has overflowed its banks or changed its course, cutting these winding channels through the wild, open country, chaparral, or forest. Some of these channels were dry or nearly so, but most of them contained more or less water below their gently sloping grassy borders. Above the banks were dense forests of large trees, huisache, ebony, hackberry, and mesquite, with a thick undergrowth of thorny shrubbery, tangles of vines, and an occasional palmetto or palm tree. In other places almost impenetrable thickets of chaparral lined the banks, with its forbidding tangle of thorny shrubs of various kinds, numerous cactuses and yuccas. These forests and thickets were teeming with bird life. Along the edges of the watercourses the pretty little Texas kingfishers were seen flying over the water or perched on some dead snag. In some small trees overgrown with *Usnea* moss the dainty little Sennett's warblers were flitting about, reminding me of our northern parulas. Handsome green jays were sneaking about in the larger trees, surprisingly inconspicuous in spite of their gaudy colors. Brilliant Derby flycatchers proclaimed their noisy presence in loud, clamorous notes from the tree tops. Sennett's thrashers scolded us in the thickets, and the confiding little Texas sparrows hopped about on the ground near us, scratching among the dead leaves. Many other birds were seen, but the most conspicuous of all were the doves; the woods and the thickets almost constantly resounded with the deep-toned notes of the white-fronted, the tiresome *who cooks for you* of the white-winged, and the soft cooing of the mourning dove. Such is the home of the chachalaca with some of its neighbors.

As we entered the chaparral before sunrise we heard the warning cry of the chachalaca on all sides; the woods fairly resounded with its cries, some of which sounded like a watchman's rattle, more wooden than metallic in quality; the birds were very shy and seldom seen; occasionally we saw one, perched on some small tree top and giving its challenge or battle cry; but as soon as it realized that it was observed, it would sail down into the thicket and keep still.

Much of the following information is taken from some very full notes obtained from Mr. Farmer. He promised to send me some notes, but unfortunately he has now gone to the "happy hunting grounds." My friend Frederic H. Kennard was more successful and has very kindly placed these notes at my disposal. According to Mr. Farmer's personal knowledge, the chachalaca occurs in Cameron, Hidalgo, Starr, Zapata, Willacy, and Kenedy Counties in Texas; the birds are never permanently located more than a mile from water; if the water dries up all about, they move; otherwise they stick to one place throughout the year.

Courtship.—According to Mr. Farmer, courtship begins about March 20 in ordinary seasons, with the *chachalac* challenge calls of the male, perched in the tops of the highest trees in the chaparral; other males answer from every direction in competition, each trying to "outholler" the others. The females can make a similar call, but it is on a higher key and less in volume. The concert begins at about sunrise or a little before. The male's call to the female is like the challenge, but it is less harsh and ends with a soft note. The females may climb up into the tree beneath the male, but in a less conspicuous place, generally keeping under cover and answering the male in their own way chattering, talking, and scolding. After the male has "hollered himself out" in the tree, he comes down to the ground and devotes himself to the females, walking about and strutting with head erect and making a low call hardly to be heard a short distance away. If another male appears he is promptly chased off. In Mr.

Farmer's experience with the birds there have always seemed to be two females to one male in this courtship. The male seems strictly impartial. Mr. Farmer has had much experience in rearing young chachalacas, and says that, so far as he has observed, they always seem to hatch out in the ratio of two females to one male. He also has, many a time, watched the courtship performance of the male with two females, no other male being tolerated in the vicinity.

He says that the males fight a great deal at this season. They have no spurs, but fight with bills, feet, and wings, jumping over one another and pecking at one another's backs rather than at the heads, their wing strokes, however, being directed at the head. At this season of the year the males frequently appear with most of the feathers pecked off their backs.

Mr. Kennard writes to me as follows:

It is a matter of common report on both sides of the Rio Grande that the chachalacas are used for crossing with game chickens for fighting purposes, the resulting cross being much quicker on its feet than the ordinary game fowl. These reports have, however, never been actually verified by either Mr. Camp or Mr. Farmer and are to be doubted.

A letter from Mr. Camp confirming this states:

I do not agree with any of the statements concerning the crossing of the chachalaca with the domestic fowl. I have traveled hundreds of miles and investigated dozens of cases both on this and the other side of the Rio Grande, endeavoring to verify reported hybrids, and at no time have I found a specimen that I would acknowledge was a cross. Last year I investigated quite extensively among the natives in the district 125 miles from here, where the chachalacas are very abundant and tame for wild birds. Most all Mexican colonies in the district had semidomesticated chachalacas running with their barnyard fowl, but none of the natives would acknowledge that he had ever seen a hybrid.

Nesting.—The only two nests of the chachalaca that I ever saw were found near the Resaca de la Palma referred to. On May 27, 1923, they both held sets of three eggs each, heavily incubated. Mr. Simmons found the first one; he had been standing under the tree for some time, when he heard the bird fly off from the thick foliage over his head and found the nest 8½ feet above the ground. I found the other by seeing the bird fly off, and I had to climb up into the very tops of several slender trees to reach the nest, which was about 18 feet up and well concealed in the leafy tops. Both nests were very small, frail structures, made of sticks and leaves and lined with a few green leaves; they were barely large enough to hold the eggs.

According to Mr. Farmer, nest building is started soon after courtship has begun. He has seen two birds, probably the male and one female, at work on the nest, while the other female was sitting about near by, perhaps helping or perhaps only watching. The nest is a scraggly but strongly built structure of short, stout twigs, so well interlaced as to stand a lot of handling; a particularly well-preserved previous year's nest measured about 24 inches across and 10 inches deep. The nest is usually built in an ebony tree, but sometimes in a mesquite or other tree, and usually between 5 and 15 feet above the ground, occasionally as high as 25 feet or as low as 2 feet but never in a hollow tree. He says that he has seen at least 1,000 nests, and that the nest is usually near the edge of the chaparral and near a resaca, never more than 200 yards from water and always near a supply of the berries on which the birds feed their young. The nest is generally built out on a limb, but sometimes in a crotch or where the limb of a tree is interlaced with vines.

There are two sets of eggs in Col. John E. Thayer's collection, said to have been taken from nests on the ground; one was "in a cane brake, composed of grass, weeds, and other litter," and the other was "on the ground among heavy grass." These were taken for Thomas H. Jackson, probably by Frank B. Armstrong or his Mexican collectors, who were known to be careless in the make-up of sets and data. Messrs. Farmer and Camp do not mention any ground nests, and George B. Sennett (1878) says:

The nest of this species is never found on the ground, but in trees and bushes varying in height from four to ten feet. The structure varies in composition and size according to its location. If it is in a large fork close to the body of the tree, a few sticks, grasses, and leaves are sufficient, and the structure will not equal in size or strength that of a Mockingbird. This small size is by far the most frequent, but I have a nest built upon a fork of two small branches, composed entirely of Spanish moss. It is bulky and flat, being a foot in diameter and four inches deep, with a depression four inches wide and two deep.

Major Bendire (1892) quotes J. A. Singley as saying:

All the nests I found were in mesquite stubs, where the limbs had been cut off to make brush fences. These limbs are never cut close to the tree and, being close together, form a cavity; leaves and twigs will fall in this and accumulate, and the bird occupies it as a nesting site. I did not find a nest that I could say was built by the bird. When the nest is approached the bird quietly flies off, rarely remaining in sight, and soon calls up its mate.

Eggs.—Mr. Farmer says that in his experience the eggs have been invariably three in number, and most of the other observers say three or rarely two. The larger sets in Mr. Jackson's collection probably came from Armstrong and may be made-up sets, though perhaps sets of four occasionally occur. The eggs of the chachalaca are ovate, short ovate, or elongate ovate. The shell is thick, tough, and roughly granulated. The color is pale creamy white or dull white. The measurements of 56 eggs average 58.4 by 40.9 millimeters; the eggs showing the four extremes measure **65.5** by **47**, **53.3** by 40.6, and 58.2 by **37.6** millimeters.

Young.—Mr. Farmer has hatched a number of eggs under hens and found that 22 days was the longest period of incubation. He says that the "sitting bird often leaves the nest to go off and feed," but he doubts whether the male helps in incubation. "The female sits fairly close if one does not make too much fuss in approaching and does it quietly and indirectly, pretending to look for something else. When she flies she 'eases' off the nest quietly and disappears into the brush, going off and hiding somewhere where you can not see her, but where she can see you, whence she 'quarrels and scolds' until you leave the vicinity. Often other birds will join in the rumpus. If you have a dog along the birds will be much more noisy, objecting to the dog even more than they do to you."

The young are precocial, leaving the nest just as soon as the down is dry. The female carries the young down to the ground clinging to her legs, one at a time, according to Mr. Farmer's observation. This particularly peculiar habit of the young clinging to its mother's legs has been verified time and again by both Mr. and Mrs. Farmer when lifting the mother hen off the nest to inspect the brood on which she was sitting, and sometimes two at a time clinging to the mother. Once in a drizzling rain Mr. Farmer heard a chachalaca chatter and went to investigate. He found a female with young, and, while he watched, saw her carry, one at a time, all three young up into an ebony tree and leave them perched in a line on a limb about 15 feet above the ground. When about 2 weeks old they can fly perhaps 100 feet, but when they are a week old they can flutter 8 or 10 feet, and even at this age they are almost impossible to catch, flitting from bush to bush among the underbrush as they do. Mr. Farmer believes that there is only one brood. He has found young birds as late as September or October, and thinks they are the result of the first nest being disturbed.

Plumages.—In the downy young chachalaca the center of the crown and the occiput are black, tinged with "russet," and there is an isolated black spot on the forehead; the sides of the head and neck are "cinnamon-buff," tinged with "cinnamon" on the neck, and finely mottled with black; the chin and throat and lower underparts are white, with a broad band of "cinnamon-buff" across the chest; the upperparts are mottled with sepia and "cinnamon-buff." Another specimen is similar, but the sides of the crown are "pale mouse gray," and the back is tinged with "russet" in the central black stripe and with "ochraceous-tawny" on the mantle. In both chicks, one of which was known to be only 4 days old, the wings are well started and already reach beyond the tail. The wings and tail in another young bird, about 9 inches long, are so well developed that it could probably fly.

The wings and tail of the juvenal plumage are the first to appear, and they grow so rapidly that the flight stage is reached at an early age. These and the upper parts in general are "Saccardo's umber"; the center of the back and rump are barred with "tawny"; the wing coverts are barred with "cinnamon-buff"; the remiges are tipped, and mottled on the outer edge, and the rectrices are tipped with "cinnamon-buff"; the rectrices are decidedly pointed. In the 9-inch specimen referred to above, the head, neck, center of breast, and belly are still downy, evidently the last parts to be feathered; the juvenal plumage coming in on the sides of the breast is "Saccardo's umber," while that on the flanks and belly is "cinnamon-buff"; these two colors are sharply defined in the juvenal plumage and not intergraded or blended, as in the adult.

Apparently the juvenal plumage is worn only a very short time, and a complete molt soon produces a practically adult plumage. I could find no traces of juvenal feathers in fully grown fall or winter birds. I have seen evidences of a complete molt in adults in August and September. Mr. Farmer says that they molt only once a year, in September and October. He also says that the naked places showing on each side of the chin are of a grayish flesh color, alike in male and female, except during courtship in spring, when the male's patches become red.

Food.—Mr. Farmer says that the food of the chachalaca consists principally of berries, though they do catch bugs, and sometimes in spring when the buds are tender they "bud" hackberry or other trees. In captivity the tame birds eat bread and crackers or chopped-up meat, and they are especially fond of milk, particularly when young, and will relish any kind of fruits, particularly apples and bananas. They are especially fond of raw, chopped-up rabbit.

Behavior.—Mr. Kennard has seen chachalacas fly silently and swiftly over the tops of the chaparral, alight heavily in the top of a tree, and hop down from limb to limb without opening the wings. Mr. Farmer once came upon a bunch of 9 or 10 chachalacas following an opossum, teasing it and attacking it, but paying no attention to him even after he had killed some of them.

Sennett (1878) writes:

> Several times, when well concealed, I have noticed a pair spring from a thicket into a large tree, jump from limb to limb close to the body until they reached the top, when they would walk out to the end of the branch and begin their song. They roost in trees, and hunters frequently get them at night. Rarely did I see them on the ground. Once, while resting in a mesquite grove which looked very much like a peach-orchard on a well-kept lawn, I saw a Chachalaca trot out from a neighboring thicket in full view. He seemed looking for food on the ground. He discovered me and we eyed each other for a moment, when it turned, ran a short distance, sprang into the lower branches of a tree, and hopping along from tree to tree disappeared into the thicket about five feet from the ground.

Voice.—The remarkable vocal performances of this species are its most interesting and striking habits. They are difficult to describe, but once heard they can never be forgotten. Dr. J. C. Merrill (1878) writes:

During the day, unless rainy or cloudy, the birds are rarely seen or heard; but shortly before sunrise and sunset they mount to the topmost branch of a dead tree and make the woods ring with their discordant notes. Contrary to almost every description of their cry I have seen, it consists of three syllables, though occasionally a fourth is added. When one bird begins to cry the nearest bird joins in at the second note, and in this way the fourth syllable is made; but they keep such good time that it is often very difficult to satisfy one's self that this is the fact. I can not say certainly whether the female utters this cry as well as the male, but there is a well-marked anatomical distinction in the sexes in regard to the development of the trachea. In the male this passes down outside the pectoral muscles, beneath the skin, to within about one inch of the end of the sternum; it then doubles on itself and passes up, still on the right of the keel, to descend within the thorax in the usual manner. This duplicature is wanting in the female.

Sennett (1879) says:

A more intimate acquaintance with this bird enables me to give a better description of its notes than the attempt in my former memoir. The notes are loud and uttered in very rapid succession, and those of the female follow the male's so closely, while so well do they harmonize, although in different keys, that I mistook the first note of one for the last note of the other. It really utters but three syllables, thus: *Cha-cha-lac,* instead of four, *cha-cha-lac-ca,* as given before. It also has a hoarse, grating call or alarm note, uttered in one continuous strain and without modulation, something like *kak-kak-kak.*

Dr. T. Gilbert Pearson told me that when he went out to camp one night with Mr. Farmer in the heart of the chachalaca country, the latter had told him that there would probably be within earshot of their camp at least 500 chachalacas, a statement about which he was very skeptical. About sundown the concert began, increasing in volume until the din became almost indescribable. Doctor Pearson was convinced, and finally suggested to Mr. Farmer that he call it 5,000 instead of 500 birds.

Game.—The chachalaca has figured largely as a game bird in the Brownsville market. Its flesh is said to make delicious eating. It hardly comes up to a sportsman's idea of what a game bird should be, though one must have a thorough knowledge of its haunts and habits to be successful in hunting it. It has been quite extensively domesticated on many Mexican ranches, lives contentedly with domestic poultry, and becomes very tame and makes a good pet, although often so familiar as to be troublesome.

DISTRIBUTION

Range.—Lower Rio Grande Valley in Texas and northeastern Mexico.

The chachalaca occupies a limited range extending **north** to southern Texas (Rio Grande City, Fort Ringgold, Lomita Ranch, Hidalgo, and Brownsville). **East** to Texas (Brownsville); eastern Tamaulipas (Matamoras, Jimenez, and Aldama); and northeastern Vera Cruz (Tampico). **South** to northeastern Vera Cruz (Tampico); and southeastern San Luis Potosi (Valles). **West** to southeastern San Luis Potosi (Valles); western Tamaulipas (Xicotencatl, Ciudad Victoria, and Rio Pilon); and southern Texas (Rio Grande City).

Egg dates.—Texas and Mexico: 73 records, March 21 to August 16; 37 records, April 27 to May 27.

Order COLUMBIFORMES

Family COLUMBIDAE, Pigeons, Doves

COLUMBA FASCIATA FASCIATA Say

BAND-TAILED PIGEON

HABITS

It was a bright, sunny, cold morning, after a frosty night in February, when I first made the acquaintance of this fine bird, the band-tailed pigeon. The sun was shining full of genial warmth on the tops of the tall sycamores and eucalyptus trees, which grew in a deep arroyo in southern California, but it had not yet penetrated to its shady depths, which still sparkled with white frost. I did not at first recognize the large plump birds, 33 of them I counted, perched in the tops of two tall sycamores, evidently enjoying their morning sun bath. But with a glass I soon recognized them as pigeons, saw the white crescent on the neck, and, as they flew, marked them as bandtails, from the bands on the broad square tails. This was one of the wandering, restless flocks which travel about during the winter, moving from one place to another as food or fancy leads them. Later I saw them flying down the arroyo in a detached flock high in the air. And almost daily for some time I saw more or less of them in the same arroyo on the outskirts of Pasadena. They remained in the vicinity off and on until the latter part of April.

This is one of the birds that was being rapidly killed off, as it was a favorite game bird in the Pacific Coast States. It was even verging toward extinction. But, fortunately, protection came in time to save it and it has made a wonderful recovery. E. A. Kitchin writes to me of conditions in Washington:

The Federal protection of these birds in recent years has made wondrous changes in the Puget Sound country, which always was a natural breeding ground. Before this protection the pigeons had become very scarce, so much so that it was even an event to see one, and these were only seen in the more isolated parts. The large gulches so numerous on Puget Sound, covered at the bottom by thick alder and on the sides by small firs, form the natural breeding grounds for these birds. Now, thanks to Government protection, there is hardly a gulch that does not contain 50 or more pairs of breeding birds.

Courtship.—The only note we have on the subject of the courtship of the bandtail is the following by Harry S. Swarth (1904):

During the breeding season the male bird is fond of sitting in some elevated position, usually the top of a tall dead pine, giving utterance, at frequent intervals, to a loud *coo*, more like the note of an owl than a pigeon, which can be heard at a considerable distance; while occasionally he launches himself into the air with wings and tail stiffly outspread, describes a large circle back to his starting point, uttering meanwhile a peculiar wheezing noise impossible of description. I had supposed that this noise was made by the outspread wings, but a male bird which Mr. Howard had in his possession for some time gave utterance to the same sound whenever angered or excited, evidently by means of his vocal organs, as we had ample opportunity of observing.

Nesting.—Mr. Kitchin, who has had considerable experience with this pigeon in Washington, contributes the following:

The flocks arrive the latter part of March or early in April and at once seek the gulches, where they feed on the seeds of the alder. They apparently have a rather prolonged breeding season, lasting from April through June. The nesting sites are mainly in the dark fir trees, where their nest of dead fir twigs is placed near the trunk and generally in the lower branches, averaging probably 20 feet from the ground. Occasionally, however, the nest is found in an alder and sometimes on the top of a thick birch overhanging the hillside.

When approached the brooding bird has a habit of standing erect in the center of the nest and by doing so becomes very conspicuous. I have read of these birds carrying their egg when disturbed, and although I have flushed many birds none showed any inclination to take the egg with it. Knowing where an occupied nest was, I have approached quietly to perhaps 6 feet before the bird flushed, and at other times I have rushed, in a startling way, taking her by surprise, but in neither case was the egg removed.

The nest is somewhat loosely made and entirely of dead twigs and, though it is not in any way cupshaped it certainly is saucershaped, and the roughness of the twigs prevents the egg from rolling in the nest. One can tell from below whether the nest contains an egg or a squab, as the brooding bird will stand in the center of the nest, astride the egg, while if a squab she will be standing on the rim.

The birds are fond of their old nesting sites and are insistent in using the site selected. They not only come back to the same tree but will use the same limb as that used the previous year, even if the first nest has been disturbed. On one occasion the bird selected a hanging bush on the hillside and built her nest near the top. The egg and nest were taken and she at once built another on the same site and raised her young. The following year she was again in the bush, sitting on a slightly incubated egg. This set I took, and by July she had another nest and egg, which were taken, and I was much surprised, in passing later in the season, to find a third nest in which she had probably raised her young to maturity.

Grinnell, Bryant, and Storer (1918) give the most comprehensive account available of the nesting habits of this bird in California, from which I quote as follows:

Nearly all authentic reports from California agree in stating that the Band-tailed Pigeon nests in trees—almost invariably in black or golden oaks—at heights ranging from eight to thirty feet above the ground. As exceptions, Littlejohn (MS) found a nest in San Mateo County in a Douglas spruce; and in Marin County, J. Mailliard found a nest in a California lilac (*Ceanothus thyrsiflorus*) overhanging a steep slope. Some early reports from this State have mentioned ground nests, as have several more recent, but scarcely trustworthy, accounts from Oregon and Washington; but there is no late evidence of the ground nesting habit in California. In a general way the nest resembles that of the Mourning Dove, save that it is considerably larger, and sometimes proportionately thicker. It is a crude structure, a mere pile of oak and other twigs, so loosely arranged that attempts to remove the mass often result in its falling to pieces. The average diameter is six or eight inches, while the thickness in two recorded instances was one and four inches, respectively. Sometimes as few as sixteen or eighteen twigs are all that go to make up the nest and again there may be more than a hundred. The twigs range from a sixteenth to a quarter of an inch in diameter and are of various lengths. They are laid across one another, with little or no weaving, forming a platform with numerous interstitial spaces. A slight lining of pine needles was found in one nest. As Gilman well says, it is a marvel how an egg can be kept warm enough to hatch while resting on such an airy platform in the cool air of a high altitude. The nest site, which is almost always on top of a large horizontal limb, seems to be so selected that the incubating bird may flush directly and rapidly from the nest when danger threatens.

There is some evidence to indicate that the band-tailed pigeon occasionally nests on the ground. Major Bendire (1892) quotes O. B. Johnson, referring to Oregon, as saying: "They nest in various situations, much like the common Dove, *Z. carolinensis*. I found one of leaves and moss beside a tree, placed on the ground between two roots; another one upon an old stump that had been split and broken about 8 feet from the ground; another was in the top of a fir (*A. grandis*), and was built of twigs laid upon the dense flat limb of the tree, about 180 feet from the ground." This statement, he says, is confirmed by Doctor Cooper, as follows: "In June they lay two white eggs, about the size of those of the house pigeon, on the ground, near streams or openings, and without constructing any nests." A similar statement is made by Mrs. Irene G. Wheelock (1904).

Although the band-tailed pigeon usually nests in widely scattered pairs, the following account of an Arizona colony, by F. H. Fowler (1903), is interesting:

When the breeding season draws near, they betake themselves to sheltered places among the lower mountains, and nest in scattered communities, or as I have seen in several cases, a pair will nest apart from the others. One of the largest breeding communities I noted was in a little pocket in the mountains, about five miles south of Fort Huachuca; this little place was at the head of a short canyon, and was indeed an ideal spot for birds, as it was well wooded and watered. Here a flock of about thirty-five pairs of band-tails nested in a scattered rookery, probably not averaging a nest to every three or four acres at the most thickly populated part; and a great majority of the nests were even farther apart than this. The nests in this colony were all placed on the forks of low horizontal limbs of live oaks usually not more than twelve feet up or less than nine, and in no case did I find more than one egg or squab in a nest. The nests were all of that very simple dove-like construction consisting of a few sticks placed on a fork of a branch.

Francis C. Willard (1916) made the following observations in the Huachuca Mountains, Ariz.:

There were a few pigeons nesting in the vicinity, and one pair near camp was watched quite closely from the time the nest was begun until the egg was laid. Nest building was carried on only in the early morning hours, from sunrise till about 8 o'clock. Both birds were present, but the female alone seemed to be engaged in the actual construction of the nest, which she went about in a very lackadaisical manner. The pair would sit together on the few sticks already in place for many minutes; at last the female seemed to remember that she was nest building and flew up the mountain side followed by the male. Considerable time was spent on every trip after material, so very few sticks were added each day, and it was not until six days had elapsed that the flimsy platform was completed and the egg laid.

Eggs.—Most authorities agree that this pigeon lays, almost invariably, only one egg; but there are a number of apparently authentic records of two eggs in a nest. The egg is elliptical ovate, generally somewhat pointed and pure white. The shell is smooth and slightly glossy. The measurements of 19 eggs average 39.7 by 27.9 millimeters; the eggs showing the four extremes measure **43.5** by 30, 39.3 by **30.2**, and **36.8** by **25.9** millimeters.

Young.—According to Major Bendire (1892) the period of incubation is "from 18 to 20 days, both sexes assisting. The young grow rapidly, and are able to leave the nest when about a month old." The nesting season is prolonged through 10 months of the year, and the evidence shows that probably several young are raised during the season. Clinton G. Abbott (1927) has published some notes from Albert E. Stillman, who had good opportunities for studying this species in San Diego County. In his notes for September 17, 1922, he writes:

That day the female left the oak tree in the early morning and returned at twilight; after quickly feeding the young she left again. Next day she left at daybreak and returned at sundown. For more than a week after that I kept the youngster with me during the day, letting him perch on my finger or hop about on the cabin floor, returning him to his nest before evening. On the morning of October 2 I found that the young bird had climbed from the nest and was sitting on a branch of the oak tree, where he remained until late in the afternoon. That night he roosted on the high limb of a nearby pine tree. On October 4 he left the neighborhood and I did not see him again.

That successive young pigeons are sometimes raised in one nest the same season was proved by Bushnell in 1925. He found a nest on March 8 containing one egg, from which the squab hatched and grew up. Then the pigeon laid an egg in the same nest and started to incubate. The second young bird hatched about the middle of May and lived to leave the nest.

Dr. Joseph Grinnell (1928a) watched a pigeon feeding its young in the Yosemite region on September 29, 1927, of which he says:

Soon an old bird alighted, coming up the same steep course as the first one, at mid-height of the trees through the forest, and alighted on a branch of the nest tree, on a level with the nest but on nearly the opposite side of the trunk.

After remaining perched quietly for awhile, the old bird then walked along the branch lengthwise to the trunk, hopped across, fluttering some, to the base of the nest branch and walked out on it to the nest.

Immediately a commotion began—the young one fluttering its wings spasmodically, the old one, not plainly seen because of intervening foliage, evidently feeding it. The process lasted fully three minutes, when the old bird flew directly off from the nest, out into space from a cliff base, and circling, was seen to alight at far distance on a middle branch of a dead tree. We would have timed the feeding process if we had had any notion of its lasting so long. After feeding, the youngster crouched down motionless and could be seen plainly no more. When being fed, its upraised, fluttering wings showed the quills to be only an inch or so long; it could have been no more than ten days old.

Mrs. Irene G. Wheelock (1904) says that the squab is " fed on a thin milky fluid, by regurgitation, for 20 days."

Plumages.—I have never seen a nestling band-tailed pigeon, but Mrs. Wheelock (1904) says that its " yellow skin is covered with the sparse, cottony, white down." The juvenal plumage is much like that of the adult, but it lacks the white collar and the iridescent metallic colors on the neck; the vinaceous tints are wholly lacking; and the feathers of the breast and wing coverts are narrowly edged with whitish, giving a slightly scaly effect. Molting birds are scarce in collections, but apparently young birds have a complete molt during the first fall, which produces a practically adult plumage. I have been unable to trace the molts of adults, but a complete molt probably occurs, as it does in European pigeons, during summer and fall; this may begin as early as May or June and end as late as October or November.

Food.—Grinnell, Bryant, and Storer (1918) have published a very full account of the food of the band-tailed pigeon, from which I can include only a condensed summary. As their food consists mainly of nuts and berries, which are intermittent crops, the pigeons find it necessary to wander about considerably, congregating in large numbers where food is abundant and deserting these same localities during seasons of scarcity. Acorns seem to be their chief food; probably all the oaks are patronized, but mainly the live oaks, golden oak, and black oak; the acorn crop lasts through a long season in fall and winter. The acorns are swallowed whole and form an attractive food supply in the fall. They resort at times to the apple-like fruits of certain species of manzanita, eating them from the time they are first formed and green until late in fall, when they are fully ripe. Early in the fall they feed on the fruit of the coffeeberry, elderberry, and chokecherry. In winter they have the toyon, or Christmasberry, and when the nut and fruit crops become exhausted they feed on the flower and leaf buds of the same plants, such as manzanita and oak buds. Early in spring sycamore balls are frequently eaten; as many as 35 have been counted in the crop of one pigeon. Fruits of dogwood, wild peas, pine seeds, and other seeds have been found in their crops. Considerable cultivated grain is eaten; this is mainly waste grain, picked up in stubble fields of barley, oats, and corn; but pigeons have been known to do some damage by pulling up newly sown seed barley; such records are scarce, however. P. A. Taverner (1926) says that in western Canada, "they are especially partial to peas and are said to pull up the sprouting seeds. The flocks so engaged are described as being numerous enough to turn the colour of the fields they alight upon from brown to blue. As they are large birds, each one intent on filling a capacious crop, their power for damage is not small. In the autumn they alight on the stooked grain and may take a considerable toll of it."

Other observers have noted in their food hazel, pinyon and other nuts, wild grapes, wild cherries, wild mulberries, blueberries, blackberries, raspberries, juniper, cascara, salmon and salal berries, and grasshoppers and other insects. Their method of feeding on manzanita berries is thus described by Laurence M. Huey (1913):

Some boys there told me that for the past two weeks a bunch of about one hundred pigeons had been feeding on green manzanita berries in a near-by thicket, and I was much pleased when they offered to take me to the place. It proved to be about one and one-half miles north of their ranch, due south of Volcan Mountain, and was the only thicket thereabout having a large crop of berries. In the morning the birds would begin to arrive a little after sunrise, leaving between eight and nine o'clock; in the evening they returned about four and stayed until dark. They seemed always to come from and return to the same place, at the top of Volcan Mountain among the pine trees.

The pigeons seen were apparently always the same bunch, as one bird noted with a few secondaries missing on the left wing was seen on three out of four occasions when the flock was encountered. It was interesting to watch them trying to alight on the clusters of berries, far too weak to support them, making many futile attempts and finally succeeding in reaching the berries only by settling on a stronger perch and then walking out to the cluster. But how they did gorge and stuff when they finally got at them.

Mr. Willard (1916) writes:

A few days later a flock was observed feeding on acorns in a group of large oak trees (*Quercus emoryi*). The antics of these birds were more like the acrobatic stunts of parrots than of pigeons. They would walk out on the slender branches till they tipped down, then, hanging by their feet, would secure an acorn and drop off to alight on a branch lower down. In spite of their large size, pigeons are surprisingly inconspicuous when thus engaged in feeding among the leaves.

M. French Gilman (1903) says of their feeding habits in the grainfields:

In March, 1901, great flocks of the pigeons poured into San Gorgonio Pass and fed in the barley fields. For about two weeks there were hundreds of them but they all left as suddenly as they had appeared. Their method of feeding was peculiar. Instead of spreading out they kept together, alternately walking and flying. Those behind would fly a few feet ahead of the advance line, alight, and walk along picking up grain until other rear ones would fly ahead and it came their turn again. In this way the flock advanced, some in the air all the time and ground was covered quite rapidly.

Behavior.—Except during the nesting season, band-tailed pigeons are decidedly gregarious, flying about in large, open, or scattered flocks, formerly in flocks of hundreds, but now more often in flocks of dozens. They are fond of perching for long periods in the tops of tall trees; in the leafless sycamores in winter the flocks are very conspicuous, but among the thick foliage of live oaks or eucalyptus they are well hidden. If approached too closely, they will begin to leave, a few birds at a time, with loud flapping of wings, and there are usually a few laggards that slip away from some unseen spot at the last moment. Their flight is strong, direct, and very swift, reminding one of domestic pigeons. According to Grinnell, Bryant, and Storer (1918), "in passage down a mountain side, the flight is inconceivably swift, the wings being held close in to the sides, beating only at long intervals, and the body veering slightly from side to side in its arrow-like course. This headlong flight produces a rushing noise as of escaping steam."

Voice.—The cooing notes of the band-tailed pigeon are much like those of the domestic pigeon. Doctor Grinnell (1905) says that "their deep monotonous *coo'-coo*, *coo'-coo*, *coo'-coo*, or *tuck-oo'*, *tuck-oo'* was a frequent sound on Mount Pinos." Mrs. Florence M. Bailey (1902) writes:

If you follow the pigeons to their breeding-grounds in some remote canyon you will be struck by the owl-like hooting that fills the place, and you will locate the sound here and there along the sides of the canyon at dead treetops, in each of which a solitary male is sunning himself, at intervals puffing out his breast and hooting. The hooting varies considerably. Sometimes it is a calm *whoo'-whoo-hoo, whoo'-hoo-hoo*, and others a spirited *hoop'-ah-whoo'*, and again a two syllabled *whoo'-ugh*, made up of a short hard hoot and a long coo, as if the breath was sharply expelled for the first note and drawn in for the second.

The method of uttering the notes is described in detail by Joseph H. Wales (1926) as follows:

When the male pigeon starts this performance he usually maneuvers around for a firm footing and perhaps opens his bill slightly once or twice. Next he stretches his neck out in a line parallel with the axis of his body, and bends his head down to a right angle. With his bill open a crack he gives one gasp which fills out the skin of his neck until about three times natural size, and at the same time utters a very faint *oo* which is not usually audible over twenty feet. All of these are preliminary actions, as directly following the first sound comes the *whoo-oo*. This hoot is made by a quick expelling of the air from the bird's lungs, and is accompanied by a slight downward push which seems to give abruptness to the first note. The swelled neck skin is not reduced, as the bill is opened and the lungs are refilled for the following *coo*. There are usually about seven or eight of these hoots in a series, but sometimes as many as eleven. When finished, the male pigeon brings his neck back into its natural position and allows the air to escape from under the neck skin. This performance is repeated at irregular intervals through the early morning and the latter part of the afternoon.

Fall.—Illustrating the heavy fall flights that formerly occurred in California, W. Leon Dawson (1923) writes:

In the fall and winter of 1911–12, lured by an unusual crop of acorns, and impelled, no doubt, by corresponding "crop" failures elsewhere, immense numbers of Band-tailed Pigeons appeared in the interior valleys of Santa Barbara County, centering about the town of Los Olivos. It is probable that practically the entire summer population of California north of the Tehachipe, Oregon, Washington, and British Columbia concentrated at this point. It is not surprising, therefore, that "millions" of birds should have been reported in this section, although half a million would probably be much nearer the truth. What followed on this occasion was a humiliating example of what human cupidity, callousness, and ignorance, when unrestrained, will accomplish toward the destruction of birds. Reports of the birds' abundance spread rapidly. The "Wild Pigeon" of the East had unexpectedly turned up in the West. Hunters from the outside flocked to the scene. Every gun was put into commission. By automobiles and trainloads they came. The country was aroar with gunfire. The ammunition business jumped in a dozen towns. Enterprising dealers organized shipments to the San Francisco and other markets. W. Lee Chambers, writing for The Condor, reports a Sunday excursion of hunters from San Luis Obispo which brought home 1,560 birds. Another man, hunting for the San Francisco market, killed 280 pigeons under one oak in one day. The stupid birds, knowing nothing of their offense, flew miserably from one part of the valley to another, but would not, or could not, forsake their food. How great the destruction of that winter really was is a matter of merest conjecture, but it must have been a very sensible proportion, possibly more than half the entire species. I passed through this section of the country on the 1st of the following April and saw only 28 pigeons, but the sides of the road in many places were so covered with paper waste from cartridge boxes that I was reminded of a street in Chinatown on the morning after New Year's. Fortunately, this destruction and the agitation which ensued prompted the Government to declare a five-year closed season on Band-tailed Pigeons.

Game.—Grinnell, Bryant, and Storer (1918) say:

The value of the Band-tailed Pigeon as a true game bird is to be conceded without argument. Its pursuit is of a different type from that offered by any other game species. An anonymous writer in southern California, who signs himself "Stillhunter," says that the best place for hunting pigeons there is near a dead tree where the birds are known to alight. For such a situation he advises using a .22 or 25–20 rifle; then single birds may be secured without frightening away others in the flock. For sneaking up on birds a "duck gun" is recommended. Ten pigeons are considered a good day's bag. If the flesh has become "strong" by reason of the birds' acorn diet, soaking in brine flavored with vinegar or lemon will remove the disagreeable taste. After such treatment the birds should be broiled, or baked in a pot pie.

Enemies.—Band-tailed pigeons apparently have few natural enemies, and these have proved of little consequence in reducing their numbers. In spite of their slow rate of increase they could hold their own against natural enemies, but they could not long resist the terrible slaughter by man when congregated in their winter quarters. Fortunately, they are now protected against this.

Mr. Willard (1916) says that in Arizona " the Prairie Falcon and Cooper Hawk take considerable toll from the flocks. These two terrors of the air will dash into a tree and grab a pigeon off a branch, rarely making an unsuccessful raid. The Prairie Falcon is the chief offender." Mr. Kitchin tells me that, in Washington, " apparently the only enemy these birds have during the breeding season is the local gray squirrel, which I know on more than one occasion has taken possession of the nest, using it as a foundation and adding to it to suit himself, and once I found the egg of the pigeon buried under the structure that the squirrel had added."

Winter.—Grinnell, Bryant, and Storer (1918) say that—

north of the northern boundary of California the Band-tailed Pigeon is wholly migratory. It seems inevitable that this northern-bred contingent should move south *into California* for the winter season, and there is, therefore, little reason to doubt the inference that the birds which concentrate in winter in west-central and southern California, represent the entire pigeon population of the Pacific coast region. If this be true, it is of course apparent that, as far as the whole Pacific coast region is concerned, California alone is, in winter, responsible for the existence of the species.

DISTRIBUTION

Range.—British Columbia, the Western United States, and Central America.

Breeding range.—The breeding range of the band-tailed pigeon extends **north** to southwestern British Columbia (Courtenay and Chilliwack). **East** to British Columbia (Chilliwack); Washington (Everett, Seattle, Tacoma, and Kalama); Oregon (Beaverton); northeastern California (Lyonsville and Stirling City); Colorado (Estes Park and the Wet Mountains); New Mexico (Tres Piedras, Pecos Baldy, Sandia Mountains, Capitan Mountains, Sacramento Mountains, and Guadalupe Mountains); western Texas (Dog Canyon, Fort Davis, Marfa, and Chisos Mountains); and Puebla (Las Vegas). **South** to Puebla (Las Vegas); Durango (Otmapa Ranch); and Lower California (Cape San Lucas). **West** to Lower California (Cape San Lucas, Victoria Mountains, and El Sauz); California (Laguna Mountains, Cuyamaca Mountains, Pine Mountain, San Jacinto Mountains, Mount Wilson, Mount Pinos, Lopez Canyon, San Jose, Lagunitas, Gualala, Eureka, and Crescent City); Oregon (Lookingglass, Newport, and Astoria); Washington (Granville, La Push, and Neah Bay); and southwestern British Columbia (Lake Cowichan and Courtenay).

Winter range.—In winter the species occurs regularly **north** to California (East Park and Alta); Arizona (Salt River Bird Reservation); and southern New Mexico (Haut Creek). **East** to New Mexico (Haut Creek and Silver City); and Guatemala (Volcano Toliman). **South** to Guatemala (Volcano Toliman); Chiapas (Pinabete); and Lower California (Mount Miraflores). **West** to Lower California (Mount Miraflores, El Sauz, La Laguna, Pierce Ranch, and Guadalupe Valley); and California (El Cajon, Los Angeles, Carpentaria, Fremont Peak, and East Park).

The range above described is for the entire species and is occupied chiefly by the typical race, *Columba f. fasciata.* Viosca's pigeon (*Columba f. vioscae*) is confined to southern Lower California and is apparently nonmigratory.

Spring migration.—Early dates of spring arrival are: Colorado, Beulah, May 7, Moraine Hill, May 25, and Gold Hill, June 2; Oregon, Mercer, March 5, Corvallis, March 14, Beaverton, April 4, Southerlin, April 8, North Bend, April 10, and Tillamook, April 14; Washington, Clallam Bay, April 9, Menlo, April 12, Vancouver, April 17, and Everett, April 26; and British Columbia, Courtenay, March 31, Sumas, April 4, Burrard Inlet, April 5, Chilliwack, April 13, and Hastings, April 26.

Fall migration.—Late dates of fall departure are: British Columbia, Courtenay, October 5, and Chilliwack, October 29; Washington, Fort Steilacoom, September 25, Clallam Bay, October 15, Argyle, October 20, and Cascades of the Columbia, October 29; Oregon, Forest Grove, October 3, Tillamook, October 10, North Bend, October 24, and Newport, October 28; and Colorado, Ouray, September 8, Del Norte, September 20, and Forks of the Rio Grande, September 26.

A vertical migration from the higher mountains in California also is occasionally noted (Escondido, 1920).

Casual records.—Band-tailed pigeons are rarely taken outside of their normal range. Patch (1922) noted them at Tow Hill, Graham Island, of the Queen Charlotte Group, on July 28, 1919, and states that there is one record from Bella Coola, British Columbia, indicating that they may at times breed farther north than is now known. One was taken in 1905 near Crescent, Okla., and another on June 2, 1912, at Englevale, N. Dak.

Egg dates.—Washington and Oregon: 6 records, May 3 to July 12. California: 46 records, March 6 to September 24; 23 records, May 10 to July 1. Arizona and New Mexico: 32 records, April 23 to October 4; 16 records, June 16 to July 14. Southern Lower California: 35 records, January 22 to December 26; 18 records, June 21 to July 28.

COLUMBA FASCIATA VIOSCAE Brewster

VIOSCA'S PIGEON

HABITS

This pale race of the band-tailed pigeon was described by William Brewster (1888), based on a study of a series of more than 100 specimens collected in southern Lower California and named in honor of Mr. Viosca, the United States consul at La Paz. He gives it the following characters: "Similar to *C. fasciata* but with the tail band wanting or only faintly indicated, the general coloring lighter and more uniform, the vinaceous tints, especially on the head, neck, and breast, much fainter and more or less replaced by bluish ash."

As to its distribution (1902), he says:

This pigeon seems to be strictly confined to the Cape Region, for neither Mr. Bryant nor Mr. Anthony has succeeded in finding it in the central or northern portions of the Peninsula where true *fasciata* is also apparently wanting.

Chester C. Lamb (1926) writes:

The Viosca Pigeon is, with one exception, known to occur only in the Victoria Mountains, sometimes known as the Sierra de la Laguna, or in the adjacent foothills. The exception is Brewster's statement that Mr. Frazar saw large numbers in San Jose del Cabo in September "passing southward." During my own two years' residence in the Cape district, however, this bird was not seen outside the mountainous district above indicated. I very much doubt the pigeons leaving Lower California at all, as implied by Brewster on Frazar's report.

I became acquainted with the Viosca Pigeon July 5, 1923, when I made my first trip to the Laguna Mountains, and in the next month found them abundant. The following year parts of four months were spent in their range, and I had ample opportunity to study and observe this isolated race of pigeon. It was common throughout the mountains, ranging from an altitude of 1,500 feet to the tops, some 6,500 feet. At the lower levels the birds are found in the canyons, where wild grapes and another native fruit grow; but the type of country they like best, and their real home, is the live-oak region of the higher valleys and canyons. These birds are swift and powerful fliers and it would not take them long to travel for their food, either to the pinyon pines above or to the wild grapes and figs below, whenever they might wish to vary their acorn diet.

Nesting.—The nesting season is very variable or very much prolonged. Reliable observers have found this pigeon nesting in January, February, April, May, June, July, August, September, and December. Mr. Lamb (1926) says of its nesting sites:

The majority of the numerous nests I examined were in live-oak trees, usually situated on the forks of the larger horizontal limbs, and placed from 10 to 20 feet above the ground. Some nests were also found placed among the smaller branches and near their extremities, but this was exceptional. A very few nests were found in a small species of white-oak tree that grows on the hillsides. This oak is peculiar in that in the dry season the leaves turn brown and appear dead, but a few days after the first rain, the leaves gradually grow green again.

There are a few pine trees, mostly pinyons, scattered among the oaks in some parts, but only in one instance did I find a pigeon's nest in a pine. This was a well built nest six feet above the ground, against the trunk where a horizontal limb grew out. One nest was found on a frond of a leaning fan palm tree. The nest is as a rule carelessly made, of a few coarse twigs, with no nest lining.

A nest collected for me by W. W. Brown, in the Sierra de la Laguna, on June 14, 1913, containing one egg, was described as a frail platform-like structure of sticks, built near the extremity of a branch, in a pine tree about 40 feet from the ground. An egg in the United States National Museum, taken by M. A. Frazar, near Pearce's ranch, on July 18, 1887, was presented by Mr. Brewster; it was taken from a nest composed of a few sticks, 18 feet up, on a broken upright branch of a giant cactus.

Eggs.—One egg seems to be the almost invariable rule with Viosca's pigeon. If two eggs are ever laid, it must be very rarely, for in more than 25 nests examined by Mr. Lamb and 8 or 10 by Mr. Brown, only one egg or young was found. The egg is pure white, like that of the band-tailed pigeon. The measurements of 25 eggs average 39.7 by 27.5 millimeters; the eggs showing the four extremes measure **43.2** by 28.5, **41.4** by **29**, **36.7** by 26.9, and 38.1 by **26.4** millimeters.

Food.—Mr. Lamb (1926) says: "Acorns, wild grapes, pinyon nuts, and a sort of wild fig were, in my experience, their only food in the summer."

Behavior.—Again Lamb writes:

At one of my camps in the Victoria Mountains, my work table was placed directly under a large live-oak tree which bore an abundant crop of acorns. This was a great attraction to the pigeons as well as to numerous Narrow-fronted Acorn-storing Woodpeckers. It was a marvel to me how such a large bird as a pigeon could alight in this tree, even on its slenderest branches, without the least audible flapping of its wings; often I would be unaware of a pigeon's presence until it was made known to me by the woodpeckers. The pigeons and woodpeckers, it appears, are inherent enemies. Let a pigeon alight in this tree, and if a woodpecker is near-by, the latter immediately, with loud cries, sets upon and drives the pigeon away, which departs with a great flapping of wings. In no case have I seen a pigeon try to defend itself, and one was never seen to take the part of the aggressor. When attacked, a pigeon flies to a near-by tree and often, as soon as the woodpecker's back is turned, so to speak, the pigeon is back again in the oak tree, only to have the same thing happen again. It is lucky for the pigeons that woodpeckers are not always on guard, else they would get but few acorns.

Voice. Again he writes:

The first bird voices one hears in the early morning in the live-oak region are those of the Narrow-fronted Woodpeckers, closely followed by the Viosca Pigeons, whose mellow *whoo-whoo* (first note short, second long and slightly lower) sounds almost human, as if someone were trying to attract attention.

From the specimens taken I learned it was only the males that make this sound. At this time the birds perch upon some dead or bare limb, usually at some elevation. They are frequently seen fluttering spirally with short wing-beats or sailing slowly over some clearing, and then an entirely different note is uttered, at short intervals, hard to describe, but which could be called a sort of hoarse, guttural croak, sounded for a sustained period.

Fall.—Brewster (1902) says:

At San José del Cabo large flocks were observed in September passing southward. Mr. Frazar believes that the majority left Lower California that season before winter set in, although he saw a few on November 15 along the road between San José and Miraflores and others at San José del Rancho December 18–25. None were found on the Sierra de la Laguna between November 27 and December 2.

COLUMBA FLAVIROSTRIS FLAVIROSTRIS Wagler

RED-BILLED PIGEON

HABITS

The large, handsome red-billed pigeon, locally known as "blue pigeon," is a Central American species that extends its range into the United States only in a narrow belt of heavily wooded bottomlands along the valley of the Rio Grande in Texas and perhaps as far west as the Graham Mountains in southern Arizona, where Major Bendire (1892) reports the capture of three specimens. I first met this pigeon in the heavy timber near the Resaca de la Palma, near Brownsville, Tex. This forest of ebony, huisache, mesquite, and hackberry trees, with its thick undergrowth of thorny shrubs and its tangles of vines, has been more fully described under the chachalaca. This pigeon is a resident in this region for about 10 months each year. George B. Sennett (1879) quotes Dr. T. M. Finley as saying that at Hidalgo these pigeons were " first noticed on January 24th in flocks; about the middle of February they were seen in the woods in pairs, and cooing. The last seen of them in 1877 was the latter part of November. These Pigeons were seen several times consorting with tame Pigeons in the ebony-trees in the neighborhood of the village of Hidalgo." Dr. J. C. Merrill (1878) says: "This large and handsome pigeon is found in abundance during the summer months, arriving in flocks of fifteen or twenty about the last week in February. Though not very uncommon about Fort Brown, it is much more plentiful a few miles higher up the river, where the dense woods offer it the shade and retirement it seeks."

Nesting.—On May 27, 1923, near Resaca de la Palma, we found two nests of the red-billed pigeon in the heavily timbered thickets. One was about 10 feet up in a tangle of vines and saplings; it was a small, frail nest of small twigs, barely strong enough to support the weight of the single, fat squab that it held. Another similar nest, in much the same kind of a location, held a single fresh egg.

Three typical nests are described by Sennett (1878) as follows:

The locality was a grove of large trees, with undergrowth, and clumps of bushes matted with vines. While prying about the thick vines I flushed the bird off its nest, and it alighted in one of the tall trees near by. It took me but a moment or two to examine the nest and shoot the bird. In less than ten minutes' time I had also its mate. The nest was only eight or nine feet from the ground, and set upon the horizontal branches of a sapling in the midst of the vines. It was composed of sticks, lined with fine stems and grasses, had a depression of an inch or more, and was about eight inches in outside diameter by two and one-half inches deep. It contained one egg, with embryo just formed. Dissection of the bird showed that she would have laid no more.

On May 8th, at Lomita Ranche, a few miles above Hidalgo, in the fine grove of ebonies in the rear of the buildings of the ranche, I found two nests. Both were well up in the trees, one about twenty-five feet and the other about thirty. The nests were situated close to the body of the trees, on large branches, and were composed of sticks and grasses, with an inside depth of about two inches. One contained a single egg, far advanced; in the other also lay a solitary egg, from which a young chick was just emerging. The parents persisted in staying about. notwithstanding we were making a great disturbance, even shooting into the same trees. Whenever we would go off some distance they would immediately go on their nests and seemed loth to leave them at our return. These were the only ones seen breeding so near habitations. The grove was a common resort for man and beast, besides being the place where wagons, tools, etc., were kept and repaired.

On May 11th, I obtained my fifth and last nest. I found it in the woods at the fork of two roads, a mile or so from the village, down the river. This nest I had discovered a week or so before, complete, but empty. It was situated about ten feet from the ground, in one of a thick clump of small trees, at the junction of several small branches. It was composed of twigs and rootlets, without grasses, and had a depression of one and one-half inches. The bird was flushed from the nest and shot. Upon examination, the solitary egg showed that incubation had begun, and dissection of the bird proved that no other eggs were developed for laying.

Again (1879) he says:

This bird breeds irregularly and lays several times in a season. I found nests during the whole time of my stay, containing eggs and young in all stages of development, but in no case did a nest contain more than one egg or young. The parents are fond and affectionate, and both assist in incubation.

A. J. Van Rossem has sent me the following notes on its nesting habits in Salvador:

The breeding season apparently extends throughout the year, for males in breeding condition were taken in January, April, May, July, September, November, and December, and females either laying or about to do so in July, September, and November. A nest was found at Lake Olomega on April 11, 1926. It consisted of only a few twigs, barely sufficient to keep the egg from rolling about, placed on top of two crossing fronds about 6 feet from the ground. The very sheltered location was the only thing that prevented the haphazard collection of twigs from falling to the ground with the first passing breeze. The male was on the nest and did not fly off until the hunter was directly beneath the nest.

Eggs.—Apparently the red-billed pigeon usually lays but one egg, though possibly very rarely two. The shape is oval or elliptical oval, the shell is smooth and slightly glossy, and the color is pure white. The measurements of 33 eggs average 38.6 by 27.3 millimeters; the eggs showing the four extremes measure **41.4** by 27.9, 39 by **29.5, 34** by 26.5, and 36.8 by **25** millimeters.

Plumages.—The squab, which I found in the nest mentioned above, is described in my notes as nearly naked, the dark, reddish-brown skin being only sparsely covered with short, black pinfeathers. Sennett (1879) says:

The young from the egg have the upper parts plumbeous and sparsely covered with dark hair-like feathers. Under parts are pale and naked. The half-grown young have plumage on the body like the adult. Head and flanks do not become feathered until bird is nearly fledged, and in half-grown young just commences to show.

Food.—There is a brief statement by Sennett (1879) that "their food when" he "saw them was chiefly the hackberry fruit."

Col. Andrew J. Grayson (1871) says:

This is the largest of our pigeons, and abundant in the Marias, as well as in some localities on the mainland. It is gregarious and frequents large forests, feeding upon various kinds of berries, acorns, etc., etc. It migrates from one part of the country to another in small flocks. In some seasons of the year the flesh of this bird has a bitter, disagreeable taste, caused by some species of berry or small bitter acorn upon which it subsists.

Dr. E. W. Nelson (1899), writing of the birds of the Tres Marias Islands, states:

On Maria Magdalena they were numerous in some trees near a group of deserted houses and in old clearings a short distance back from the shore. They came to these trees to feed upon the ripening fruit, but were rather shy. When one becomes startled and takes wing it makes a loud flapping noise that alarms its companions, and then all dash swiftly away. They were less confiding than most of the birds on the islands, but were not so shy as their representatives on the mainland. Wild figs and the small fruit of a tree, probably a species of *Psidium*, or wild guava, were favorite articles of food. Their loud cooing note is uttered at short intervals and is one of the characteristic sounds in the forests they frequent. They are essentially arboreal in habits and are rarely seen near the ground.

Behavior.—The red-billed pigeon reminds one of a domestic pigeon in its swift, strong, steady, and direct flight and in its similar but louder cooing notes.

Sennett (1878) writes:

Like all the Pigeons, it is fond of the water. Any morning will find numbers of all the different species going to and coming from the sand-bars in the river, where they are in the habit of drinking and bathing. The cooing of this bird is clear, short, and rather high-pitched. It is more secluded in its habits than any of the others, except the one I have lately found new to our fauna, *Aechmoptila albifrons*. In point of numbers it is much less numerous than the Carolina and the White-winged Doves; still, it is quite extensively shot for market.

George N. Lawrence (1874) quotes Colonel Grayson as follows:

This fine species in some localities of Western Mexico is quite abundant, particularly in the region of Mazatlan River. It frequents the larger forests, and feeds upon various kinds of berries, acorns, and the tender buds of some trees. It is partially gregarious but is often seen solitary and in pairs. Small flocks of from twenty to fifty migrate from one part of the country to another in search of its favorite food. I have found it at a considerable height on the western slope of the Sierras Madres, feeding upon acorns, that are there in abundance in some seasons. The flesh of this pigeon is tough, and sometimes bitter to the taste, caused by the bitter acorn, and also by an astringent kind of berry, upon which they may be subsisting at the time. As a game bird it is inferior to most of our pigeons for the table, but, being a large and handsome bird it is sure to attract the attention of the gunner. They are not easily approached, however, being very shy, and without the strictest caution the hunter would not be able to fill his bag with this game in a long day's tramp.

DISTRIBUTION

Range.—Central America north to the lower Rio Grande Valley; nonmigratory.

The range of the red-billed pigeon extends **north** to southern Sonora (Sierra de Alamos); northeastern Sinaloa (El Toro); northern Nuevo Leon (Alamo and Trevino); and southern Texas (Carrizo, Lomita, Hidalgo, and Brownsville). **East** to Texas (Hidalgo and Brownsville); Tamaulipas (Matamoras, Rio Pilon, Rio Cruz, Ciudad Victoria, Aldama, and Altamira); Vera Cruz (Tampico, Misantla, Cordoba, and Alvarado); Yucatan (Tunkas and Chichen Itza); and Costa Rica (Guayabo, Tius, and Dota). **South** to Costa Rica (Dota, Naranjo, and Bolson); Salvador (Volcano San Miguel); Guatemala (Esquintla, San Jose, Retalhuleu, and El Naranjo); Chiapas (Tonala); Oaxaca (Topana and Tehuantepec); Guerrero (Acapulco); and Jalisco (Las Penas). **West** to Jalisco (Las Penas); Tepic (Cleofas Island and Tres Marias Islands); southwestern Sinaloa (Escuinapa, Presidio, and Mazatlan); and southern Sonora (Sierra de Alamos).

Casual records.—Bendire (1892) reports three specimens taken near Fort Grant, in the foothills of the Graham Mountains, Ariz., July 25, 1886.

Egg dates.—Texas and Mexico: 79 records, March 1 to August 8; 40 records, April 23 to June 6.

COLUMBA LEUCOCEPHALA Linnaeus

WHITE-CROWNED PIGEON

HABITS

The fine white-crowned pigeon is a permanent resident in the Bahamas and West Indies and occurs in the Florida keys as a summer resident only. Audubon (1840) writes:

The White-headed Pigeon arrives on the Southern Keys of the Floridas, from the Island of Cuba, about the 20th of April, sometimes not until the 1st of May, for the purpose of residing there for a season, and rearing its young. On the 30th of April, I shot several immediately after their arrival from across the Gulf Stream. I saw them as they approached the shore, skimming along the surface of the waters, flying with great rapidity, much in the manner of the common house species, but not near each other like the Passenger Pigeon. On nearing the land, they rose to the height of about a hundred yards, surveyed the country in large circles, then with less velocity gradually descended, and alighted in the thickest parts of the mangroves and other low trees. None of them could be easily seen in those dark retreats, and we were obliged to force them out, in order to shoot them, which we did at this time on the wing.

In creeping among the bushes to obtain a view of them whilst alighted, I observed that the more I advanced, the more they retired from me. This they did by alighting on the ground from the trees, among which they could not well make way on wing, although they could get on with much ease below, running off and hiding at every convenient spot that occurred. These manoeuvres lasted only a few days, after which I could see them perched on the tops of the trees, giving a preference perhaps to dry branches, but not a marked one, as some other species are wont to do.

Of their haunts there he says:

The key on which I first saw this bird, lies about twenty-five miles south of Indian Key, and is named Bahia-honda Duck Key. The farther south we proceeded the more we saw, until we reached the low, sandy, sterile keys, called the Tortugas, on none of which did I see a Pigeon of any kind. On our return from the Tortugas to Key West, our vessel anchored close to a small key, in a snug harbour protected from the sea winds by several long and narrow islands well known to the navigators of those seas. Captain Day and myself visited this little key, which was not much more than an acre in extent, the same afternoon. No sooner had we landed than, to our delight, we saw a great number of White-headed Pigeons rise, fly round the key several times, and all realight upon it. The Captain posted himself at one end of the key, I at the other, while the sailors walked about to raise the birds. In less than two hours we shot thirty-six of them, mostly on the wing. Their attachment to this islet resulted from their having nests with eggs on it. Along with them we found Grakles, Red-winged Starlings, Flycatchers, and a few Zenaida Doves.

The next morning we thought of calling at this little key on our way, and were surprised to find that many new comers had arrived there before us. They were, however, very shy, and we procured only seventeen in all. I felt convinced that this spot was a favourite place of resort to these birds. It being detached from all other keys, furnished with rank herbaceous plants, cactuses, and low shrubs, and guarded by a thick hedge of mangroves, no place could be better adapted for breeding; and, at each visit we paid it, White-headed Pigeons were procured.

On Jamaica, P. H. Gosse (1847) says:

This fine dove is common in almost all situations, but chiefly affects the groves of pimento, which generally adorn the mountain pens. The sweet aromatic berries afford him abundant and delicious food during the pimento season; the umbrageous trees afford him a concealment suited to his shy and suspicious character; and on them his mate prefers to build her rude platform-nest and rear her tender progeny.

Dr. Alexander Wetmore (1916) writes:

The white-crowned pigeon was formerly one of the most abundant species in Porto Rico, but now is found only in a few localities. Gundlach spoke of it as very common in the seventies, but its numbers have undoubtedly greatly decreased. The birds occur mainly near the coast, usually in dense swampy growths, though one was seen near Aibonito; and the few small areas of forest remaining in the lowlands may account for their diminution in numbers. Around Punta Picua, north of Mameyes, they were found preparing to breed in the swamps, where the growth was so dense that it was hard to get near them. They usually came out into the more open portions late in the evening to feed on the fruit of the icaco (*Chrysobalanus* sp.), but even then kept well concealed in the thick leaves.

Courtship.—Audubon (1840) describes its courtship as follows:

The White-headed Pigeon exhibits little of the pomposity of the common domestic species, in its amorous moments. The male, however, struts before the female with elegance, and the tones of his voice are quite sufficient to persuade her of the sincerity of his attachment. During calm and clear mornings, when nature appears in all her purity and brightness, the cooing of this Pigeon may be heard at a considerable distance, mingling in full concord with the softer tones of the Zenaida Dove. The bird, standing almost erect, full plumed, and proud of his beauty, emits at first a loud *croohoo*, as a prelude, and then proceeds to repeat his *coo-coo-coo*. These sounds are continued during the period of incubation, and are at all times welcome to the ear of the visitor of these remarkable islands. When approached suddenly, it emits a hollow, guttural sound, precisely resembling that of the Common Pigeon on such occasions.

Nesting.—The same gifted author says:

The nest is placed high or low, according to circumstances; but there are never two on the same tree. I have found it on the top shoots of a cactus, only a few feet from the ground, on the upper branches of a mangrove, or quite low, almost touching the water, and hanging over it. In general the nest resembles that of the *Columba migratoria*, but it is more compact, and better lined. The outer part is composed of small dry twigs, the inner of fibrous roots and grasses.

In the Bahamas, according to Dr. Henry Bryant (1861)—

It breeds in communities, in some places, as at Grassy Kays, Andros Island, in vast numbers; here the nests were made on the tops of the prickly pear, which cover the whole kay; at the Biminis and Buena Vista Kay, Ragged Island, on the mangroves; and at Long Rock, near Exuma, on the stunted bushes. I do not think they ever select a large kay for their breeding place.

C. J. Maynard (1896) describes an abandoned nesting colony in the Bahamas as follows:

One of the most remarkable sights that I ever witnessed as regards numbers of birds' nests was on one of the Washerwomen Keys off the South shore of Andros. These are small, rocky islets, lying on the barrier reef, and are some twenty-five feet high. On one of these little keys, which did not contain over an acre of land, there were at least ten thousand nests of the White-headed Pigeon. The rocks were mostly covered with a scanty growth of low bushes and with a more luxuriant growth of cacti, and upon both plants and bushes the birds had placed their nests, and some were upon elevated portions of rock, while a few were placed upon the naked ground. So completely covered was the southern and northern portion of the key that the nests were nowhere over two feet apart and often nearer together than that. Unfortunately, however, all of these nests were of the previous year, only a single dove being seen. My boatmen informed me that this rookery was occupied by many thousand birds during the past year, and that the spongers were accustomed to visit the place at night and capture the sitting birds. This statement was confirmed by the remains of torches which were scattered about the island. Many nests contained eggshells, the contents of which had been removed by Buzzards, Man-of-war birds or Gulls. The time of this visit was May 8th, 1884.

On the Isle of Pines, W. E. Clyde Todd (1916) says that "the nest is usually built in the top of a royal palm, but along the Los Indios River the birds were found nesting in the mangroves, rather low down."

Dr. Paul Bartsch writes to me that the nests he found on San Salvador

were all placed in mangrove clumps such as are shown in the habitat photograph. These clumps stood out in the lakes at some distance from shore and furnished splendid protection. Furthermore, as a rule, there was a gray kingbird's nest in the top branches of these clumps, and the kingbird served as an alarmist. The birds were exceedingly shy, regardless of whether they were incubating eggs or taking care of young.

Eggs.—The white-crowned pigeon ordinarily lays two eggs, but sometimes only one. The eggs are elliptical oval or nearly oval, pure white, smooth, and quite glossy. The measurements of 35 eggs average 36.8 by 27 millimeters; the eggs showing the four extremes measure **40.2** by 28.4, 39.4 by **29.5**, **32.3** by 26.2, and 36.8 by **25.2** millimeters.

Plumages.—I have not seen the downy young of this pigeon, but Doctor Bartsch tells me that the down is of a "light buff color." Audubon (1840) says:

The young birds are at first almost black, but have tufts of a soft buff-coloured down distributed mostly over the head and shoulders. While yet squabs they have no appearance of white on the head, and they take about four months before they acquire their perfect plumage. Smaller size, and a less degree of brilliancy, distinguished the female from the male.

Wilson (1832) says:

The young are distinguished by duller tints, and the crown is at first nearly uniform with the rest of their dark plumage; this part, after a time, changes to grey, then greyish white, and becomes whiter and whiter as the bird grows older.

Gosse (1847) writes of some squabs that he raised in captivity:

Both were exceedingly ugly; long-necked, thin-bodied, the head not well rounded, the fleshy part of the beak prominent, and its base unfeathered. The whole plumage was blackish ash-colored, each feather slightly tipped with paler, and the feather of the head terminating in little curled grey filaments, which added to the uncouth appearance of the birds. In a week or two I perceived these filaments were gradually disappearing, and about the beginning of October the small feathers began to clothe the base of the beak; these feathers were greyish-white, and at the same time the grey hue was beginning to spread up the forehead, I believe, by the dropping of the black feathers and their immediate replacement by the white ones. About this time also the general plumage began to assume the blue hue of the adult, in patches; and on the 12th of October, I first observed the beautiful iridescent feathers of the neck, but as yet only on one side. These notes refer to the elder; the other was about two weeks more backward. On the 16th, I first heard it coo; for some time it had now and then uttered a single note, but on this day it gave the whole *Sary-coat-blue*, but short, and in a low tone; and that only once. By the end of November the white had spread over the whole crown, as in the adult; but was not yet so pure or so smooth.

I have seen several young birds, both males and females, apparently in first-winter plumage, taken at various dates from December 14 to May 16, that had evidently matured more slowly or were, perhaps, hatched later. They all had dull gray or dirty white crowns and were otherwise like adult females, except that they had no scaly markings on the neck; the neck and mantle were dull brown, with darker edgings; the wing coverts and scapulars had narrow light tips. There was a mixture of new plumbeous, adult plumage in the back, and the wings and tail were either molting or had been recently renewed, showing that young birds at least have a complete molt in winter and spring. I have been unable to learn anything about the molts of adults.

Food.—Gosse (1847) gives a very good account of the feeding habits of the "baldpate," as he calls this pigeon; he writes:

When the pimento is out of season, he seeks other food; the berries of the sweetwood, the larger ones of the breadnut, and burn-wood, of the bastard cedar, and the fig, and the little ruddy clusters of the fiddle-wood, attract him. He feeds early in the morning, and late in the afternoon; large numbers resort to a single tree (though not strictly gregarious), and when this is observed, the sportsman, by going thither before dawn, and lying in wait, may shoot them one by one, as they arrive. In September and October they are in fine condition, often exceedingly fat and juicy, and of exquisite flavour. In March the clammy-cherry displays its showy scarlet racemes, to which the Bald-pates flock. The

Hopping Dick, Woodpecker, and the Guinea-fowl feed also upon it. In April, Sam tells me he has seen as many as thirty, almost covering a tree, feeding on berries which he believes were those of the bully-tree. Laté in the year they resort to the saline morasses, to feed on the seeds of the black-mangrove, which I have repeatedly found in the craw; I have even seen one descend to the ground beneath a mangrove, doubtless in search of the fallen seeds. In general, however, the Bald-pate is an arboreal pigeon, his visits to the earth being very rare. He often feeds at a distance from home; so that it is a common thing to observe, just before nightfall, straggling parties of two or three, or individuals, rushing along with arrowy swiftness in a straight line to some distant wood.

Doctor Wetmore (1916), on Porto Rico, found that "five stomachs examined contained vegetable matter only, composed of drupes and fruits of fair size. The icaco and berries of various palms (palmo real and lluma) are favorites with these birds, while a tree known as palo blanco (*Drypetes* sp.) is said to furnish them food in season. No cultivated crop is injured, the bird depending wholly upon wild fruits for its sustenance." Mr. March, as quoted by Baird, Brewer, and Ridgway (1905), states that "they commit serious depredations on the Guinea-corn fields, not only by the quantity they devour, but by breaking down the brittle corn-stalks with the weight of their bodies."

Behavior.—Maynard (1896) writes:

The White-headed Pigeons are thoroughly at home among the thick branches of the trees and shrubbery of the Bahamas, moving about among them as easily as do the smaller-perching birds, and they make very little noise. When surprised by an intruder they will remain perfectly quiet until approached within a few yards, when they will spring rapidly into air, rise to the tips of the woodland, and dart off with an exceedingly rapid flight; in fact, few, if any birds, can fly any more quickly than do these Pigeons. I have shot several in air, as they rose from the bushes and darted away, but I never attempted to shoot one as it passed me at full speed at right angles. When dashing along at this headlong speed they will suddenly alight upon a branch or on the ground, without the beating or fluttering of the wings, which usually attends a similar abrupt stoppage in most birds of a similar size and which is so noticeable in our domestic pigeon.

Dr. Thomas Barbour (1923) says:

The White-crowned Pigeon is of irregular appearance in any given locality, its presence depending on the abundance of the fruits upon which it feeds. It is essentially a coastal form, and one which is always gregarious. It roosts in great hordes, usually on some mangrove islet, and bands sally forth each morn to feed, returning from their distant foragings at dusk. Then they rush and swirl into the greater resorts, or *palomares*, in incredible hosts. Famous roosts are Moraine Cay north of Grand Bahama, where I have shot, and Green Cay, south of New Providence. Gundlach speaks of their seldom being seen in Cuba except when nesting, which they do at various seasons of the year. This intermittent appearance is noticed everywhere. They are in the Florida Keys in summer only, but not every summer in equal numbers; in certain of the Bahamas they abound at one season, elsewhere at others.

The fact is, the individual bands are capable of long flights, and move far and wide as food supplies dictate. Great numbers are slaughtered by hunters, who build an ambush near roost or rookery and kill the returning birds as they fly in just before dark. Unfortunately, this leaves many young birds to starve.

Voice.—In addition to the notes mentioned under courtship, I might quote what Maynard (1896) says:

The notes of this Pigeon are very loud and characteristic, sounding something like *wof, wof, wo, co-woo.* The first three notes are repeated several times, then the *co-woo* is long drawn out; all being in as low a key as the hoot of an owl. The entire cry is cleverly imitated by the Creoles when they wish to decoy the bird within gun shot, but there is a certain tremulousness in the real notes which cannot be imitated by the human voice.

Again Gosse (1847) gives a still different wording of it as "*sary-coat-blue*, uttered with much energy, the second syllable short and suddenly elevated, the last a little protracted and descending."

Game.—Nearly everyone who has written about the birds of the Bahamas or the West Indies has referred to the white-crowned pigeon as one of the finest game birds on these islands. Its game qualities are excellent, as it is rather wild and a very swift flier; and its flesh is delicious on the table. It was formerly much more abundant than it is now, for it has been shot in enormous numbers while flying to or from its breeding, feeding, or roosting grounds. As the pigeons were shot most easily on or about their breeding grounds, many young were left to starve in the nests and many older squabs were taken for food or to be raised in captivity. With such wholesale slaughter the species is rapidly disappearing and is badly in need of protection. Maynard (1896) relates the following tale:

About the first week in July, previous to 1884, sportsmen from Nassau had been in the habit of visiting Green Key and shooting the breeding Pigeons as they flew from their nests to cross to Andros Island, some fifteen miles distant, where they are said to go daily for food and water. Many of the nests of the previous season which I had examined on Green Key contained broken eggs that contained the remains of half-formed young, and in some of the nests were the skeletons of newly hatched young; the parents of both eggs and young had doubtlessly been killed as they left the nests. This sight was a most piteous appeal to humanity. I was informed by one of my boatmen, who had accompanied hunting parties to the key, that so great was the slaughter of Pigeons that many more were killed than were needed, and that he had frequently seen hundreds of birds buried in the sand of the beach near where they were shot. Upon my return to Nassau I promptly stated the facts as I had observed them to the Governor, Sir Henry A. Blake, and, as I have elsewhere stated in this work, through his ready and sympathetic cooperation a law was enacted protecting these Pigeons during the breeding season.

Winter.—The wanderings of the white-crowned pigeons have been referred to in the foregoing quotation from Doctor Barbour. These are all winter, or between breeding seasons, wanderings. The birds are evidently absent from the Florida Keys in winter, and Mr. Todd (1916) says of their exodus from the Isle of Pines:

> This is a common species everywhere, except in the Cienaga, appearing in flocks late in February, and remaining until the last of September. Although a few stragglers may be seen through the winter months, the vast majority of the individuals withdraw at that season from their usual range and according to native report resort to the "south coast," in great numbers. It is one of the most numerous birds of the various mountain ridges in the interior of the island during the breeding-season, which begins in May.

This last movement is northward to the larger land area of Cuba, probably to find a better food supply. The birds that breed on the Florida Keys are probably those that migrate northward to the mainland of Florida in winter. Recent information indicates that these pigeons migrate to extreme southern Florida occasionally, perhaps regularly, in winter. Gilbert R. Rossignol writes me that he saw some of these pigeons between Flamingo and Coot Bay five different times between December 30, 1928, and February 24, 1929. He says: "I recall seeing three at one time and a pair here and there between the first and third bridges, but mostly around the second bridge, where there is considerable open country due to some farming and a burnt district." Frank N. Irving, who was with Mr. Rossignol on some of these trips, tells me that he saw the white-crowned pigeons in the same region during March, 1928, and January and February, 1929. I wrote to Harold H. Bailey for his experience, and he replied that he has taken several on the mainland at Cape Sable and has seen flocks there of 20 or 30 birds, or more, many times; they were feeding in the higher foliage of the Florida holly and other berry-bearing trees. He has not found a nest on the mainland and thinks they come there to feed only and spend the winter near abundant food. He says that a similar movement takes place in the Bahamas, where the pigeons desert their breeding grounds on the outlying keys and come to Andros Island to spend the winter.

DISTRIBUTION

Range.—Southern Florida, the West Indies, and locally in central Central America.

The range of the white-crowned pigeon extends **north** to southern Florida (probably Dry Tortugas, Key West, Bahia Honda, Cape Sable, Coconut Grove, and Indian Key); and the Bahama Islands (Abaco Island). **East** to the Bahama Islands (Abaco, New Providence, Green Cay, and Mariguana Island); and the Lesser Antilles (Barbuda Island and Antigua Island). **South** to the Lesser Antilles (Antigua and St. Croix Islands); Porto Rico (Vieques Island, Punta Picua, and Mona Island); Haiti (San Domingo and Jacmel);

Jamaica (Spanishtown); Nicaragua (Great Corn Island); Honduras (Ruatan Island); British Honduras (Half Moon Cay, Turneff Island, and Belize); and Oaxaca (Salina Cruz). **West** to Oaxaca (Salina Cruz and Tehuantepec); Yucatan (Cozumel Island and Buchotz); and southern Florida (probably Dry Tortugas).

Scott (1889) records a specimen taken at Punta Rassa, Fla., August 16, 1886, and Cory (1891) noted them in the winter of 1891 at Caicos Islands of the eastern Bahamas.

Although given to considerable wandering, the white-crowned pigeon does not appear to have a regular migration, at least in the main part of its range. Audubon (1840) stated that they arrived on the Florida Keys from April 20 to May 1, while Maynard (1896) reported their arrival in this region about the 1st of June and their departure late in October. Some appear to winter regularly in southern Florida.

Egg dates.—Bahamas: 48 records, May 21 to December 8; 24 records, June 18 to 29.

COLUMBA SQUAMOSA Bonnaterre

SCALED PIGEON

HABITS

The scaled pigeon is a West Indian species, which is included in our list as an accidental straggler to Key West, Fla., and therefore extralimital. John W. Atkins (1899) records the incident, as follows:

On October 24, 1898, an adult female of this species was shot on the Island of Key West, and brought to me in the flesh, by a young collector in my employment, who found it among some Doves in the possession of a dove hunter, who had shot it from a wild fig tree on the outskirts of the town.

Dr. Thomas Barbour (1923) says of its haunts in Cuba:

In western and central Cuba this beautiful Pigeon is by no means common at the present time. It is a highland bird but not exclusively confined to mountain ranges. One finds the Torcaza Morada usually perched high on the dead branches of some towering tree, most often on cliffs or steepish slopes. The birds seem sluggish and make short flights, booming their heavy, sonorous call through the heat of the day. Attempt to approach, and the bird is off, for no Pigeon is more alert. Its flesh is excellent, and the body is heavy beyond other local species. In appearance in the field it is larger and darker than a domestic pigeon, and it has a patch of brilliant metallic feathers on each side of the neck. It is never terrestrial. Ramsden has given an excellent account of the persecution it suffers in Oriente, where it appears at intervals in great numbers. Ramsden also recalls breeding rookeries which Gundlach never found. This gregarious habit is beyond a doubt confined to the wild Eastern Province, where the Scaly-naped Pigeon still is more abundant than elsewhere. For never elsewhere have I met with numbers which would allow of killing five-thousand individuals in a couple of weeks in one locality. It still occurs in the regions mentioned by Gundlach, in the mountains of Vuelta Abajo and Trinidad, but in both these highlands it may today be seen regularly only in pairs, trios, or small bands, and probably never over a few dozen in a day—and many days far fewer would be seen. Slaughter for food and sport has already very greatly reduced this splendid species, and it needs now protection, which probably will not be granted to it, and which, if granted, can not be enforced.

In Porto Rico, in 1912, Dr. Alexander Wetmore (1927) "found the species common in the hills and mountains, where small tracts of natural forest and extensive coffee fincas offered it shelter, but saw it seldom on the coastal plain, where there was in the main little shelter to attract it."

Nesting.—As to its nesting habits he says:

It is a common belief in Porto Rico that the scaled pigeon is only a migrant on the island—a belief promulgated, it may be said, by gunners who desire an open season during the entire year. That this is erroneous was proven on March 8, 1912, when without special search I found three nests on El Yunque, while there was no doubt whatever that the dozens of birds flushing on every hand were breeding. The three nests definitely located were made of sticks loosely piled together and placed about fifteen feet from the ground on horizontal limbs, or on refuse piled on large air plants. Two were empty, while one contained a single egg, plain white in color with a slight gloss, which was collected. This egg had had about five days' incubation. It measures 34.8 by 26.7 mm. At Maricao, on June 1, a native brought me a young bird about two-thirds grown, and said it was the only one in the nest. Gundlach has said that two eggs are laid, but from these instances it would seem that a single egg in a set is not unusual.

A nest found by Harry A. Beatty (1930) on St. Croix "was a very frail platform of coarse sticks, situated 25 feet up on the forked branch of a mahogany tree. I could see plainly the two glossy white eggs through the nest from below."

Eggs.—The egg, taken by Doctor Wetmore, referred to above, is oval in shape, slightly elliptical, rather glossy, and pure white. The measurements of seven eggs average 36 by 26.7 millimeters; the eggs showing the four extremes measure 37.6 by 27.4, 35.5 by 27.5, 31.8 by 26.7, and 34.5 by 25.7 millimeters.

Plumages.—I have not seen the downy young. Young birds in juvenal plumage, in July, are like the adult, but lack the richly colored, scaly markings on the hind neck; the head and neck are "warm blackish brown," shading off to "dark plumbeous" on the back and to "walnut brown" on the upper breast, the latter feathers having darker edgings; the wing coverts are edged with "walnut brown." I have seen adults showing wing molt from September to December, but I can learn nothing further from the scanty material available.

Food.—Doctor Wetmore (1927) says that "it feeds on wild berries and fruits, with occasional succulent leaves or shoots. The berries of various palms, wild figs, the moral (*Cordia*), and jagua (*Genipa americana*), with various wild legumes, are eaten extensively." He says elsewhere (1916):

All of the smaller wild fruits in season appear to furnish food, and these are so abundant that cultivated fields are not molested. The fruits eaten, though sometimes of comparatively large size and with hard stony pits, are swallowed entire. The strong muscular gizzard of the bird has a tremendous triturating power, however, and the fruits are easily crushed and the meaty centers opened to the processes of digestion.

Behavior.—Referring to the habits of the scaled pigeon in Porto Rico, Doctor Wetmore writes:

The dense forests covering the slopes of El Yunque de Luquillo, in the northeastern part of Porto Rico, harbored great numbers of these birds, which ranged commonly up to 2,500 feet above the sea. In late afternoon and evening, near the Hacienda Catalina, it was a common sight to see them circling about high in the air. In spite of their large size, they were difficult to see in the trees, even in the thin foliage of the cacao rosetta (*Sloanea berteriana*). Thus it often happened that bird after bird flew out from amid the limbs, with loudly clapping wings, yet failed to offer a shot, while I peered vainly upward in search for their hidden companions. When one of the big males chanced to drop in near another, a great flapping of wings ensued until one was forced to take flight. The ordinary call note was a loud, strongly accented *who-hoo-hoo*, while a burring guttural *hoo-o-o-o*, given with a throaty rattle, was almost startling when heard from directly overhead. Many birds descended to feed amid the tall trees fringing small streams at the foot of the mountain, and some were encountered in the dense, swampy forests near Punta Picua, beyond Mameyes. Males rest and call at times in the tops of tall, dead trees.

Fall.—Doctor Wetmore says that

The paloma turca is said to occur in large flocks during fall, and to gather in numbers where wild fruits are ripening, at which time many are killed. It is common belief that these flocks are entirely migratory, but there can be no doubt that they come mainly from the forests on El Yunque and elsewhere in the interior.

Game.—The same writer states:

The species is the only game bird of importance in the inland region of Porto Rico and affords excellent sport, as it is wary, strong on the wing, and is found only in the wildest, roughest country. It should be protected from February 1 to October 15 each year, if not longer, to permit it to breed, as otherwise it cannot maintain its status.

Austin H. Clark (1905), referring to the southern Lesser Antilles, writes:

This is the chief game bird of these islands, and is much hunted. The flight is rapid and powerful, and the birds regularly cross over from one island to another to feed, returning at night to roost on the smaller keys. Formerly numbers could be shot any evening about four o'clock from Clifton House, Union Island, as they flew from that island over to Prune to spend the night. They could be obtained at Hermitage House, Carriacou, in the same manner, as they flew past, going to one or other of the small keys near by.

DISTRIBUTION

Range.—The West Indies; accidental at Key West, Fla.; non-migratory.

The range of the scaled pigeon extends **north** to Cuba (Nueva Gerona and Guantanamo); northern Haiti (Massif du Nord, Morne Salnave, and Catare); northern Porto Rico (Desecheo Island, Culebra Island, and Culebrita Island); and the Lesser Antilles (St. Eustatius and Guadaloupe). **East** to the Lesser Antilles (Guadeloupe, Dominica, Martinique, Santa Lucia, St. Vincent, the Grenadines, Carriacou, and Grenada). **South** to the southern Lesser Antilles (Grenada); southern Haiti (Selle Mountains and Hatte Mountains); and the Isle of Pines, Cuba (Nueva Gerona). **West** to the Isle of Pines (Nueva Gerona).

Egg dates.—West Indies: 2 records, March 8 and May 26.

ECTOPISTES MIGRATORIUS (Linnaeus)

PASSENGER PIGEON

HABITS

Contributed by Charles Wendell Townsend

The passenger pigeon, or wild pigeon, as it was often called, is generally believed by ornithologists to be extinct. Of the mighty hosts of this splendid bird that swarmed over the country, not a single individual is thought to remain alive to-day, and yet within the memory of men not yet old, the bird was well known, and the possibility of its extinction was far from their thoughts. Indeed, whenever laws were proposed for conserving the bird, the cry at once went up that it needed no protection, for its numbers and the extent of country over which it ranged were both so huge that protection seemed unnecessary. Even the tardy protective laws passed by some States were largely disregarded.

At last, in 1910, 1911, and 1912, when it was too late, attempts were made to save the bird, and rewards that totaled more than $1,000 were offered for evidence that it was living and nesting—the live bird, not the dead one was sought. But it was all in vain. The passenger pigeon appears to have gone the way of the dodo and the great auk.

James H. Fleming, who has made the most complete and critical studies of the recorded specimens of the passenger pigeon, believes (1907) that " for all practical purposes the close of the nineteenth century saw the final extinction of the passenger pigeon in the wild state and there remained only the small flock, numbering in 1903 not more than a dozen, that had been bred in captivity by Prof.

C. O. Whitman of Chicago. These birds, the descendants of a single pair, had long before that ceased to breed." The last of this group, a female in the Cincinnati Zoo, died of old age in September, 1914.

The last records of wild birds that are based on specimens about which there is no doubt appear all to have been taken in 1898—one, an adult male taken at Lake Winnipegosis, Manitoba, on April 14 (Fleming, 1903 and 1907); an immature male at Owensboro, Ky., on July 27, 1898 (Fleming, 1907), now in the Smithsonian Institution; and an immature bird taken at Detroit, Mich., on September 14, 1898 (Fleming, 1903 and 1907). Fleming (1907) adds in a footnote: "There is a mature female in the collection of the Carnegie Institution of Pittsburg, Pa., marked 'Pennsylvania, August 15, 1898,' but without further locality." Still another specimen, the fifth for 1898, was a young male shot by Addison P. Wilbur at Canandaigua, N. Y., on September 14, 1898, and is recorded by E. H. Eaton (1910), who states that he saw it killed, and that it "was unquestionably reared in the spring of 1898, as it was just assuming the adult plumage."

L. E. Wyman (1921), of the Museum of History, Science, and Art, Los Angeles, Calif., records the following: "A mounted specimen of passenger pigeon acquired by the late F. S. Daggett, in January, 1920, and now in the Daggett collection, deposited in this museum, bears the following label: 'Passenger Pigeon, male, No. 315, Coll. of Geo. S. Hamlin. Shot by a Swede, North Bridgeport, Fairfield Co., Conn., Aug., 1906.'" This seems conclusive, but under date of September 19, 1930, Mr. Fleming writes me, "I suggest that August, 1906, is the date the bird was acquired by Hamlin and who then wrote the label." Mr. Fleming calls attention to the fact that in "The Birds of Connecticut," by Sage, Bishop, and Bliss (1913), the last authentic specimen for the State is recorded for October 1, 1889, and, what is very significant, in notes communicated by Hamlin there is no mention of a specimen taken later than 1892.

A specimen in the Cornell University Museum, recorded by S. C. Bishop and A. H. Wright (1917), was shot at Clyde, N. Y., by J. L. Howard, who, when more than 80 years of age in 1915, stated from memory that it was taken about 6 years previous, that is, in 1909. This date is rendered extremely doubtful from the fact that on the bottom of the mount is the date July 5, 1898, although this may have been the record of another bird on the same mount. Mr. Howard gives a circumstantial account of his shooting the bird, and states he had not seen any passenger pigeons before this for about 15 years.

It is true that since this time there have been many sight records of the bird reported, some of which, at least some of the earlier ones, are doubtless authentic, and at the present day there seems to be a recrudescence of the belief that the bird still exists. In Science of February 14, 1930, Prof. Philip Hadley (1930) notes on the authority of others—not ornithologists—sight records of these birds in northern Michigan. Unfortunately, the mourning dove is often mistaken for the passenger pigeon, and in the West the band-tailed pigeon has been similarly mistaken. The distinguishing field marks of these birds will be discussed later, but it seems to be a common idea that the passenger pigeon is easily recognized by its size, which is larger than that of the mourning dove. All ornithologists know, however, in the absence of direct comparison, how deceptive difference in size may be, and they are well aware that not only to an uncritical observer but even to an expert ornithologist, a mourning dove may often look as large as a passenger pigeon. The fact that the observer has seen thousands of passenger pigeons in years before, and has handled and plucked them, does not necessarily mean he is a good judge of the bird. The wish is father to the thought, especially in the unscientific, and it has been proved over and over again that birds reported as passenger pigeons have turned out to be mourning doves.

It may be worth while to enter here the following example among many of the manner in which the will to believe in the existence of the passenger pigeon surmounts all obstacles and all evidence. Dr. C. F. Hodge (1912), who had charge of the offer of $1,000 for the location of a nesting pair of passenger pigeons, received a letter from a man in Maine that he had shot a bird "that proved to be a passenger pigeon," and that he had had it mounted. Doctor Hodge sent him descriptions and colored plates of the passenger pigeon and mourning dove, and he underscored in red ink the comparative lengths of these two birds, and he asked the man if, after study of these, he still thought the bird a passenger pigeon to send it to him. The specimen arrived and proved to be a mourning dove. In a recent newspaper article, an old-time pigeon trapper is quoted as saying: "No one familiar with this bird could possibly make a mistake. The mourning dove, although it has the same coloration, is smaller. I can tell a mourning dove from a pigeon as far as I can see it." With this belief it is easy to see passenger pigeons! This, however, is not conclusive that all the reports are erroneous, and, although the evidence points strongly to the extinction of the passenger pigeon, it is proper to keep an open mind on the subject and investigate plausible clues.

There is a popular idea that the passenger pigeon mysteriously disappeared, and that, while still enormously numerous, it suddenly ceased to exist. Its annihilation has been popularly attributed to various natural phenomena, and it has even been rumored that the bird has migrated to South America, but the real cause for its extinction—man—is not mentioned. The natural phenomena supposed to be causative of the extinction are: Epidemics, tornadoes, early deep snowstorms, forest fires, strong winds, while the birds were crossing large bodies of water, causing exhaustion and death by drowning. Some of these were reported while the bird was still common in other localities. Circumstantial accounts were published of immense numbers drowned in the Gulf of Mexico, a region beyond the usual range of the bird. The destruction of the forests by ax and by fire undoubtedly has been a large detrimental factor in the life history of the pigeons, for the forests supplied their principal food as well as roosting and nesting places. These natural causes had acted for countless ages but the passenger pigeon survived, but when the white man arrived on the North American Continent, and especially after the pigeon became a commercial asset, its destruction, unless some curb was put on the slaughter, was ordained.

The evidence that man is responsible for the enormous destruction is voluminous and convincing, but why, it may be asked, did not a few escape and continue the race when the numbers became so small that the pursuit of the bird was no longer commercially profitable? It is believed by some that the bird still exists in small numbers in remote parts of northern Michigan and in Canada. In answer to this it may be said that a bird accustomed for ages to living together in large numbers and close ranks, whether in feeding, migrating, roosting, or nesting, might find it impossible to continue satisfactorily these functions with greatly reduced and scattered ranks. It is probably no mere figure of speech to say that under these circumstances such a communistic bird would "lose heart," nor is it fanciful to suppose that sterility might in consequence affect the remnants. The ease and thoroughness with which the squabs were destroyed, even in preference to the adults, would alone account for the extinction. Just as a forest finally dies of disease, accident, and old age if all the seedlings are destroyed by intensive grazing, so any species of bird is doomed in the same way to extinction if its offspring are annihilated before they reach maturity.

The probable final stages in the disappearance of the passenger pigeon are well portrayed by W. B. Barrows (1912) as follows:

In the opinion of the writer the most probable cause for the disappearance of the pigeon lies in the fact that, through the clearing of the forests and the increasing persecution by man, the birds were driven from one place to another and gradually compelled to nest farther and farther to the north, and under conditions successively less and less favorable, so that eventually the larger part of the great flocks consisted of old birds, which, through stress of weather and persecution, abandoned their nesting places and failed to rear any considerable number of young. Under such conditions they would naturally become weaker, or at least less resistant, each year, and in the attempt to find nesting places in the far north they may have been overwhelmed by snow and ice during one or two of the unusually severe summers that occurred between 1882 and 1890.

Dr. Thomas S. Roberts (1919) sums up the whole situation when he says of this bird in Minnesota:

Formerly an abundant summer resident. Rapidly diminished in numbers between the years 1878 and 1885, finally disappearing entirely between 1890 and 1900. It is now extinct everywhere. All other theories to the contrary, the extermination of this bird was the result of ruthless and wholesale destruction by man.

Courtship.—The early writers describe a courtship very much after the fashion of the domestic pigeon. Thus Audubon (1840) says:

The male assumes a pompous demeanor, and follows the female, whether on the ground or the branches, with spread tail and drooping wings, which it rubs against the part over which it is moving. The body is elevated, the throat swells, the eyes sparkle. He continues his notes, and now and then rises on the wing, and flies a few yards to approach the fugitive and timorous female. Like the domestic pigeon and other species, they caress each other by billing, in which action the bill of the one is introduced transversely into that of the other.

All that Wilson (1832) has to say on the subject is that "they have the same cooing notes common to domestic pigeons, but much less of their gesticulations."

Wallace Craig (1911a) believes that Audubon's description of the courtship and voice "came largely by reasoning by analogy from the domestic pigeon and from the author's charming but somewhat unscientific imagination," because Craig's careful studies of captive passenger pigeons in the aviary of Prof. C. O. Whitman showed various peculiarities and characteristics, unlike those of the domestic pigeon, characteristics that "all seem connected, directly or indirectly, with the extreme gregariousness, the breeding in vast colonies." Thus he found no bowing or strutting or charging as in other species. The male emits a loud *kek* and chattering notes and waves his wings in a single sweep, or flaps the wings repeatedly, holding tight to the perch the whole body, head, and tail rising and falling with each stroke.

When close beside the female, the male *Ectopistes* had a way all his own of sidling up to her on the perch, pressing hard upon her, sometimes putting his neck over her neck, "hugging" her as Professor Whitman expressed it. * * * When the female becomes amorous, instead of edging away from the male when he sidles up to her, she reciprocates in the hugging, pressing upon the male in somewhat the same manner that he presses upon her. * * * The act of billing, which occurs in all pigeons before copulation, is in *Ectopistes* reduced to a mere form. * * * The bills are quickly clasped, shaken for a fraction of a second, and as quickly separated; the performance is precisely like a brief, quick handshake. It is probable that there is no passing of food from one mouth to the other.

The passenger pigeon was essentially an arboreal bird and conducted its courtship in the trees. Chief Pokagon (1895), the last Pottawattomie chief, says that about the middle of May, 1850, at a great nesting place in Michigan, "the trees were filled with them sitting in pairs in convenient crotches of the limb, now and then gently fluttering their half-spread wings and uttering to their mates those strange, bell-like waving notes which I had mistaken for the ringing of bells in the distance."

During the courtship period much fighting occurred among the males in their crowded quarters, but little injury resulted. Craig (1911a) says:

The male *Ectopistes* was a particularly quarrelsome bird, ever ready to threaten or strike with his wings (though perhaps not quite so ready with his beak), and to shout defiance in his loud strident voice. * * * He was an aggressive, violent threatener, but not a real fighter.

Nesting.—As the passenger pigeon approached extinction it nested in small companies or singly, but until its numbers were greatly reduced the nestings took place in communities of great size. The accounts of these nesting communities by various authors are almost invariably accompanied with a description of the means used for the capture and slaughter of the adults and squabs. These will be quoted later under enemies. There are, however, a few descriptions of the nestings alone. Wilson (1832) says they

are generally in beech woods, and often extend in nearly a straight line across the country for a great way. Not far from Shelbyville in the state of Kentucky, about five years ago, there was one of these breeding places, which stretched through the woods in nearly a north and south direction, was several miles in breadth, and was said to be upwards of forty miles in extent! In this tract almost every tree was furnished with nests wherever the branches could accommodate them."

More than 100 nests in one tree alone were not uncommon. Mershon (1907) says a game dealer in Detroit had seen a nesting place in Wisconsin that extended through the woods for 100 miles.

In the spring of 1885 William Brewster and Jonathan Dwight went to Michigan hoping to see a nesting of the passenger pigeon. In this they were unsuccessful as they found only one nest, but they collected a great deal of valuable information. Brewster (1889) quotes S. S. Stevens of Cadillac, "a veteran pigeon netter of large experience—a man of high reputation for veracity and carefulness of statement," as follows:

The last nesting in Michigan of any importance was in 1881, a few miles west of Grand Traverse. It was of only moderate size, perhaps eight miles long. The largest nesting he ever visited was in 1876 or 1877. It began near Petosky and extended northeast past Crooked Lake for twenty-eight miles, averaging three or four miles wide. Nestings usually start in deciduous woods, but during their progress the pigeons do not skip any kind of trees they encounter. The Petosky nesting extended eight miles through hard-wood timber, then crossed a river bottom wooded with arbor-vitae, and thence stretched through pine woods about twenty miles. For the entire distance of twenty-eight miles every tree of any size had more or less nests, and many trees were filled with them None were lower than fifteen feet above the ground.

Mr. Stevens also stated that "so rapidly did the colony extend its boundaries that it soon passed literally over and around the place where he was netting, although when he began, this point was several miles from the nearest nest."

The nests were simple frail structures, composed of sticks and twigs crossing one another and supported by forks of the branches at a height of 10 to 50 feet or more. The nest was often so loosely made that the egg or squab could be seen through it from below.

Ruthven Deane (1896) observed the habits of captive birds belonging to David Whittaker, of Milwaukee. These birds had increased from two pairs procured in 1888 to six males and nine females in 1896.

When the pigeons show signs of nesting, small twigs are thrown on the bottom of the enclosure, and on the day of our visit, I was so fortunate as to watch the operations of nest building. There were three pairs actively engaged. The females remained on the shelf, and at a given signal which they only uttered for this purpose, the males would select a twig or straw, and in one instance a feather and fly up to the nest, drop it and return to the ground, while the females placed the building material in position and then called for more. In all of Mr. Whittaker's experience with this flock he has never known of more than one egg being deposited. * * * The eggs are usually laid from the middle of February to the middle of September, some females laying as many as seven or eight during the season, though three or four is the average.

This fact of one egg to the clutch is also confirmed by Prof. C. O. Whitman. Wilson (1832) also confirms this, "a circumstance," he adds, "in the history of the bird not generally known to naturalists." Audubon and many who have followed and copied him state, however, that two eggs form the clutch, as is the case with the domestic and other pigeons. There are, however, a sufficient number of independent observations that show that two eggs are often, or, according to others, rarely found. It is possible, in the crowded nesting places, that two females may have laid in the same nest. Brewster (1889) questioned Stevens closely on the number of eggs in the nest. "He assured me that he had frequently found two eggs or two young in the same nest, but that fully half the nests which he had examined contained only one." Mr. Brewster adds: "Mr. Stevens is satisfied that pigeons continue laying and hatching during the entire summer. They do not, however, use the same nesting place a second time in one season, the entire colony always moving from twenty to one hundred miles after the appearance of each brood of young." When food was plentiful it is believed that three or even four broods of young were raised in a season.

Eggs.—[AUTHOR'S NOTE: The passenger pigeon laid either one or two eggs in a set, probably more often only one. Most of the authentic eggs that I have seen are decidedly elongated, elliptical-ovate in shape, but this may not be the invariable rule. The shell is smooth and slightly glossy. The color is pure white. The measurements of 32 eggs, apparently authentic, average 38.2 by 27 millimeters; the eggs showing the four extremes measure **45.2** by **29.7**, **33.5** by 26, and 36.2 by **24.9** millimeters.]

Young.—The incubation period is 14 days as observed exactly in captive birds (Deane, 1896). Both birds take part in the incubation, a point that was accurately determined by netters as shown by their catches. Brewster (1889) quoting Stevens, says:

Both birds incubate, the females between two o'clock p. m. and nine or ten o'clock the next morning; the males from nine or ten o'clock a. m. to two o'clock p. m. The males feed twice each day, namely, from daylight to about eight o'clock a. m. and again late in the afternoon. The females feed only during the forenoon. The change is made with great regularity as to time, all the males being on the nest by ten o'clock a. m. During the morning and evening no females are ever caught by the netters; during the forenoon no males. The sitting bird does not leave the nest until the bill of its incoming mate nearly touches its tail, the former slipping off as the latter takes its place.

Ruthven Deane (1896) says:

During the first few days, after the young is hatched, to guard against the cold, it is like the egg, concealed under the feathers of the abdomen, the head always pointing forward. In this attitude, the parents, without changing the sitting position or reclining on the side, feed the squab by arching the head and neck down and administering the food. The young leave the nest in about fourteen days.

As the nesting of passenger pigeons often began before snowfalls had ceased, the following note by Frank J. Thompson (1881) showing their hardihood under such circumstances is of interest:

In confinement in Cincinnati early in March, 1878, two pairs began nesting, the male carrying the material while the female busied herself in placing it. A single egg was soon laid in each nest and incubation commenced. On March 16 there was quite a heavy fall of snow and on the next morning I was unable to see the birds on their nests on account of the accumulation of the snow piled on the platforms around them. Within a couple of days it had disappeared and for the next four or five nights a self-registering thermometer, hanging in the aviary, marked from 14° to 19°. In spite of these drawbacks both of the eggs were hatched and the young ones reared.

Brewster (1889) says:

The young are forced out of their nests by the old birds. Mr. Stevens has twice seen this done. One of the pigeons, usually the male, pushes the young off the nest by force. The latter struggles and squeals precisely like a tame squab, but is finally crowded out along the branch and after farther feeble resistance flutters down to the ground. Three or four days elapse before it is able to fly well. Upon leaving the nest it is often fatter and heavier than the old birds; but it quickly becomes much thinner and lighter, despite the enormous quantity of food that it consumes.

Another point about young birds was brought out by Brewster (1889) from Stevens. He writes:

On one occasion an immense flock of young birds became bewildered in a fog while crossing Crooked Lake and descending struck the water and perished by thousands. The shore for miles was covered a foot or more deep with them. The old birds rose above the fog and none were killed.

As with all the pigeons, the passenger pigeon fed its young at first with the so-called "pigeon milk," the secretions from the glandular crop, mixed with the food in the crop, and serving to digest it. It is probable that the young inserted its bill into that of the mother to obtain this "milk," although some writers state that the mother inserted her bill into that of the young. In these large colonies, communal feeding of squabs that had lost their mothers has been observed.

Plumages.—[Author's note: I have seen but one nestling of the passenger pigeon; this is quite completely, but very thinly, covered with long, soft, hairlike, "honey-yellow" down. This and two specimens showing the development of the juvenal plumage are in the Museum of Comparative Zoölogy, in Cambridge, Mass. A small juvenal female, 8 inches long, is nearly fledged, but the yellowish down filaments still adorn the head, neck, and breast; the crown and upper back are "bister" or "warm sepia," shading off to "natal brown" on the breast and to "wood brown" on the lesser wing coverts and scapulars; the feathers of the back, wing coverts, and scapulars are edged with whitish, or pinkish, buff; the greater coverts shade from "fawn color" to "French gray," and are more narrowly edged; many of the inner coverts have a large patch of "bister" on the outer web; the inner primaries are tipped and broadly edged on the outer web with "Mikado brown," the edgings gradually disappearing outwardly; the lower back and rump are "Quaker drab" to "mouse gray"; the underparts shade off from "wood brown" on the flanks to whitish on the belly and chin.

Another young bird is fully fledged in juvenal plumage; the feathers of the head, neck, and breast, now fully grown, have narrow, buffy-white edgings; many of the outer wing coverts, especially the greater, are "French gray"; the tail is shorter than the adult's, the central rectrices are browner and the lateral ones are darker gray, so that there is less contrast in the tail.

The postjuvenal molt took place during August, September, and October, earlier or later according to the date of hatching. It apparently involved all the contour plumage, the wing coverts, and the tail. In the first winter plumage, which was practically adult, the sexes were differentiated. Adults had one complete molt in August and September.]

Food.—The immense forests of North America, before they were devastated and wasted by the ax and fire of the white settlers, furnished an inexhaustible supply of food to the enormous hosts of wild pigeons. The principal food was mast—acorns, beechnuts, and chestnuts—and also the fruit or nuts of any of the forest trees, as well as wild berries and fruits, such as cherries, raspberries, blueberries, currants, pokeberries (also called pigeon berries), and strawberries, grain, seeds of weeds, and grasses. I have the records of the stomach contents of three passenger pigeons I shot at Magnolia, Mass., two in September, 1877, and one in July, 1878. All contained small pebbles and the hips and seeds of wild roses. S. C. Bishop (1924) reports that in the partially mummified remains of a passenger pigeon taken in New York State many years ago, he found in the crop 25 well-preserved seeds of the sugar maple and close to the base of each fruit the wing had been sheared off and discarded. Aughey (1878) found that six stomachs of birds, taken in Nebraska, each averaged eight locusts and two other insects and some seeds. Audubon (1840) found three entire acorns in one crop and, in the stomach, fragments of others and three pieces of quartz. Records of stomach contents of the passenger pigeon are rare. During the nesting season insects were largely eaten, especially earthworms, grubs, and grasshoppers. The pigeons were fond of salt and frequented natural salt licks as well as grounds baited with salt.

Pehr Kalm (1911) lists the following as food of the passenger pigeon: Acorns, beechnuts, seeds of red-flowered maple, of American elm, mulberries, rye, wheat, buckwheat, but not Indian corn, berries of the tupelo, as well as of other trees and plants. He also notes the fondness of the pigeons for salt.

Wilson (1832) says that besides acorns and beechnuts, "buckwheat, hemp seed, Indian corn, holly berries, hackberries, huckleberries, and many others furnish them with abundance of food at all seasons. The acorns of the live oak are also eagerly sought after by these birds, and rice has been frequently found in individuals killed many hundred miles to the northward of the nearest rice plantation. * * * I have taken from the crop of a single wild pigeon, a good handful of the kernels of beechnuts intermixed with acorns and chestnuts." He calculated that the immense number of pigeons he observed in a flock in Kentucky would eat 17,424,000 bushels of mast a day, supposing that each bird consumed a pint. His calculations of the number of birds—2,230,272,000—an incredible number, seems, as will be recorded later, to be a fair one.

Ruthven Deane (1896), speaking of observations on captive birds, says:

As soon as the young are hatched the parents are fed on earthworms, beetles, grubs, etc., which are placed in a box of earth, from which they greedily feed, afterwards nourishing the young in the usual way, by disgorging the contents from the crop. At times the earth in the enclosure is moistened with water and a handful of worms thrown in, which soon find their way under the surface. The pigeons are so fond of these tidbits, they will often pick and scratch holes in their search, large enough to almost hide themselves.

Behavior.—The passenger pigeon was such a spectacular species in its migratory flights, its roostings, and its nestings, in which such enormous numbers took part, that there are many references to it from the times of the earliest pioneers. From this mass of literature it will be well to enter here some of the important reports, omitting till later those that deal largely with the slaughter of the bird.

Higginson (1630) writing of the region about Salem, Mass., says:

Upon the eighth of March from after it was faire Daylight until about eight of the clock in the forenoon, there flew over all the towns in our Plantacions soe many flockes of Doves, each flock contayning many thousands, and soe many that they obscured the light, that passeth credit, if but the Truth should be written.

Wood (1635) in the same region says:

These Birds come into the Country to goe to the North parts in the beginning of our Spring, at which time (if I may be counted worthy, to be beleeved in a thing that is not so strange as true) I have seen them fly as if the Ayerie regiment had been Pigeons, seeing neyther beginning nor ending, length, or breadth of these Millions of Millions. The shouting of people, the ratling of Gunnes, and pelting of small shotte could not drive them out of their course, but so they continued for foure or five houres together; yet it must not be concluded, that it is thus often; for it is but at the beginning of the Spring, and at Michaelmas [September 29], when they returne back to the Southward; yet are there some all the yeare long, which are easily attayned by such as looke after them.

Pehr Kalm (1911) writing of a migration in March, 1740, in Pennsylvania says:

Their number, while in flight, extended 3 or 4 English miles in length and more than one such mile in breadth, and they flew so closely together that the sky and the sun were obscured by them, the daylight becoming sensibly diminished by their shadow.

The big as well as the little trees in the woods, sometimes covering a distance of 7 English miles, became so filled with them that hardly a twig or branch could be seen which they did not cover; on the thicker branches they had piled themselves up one above another's backs, quite about a yard high.

When they alighted on the trees their weight was so heavy that not only big limbs and branches of the size of a man's thigh were broken straight off, but less firmly rooted trees broke down completely under the load.

The ground below the trees where they had spent the night was entirely covered with their dung, which lay in great heaps.

Wilson (1832) thus describes a flock of pigeons that passed over him as he was on his way to Frankfort, Ky.:

Coming to an opening by the side of a creek called the Benson, where I had a more uninterrupted view, I was astonished at their appearance. They were flying with great steadiness and rapidity, at a height beyond gunshot, in several strata deep, and so close together, that could a shot have reached them, one discharge could not have failed of bringing down several individuals. From right to left as far as the eye could reach, the breadth of this vast procession extended, seeming everywhere equally crowded * * *. It was then half past one. I sat for more than an hour, but instead of diminution of this prodigious procession, it seemed rather to increase both in numbers and rapidity; and, anxious to reach Frankfort before night, I rose and went on. About four o'clock in the afternoon I crossed the Kentucky river, at the town of Frankfort, at which time the living torrent above my head seemed as numerous and as extensive as ever. Long after this I observed them, in large bodies that continued to pass for six or eight minutes, and these again were followed by other detached bodies, all moving in the same south-east direction, till after six in the evening.

Wilson calculated that this great mass of birds contained the incredible number of 2,230,272,000 individuals, and his method of calculation seems to be a conservative one. He assumed the flock to be a mile in breadth, although he believed it was much more. Supposing it was moving at the rate of a mile a minute, as it was four hours in passing, he estimated that its whole length would have been 240 miles. He also assumed that each square yard contained three pigeons. As the flock was several strata deep there must have been many more than this.

Wilson (1832) says:

In descending the Ohio by myself, in the month of February, I often rested on my oars to contemplate their aerial manoeuvres. A column, eight or ten miles in length, would appear from Kentucky, high in air, steering across to Indiana. The leaders of this great body would sometimes gradually vary their course, until it formed a large bend of more than a mile in diameter, those behind tracing the exact route of their predecessors. This would continue sometimes long after both extremities were beyond the reach of sight, so that the whole, with its glittering undulations, marked by a space on the face of the heavens resembling the winding of a vast and majestic river. * * * . Sometimes a hawk would make a swoop on a particular part of the column, from a great height, when almost as quick as lightning, that part shot downwards out of the common track, but soon rising again, continued advancing at the same height as before; this inflection was continued by those behind, who on arriving at this point, dived down almost perpendicularly, to a great depth, and rising followed the exact path of those that went before.

On another occasion, he says, " while talking with the people within doors, I was suddenly struck with astonishment at a loud rushing roar, succeeded by instant darkness, which, on the first moment, I took for a tornado about to overwhelm the house, and everything around in destruction. The people, observing my surprise, coolly said, 'It is only the pigeons.' "

Audubon (1840) describes similar great multitudes of passenger pigeons. In the autumn of 1813, while traveling 54 miles on the banks of the Ohio River between Hardensburgh and Louisville, he observed great flocks of pigeons flying southwest. He counted the flocks for 21 minutes in the morning and found that 163 had passed. " I traveled on," he says, " and still met more the farther I proceeded. The air was literally filled with pigeons; the light of noonday was obscured as by an eclipse, the dung fell in spots not unlike melting flakes of snow, and the continued buzz of wings had a tendency to lull my senses to repose." He reached Louisville at sunset. " The pigeons were still passing in undiminished numbers and continued to do so for three days in succession."

Audubon (1840) describes as follows a roosting place on the banks of the Green River in Kentucky:

It was, as is always the case, in a portion of the forest where the trees were of great magnitude, and where there was little underwood. I rode through it upwards of forty miles, and, crossing it in different parts, found its average breadth to be rather more than three miles * * *. The dung lay several inches deep, covering the whole extent of the roosting-place. Many trees two feet in diameter, I observed, were broken off at no great distance from the ground, and the branches of many of the largest and tallest had given way as if the forest had been swept by a tornado. Everything proved to me that the number of birds resorting to this part of the forest must be immense beyond conception.

When the birds were coming to the roost, he continues—

the noise which they made, though yet distant, reminded me of a hard gale at sea, passing through the rigging of a close-reefed vessel. As the birds arrived and passed over me, I felt a current of air that surprised me * * *. The pigeons, arriving by thousands, alighted everywhere, one above another, until solid masses were formed on the branches all round. Here and there the perches gave way under the weight with a crash, and, falling to the ground, destroyed hundreds of birds beneath, forcing down the dense group with which every stick was loaded.

A similar account, confirming that of Audubon, is given by Revoil (1869) of his experiences in 1847 of a pigeon roost near Hartford, Ky. As the sun set, the birds began to come.

Indeed, the horizon grew dark, and the noise made by the pigeons resembled that of the terrible Mistral of the Provence, engulfing itself in the gorges of the Apennines.

When the column of the pigeons passed over my head, I felt a chill, caused all at once by the astonishment and the cold, for this displacement of the air produced a strongly unusual atmospheric draught of air. * * * The pigeons arrived by millions, precipitating themselves, the ones upon the others, pressed together like the bees in a swarm that escape from the hives in the month of May. The over-loaded tree-tops of the roosting place broke, and, falling to earth, pulled down at the same time the pigeons and the branches which found themselves below them. The noise was so great that even the next neighbors could not hear each other if they cried out with all the power of their lungs.

Sutton (1928) quoting from an account of a roost by Dr. Samuel P. Bates says: "In the hot summer nights the constant flapping of their wings produced by being crowded from their perches, gave forth a sound not unlike the distant roar of Niagara." John C. French (1919) speaking of a flight from a roost in 1866 in Potter County, Pa., says: "Each morning a valley a mile wide between the hills was filled strata above strata, eight courses deep at times, for about an hour with the multitude of birds flowing westward at the rate of a mile a minute, going for food. The roar of wings was like a tornado in the tree tops and the morning was darkened as by a heavy thunder-shower." These accounts, coming from different independent sources, are so similar that we are compelled to believe what seems to pass belief.

Although some of the great flights described above are the migratory flights of spring and autumn, as for example those described by Higginson and Wood, many of them are merely flights to and from the roosts or nesting grounds to the feeding grounds. The locality of these feeding grounds, as well as the winter residence, depended on the abundance of mast, which varied from year to year in different forests. As the mast could not be picked up from the ground if it were covered with snow, the birds migrated south of the snow line. Thus Mearns (1879) notes that the pigeon was rare in winter in the Hudson Valley except in very mild weather when the ground was bare.

Beechnuts were particularly sought by the pigeons. H. J. Jewett (1918) describes as follows the activities of a flock of about 120, feeding on these nuts:

They lighted in the top of a large beech tree; and, finding the beechnuts had fallen out of the hulls, dropped in rapid succession from branch to branch till all had reached the ground. I never have seen more intense activity or seeming system in feeding than these birds displayed. They worked in a wing-shaped group, moving nervously forward in one direction around the tree, gleaning the entire nut-covered space as they went. Those falling to the rear of the flock, where the nuts were picked up, kept flopping across to the front so as to get the advantage of the unpicked ground. A few that wandered apart in search of scattered nuts kept scurrying about and tilting as they picked them up and then hurried back to the flock as if they feared that the flock would soon be through feeding and off on the wing. This restless, voracious activity continued till the flock took fright and burst into the air.

Audubon (1840) thus describes the activities of a larger flock:

When alighted, they are seen industriously throwing up the withered leaves in quest of the fallen mast. The rear ranks are continually rising, passing over the main body, and alighting in front, in such rapid succession, that the whole flock seems still on wing.

Herman Behr (1911) speaks of the birds frequenting alder marshes for food:

Here they pried under the old leaves, searching for worms or insects, scratching and digging with great energy. Throughout these operations I do not recall them using their feet once, but always they pried and scratched and dug with their bills.

Wallace Craig (1911a) made valuable and interesting studies of the expressions of emotion of captive passenger pigeons from which I have transcribed the following:

It was eminently a bird of flight; on the ground it was rather awkward for a pigeon, its legs seeming too short and its massive shoulders too heavy. The nod of the passenger pigeon was utterly different from that of the mourning dove. The specific manner of nodding seemed an integral part of the bird's general bearing. The nod consisted of a movement of the head in a circle, back, up, forward, and down, as if the bird were trying to hook its bill over something. Often two or three such nods were given with no pauses between, following each other much more rapidly than in the mourning dove, because body and tail remained all the while stationary * * *. Ordinary walking pace of the male, 12–13 steps in 5 seconds. In eating, female pecks at rate of 12 pecks in 5 seconds on an average, and as head moves through a considerable arc, its motion is very quick.

They expressed fear and alarm by beating the wings together in quick succession, making a sound like the rolling beat of a snare drum.

Their flight was direct and made with great velocity. Maynard (1896) says that in twisting and turning they surpassed the Wilson's snipe. At times the flocks swept along close to the ground, at other times they flew at a great height. Sutton (1928), quoting W. G. Hayes, says: "Usually they flew 10 or 12 feet from the ground. They rose in waves to pass over fences and trees, but sometimes they flew from 30 to 50 feet in the air without the undulating motion." James G. Suthard communicates the following from the notes of Prof. J. J. Glen: "Standing in the open, I watched their flight above in every direction. They were so close the one to the other that it seemed as if their wings would touch. They were so high above the earth that nothing short of a modern rifle could reach them."

Maynard (1896), quoting Edward A. Bowers, describes an immense congregation of pigeons at a spring of brackish water in Michigan. "In an incredibly short time the birds begin to come; first in small numbers, then increasing rapidly until, in a few moments, they come in a living avalanche, covering the trees." After a few have cautiously alighted at the mound by the spring, "others follow slowly until, at last, a perfect torrent falls upon the spot, covering it so deeply as to endanger the lives of many of them by suffocation; then the whole enormous body suddenly rises with a deafening roar and alights on the trees. This is repeated until all are satisfied."

Voice.—Craig (1911a), as a result of intimate studies of captive birds, says: "Its voice was loud and strident, the hard notes being predominant and the musical notes somewhat degenerated; this being probably the result of its living and breeding in colonies so populous that only the loudest sounds could be heard." He gives at considerable length and, often with the aid of musical notation, the various calls of the bird, which he divides into: "1. The copulation-note; 2. The *keck* (a name not used for the note of any other species); 3. Scolding, chattering, clucking (these names also peculiar to the species); 4. The vestigial *coo* or *keeho;* 5. The nest-call." The copulation note he considers to be essentially the same as that of the mourning dove. The *keck* or *kheck* is loud and harsh, generally given singly, and sometimes it is accompanied with a flapping of the wings. The scolding, chattering, and clucking notes have been written *kee-kee-kee-kee* and *tete! tete!* and have been said to resemble the croaking of wood frogs. "In expressing high excitement, it becomes loud and high-pitched, and in the excitement of fighting especially it becomes very rapid." The vestigial *coo* or *keeho* was a mere remnant of the *coo* of other pigeons. "One sees in this," says Craig, "probably an adaptation to life in a community so populous and hence so noisy that cooing could hardly be heard, and the pigeon which could best win a female or warn off an interloper would be the pigeon with the merely loudest voice." The nest call he describes as "very much blurred—more so than any other note of this species. A great mixture of high and low tones."

Herman Behr (1911) says that feeding birds hailed "newcomers with a call, peculiar to the occasion. It was a long-drawn and moderately loud repetition of one note, which sounded like *treet*, and this would cause the flying birds to alight in nearby trees, giving in their turn a low call, *tret*, *tret*, *tret*. To me these seemed to be notes of greeting, while other sounds were indicative of sex. For instance, the female call note is similar to the *treet* above, but the male response is a low *oorn*, which can not be heard farther than two or three hundred feet. My knowledge of these notes is due to the fact that I learned to imitate them perfectly, in order to call the birds up within good shooting distance." Maynard (1896) quotes August Koch as describing their note of alarm "sounding something like a laugh made with a child's trumpet."

Field marks.—The passenger pigeon may easily be confused with the mourning dove. It is considerably larger, the adult male being nearly half as long again, but in the female and young, with their shorter tails, the difference is not so great. As already remarked, and as all ornithologists know, difference in size alone without direct comparison is of little value as a field mark. The best observer, depending on this alone, may easily be mistaken. The passenger pigeon lacks the spot on the neck that is present in adult mourning doves, but is absent in juveniles. The adult male passenger pigeon has a much redder breast than the mourning dove. The iris of the adult male passenger pigeon is scarlet, that of the female orange, while the iris of a juvenile bird taken by me on September 4, 1877, is described in my notes as " hazel and gray outside." The iris of the mourning dove on the other hand is dark brown. A capital mark of the adult male passenger pigeon is its blue rump. To some extent in the female and especially in the young, however, this is obscured by a brown or olivaceous hue, and is more like the rump of the mourning dove. Unlike the mourning dove, the passenger pigeon does not make a twittering sound with its wings when it rises in flight.

It is hardly necessary to consider the band-tailed pigeon, which more nearly resembles the domestic pigeon. Its tail is short and square. Flocks of this bird in the West have, however, sometimes been mistaken for the passenger pigeon.

Enemies.—Before the arrival of Europeans in this country predatory animals and birds found in the passenger pigeon a large and easily accessible store of food, and with these the Indian joined in the feast. Yet, judged from the enormous numbers of pigeons observed by the white pioneers, the drain on the pigeons must have been insignificant. The white man from the first began destroying the pigeons excessively, and later, with the development of the railroads and the telegraph and with increasing demands of the markets, the destruction advanced by leaps and bounds. The pigeon had become a commercial asset of great value. The Indians, before they were contaminated by the whites killed no more than they could use themselves, cooking and eating them, or drying the flesh and trying out the fat of a moderate number for future use.

The following from Kalm (1911) is an interesting reflection on the difference of treatment of the pigeons by the Indians and so-called civilized white man:

While these birds are hatching their young, or while the latter are not yet able to fly, the savages or Indians in North America are in the habit of never shooting or killing them, nor of allowing others to do so, pretending that it would be a great pity on their young, which would in that case have to starve to death. Some of the Frenchmen told me that they had set out with the intention of shooting some of them at that season of the year, but that the savages had at first with kindness endeavored to dissuade them from such a purpose, and later added threats to their entreaties when the latter were of no avail.

There are numerous accounts of the slaughter of the pigeons, but a few will suffice. Wilson (1832) described the slaughter in a nesting, said to be 40 miles in extent, in Kentucky, as follows:

As soon as the young were fully grown, and before they left the nests, numerous parties of the inhabitants, from all parts of the adjacent country, came with wagons, axes, beds, cooking utensils, many of them accompanied by the greater part of their families, and encamped for several days at this immense nursery. * * * The ground was strewed with broken limbs of trees, eggs, and squab pigeons, which had been precipitated from above, and on which herds of hogs were fattening. Hawks, buzzards, and eagles were sailing about in great numbers, and seizing the squabs from their nests at pleasure; while from twenty feet upwards to the tops of the trees the view through the woods presented a perpetual tumult of crowding and fluttering multitudes of pigeons, their wings roaring like thunder; mingled with the frequent crash of falling timber; for now the axe-men were at work cutting down those trees that seemed to be most crowded with nests; and contrived to fell them in such a manner that in their descent they might bring down several others; by which means the falling of one large tree sometimes produced two hundred squabs, little inferior in size to the old ones, and almost one mass of fat. * * * It was dangerous to walk under the flying and fluttering millions, from the frequent fall of large branches, broken down by the weight of multitudes above, and which in their descent often destroyed numbers of the birds themselves; while the clothes of those engaged in traversing the woods were completely covered with the excrements of the pigeons.

Wilson also described the destruction of pigeons on migration:

In the Atlantic states, though they never appear in such unparalleled multitudes, they are sometimes very numerous; and great havoc is then made amongst them with the gun, the clap-net, and various other implements of destruction. As soon as it is ascertained in a town that the pigeons are flying numerously in the neighborhood, the gunners rise *en masse;* the clap-nets are spread out on suitable situations, commonly on an open height, in an old buckwheat field; four or five live pigeons with their eyelids sewed up, are fastened on a moveable stick—a small hut of branches is fitted up for the fowler at the distance of forty or fifty yards; by the pulling of a string, the stick on which the pigeons rest is alternately elevated and depressed, which produces a fluttering of their wings similar to that of birds just alighting; this being perceived by the passing flocks, they descend with great rapidity, and finding corn, buckwheat, etc., strewed about, begin to feed, and are instantly, by the pulling of a cord, covered with the net. In this manner ten, twenty, and even thirty dozen have been caught in one sweep. Meantime the air is darkened with larger bodies of them moving in various directions; the woods also swarm with them in search of acorns; and the thundering of musketry is perpetual on all sides from morning to night. Wagon-loads of them are poured into market, where they sell from fifty to twenty-five and even twelve cents a dozen.

Audubon (1840) describes the killing in a roost in a similar way and adds iron pots containing burning sulphur, torches of pine knots, and long poles to the instruments of destruction. He says that the next morning "the authors of all this devastation began their entry amongst the dead, the dying, and the mangled. The pigeons were picked up and piled in heaps until each had as many as they could possibly dispose of, when the hogs were let loose to feed on the remainder."

John Lewis Childs (1905) says that an old settler in Maine told him "that a common way of killing them off was to dig a long trench in which a quantity of wheat was scattered to attract the birds. When they came and settled down to feed, filling the trench to its utmost capacity, one discharge from some advantageous point of an old flint-lock musket loaded with a handful of shot would often result in the killing of as many as 75 birds." He adds that "in those days wild pigeons were hunted for three distinct reasons—as sport, as an article of food, and because they were destructive to crops."

Although this early slaughter in many places, largely for individual use, doubtless reduced the numbers of the passenger pigeon and started it on the road to extinction, the systematic destruction on a large and commercial scale, which began in the forties and reached its highest point late in the sixties and in the seventies of the nineteenth century, was responsible for the ultimate result. This perfection of extermination was brought about by the increasing development of the mail and telegraph systems, by which the professional netters, or "pigeoners" as they were called, were informed of and kept in touch with roosting places, in both winter and summer, with flights, and especially with nesting places. A spreading network of railroads enabled the pigeoner to arrive promptly in the region of his quarry and to send back to the markets many tons of the dead bodies of this beautiful bird, as well as thousands of living ones to be used in clubs for trap shooting. The decreasing area of forest suitable for the pigeons tended to concentrate the remainder and render it easier to locate them.

Prof. H. B. Roney (1879) estimated that at the time of his writing there were about 5,000 men in the United States who pursued pigeons year after year as a business. These men had perfected the methods of netting, which was generally done at a bed of muck baited with salt and sulphur, or grain. The net, about 6 feet wide and 20 to 30 feet long, could be sprung by means of ropes and a powerful spring pole from a blind by the operator, and "fliers" (captive birds thrown up into the air) or "stool pigeons" (birds tied to a pole that could be suddenly raised and lowered by a cord, making the birds flutter as if alighting) were used to lure down the birds flying overhead. Professor Roney found that 60 to 90 dozen birds a day to the net were a fair average of the numbers caught. He adds:

Higher figures than these are often reached, as in the case of one trapper who caught and delivered 2,000 dozen pigeons in ten days, being 200 dozen, or about 2,500 birds per day. A double net has been known to catch as high as 1,332 birds at a single throw, while at natural salt licks, their favorite resort, 300 and 400 dozen, or about 5,000 birds have been caught in a single day by one net.

At this time, 1879, dead birds were sold in Chicago markets for 50 to 60 cents a dozen, 35 to 40 cents at the nestings, while live birds brought $1 to $2 a dozen, and Professor Roney estimated that the pigeoner made at these times $10 to $40 a day. As long as money was obtained by the sale of the birds, men were ready to slaughter them. Mershon (1907), in his book on the passenger pigeon, gives a large number of statistics and estimates of the immense slaughter of this bird.

The last great nesting of this beautiful bird to be desecrated was in 1878 at Petoskey, Mich. Here the persecuted birds had gathered in old-time numbers, driven out from roosts and nestings in other places. The nesting was said to have been 28 to 40 miles in length and 3 to 10 miles in width. Professor Roney (1879), who visited the nesting at the time of the slaughter, thus describes it:

Scarcely a tree could be seen but contained from 5 to 50 nests, according to its size and branches. Directed by the noise of chopping and of falling trees, we followed on, and soon came upon the scene of action. Here was a large force of Indians and boys at work, slashing down the timber and seizing the young birds as they fluttered from the nest. As soon as caught, the heads were jerked off from the tender bodies with the hand, and the dead birds tossed into heaps. Others knocked the young fledglings out of the nests with long poles.

The Indians were paid a cent apiece for the slain. Within 100 rods of the nests, instead of 2 miles away as the law prescribed, netters were hard at work, one taking 984 birds in a single day. He saw men going about close to the nesting, shooting the birds as they roosted in rows on the branches and passed in clouds overhead. Scores of dead pigeons were left on the ground to decay, and the woods were full of wounded ones, while many of the squabs, deprived of their parents, starved to death. It is a terrible picture of cruelty and greed.

Roney (1879) adds:

For many weeks the railroad shipments averaged fifty barrels of dead birds per day—thirty to forty dozen old birds and about fifty dozen squabs being packed in a barrel. Allowing 500 birds to a barrel, and averaging the entire shipments for the season at 25 barrels per day, we find the railroad shipments to have been 12,500 dead birds daily, or 1,500,000 for the summer. Of live birds there were shipped 1,116 crates, six dozen per crate, or 80,352 birds. These were railroad shipments only and not including the cargoes by steamer from

Petoskey, Cheboygan, Cross Village, and other like ports, which were as many more. Added to this were the daily express shipments in bags and boxes, the wagon loads hauled away by the shotgun brigade, and the myriads of squabs dead on the nest.

The profit from these transactions was the lure that led to the annihilation of the passenger pigeon. Barrows (1912) says:

Dr. Isaac Voorheis, of Frankfort, Michigan, told the writer personally that in 1880 or 1881, when there was a large nesting in Benzie County, he took at one throw the net 109 dozen and 8 pigeons (1,316 birds), and that six catches of the net brought him $650. These birds were kept alive until a schooner load was obtained, when they were sent directly to Chicago for trap shooting.

A provision dealer at Cheboygan, Mich., is said to have shipped live pigeons in numbers up to 175,000 a year from 1864 until the end.

Kumlien and Hollister (1903) say of Wisconsin:

Mr. J. M. Blackford, now residing at Delaware, states the last large catch of nesters was in 1882. The following spring but 138 dozen were taken in the best pigeon ground in the State, and this was practically the end.

Ruthven Deane (1896) wrote in 1895 to N. W. Judy & Co., of St. Louis, Mo., dealers in poultry and the largest receivers of game in that section, as to their dealings in passenger pigeons and they replied: "We have had no wild pigeons for two seasons; the last we received were from Siloam Springs, Ark. We have lost all track of them and our netters are lying idle."

This firm, Widmann (1907) says, "handled more dead and live pigeons than any other firm in the country, and had their netters employed all the year around, tracing the pigeons to Michigan and Wisconsin in spring and to the Indian Territory and the south in winter."

This is the last word on the commercial extinction of the passenger pigeon by its greatest of all enemies, so-called civilized man.

DISTRIBUTION

Range.—Eastern North America, casual in Bermuda, the West Indies, and in Mexico.

Breeding range.—The breeding range of the passenger pigeon extended **north** to Montana ("latitude 49°," Great Falls, and the Yellowstone River); North Dakota (Fort Berthold); Manitoba (Waterhen River); Minnesota (Lake Itasca and Northern Pacific Junction); Wisconsin (Stockholm, Shiocton, and West Depere); Michigan (Petoskey, Brant, and Detroit); Ontario (Moose Factory); Quebec (Fort George and Quebec); New Brunswick (Grand Falls); Prince Edward Island; and probably Newfoundland (St. Johns). **East** to probably Newfoundland (St. Johns); Nova Scotia (Halifax and Yarmouth); New Hampshire (Conway and Webster);

Massachusetts (Plymouth); Connecticut (Portland); southeastern New York (Croton Falls); southeastern Pennsylvania (Columbia); and northeastern Virginia (Clarendon). **South** to northeastern Virginia (Clarendon); northwestern West Virginia (Marion County); Kentucky (Green County); Illinois (Mount Carmel, Hillsboro, and Quincy); and eastern Kansas (Neosho Valley). **West** to eastern Kansas (Neosho Valley); South Dakota (Vermilion and Fort Pierre); and Montana (Hellgate River, Fort Shaw, and "latitude 49°").

Passenger pigeons also were frequently recorded in summer at points well outside the breeding range above outlined. Among these records are Mackenzie (Fort Good Hope, by Alexander Mackenzie in 1789; Fort Norman, by Ross; and Fort Simpson, by Kennicott); Saskatchewan (Ile a la Crosse, by Hood in 1820; near the mouth of the South Saskatchewan River, August 2 to 5, 1858; and Fort Qu' Appelle); northern Manitoba (Island Lake, where in 1903 E. A. Preble received a report of a small flock seen three or four years previously; York Factory; and Fort Churchill, where two specimens were taken); Quebec (Tadousac, one recorded by Fleming as taken July 20, 1889; Godbout, one taken June 27, 1859; and Anticosti Island, one record); Franklin (one recorded by J. C. Ross as coming aboard their vessel in Baffin Bay at latitude 73½° N.); northwestern South Carolina (Caesars Head, two pairs seen by Arthur T. Wayne in the summer of 1882); Missouri (New Haven, large flocks observed in 1872 by Dr. A. F. Eimbeck); Wyoming (one specimen taken 40 miles west of Fort Laramie, September 8, 1857); and Nevada (West Humboldt Mountains, in September, 1867, according to Ridgway [1873]). Passenger pigeons also have been recorded from British Columbia, but the evidence seems rather unsatisfactory (*cf.* Brooks and Swarth, 1925).

Winter range.—The vast numbers of these birds and the extensive area included in their breeding grounds considered, the winter range was greatly restricted. Data bearing upon this phase of the subject, however, are not plentiful. Apparently at this season, the species extended **north** regularly to Arkansas (Rogers); southeastern Missouri (Attie); and northern South Carolina (Chester County). **East** to South Carolina (Chester County and Sineaths Station); and Florida (Amelia Island and Gainesville). **South** to Florida (Gainesville); Alabama (Greensboro); and Louisiana (Mandeville). **West** to Louisiana (Mandeville); and Arkansas (Judsonia, Fayette, Siloam Springs, and Rogers). In some seasons large numbers wintered much farther north as they were recorded as abundant in the winter of 1853–54 at Brookville, Ind.; a flock of about 300 was seen near Harrisburg, Pa., December 25, 1889; and some wintered near Hartford, Conn., in 1882–83. At this season, they also were recorded at more southern points: Cuba (a female taken at Triscornia on Havana Bay, and a male obtained in the Havana market, both specimens being preserved in the Museo Gundlach); while four specimens were taken at Jalapa, Vera Cruz, in the winter of 1872–73, and it was reported by Herrera as an occasional visitant to the Valley of Mexico (no details). A migrant specimen was killed in Bermuda on October 24, 1863.

Migration.—Despite a rather voluminous literature, actual useful data bearing upon the seasonal movements of the passenger pigeon are decidedly scarce. The modern method of recording arrival and departure was not developed until the species was so reduced in numbers that its ultimate extinction was already apparent. For this reason, many of the dates of later years have no significance as bearing upon migration, and wherever possible data obtained prior to 1890 have been used.

Spring migration.—Early dates of spring arrival are: North Carolina, Raleigh, April 18; Virginia, Highgate, April 20; West Virginia, French Creek, February 27 and April 10, and Fairview, March 19; District of Columbia (occasionally wintered), Washington, April 3; Pennsylvania (occasionally wintered), Chambersburg, February 13, Erie, March 13, Brockney, March 18, Guths Station, March 20, and Ridgway, March 22; New Jersey, Raccoon, March 3, Plainfield, March 12, and Caldwell, March 31; New York (rarely wintered), Locust Grove, March 3, Cornwall-on-Hudson, March 4, Elmira, March 7, Painted Post, March 16, Glasco, March 20, and Buffalo, March 27; Connecticut, Gaylordsville, March 13, and Saybrook, March 21; Massachusetts, Woods Hole, March 20, Ponkapog, March 31, and Amherst, April 7; Vermont, Tydeville, April 9; New Hampshire, Hollis, April 2; Maine, North Livermore, April 28; Illinois (occasionally wintered in southern part), Charleston, February 3, Virden, February 11, Lake Forest, March 4, and Carthage, March 12; Indiana, Millwood, February 18, Brookville, February 28, Jonesboro, March 1, Brown County, March 7, and Kokomo, March 13; Ohio, Hudson, March 8, West Liberty, March 10, and Fayette, March 15; Michigan, Petersburg, March 20, Locke, March 22, Battle Creek, March 27, and Plymouth, April 7; Ontario, Kingston, March 21, London, March 24, Port Rowan, March 26, and Toronto, April 13; Iowa, Iowa City, March 13, Burlington, March 15, La Porte, March 16, and Dubuque, March 22; Wisconsin, Ripon, March 8, Janesville, March 23, Racine, March 25, Shiocton, March 29, Delavan, April 1, and Green Bay, April 7; Minnesota, Lanesboro, March 22, Bradford, April 1, Minneapolis, April 3, Zumbrota, April 8, and Lake Andrew, April 9; and Manitoba, Aweme,

April 8, Lake Winnipegosis, April 14, Greenridge, April 17, and Ossowo, April 18.

Fall migration.—Late dates of fall departure are: Manitoba, Mount Royal, September 15, and Aweme, September 21; Minnesota, Lake Andrews, September 28, Lanesboro, October 5, and Zumbrota, November 15; Wisconsin, Delavan, September 8, and Kelley Brook, September 16; Iowa, Williamstown, September 27, and Keokuk, October 18; Ontario, Ottawa, September 3, and Toronto, October 22; Michigan, Ann Arbor, October 12, and Newberry, October 24; Ohio, Wayne County, September 19; Indiana (probably occasionally wintered), Bloomington, September 28; Illinois (occasionally wintered in southern part), Virden, October 5; Kentucky, Casky, October 30; Quebec, Montreal, September 15, and Valley River Rouge, October 7; New Hampshire, Acworth, October 10; Massachusetts (rarely wintered), Worcester, September 25, and Plymouth, October 16; Connecticut (rarely wintered), East Hartford, October 19; Rhode Island, Newport, October 19; New York, Ossining, October 1, Croton Falls, October 16, Locust Valley, November 4, and Orange Lake, November 17; New Jersey, Morristown, September 16, and New Providence, November 12; Pennsylvania (occasionally wintered), Linden, October 5, Monroe County, October 25, and Wayne County, November 2; Maryland (occasionally wintered), Laurel, October 11; Virginia, Dunn Loring, October 19; and West Virginia, French Creek, October 20.

Egg dates.—Saskatchewan and Manitoba: 6 records, May 21 to June 23. Minnesota and Wisconsin: 11 records, May 5 to September 10; 6 records, May 20 to 30.

ZENAIDURA MACROURA CAROLINENSIS (Linnaeus)

EASTERN MOURNING DOVE

HABITS

CONTRIBUTED BY WINSOR MARRETT TYLER

The mourning dove must have been one of the first birds that attracted the attention of the early settlers when this country was new and wild. They must have recognized the bird as not far removed from some of the Old-World species of pigeons, and its notes must have recalled to them their old home. The writers of these times speak of the bird familiarly, especially as a game bird that relieved the hardships of pioneer life.

At the present time, in the Northern States, protected as a song bird, it adds a quiet dignity to our avifauna, while in the Southern States it is a common, tame, almost a dooryard bird and a gleaner of fields, except when, during the hunting season, it is shot for food and sport. In the West it is an inhabitant both of the plains and the mountains, ranging commonly to 7,000 feet altitude.

And yet, well known and widely distributed as the bird is, it is not a conspicuous bird of the country at all. It is quiet in voice, neutral in color, and so unobtrusive in deportment that it seems little more than a part of the background; a quiet, pastoral bird, reminding us of the man in "The Bab Ballads"—"no characteristic trait had he of any distinctive kind"—or of sweet, lovable, but wholly negative Hero in "Much Ado about Nothing."

Spring.—In the parts of the country where the mourning dove spends the winter, one of the early signs of spring is when the winter flocks begin to break up and the doves separate into mated pairs. Just as the mockingbird in the Southern States bursts suddenly into song and separates winter from spring, so the male mourning dove, who has been silent through the winter, at the first hint of spring begins to coo.

As the breeding season approaches, the birds become gradually tamer and, as Wilson (1832) says, they "are often seen in the farmer's yard before the door, the stable, barn, and other outhouses, in search of food, seeming little inferior in familiarity, at such times, to the domestic pigeon"—a contrast to the wild game bird of the autumn.

Courtship.—Very little has been published on the courting actions of the mourning dove, and apparently no detailed study has been made of them. Indeed, many observers who know the bird well state that they have seen no courting at all.

Barrows (1912), who gives a careful description of a nuptial flight, points out that "although familiar with the mourning dove's habits in New England, western New York, and elsewhere we have never seen this peculiar flight except in Michigan." He says:

An individual leaves its perch on a tree, and, with vigorous and sometimes noisy flapping (the wings seeming to strike each other above the back), rises obliquely to a height of a hundred feet or more, and then, on widely extended and motionless wings, glides back earthward in one or more sweeping curves. Usually the wings, during the gliding flight, are carried somewhat below the plane of the body, in the manner of a soaring yellowlegs or sandpiper, and sometimes the bird makes a complete circle or spiral before again flapping its wings, which it does just before alighting. * * * This peculiar evolution is commonly repeated several times at intervals of two or three minutes, and appears to be a display flight for the benefit of its mate, the assumption being that only the male dove soars.

Goss (1891) speaks of the courtship thus:

During the pairing season the male often circles and sails above his mate, with tail expanded, and upon the ground struts about with nodding head, and feathers spread in a graceful manner.

Craig (1911), speaking of the "nest-calling attitude," calls attention to the display of the ornamented tail. He says:

> The male [sits] with his body tilted forward, tail pointing up at a high angle, the head so low that bill and crop may rest on the floor, or if the bird be in the nest, the head is down in the hollow. Both the voice and the attitude of the male serve to attract the female, for in all pigeons the nest-call is accompanied by a gentle flipping of the wings, ogling eyes, and a seductive turning of the head. In addition to these general columbine gestures, Zenaidura has a special bit of display of his own, for during the first note of the nest-call he spreads his tail just enough to show conspicuously the white marks on the outer feathers; soon as this first note is past, the tail closes and the white marks disappear, to flash out again only with the next repetition of the nest-call, before which there is always a considerable interval.

Forbush (1927) says that "in courtship the male mourning dove sometimes strikes his feet hard on his perch one after another."

James G. Suthard, speaking of the bird in Kentucky, says in his notes:

> During the nesting season, the female acts very much like the tame pigeon. The male prances around with his neck feathers all ruffed up, cooing and billing with the female. I have noticed that he sometimes picks up pieces of grass in his courtship antics. The intrusion of another male on one of these scenes results in a fight whereupon the female usually disappears.

Nesting.—The mourning dove uses a very wide choice in selecting a site for its nest. Perhaps the site most nearly typical is not far from the trunk on a horizontal branch of an evergreen tree—a pine or cedar—affording a firm foundation for the flimsy nest. The bird frequently nests on the ground, however, even on a clump of grass, sometimes on the stump of a tree, and there are several recorded instances where the nest has been found placed on a wooden ledge attached to an inhabited building. Indeed Gardner (1927) says that the birds in Kansas "preferred the vicinity of buildings to the wooded and secluded canyons of the back country by a ratio of at least ten to one."

The chief requisite, apparently, is a level support that will give stability to the nest, and to acquire this security the dove often makes use of the experience of another species of bird and builds its own nest on a nest (for example, that of a robin, brown thrasher, or mockingbird) that has weathered the previous winter.

Bendire (1892) cites an extreme instance of this habit. Quoting J. L. Davidson from Forest and Stream, he says:

> I found a black-billed cuckoo and a mourning dove sitting together on a robin's nest. The cuckoo was the first to leave the nest. On securing the nest I found it contained two eggs of the cuckoo, two of the mourning dove, and one robin's egg. The robin had not quite finished the nest when the cuckoo took possession of it and filled it nearly full of rootlets, but the robin got in and laid one egg.

As a rule a pair of mourning doves, in contrast to the habit of the passenger pigeon, nests well removed from the nests of other doves, but Charles R. Stockard (1905) reports in Mississippi an interesting exception to this rule. "Doves," he says, "often nested in small colonies. In a clump of about fifteen young pine trees I once found nine nests, and in an Osage orange hedge about one-half mile long twelve nests were located. But most doves nest singly, or with the nests too far apart to suggest any gregarious nesting habit."

Most commonly the nest is made of sticks and is lined with finer twigs. A. D. DuBois, however, in his notes records the use of grass, weed stalks, roots, and a lining of leaves and mentions one nest made "almost entirely of rootlets and stems lined with finer rootlets (a shallow affair)."

The nest, oftenest, perhaps, just a platform of sticks, but firm enough to withstand usage for 30 days, is made apparently entirely by the female bird. Frank F. Gander in his notes states this to be a fact in the case he describes, and he demonstrates the aid that the male bird gives to his mate. He says:

The bringing of the material was accomplished by the male, who flew to the ground and searched about until a suitable stick was found. In selecting material the male was very careful and tested the sticks by shaking them vigorously. Perhaps this was as much to test his hold upon the stick as to test the stick itself, as many times sticks were shaken from his beak. So much time was consumed in this choosing of a twig that his trips to the nest averaged about one every two minutes. He always approached the nest by the same route, alighting upon a protruding branch, hopping from this to another, and walking along the latter to the nest. Reaching the nest, he turned the material over to the female, who reached up her beak to receive it. Sticks were frequently dropped during this exchange. The female placed the sticks under and about her to construct the nest.

Building did not continue uninterruptedly, as the female frequently left the nest when the male would pursue her and peck at her until she returned. Work for the day was stopped at about 11 a. m. The nest building was taken up again on the following morning and carried on until about 10 a. m.

Continuing, he shows the division of labor during the incubation period of 15 days:

The male took up his duties at the nest at about 10 in the morning and was relieved again at about 3 in the afternoon. The male often left the eggs unguarded for a few minutes about noon while he flew to a near-by watering place to drink.

Margaret M. Nice (1922), referring to the building of the nest, corroborates the observation quoted above. She says:

Nest building as a rule takes place in the early morning. The male mourning dove gathers the materials and carries them to his mate who arranges them. He takes one piece at a time, and if he happens to drop it, he does not stop but continues his journey to the tree and then starts over again.

Mrs. Nice's articles form an exhaustive study of the nesting habits of the mourning dove and contain many statistical data of the utmost interest. Readers are referred to these valuable articles for detailed information.

That the division of labor, with a well-ordered time for relieving each other on the nest, continues through the incubation period and during the 15 days that the young spend in the nest, is shown by the following extracts. Wallace Craig (1911) says:

> Male and female take regular daily turns in sitting on the eggs or young: the female sits from evening till morning, the male from morning till evening, the exchanges taking place usually about 8.30 a. m. and 4.30 p. m. This arrangement is very regular if there is nothing to disturb the birds; but if interloping birds come about, this arouses the anger of the male and he leaves the nest in order to attack them.

Mrs. Nice's (1923) experience corroborates Doctor Craig's observation. She says:

> The male incubates and broods during the day while the female does the same during the night, early morning, and late afternoon; and both parents regurgitate "pigeon milk" for the young. Two striking differences between the nest behavior of mourning doves and most passerine birds is the almost constant brooding of the young till near the end of the nest life and the lack of any sanitary care of the nest.

And further (1922):

> As a rule one or the other parent is continuously on the nest from the time the first egg is laid until the young are fairly well grown.

I have approached the nest of a mourning dove and come almost within arm's reach of the bird before it flew quietly away, but there is plenty of evidence that this behavior is not invariable for frequently the bird is reported to flop from the nest and resort to the ruse of the broken wing.

The breeding season is very long; in the Middle States it lasts from May to August and rarely to early September. The birds commonly rear two broods in a season, and Miss A. R. Sherman believes that they probably rear three sometimes. In her notes Miss Sherman says in substance:

> The doves are so numerous and so secretive in their ways that it is not possible to say whether a pair of birds, which has nested in May or June, breed again late in June or in July. When a nest is used twice in the same season, however, the assumption is that a pair of birds is using their own nest a second time.

[AUTHOR'S NOTE: On April 19, 1907, while hunting for hawks' nests in a grove of tall pines, I rapped on a tree containing a likely looking old nest and was surprised to see a mourning dove fly from it. I climbed to it, more than 40 feet up in a white pine, and found it to be an old hawk's nest, that had since been used by squirrels, as it was full of pine needles and soft rubbish, such as squirrels use. It was quite a large nest, measuring 25 by 15 inches. The doves had scraped out a hollow in the pine needles and laid one egg. I visited the nest again on April 28, when the dove flew off, as before, and the nest held two eggs. I photographed the nest and collected the eggs. As a pair of great-horned owls and a pair of Cooper's hawks were nesting in these woods, the doves stood a poor chance of raising a brood, or even escaping with their own lives.

Eggs.—The mourning dove lays almost always two eggs, but there are a few records of three, or even four. In shape they vary from elliptical oval, the commonest shape, to elliptical ovate or ovate. The shell is smooth with very little gloss. The color is pure white. The measurements of 47 eggs average 28.4 by 21.5 millimeters; the eggs showing the four extremes measure 31 by 22, 29.5 by 23, 26 by 20.5, and 28.5 by 20 millimeters.]

Young.—The young of the mourning dove are helpless when hatched, and during the two weeks they remain in the nest they require constant care from their parents. They are fed by regurgitation during most of their nest life, but solid food, such as insects and seeds, is gradually substituted, and at the time of leaving the nest it largely replaces the "pigeon milk." The contents of the crop of a young bird, examined at the end of its nest life, consisted almost entirely of seeds (principally grass seeds) and less than 2 per cent "pigeon milk." [See Townsend (1906).]

Gabrielson (1922), who studied the nest life from a blind, clearly describes the process of regurgitation thus:

At 7.30 a. m. a squab backed toward the blind and getting from beneath the parent raised its head and mutely begged for food. The adult (presumably the female) responded immediately by opening her beak and allowing the nestling to thrust its beak into one corner of her mouth. She then shut her beak on that of the nestling and after remaining motionless for a short time began a slow pumping motion of the head. The muscles of her throat could be seen to twitch violently at intervals, continuing about a minute, when the nestling withdrew its beak. The other nestling then inserted its beak and the process was repeated, 15 seconds elapsing before its beak was removed. With intervals varying from 5 to 10 seconds (watch in hand) four such feedings, two to each nestling, occurred. The nestling not being fed was continually trying to insert its beak in that of the parent and at the fifth feeding both succeeded in accomplishing this at the same time. The nestlings' beaks were inserted from opposite sides of the parent's mouth and remained in place during the feeding operation although I could not say whether or not both received food. While being fed the nestlings frequently jerked the head from side to side and also followed the motion of the parent's beak by raising and lowering themselves by the use of the legs. They were not more than five days old but had better use of their muscles than the young of passerine birds at from eight to ten days of age. The entire process described above occupied about six minutes, after which the nestlings crawled back beneath the parent.

Miss A. R. Sherman, who has had ample opportunity to study the mourning dove and a wide experience as a field observer, gives in her notes the period of incubation definitely as 15 days.

Plumages.—[AUTHOR'S NOTE: The young squab is fat and unattractive, scantily covered with short, white down, through which the yellowish skin shows. The stiff quills of the juvenal plumage soon appear, giving the young bird an ugly, spiny appearance. The juvenal plumage is well developed before the young birds leave the well-filled nest. In this plumage the upper parts are " buffy brown " to " snuff brown," with faint, whitish edgings on the back and wing coverts; the scapulars and some of the inner wing coverts have large black patches; the underparts are from " pinkish cinnamon " to " light vinaceous-cinnamon," paler on the belly and grayer on the flanks. A postjuvenal molt of the contour plumage and tail, during fall, produces a first winter plumage, which is like the adult but somewhat duller. Adults have a complete molt in fall.]

Food.—Adult mourning doves are essentially seedeaters. Wheat and buckwheat are said to be their favorite grains, but they consume such enormous numbers of weed seeds that they prove to be a highly beneficial species, as the following quotation from Dutcher (1903) shows:

The examination of the contents of 237 stomachs of the dove shows over 99 per cent of its food consists wholly of vegetable matter in the shape of seeds, less than 1 per cent being animal food. Wheat, oats, rye, corn, barley, and buckwheat were found in 150 of the stomachs, and constituted 32 per cent of the total food. However, three-fourths of this amount was waste grain picked up in the fields after the harvesting was over. Of the various grains eaten, wheat is the favorite, and is almost the only one taken when it is in good condition, and most of this was eaten in the months of July and August. Corn, the second in amount, was all old damaged grain taken from the fields after the harvest, or from roads or stock yards in summer. The principal and almost constant diet, however, is the seeds of weeds. *These are eaten at all seasons of the year.* In one stomach were found 7,500 seeds of the yellow wood-sorrel (*Oxalis stricta*), in another, 6,400 seeds of barn grass or fox tail (*Chaetocloa*).

Wayne (1910) records this interesting observation on the dove's feeding habits:

Although this species is supposed to feed upon the ground, this is by no means always the case as the birds resort to the pine woods for weeks at a time to feed upon the seeds of these trees, which they obtain by walking out on the limbs and extracting them from the cones. The flesh at this time is very strongly impregnated with a piney flavor.

Behavior.—Although mourning doves spend a large part of the year in flocks, they have a strong tendency in spring to separate into pairs and scatter over the country to nest. Doubtless they owe their present status, perhaps even their existence, to this habit, for, had they bred in colonies as the passenger pigeon did, the doves would have been subjected at their nests to the wholesale slaughter that exterminated the pigeon.

As we watch a number of doves feeding in a stubble field we soon see that there is no very strong tie binding together the members of the company—no such bond as holds together a flock of sandpipers and suppresses individual action. The doves are spread out over the ground, each walking off by itself and feeding more or less alone, like grazing cattle. When we walk toward them they start into the air, but not all together; a few, very often only two, fly away; then, after a moment, a few more take flight and go off, very likely in another direction. The flock when alarmed, instead of moving off as a unit, breaks up, and the birds retreat individually or in pairs. Thus even when the doves are assembled in numbers there is a tendency to segregate into pairs—a characteristic of the breeding season.

The birds leave the ground very quickly, gaining speed rapidly with strong, sweeping wing beats and fly with whistling wings, suggesting the whistling flight of the golden-eye.

In eastern Massachusetts, where since 1910 the birds have become well established, they frequent the dry, sandy, sparsely wooded hillsides characteristic of this glaciated country, and retire to nest in the near-by pine woods, where they seem much at home, walking easily among the branches.

Doves often visit gravelly roads and are sometimes seen on the sea beaches. On the dry plains of western Texas (Merriam, 1888) they were found 3 to 5 miles from the nearest water, and Merriam (1890) describes thus the coming of the doves at dusk to drink:

> Common from the Desert of the Little Colorado to the upper limit of the pine belt. Every evening they assemble at the springs and water holes, coming in greatest numbers just at dark, particularly about the borders of the Desert where water is very scarce. On the evening of August 20 we camped for the night at a small spring about 5 miles west of Grand Falls. At dusk hundreds of doves came to drink, and continued coming until it was so dark that they could not be seen.

Voice.—The mourning dove takes its name from its common note, a low-toned, moaning *coo*. This is one of the bird notes that, while fairly loud and perfectly distinct, does not readily attract the attention of one who is not familiar with it. In this respect it resembles the diurnal hooting of the screech owl; both of these notes in some strange way are disregarded by the ear until it is trained to detect them. We then recognize them both as familiar sounds of the countryside.

A. A. Saunders in his notes describes a typical song thus:

> The sound is well imitated by a low-pitched whistle, but some birds strike notes lower than I can whistle. The song consists of four notes. The first is usually twice as long in time as the others, and slurred first upward, and

then downward. The other three notes are either all on one pitch or slightly slurred downward. Lengths of songs vary from 3 to 4⅗ seconds. Usually one bird sings a single song and repeats it over and over in just the same manner; but I once recorded five different songs from a single bird.

Craig (1911) has reported the results of an exhaustive study of the bird in confinement. The following are the outstanding features of his article. He describes the song, which he terms the "*perch coo*" essentially as Saunders does, and adds:

When delivering his song, the mourning dove does not perform any dance or gesture, as some birds do. He invariably stands still when cooing; even when he coos in the midst of pursuing the female he stops in the chase, stands immovable until the coo is completed, and then runs on. His attitude is, to be sure, very definite, the neck somewhat arched and the whole body rigid; but the impression it gives one is, not that the bird is striking an attitude, but that he is simply holding every muscle tense in the effort of a difficult performance.

The female also utters the perch-coo, though less often than the male, and in a thin, weak voice and staccato tones, which, as compared with the male's song, form so ludicrous a caricature that on first hearing it I burst out laughing.

To this commonly heard note he adds two others; the nest call of which he says: "This call is much shorter than the song, and much fainter, so that the field observer may fail ever to hear it. Its typical form is of three notes, a low, a high, and a low, thus somewhat resembling the first bar of the song, but differing in that the three notes do not glide into one another, there being a clear break from each note to the next"; and the copulation note, which "is given by both the male and female, immediately after coition; in the mourning dove it is a faint growling note, repeated two to four times with rests between. So far as I have seen, the mourning doves, throughout the utterance of these sounds, keep the bill wide open."

Field marks.—The mourning dove in the field appears as a long, slim, gray-brown bird with small, nodding head, whistling wings, and long, pointed tail. The sparrow hawk resembles the dove very closely in flight, but it has strong, heavy shoulders, a larger head, and squarely tipped tail. The little ground dove of the Southern States is instantly distinguished from the mourning dove by its stumpy tail and the flash of bright color under the wing.

Enemies.—Harold C. Bryant (1926) speaks thus of the dove's enemies:

Apart from man, the dove has other enemies. The duck hawk is swift enough to overtake the dove, and this bird is probably the dove's most dreaded avian foe. Other predatory species take a toll during the nesting season. Its rapid flight frequently brings the dove in violent contact with telephone wires, and many birds die annually from this cause. Rodent-poisoning operations have in recent years been responsible for the death of many doves; for, unlike the quail, the mourning dove and the band-tailed pigeon are both susceptible to strychnine.

C. S. Thompson (1901) notes a very long tapeworm wound round and round in the intestines of an emaciated bird, and Lloyd (1887), writing of the bird in western Texas, speaks of owls as enemies in winter, "when they frequently change their roosting place, as a friend (Mr. Loomis) suggests, in consequence of being disturbed by the numerous owls."

The cowbird not infrequently selects the dove as a host for its young.

Game.—Bryant (1926) shows why, aside from its desirability as a table delicacy, the dove is a popular game bird, affording a rapidly moving target that demands the utmost skill on the part of the hunter. He says:

Unless favorably located near a watering place, one bird in three or four shots makes a good average for all but the most experienced hunter. The small size and great speed make the bird a difficult target. The variety of shots possible is almost endless. Quartering and side shots are most difficult because of the speed of the birds in flight. Then come shots at towering or descending birds, often dependent on whether they are coming or going. The easier straight-away shots are to be expected less often in dove shooting than in quail shooting.

Thus it will be seen that dove hunting gives the best of practice to the lover of wing shooting. No finer test of skill is afforded unless it be in snipe shooting.

Fall and winter.—In regions where mourning doves are common, they begin to resume their gregarious habits soon after the breeding season is over.

J. A. Spurrell (1917), speaking of the bird in Iowa, says:

From the latter part of July until the doves depart on their fall migration in late October they select common roosting places, one of which happens to be our orchard. Toward sunset the doves visit some place to drink and then fly to the roosting place from all directions until between five and six hundred are roosting there. They depart again just as it becomes light in the morning, spending the day far away in pastures and grain fields. During the month of August they may be commonly found about salt troughs for cattle, seeming to eat the salt.

Stockard (1905) speaks thus of the dove's habit of roosting in Mississippi, where they remain during the winter:

Late in summer they begin roosting in company, and many hundreds come about sunset to their chosen place for the night. During this season they are shot in large numbers while flying to the hedge or small wood that has been selected as a roosting place.

Throughout the winter in the Southern States we see the doves, in companies of a dozen or more, feeding quietly in the stubble and pea fields, from which, as we approach, they flush rather wildly and, scattering, retreat in twos and threes beyond the surrounding pine trees.

[AUTHOR'S NOTE: A comprehensive and interesting study of the fall migration of the mourning dove has been published recently by William B. Taber, jr. (1930), to which the reader is referred. F. C. Lincoln (1930) proves the migratory status of the mourning doves from evidence obtained from returns from banded birds.]

DISTRIBUTION

Range.—North America, Central America, and the West Indies.

Breeding range.—The mourning dove breeds **north** to British Columbia (Chilliwack and Okanagan Landing); Saskatchewan (Qu'Appelle); Manitoba (Aweme, Portage la Prairie, and Winnipeg); Michigan (Newberry, Mackinac Island, Hillman, and Zion); Ontario (Sarnia, Plover Mills, Guelph, Toronto, and Brighton); Vermont (Burlington); and New Hampshire (Concord and Hampton Falls). **East** to New Hampshire (Hampton Falls and Seabrook); Massachusetts (Boxford, Boston, North Truro, and Woods Hole); New York (Shelter Island and Yaphank); New Jersey (Red Bank, Spring Lake, and Sea Isle City); Delaware (Lincoln); Maryland (Cambridge); Virginia (Wallops Island and Spottsville); North Carolina (Beaufort and Fort Macon); South Carolina (Waverly Mills, Mount Pleasant, St. James Island, and Hilton Head); Georgia (Savannah, MacIntosh, Cumberland, and St. Marys); Florida (St. Augustine, Fruitland Park, Longwood, and Fort Myers); the Bahamas (Abaco Island, Eleuthera Island, Long Island, and Bird Rock); and Haiti (Monte Cristi, La Vega, Mahiel, and San Cristobal). **South** to Haiti (San Cristobal and Port au Prince); Cuba (Trinidad, La Ceiba, and McKinley); and central Mexico (Aquas Calientes and Las Penas). **West** to Mexico (Las Penas and probably San Blas); Lower California (Comondu, San Quintin Bay, and probably Todos Santos Island); California (San Clemente Island, Santa Catalina Island, Santa Cruz Island, San Francisco, Napa, Cahto, and Eureka); Oregon (Elkton, Eugene, Corvallis, and Portland); Washington (Yakima, Seattle, and Bellingham); and British Columbia (Chilliwack).

Winter range.—In winter doves are found with regularity **north** to California (San Geronimo); central Arizona (Fort Verde); Colorado (Navajo Springs and Pueblo); Nebraska (Lincoln); Iowa (Wiota, Ames, and Sabula); Illinois (Rantoul); Indiana (Camden); southern Michigan (Manchester); Ohio (Oberlin and Cleveland); Pennsylvania (Philadelphia); and New Jersey (Morristown and Englewood). **East** to New Jersey (Englewood and Newfield); Virginia (Wallops Island and Bowershill); Bermuda; North Carolina (Raleigh and New Bern); South Carolina (Summerville, Charleston, and Beaufort); Georgia (Savannah and Darien); Florida (Daytona, Titusville, St. Lucie, and Royal Palm Hammock); the Bahamas (Abaco Island, Eleuthera Island, Long Island, and Bird Rock); Haiti (Monte Cristi, La Vega, and San Cristobal); and Panama (Calobre). **South** to Panama (Calobre and Volcano de Chiriqui); Costa Rica (Azahar de Cartago and San Jose); Guatemala (Duenas and Quezaltenango); Oaxaca (Tehuantepec); Jalisco (Zapotlan); and Lower California (La Laguna). **West** to Lower California (La Laguna and Triunfo); and California (Santa Barbara, Paicines, Gilroy, San Francisco, and San Geronimo).

In addition to this normal winter range, individuals or small numbers of mourning doves will occasionally spend the winter north almost to the limits of the breeding range. Among such cases are: British Columbia (Okanagan Landing); Idaho (Emmett and Gray); Minnesota (Lanesboro); Wisconsin (Beloit); Ontario (Plover Mills); New York (Rochester and Rhinebeck); and Massachusetts (Danvers and Barnstable).

The range as described is for the entire species, which has, however, been divided into several races. True *macroura* is restricted to the Greater Antilles, although occurring in winter along the coast of Central America. It has no record in North America. The eastern mourning dove (*Zenaidura m. carolinensis*) occurs from the Atlantic seaboard west to the eastern edge of the Great Plains. It is found also in the Bahamas and on the Gulf and Caribbean coasts of Mexico and Central America. The western mourning dove (*Z. m. marginella*) occupies the territory from the Great Plains to the Pacific coast, and south through Mexico to Panama.

Spring migration.—Early dates of spring arrival are: New York, Orient, March 1; Canandaigua, March 4; Hamburg, March 6; Ithaca, March 8; Buffalo, March 9; Rochester, March 14; and Rhinebeck, March 23; Connecticut, Fairfield, March 10; Jewett City, March 12; Norwalk, March 14, New Haven, March 18; Hadlyme, March 21; and Saybrook, March 26; Rhode Island, Providence, March 31. Massachusetts, Harvard, March 12; Taunton, March 12; Amherst, March 17; Somerset, March 21; and Danvers, March 29; Vermont, Rutland, April 4; Castleton, April 8; Bennington, April 10; and Wells River, April 13; New Hampshire, Charlestown, April 14; Manchester, April 16; and Concord, April 20; Maine, Bucksport, March 21; Popham Beach, March 30; and Machias, April 6; Ontario, Port Dover, March 9; Point Pelee, March 17; London, March 17; Brighton, March 19; Harrow, March 20; Preston, March 24; and Guelph, March 25; Wisconsin, Janesville, March 10; Beloit, March 16; Prairie du Sac, March 18; Milwaukee, March 20; North Freedom, March 21; Madison,

March 22, and Menomonie, March 23; Minnesota, Redwing, March 12; Lanesboro, March 15; Hutchinson, March 23; St. Cloud, March 24; Elk River, March 25; and Wilder, March 26; South Dakota, Yankton, March 14; Vermilion, March 23; Sioux Falls, March 25; and Dell Rapids, March 28; North Dakota, Fargo, March 18; Bismarck, April 3; Larimore, April 5; Wahpeton, April 6; and Grand Forks, April 12; Manitoba, Margaret, April 7; Aweme, April 8; Treesbank, April 10; and Pilot Mound, April 21; Saskatchewan, Eastend, April 23; Indian Head, April 29; Muscow, May 4; and Wiseton, May 16. Utah, Clinton, April 18, and Salt Lake, April 26; Wyoming, Sheridan, April 13; Cheyenne, April 17; and Yellowstone Park, May 9; Idaho, Deer Flat, April 3; Meridian, April 6; and Pocatello, April 22; Montana, Knowlton, April 25; Bozeman, April 26; Corvallis, April 29; and Billings, May 3; Alberta, Banff, May 24; Nevada, Carson City, April 23; Oregon, Albany, March 18; Beaverton, March 30; Portland, March 30; and Mulino, April 15; Washington, North Yakima, March 21; Camas, April 7; and Pullman, April 16; and British Columbia, Okanagan Landing, April 20; Burrard Inlet, May 7; and Edgewood, May 15.

Fall migration.—Late dates of fall departure are: British Columbia, Chilliwack, October 2; Courtenay, October 9; and Okanagan Landing, November 8; Washington, Tacoma, October 1, and Pullman, October 2; Oregon, Cold Spring Bird Reservation, October 14, and Portland, November 19; Nevada, Winnemucca, September 9; Alberta, Red Deer River, September 15; Montana, Bozeman, October 2, and Glacier Park, October 30; Idaho, Minidoka, September 29, and Meridian, November 4; Wyoming, Yellowstone Park, October 2, and Sundance, October 4; Utah, Clinton, September 10; Saskatchewan, Eastend, September 15, and Indian Head, October 1; Manitoba, Killarney, October 14; Ninette, October 15; Aweme, October 19; and Margaret, October 20; North Dakota, Harrisburg, October 11; Inkster, October 14; Fargo, October 15; and Grafton, November 12; South Dakota, Wall Lake, October 13; Harrison, October 18; and Rapid City, October 24; Minnesota, Hutchinson, October 22; Twin Valley, October 25; Elk River, October 26; and Minneapolis, November 10; Wisconsin, Delavan, October 31; Shiocton, November 4; Trempealeau, November 17; Menomonie, November 20; Elkhorn, November 22; and Meridian, November 23; Ontario, Guelph, November 3; Port Dover, November 4; Plover Mills, November 14; Harrow, November 21; Windsor, November 27; and Point Pelee, December 4; Maine, Gorham, September 26; Mount Desert, October 16; and Machias, November 10; Vermont, Bennington, September 1; Massachusetts, Harvard, October 12; Lunenburg, October 22; Amherst, October 28; Braintree, November 7; and Marthas

Vineyard, November 21; Rhode Island, Block Island, October 22, and South Auburn, November 20; Connecticut, Fairfield, October 12; Meriden, October 20; Hartford, October 24; New Haven, October 30; and Portland, November 30; and New York, Rochester, October 5; Rhinebeck, October 6; Geneva, October 12; Collins, October 19; and Howard, October 24.

From the fact that mourning doves winter so far north it might be assumed that they do not have an extensive migration. The returns from banded birds, however, indicate that the majority of these birds move in winter well to the south. For example, one banded at Wauwatosa, Wis., on June 6, 1929, was shot in Dale County, Ala., on January 3, 1930, and another banded at Madison, Wis., on April 12, 1929, was recovered at Jennings, La., on December 2, 1929.

The region from Texas east to Georgia is the favored winter home of the species. Eight birds banded at points in Nebraska, Kansas, Minnesota, Missouri, Illinois, Indiana, and Ohio were retaken in Texas; 7 banded in Illinois and Indiana were shot in Louisiana; 10 banded in Illinois, Indiana, Michigan, and Ohio were recovered in northern Florida; 4 banded in Illinois, Indiana, and Ohio were shot in Alabama; and 14 banded in Illinois, Indiana, Michigan, New York, New Jersey, Pennsylvania, and Virginia were recovered in Georgia. (See Lincoln, 1930.)

Casual records.—Casual occurrences of the mourning dove have been chiefly north of the regular range, where it appears they may sometimes breed. Two young barely able to fly were noted on the Stikine River, Alaska, by Willett, who also records one from Hydaburg, Prince of Wales Island, on September 1, 1926, and one taken at Sitka, Alaska, on September 14, 1912. Kermode (1913) records a specimen from Telegraph Creek, British Columbia. There are several records for Quebec, among which are: Godbout, October 10, 1881, and June 6, 1882; once at St. Joachim (date ?), and one in the fall of 1887 near Quebec City (Dionne). Newfoundland has one record on October 16, 1890, but there are a few others on the Labrador coast: Spotted Islands, October 17, 1912; Battle Harbor, October 20, 1912; and Red Bay, September 7, 1891. The species has been recorded from New Brunswick at Fredericton, October 14, 1899; Hampton, June, 1880; Rothesay, September 30, 1881; and Milkish, October 17, 1881.

Egg dates.—Southern New England and New York: 20 records, April 6 to August 8; 10 records, May 20 to June 15. New Jersey and Pennsylvania: 72 records, April 6 to July 8; 36 records, April 26 to May 29. Florida: 10 records, March 11 to July 10. Michigan to the Dakotas: 18 records, April 27 to August 9; 9 records, May 14 to June 5. Indiana to Iowa: 58 records, April 4 to September 1; 29 records, April 30 to June 7. Texas: 89 records, February 20 to September 24; 45 records, April 21 to May 28. Arizona and New Mexico: 54 records, April 1 to September 2; 27 records, May 1 to June 6. California: 170 records (every month but October and November), January 18 to December 5; 85 records, May 6 to June 19.

ZENAIDURA MACROURA MARGINELLA (Woodhouse)

WESTERN MOURNING DOVE

HABITS

The habits of the western mourning dove are so much like those of the eastern race that there is very little to be added to the excellent life history of the latter contributed by Doctor Tyler.

The western race ranges from Manitoba and Oklahoma westward. Being largely an inhabitant of the open plains and more arid regions, it averages slightly paler in coloration, with the upper parts more grayish, and slightly larger than the eastern race.

Courtship.—Frank F. Gander has noted some aggressiveness on the part of the female, on which he has sent me the following notes:

On April 7, 1927, I watched a pair of doves mate. The male flew to a bare limb and was closely followed by the female, who pressed close to him and reached up with her beak as if begging for food. She also fluttered her wings a little and squatted low to the branch. The male seemed inattentive at first, but gradually began to pay more attention, billing with her as if feeding, but there was none of the straining of regurgitation. After a few moments of this, with the female continuing to beg and remaining in a squatting posture, the act of copulation took place. The wings of the male were slightly raised and used to maintain his balance. After mating, there was no strutting flight as in the pigeon, but the two birds calmly resumed their respective places on the limb and, beyond a slight craning of necks and peering about, showed no signs of having been affected. In a few moments the female again made advances toward her mate and the billing took place, but the act of mating was not repeated before the birds flew away.

Nesting.—Mr. Gander has found nests as high as 40 feet in cottonwood and eucalyptus trees, and he says in his notes:

Many years ago I found a set of eggs on a piece of loose bark that had become lodged in the branches of a large tree. The only additions to this site were two sticks, which were crossed, and the eggs were in the angle thus formed. In southern California, especially, the doves use a nest over and over again, and as some additional material is carried before each set of eggs, it often grows to a rather substantial nest.

On April 27, 1927, I saw an adult dove brooding a set of two eggs in the same nest with a well-grown young one. I have observed this on other occasions but have no further data.

George Finlay Simmons (1925) says that in Texas the nest is placed " rarely on leaning corn-stalks, rail fences, tops of rock fences, or ledges in cliffs. Occasionally old nests of the gray-tailed cardinal or the western mockingbird are repaired and used; one bird with eggs was found occupying an old nest of the Audubon caracara."

Eggs.—The 2 eggs, rarely 1 or 3, of the western mourning dove are pure white like those of the eastern bird. The measurements of 48 eggs average 28 by 20.9 millimeters; the eggs showing the four extremes measure **31** by 21.5, 30 by **22**, and **25** by **19.5** millimeters.

Behavior.—Simmons (1925) writes:

In desert country, from Austin westward, water holes and the cool shade of the narrowly timbered prairie creeks can be located by watching during mid-day the direct lines of flight of the swiftly moving doves, which come from miles across the prairie, since they frequently nest a long distance from water. Hunters make use of this water-seeking habit of the doves by hiding near a water hole at dawn or dusk and shooting the birds as they drop in to water preparatory to leaving or roosting in the nearby trees where they spend the night.

ZENAIDA ZENAIDA ZENAIDA (Bonaparte)

ZENAIDA DOVE

HABITS

On April 24, 1903, while crawling on all fours under the thorny tangle of tropical shrubs and vines on Indian Key, one of the lower Florida Keys, I saw several small doves, with white in the wings and tail, flitting along ahead of me near the ground or breaking out and flying away over the tops of the bushes. The vegetation was too thick to shoot them or even to get a good look at them, but I have always suspected that they were Zenaida doves, as this is the key where Audubon (1840) mentions finding them. Here is what he says about it:

The Zenaida Dove is a transient visitor of the Keys of East Florida. Some of the fishermen think that it may be met with there at all seasons, but my observations induce me to assert the contrary. It appears in the islands near Indian Key about the 15th of April, continues to increase in numbers until the month of October, and then returns to the West India Islands, whence it originally came. They begin to lay their eggs about the first of May. The males reach the Keys on which they breed before the females, and are heard cooing as they ramble about in search of mates, more than a week before the latter make their appearance. In autumn, however, when they take their departure, males, females, and young set out in small parties together.

Dr. Thomas Barbour (1923) says of its haunts in Cuba:

This wide-ranging Pigeon is more shy and retiring than the Rabiche, and more solitary. Nevertheless it is found in varying numbers throughout the island. Its noisy flight is often startling. It is found rarely in deep forest, though Brooks and I have taken it in the high woods about the Cienaga. It is far more characteristic of open savanna lands and the shady second-growth *manigua* along water-courses in pastures and the outer boundaries of cultivated fields. It shuns habitations, and is seldom seen in cultivated land; in fact, it feeds but little on the ground. Its flesh is excellent.

P. H. Gosse (1847) says that in Jamaica:

The open pastures, or the grassy glades of pimento pens, are the favourite haunts of this pretty Dove, where it walks on the ground singly or in pairs. In such open situations, it can discover, and mark the motions of an intruder, and long before he is within gun-range it is upon the wing. When the rains have ceased, the increasing drought renders these, as it does many other birds, more familiar; and they may be seen lingering on the borders of streams and ponds. Indeed they seem, of all our Doves, to haunt most the vicinity of water; particularly those dreary swamps or morasses which are environed by tall woods of mangrove. In the winter months, when the pastures are burnt up with drought, we may hear all day long their plaintive cooing, proceeding from these sombre groves, though it is not much heard in any other situation.

Nesting.—Audubon (1840) writes:

Those Keys which have their interior covered with grass and low shrubs, and are girt by a hedge of mangroves, or other trees of inferior height, are selected by them for breeding; and as there are but few of this description, their places of resort are well known and are called *Pigeon* or *Dove Keys.* It would be useless to search for them elsewhere. They are by no means so abundant as the White-headed Pigeons, which place their nest on any kind of tree, even on those whose roots are constantly submerged. Groups of such trees occur of considerable extent, and are called Wet Keys.

The Zenaida Dove always places her nest on the ground, sometimes artlessly at the foot of a low bush, and so exposed that it is easily discovered by anyone searching for it. Sometimes, however, it uses great discrimination, placing it between two or more tufts of grass, the tops of which it manages to bend over so as completely to conceal it. The sand is slightly scooped out, and the nest is composed of slender dried blades of grass, matted in a circular form, and imbedded amid dry leaves and twigs. The fabric is more compact than the nest of any other Pigeon with which I am acquainted, it being sufficiently solid to enable a person to carry the eggs or young in it with security. The eggs are two, pure white, and translucent. When sitting on them, or when her young are still small, this bird rarely removes from them, unless an attempt be made to catch her, which she, however, evades with great dexterity. On several occasions of this kind, I have thought that the next moment would render me the possessor of one of these Doves alive. Her beautiful eye was steadily bent on mine, in which she must have discovered my intention, her body was gently made to retire sidewise to the farther edge of her nest, as my hand drew nearer to her, and just as I thought I had hold of her, off she glided with the quickness of thought, taking to wing at once. She would then alight within a few yards of me and watch my motions with so much sorrow, that her wings drooped, and her whole frame trembled as if suffering from intense cold. Who could stand such a scene of despair? I left the mother to her eggs or offspring.

Gosse (1847) describes its nesting habits in Jamaica quite differently, as follows:

The nest is, as usual, a loose platform of twigs interlaced, with scarcely any hollow, and no leaves; it is often built in an orange, or a pimento, and contains two eggs of a drab hue. Near the end of March we started a Pea-dove from the centre of a lofty Ebby palm (*Elais*) in Mount Edgecumbe; it immediately alighted on the ground just before my lad, and began to tumble about in a grotesque manner, affecting inability to fly. Sam was not to be caught, however; but calling my attention to the circumstance, we began to peer among the fronds of the tree, where we presently discerned the projecting ends of the twigs that constituted her nest, the centre of her fears and anxieties. It was inaccessible, however, when discovered.

Doctor Barbour (1923) says that "Gundlach found nests from April to July," in Cuba, "the usual shabby platform with two eggs, on bunches of epiphytic bromeliads or on some horizontal limb." Stuart T. Danforth (1925) found the Porto Rican form nesting still differently; he says:

At the lagoon a few of these doves forsake their usual nesting sites and nest in the cat-tails. I found two such nests in 1924. The first was a crude platform of twigs on some bent cat-tail leaves in a dense clump on Las Casitas. It was three feet above the water, which was two feet deep at that place. On May 13 it contained two white eggs, but the nest was so frail it fell to pieces before the eggs had time to hatch. The second nest was built of dry cat-tail leaves and placed in the cat-tails at a height of four feet from the water, which was two inches deep at that time. When I found it on May 27 it contained two egg shells from which the young had hatched but the young birds were nowhere to be seen.

And all the nests that Dr. Henry Bryant (1861) found in the Bahamas "were made in holes in the rocks, and consisted, as is always the case in this family, of but a few sticks."

Eggs.—The eggs of the Zenaida dove are two in number, oval, but more rounded than doves' eggs usually are, and pure white. The measurements of 17 eggs average 29.6 by 22.8 millimeters; the eggs showing the four extremes measure **34** by 23, 31 by **24.5**, and **25.2** by **19.8** millimeters.

Plumages.—I have not seen enough specimens in immature or transition plumages to say much on this subject. Young birds in juvenal plumage seem to be similar to adults, but paler or duller, with more whitish on the chin and without the iridescent colors. An adult, taken July 10, is completing the molt of wings and tail, which indicates that the complete molt begins early.

Food.—Audubon (1840) says of the food of this dove:

The flesh is excellent, and they are generally very fat. They feed on grass seeds, the leaves of aromatic plants, and various kinds of berries, not excepting those of a tree which is extremely poisonous,—so much so, that if the juice of it touch the skin of a man, it destroys it like aquafortis. Yet these berries do not injure the health of the birds, although they render their flesh bitter and unpalatable for a time. For this reason, the fishermen and wreckers are in the habit of examining the crops of the Doves previous to cooking them. This, however, only takes place about the time of their departure from the Keys, in the beginning of October. They add particles of shell or gravel to their food.

In Porto Rico, Dr. Alexander Wetmore (1927) found that

the bulk of the food of this dove consists of seeds, including many wild legumes, euphorbias, mallows. knotweed and pigweed. Waste grain is also taken and various small wild fruits in season. The species is a valuable game bird and should be carefully protected during the breeding season.

Gosse (1847) writes:

The Pea-dove subsists on various fruits and seeds: pimento-berries, orange-pips, sop-seeds, castor-oil nuts, physic-nuts, maize, and the smaller seeds of pasture weeds are some of his resources. His flesh is white and juicy, and when in good condition is in general estimation.

Behavior.—Doctor Wetmore (1927) refers to its habits as follows:

The flight is strong and direct and it flushes with a loud clapping of wings. On the ground this species resembles the mourning dove, as it walks quickly about with nodding head, and it has a cooing note almost indistinguishable from that of the bird mentioned.

During the breeding season the males are frequently seen sailing out in circles, with the wings held stiffly, and their cooing notes come from the hillsides all day long. They are also observed at times walking rapidly about on the ground near the females, striking at each other with their wings.

Between nine and ten in the morning the Zenaida dove comes in to streams or ponds for water, usually in pairs, swiftly flying high in the air. On the gravel bars of the larger rivers they walk about quickly, quenching their thirst and picking up bits of sand and gravel. Usually they are quite wary, but sometimes prefer to hide and let an intruder pass, rather than fly.

The species frequents open country and is thus the only one of the large pigeons in Porto Rico that prospers with the clearing of the land. In some localities it is hunted constantly and then is very wild; elsewhere it is quite tame.

Audubon's (1840) account is somewhat different:

The flight of this bird resembles that of the little Ground Dove more than any other. It very seldom flies higher than the tops of the mangroves, or to any considerable distance at a time, after it has made choice of an island to breed on. Indeed, this species may be called a Ground Dove too; for, although it alights on trees with ease, and walks well on branches, it spends the greater portion of its time on the ground, walking and running in search of food with lightness and celerity, carrying its tail higher than even the Ground Dove, and invariably roosting there. The motions of its wings, although firm, produce none of the whistling sound, so distinctly heard in the flight of the Carolina Dove; nor does the male sail over the female while she is sitting on her eggs, as is the habit of that species. When crossing the sea, or going from one Key to another, they fly near the surface of the water; and, when unexpectedly startled from the ground, they remove to a short distance, and alight amongst the thickest grasses or in the heart of the low bushes. So gentle are they in general, that I have approached some so near that I could have touched them with my gun, while they stood intently gazing on me, as if I were an object not at all to be dreaded.

Voice.—C. J. Maynard (1896) gives an elaborate description of the sound-producing apparatus of the Zenaida dove, and then describes its notes, as follows:

This dove, as might be expected from such a musical apparatus, has a singular note. The male perches upon a limb of a tree, swells out his throat, and utters his cooing song, which he repeats at rather regular, but protracted intervals.

This song begins with two notes, the first uttered with a falling inflection, the second with a rising. The second follows the first rather quickly and is not as prolonged. Both are in a low key. Then follows three other notes, sounding like "Who, who, who," but there is a decided pause between the first two; the last three are given in the same time and in the same key. The notes are all loud, but when softened by distance have a singularly mournful effect.

The notes of this bird which I have described above are very loud and on a still morning can be heard for a long distance. The males begin to coo with the first indication of the dawn, and begin to fly about some time before sunrise. They also coo at sunset and continue to utter their mournful notes until darkness fairly begins.

Enemies.—One of the worst enemies of this and other ground-nesting birds is the mongoose, but fortunately the dove has developed the tree-nesting habit in certain places, perhaps as a result of predatory attacks. Several writers speak of it as an important and popular game bird and much in need of protection. Mr. Danforth (1925) says:

This species is probably preyed upon to some extent by the Mongoose, although its nesting habits make it more immune to attack than the Ground Dove. The Zenaida Dove is one of the most-sought-for game birds on the Island, and justly so, as its meat is of good flavor, and it is so wary that one has to be an expert marksman to obtain many. But it has a disadvantage in that it frequently flies long distance after being shot before it falls, making the recovery of the birds shot very difficult. Often but a small percentage of the birds killed are recovered, especially when one is hunting in rough country. The few that I have shot were lost in this manner.

DISTRIBUTION

Range.—The West Indies, including the Bahama Islands, and the coast of Yucatan; casual visitor on the Florida Keys. Generally nonmigratory.

The range of the Zenaida dove extends **north** to northern Yucatan (Progreso and Holbox Islands); northwestern Cuba (San Cristobal and Habana); southern Florida (Indian Key); and the Bahama Islands (Great Bahama and Stranger Cay). **East** to the Bahama Islands (Stranger Cay, Moraine Cay, Abaco Island, Nassau, Cat Island, Watlings Islands, Acklin Island, North Caicos, Grand Caicos, and East Caicos); and the Lesser Antilles (Sombrero Island, Barbuda Island, Antigua Island, Grande Terre, Barbados Island, and Grenada). **South** to the Lesser Antilles (Grenada and St. Croix Island); southern Porto Rico (Comerio, the Cartagena Lagoon, and Mona Island); southern Haiti (San Domingo); Jamaica (Spanishtown, Port Henderson, Cayman Brac, Little Cayman, and Grand Cayman); and the coast of Yucatan (Mujeres

Island, La Logartos, and Progreso). **West** to the coast of Yucatan (Progreso).

The Zenaida dove is apparently only a visitor to the Florida Keys, arriving about the middle of April and remaining until about October, when they return to the islands of the West Indies.

Two birds were reported by Pangburn (1919), as seen at Passagrille, Fla., February 11, 1918, but as no specimen was obtained, the record is considered doubtful.

Egg dates.—Bahamas and West Indies: 10 records, April 6 to December 8; 5 records, May 13 to June 12.

LEPTOTILA FULVIVENTRIS ANGELICA Bangs and Penard

WHITE-FRONTED DOVE

HABITS

The white-fronted dove was first discovered as an inhabitant of the United States by George B. Sennett (1878) in the valley of the Rio Grande in southern Texas, of which he writes:

On April 18th, I obtained my first in a tract of timber a mile below Hidalgo, near the bank of the river. It was shot from the upper branches of the tallest trees. Scattered about the woods in pairs were *Columba flavirostris*, Red-billed Pigeon, and *Melopelia leucoptera*, White-winged Dove. On the 19th, another was shot in the same locality. Five specimens were secured up to the time of leaving, and a number of others seen and heard. It is more secluded than the other Pigeons, and only found among the tallest timber. Seen in the woods, it resembles *M. leucoptera* both in size and shape of tail, but can be recognized from it at sight by the absence of the large, white wing-patch.

Recently the northern form of this dove, found in northern Mexico and southern Texas, has been separated as a distinct subspecies by Bangs and Penard (1922), given the name *angelica*, and characterized as "similar to *Leptotila fulviventris fulviventris* Lawrence, but under parts less buffy, the under tail-coverts almost pure white; forehead more grayish, less vinaceous; neck and chest less vinaceous."

Spring.—Sennett (1879) writes:

Dr. Finley reports the arrival of this Pigeon at the vicinity of Hidalgo and Lomita about the middle of February, its departure having taken place in November. Although it is less numerous than the Red-billed Pigeon, yet, by its peculiar note, it is easily distinguished from all other species, and can thus be readily obtained. We heard it daily. It is so much more retiring in its habits than other Pigeons, that were it not for the peculiarity we mention, it would be met with very seldom. It frequents the dense and heavy growth of timber, and long and frequent were our endeavors to find its nest.

Nesting.—On May 27, 1923, we found the white-fronted dove quite common and breeding in the dense forest around the Resaca de la Palma, near Brownsville, Tex. These and the mourning and white-winged doves were especially abundant and much in evidence among the varied bird life of that interesting region, more fully described under the chachalaca. Their characteristic notes were almost constantly heard, and we found several nests of each of these species. The white-fronted and white-winged doves looked much alike, but could be easily recognized by the deep-toned notes of the former and the white wing patches of the latter. The nests and the eggs of these two species were also much alike, but could usually be identified by seeing the birds leave them. The only set of eggs of the white-fronted dove that I collected was taken from a small, frail nest of sticks, 10 feet above the ground, on a horizontal limb of a small tree, in the dense underbrush of the forest. I attempted to photograph the bird on a similar nest, not so high up, but something destroyed the eggs before the bird returned.

Sennett (1879) had a nest brought to him (with the parent bird) that " was situated in the forks of the bushes, about five feet from the ground, was flat and quite large for a Pigeon's nest, and composed of the dead branches, twigs, and bark of pithy weeds." Dr. J. C. Merrill (1879) took a nest that " was about seven feet from the ground, supported by the dense interlacing tendrils of a hanging vine growing on the edge of a thicket." Major Bendire (1892) says:

Mr. William Lloyd writes me that this Pigeon breeds abundantly in the Sierra Madre, from southern Chihuahua to Beltran, Jalisco, Mexico, at an altitude of from 1,100 to 2,200 feet. The nests, usually placed in thorny shrubs, *Huisache, Acacia farnesianna,* 10 to 12 feet from the ground, are substantially made of straw. He found eggs as early as May 10, and up to June 13, when they were much incubated. It frequents deep arroyas mostly during the breeding season.

According to George N. Lawrence (1874) the white-fronted dove of western Mexico nests on the ground. He quotes Colonel Grayson as saying that " differing from all our American doves, it deposits its eggs upon the ground, forming scarcely any nest; the eggs are two, and white; the young soon follow the mother, before being able to fly, like some of the gallinaceous birds."

Eggs.—The white-fronted dove lays two eggs, which are easily recognized, when first collected, by their color. They are elliptical oval in shape and somewhat glossy. The color is " cream-buff " at first, but it soon fades to " cartridge buff " or dull white. The measurements of 49 eggs average 30.6 by 22.9 millimeters; the eggs showing the four extremes measure **33.7** by 23.5, 29.2 by **24.1**, **28** by 22.2, and 30.2 by **21.4** millimeters.

Plumages.—I have seen no specimens of the downy young. Specimens in juvenal plumage, in June, have the crown, mantle, wings, and tail " sepia "; the tail and wing feathers and scapulars are narrowly edged with " cinnamon-buff " or " pinkish cinnamon "; the greater and lesser wing coverts are more broadly edged with " cinnamon-rufous"; the breast is "wood brown," with "pinkish buff" edgings, shading off to grayish or buffy white on the belly.

Material in collections is too scanty to illustrate subsequent molts and plumages. A young bird, taken July 2, is fully grown and has nearly completed the body molt into a plumage that is practically adult, but the flight feathers and wing coverts are still juvenal. Apparently both young and old birds have one complete molt during a prolonged period in summer and fall.

Food.—Little seems to be known about the food of these doves, which probably consists mainly of seeds, fruits, and berries. Austin P. Smith (1910) says: "They feed almost entirely on small herb and grass seed, rarely partaking of the mesquite or ebony bean."

Behavior.—Smith (1910) says:

> This dove approaches the true pigeons in bulk, but is more eminently terrestrain than any of the several pigeons I am acquainted with. The White-fronted Dove is a slow-moving bird on the ground and quite unsuspicious; and as it generally prefers to feed under growth of some sort, proves an easy target for the pot-hunter.

Lawrence (1874) quotes from Colonel Grayson's notes as follows:

> This "ground dove," as its name indicates, is usually met with upon the ground in search of its food, or sometimes resting upon low branches, or old logs, and always in the thickest woods, out of which they are seldom seen. It walks and runs with great facility upon the ground, whilst its flight is always low amidst the bushes or underbrush as if to conceal itself, and not long continued, usually alighting upon the ground beneath a massive canopy of underbrush, where it continues to walk or run to elude pursuit, or search for its food. When suddenly started from its retreat, the wings whir, accompanied by a whistling sound, very similar to that of the wood-cock. Its habits are solitary, never congregating into flocks, and only during the breeding season do we ever find a pair together.

Sennett (1879) considered it less of a ground dove, for he writes:

> During both seasons that I passed on the Rio Grande, I saw this bird upon the ground but once, and it was then feeding upon some corn that was scattered in the roadway, and, so far from its remaining near or on the ground, its habit is to frequent the high branches of tall trees; indeed, on this account almost every specimen shot was more or less injured in falling, a number being too much so to save.

Voice.—The notes of the white-fronted dove are characteristic; I recorded them in my notes as deep-toned. Sennett (1878) says:

> Its note is somewhat prolonged, ends with a falling inflection, and is exceedingly low in pitch. Most of my birds were obtained by following the sound of their notes until within range; all were seen sitting quietly in secluded places; all are males, and injured considerably by falling from great heights.

Winter.—From November to February this dove is apparently absent from the northern portion of its range.

DISTRIBUTION

Range.—Central America, north to the lower Rio Grande Valley in Texas; nonmigratory except that in the northern part of the range there is probably a slight withdrawal from November to February.

The range of the white-fronted dove extends **north** to Tepic (Santiago); northern Nuevo Leon (Rio Salado); and southern Texas (Rio Grande City, Lomita, and Sause Ranch). **East** to Texas (Sause Ranch, Santa Rita Ranch, and Brownsville); southern Nuevo Leon (Montemorelos); Pueblo (Tecali); eastern Guatemala (San Geronimo); and northwestern Nicaragua (Chinandega). **South** to northwestern Nicaragua (Chinandega); Salvador (La Libertad); western Guatemala (San Jose, Duenas, and Retalhuleu); Chiapas (Tonala); Oaxaca (Santa Efigenia); and Colima (Santiago). **West** to Colima (Santiago); and Tepic (San Blas and Santiago).

A specimen in the United States National Museum from Tres Marias Island, Tepic, can be considered only as a straggler from the mainland.

Egg dates.—Texas: 42 records, March 30 to July 25; 21 records, May 6 to 27. Mexico: 18 records, April 13 to June 28; 9 records, May 20 to June 24.

MELOPELIA ASIATICA ASIATICA (Linnaeus)

EASTERN WHITE-WINGED DOVE

HABITS

The type name of the white-winged dove is now restricted to the birds found in the West Indies and the eastern part of the range of the species. This eastern form is much less numerous than the western form and is not nearly so well known. For these reasons it has seemed best to write a full life history of the western form only.

Baird, Brewer, and Ridgway (1905) say:

This species is abundant in Jamaica, where, according to Mr. March, it is more a lowland than a mountain dove. They are said to be gregarious, usually keeping in flocks of from 10 to 20, but in January and in February, in the Guinea-corn season, and at other times when the *Cerei* are in fruit, they congregate in large flocks, often of several hundreds. Their food is principally grain and seeds, but they are equally fond of the ripe fruit of the different species of *Cereus* abounding on the savannas and salines during the summer. Inland, the White-Wings, in the same manner as the Baldpate, breed in solitary pairs; but in the mangrove swamps, and in the islands along the coast, they breed in company, many in the same trees. The nest is a frail platform of sticks, with a slight hollow lined with leaves and bark, and sometimes a few feathers.

P. H. Gosse (1847) writes:

In the early months of the year, when the physic-nut (*Jatropha curcas*) is ripening, and oranges come in, the Whitewing becomes plentiful in open pastures, and the low woods in the neighbourhood of habitations; the seeds of these fruits, and the castor-oil nut, forming the principal part of their food. At this time they are very easily shot, as they walk about on the ground. From the ease with which they are procured, they are a good deal eaten, though seldom fat, and rather subject to be bitter.

When the rains fall, we see the Whitewings but seldom; they betake themselves to the deep woods and impenetrable morasses, when their presence is indicated by their loud stammering coo.

Farinaceous and pulpy berries are found in the woods at all seasons, so that the Pigeons and other frugivorous birds have not only abundance but variety. Its nest is not very often met with. I am informed that it occasionally builds in a pimento; Robinson says that it builds also in the orange, and sea-side grape, in May, a very slight and narrow platform of rude twigs, and lays two eggs, of a pale drab hue.

We found this dove abundant in Hidalgo and Cameron Counties in southern Texas, where it evidently was the most numerous bird, next to the omnipresent great-tailed grackle, in the forests and thickets about Brownsville. We found a few nests in the chaparral and in the dense forests around the resacas, which I have already described under the chachalaca. The nests were in low trees or bushes and were made of small twigs, grasses, and weeds. George B. Sennett (1878) found one nest made of Spanish-moss. The eggs are like those of the western form, but average a little smaller. The measurements of 33 eggs average 29.8 by 22.1 millimeters; the eggs showing the four extremes measure **33** by 23, 31.5 by **24.5**, **26.5** by 20.5, and 28 by **19.5** millimeters.

DISTRIBUTION

Range.—Southern United States, the West Indies, and Central America.

The range of the white-winged dove extends **north** to southern California (Brawley and Palo Verde); Arizona (Little Meadows, Big Sandy Creek, Congress Junction, New River, Roosevelt, and Graham Mountains); New Mexico (Hidalgo County, Cloverdale, Mesilla Park, and Cliff); southern Texas (Del Rio, Uvalde, Castroville, San Antonio, Beeville, and probably High Island); the Bahamas (Great Inagua Island); and the Lesser Antilles (St. Bartholomew Island). **East** to the Lesser Antilles (St. Bartholomew Island). **South** to the Lesser Antilles (St. Bartholomew Island); Haiti (Mount La Laguneat and Gonave Island); Jamaica (Spanishtown and Port Henderson); Costa Rica (La Palma); Nicaragua

(San Juan del Sur); Salvador (La Libertad); Guatemala (Volcano Agua and Duenas); Oaxaca (Tehuantepec); Tepic (Las Penas Island and San Blas); and Lower California (Cape San Lucas). **West** to Lower California (Cape San Lucas, San Jose del Cabo, Santa Anita, Triunfo, La Paz, and Comondu); western Sonora (Guaymas); and southern California (Brawley).

Migration.—Although white-winged doves are found in winter more or less throughout their breeding range, migrating birds have been observed to arrive in Arizona as follows: Sabina Canyon, April 6, Otero Creek Canyon, April 12, Oracle, April 15, and Tombstone, April 15. Similarly, fall migrants have been observed at Phoenix, October 1, and Tombstone, October 21.

The range as above outlined is for the entire species, which has, however, been separated into two subspecies. The eastern white-winged dove (*Melopelia a. asiatica*) is found from Texas, eastern Mexico, and Costa Rica, east to the West Indies, and casually southern Florida. The western white-winged dove (*M. a. mearnsi*) occurs in the Southwestern United States and western Mexico.

Casual records.—The occurrence of white-winged doves north of their normal range has been noted on numerous occasions. Among these records are California, a specimen at Escondido, about September 25, 1911, one "heard" at Needles (Stephens, 1903), and one seen at Santa Barbara, November 8, 1922; Washington, one taken at Puyallup, November 11, 1907; British Columbia, two seen and one taken at Sherringham Point, Vancouver Island, in July, 1918; Colorado, one shot in the Wet Mountains in September, 1899; Texas, the most northwestern record being a specimen at Kerrville, November 25, 1910; Louisiana, one of a pair taken at Venice, about November 20, 1910, and Grand Isle, May, 1894, and August, 1895; Mississippi, one in Jackson County, on November 13, 1915; Alabama, one taken at Point Clear, about December 23, 1916, one taken at Daphne, about December 2, 1916, and another, also in Baldwin County, exact date and locality unknown (A. H. Howell, 1928); Florida, three specimens at Key West on November 14, 1888, November 20, 1895, and November 28, 1895, Kissimmee, November, 1896, and one taken near Orlando in the winter of 1908–9; Georgia, a specimen at Hoboken, January 6, 1917; and Maine, one taken at Lincoln, November 5, 1921.

Egg dates.—Texas: 108 records, March 30 to July 14; 54 records, May 12 to 29. Arizona: 68 records, April 2 to August 2; 34 records, May 18 to June 2. Mexico: 20 records, April 20 to August 5; 10 records, May 18 to June 20.

MELOPELIA ASIATICA MEARNSI Ridgway

WESTERN WHITE-WINGED DOVE

HABITS

The name *mearnsi* was applied to the white-winged dove of the Southwestern United States and Mexico by Robert Ridgway (1915), who described it as "similar to *M. a. asiatica* but averaging decidedly larger, and coloration paler and grayer, the foreneck and chest light drab to hair brown instead of fawn color, the back, etc., hair brown to deep drab."

In the regions where I have met with the white-winged dove, I found it to be one of the commonest and decidedly the noisiest of birds. Its monotonous cooing and hooting notes were heard almost constantly in the chaparral and forests, especially early in the morning and toward night.

We found it very abundant in certain parts of southern Arizona. We camped for several days on the edge of the mesquite forest, on the Santa Cruz River, south of Tucson, where we were lulled to sleep at night, or awakened in the morning, with the monotonous notes of the white-winged doves ringing in our ears. They were most noisy at morning and evening, but could be heard at all hours of the day and sometimes during the night. These doves were also common in nearly all the canyons, and a few were found in the more fertile valleys of the San Pedro River.

Harry S. Swarth (1920) says:

Throughout the valleys of southern Arizona the white-winged dove, or Sonora pigeon as it is generally known, is an abundant summer visitor. Mesquite-grown bottom lands form the favorite breeding resort, and it is there or in cultivated fields that the white-wings are to be found in numbers. Anywhere on the desert, however, one is apt to see them, passing overhead, feeding, or resting on the giant cactus or in the shade of the thicker bushes. They also invade the towns to some extent, and may frequently be observed in garden shrubbery or perched on fences or electric wires.

Spring.—Although Major Bendire (1892) found it partially resident throughout the year in the vicinity of Tucson, Ariz., and observed it during every month of winter, they were not so abundant then as in summer, many having migrated. Mr. Swarth (1920) says: "The birds, as a rule, arrive in southern Arizona about the third week of April. Gilman gives the date of arrival at Sacaton as April 20, while I found a bird sitting on eggs near Tucson as early as April 13."

M. French Gilman (1911) writes:

Their coming is coincident with the ripening of the berries of the wild jujube, *Zizyphus lycioides*, upon which they feed greedily as long as the fruit lasts, consuming both ripe and green. They come in such great numbers that the

wheat fields suffer and the loss is considerable. The Indians try to frighten them away from the fields but do not hunt them. Probably they figure that ammunition would count up more on the debit side than would the wheat destroyed.

Referring to southern California, W. Leon Dawson (1923) says:

The White-winged Dove is a tardy migrant, and its numerous arrival in late April is quite conspicuous. Flight is conducted at low levels, and occupancy is effected by a progressive invasion rather than by a sudden coup. The birds troop across the roads in endless desultory columns, or else rise hastily from a wayside snack; or, most likely of all, gather upon exposed branches to mark with curious wooden detachment the passing of the intruder.

Courtship.—Dr. Alexander Wetmore (1920) has published a most interesting account of his extensive studies of the white-winged dove, from which I shall quote freely; regarding courtship, he writes:

In displaying before females males had a curious habit or pose in which they raised the tail high and tilted the body forward. At the same time the tail was spread widely and then closed with a quick flash of the prominent black and white markings. In the breeding colonies males at intervals flew out with quick, full strokes of the spread wings, rising until they were thirty or forty feet in the air. The wings were then set stiffly with the tips decurved, while the birds scaled around above the mesquites in a great circle that often brought them to their original perches. The contrasted markings of the wings showed brilliantly during this flight and the whole was most striking and attractive. In the cooler part of the morning males performed constantly in this manner over the rookery.

Mrs. Florence M. Bailey (1923) describes it as follows:

One was seen displaying as he gave his call, as is described by Bendire. Instead of inflating his chest pouter-pigeon style, as is done by the Band-tails, he puffed out his throat, and, as if about to launch into the air, threw up his wings as some of the ducks do in courtship display of the speculum, showing the handsome white wing crescent; and at the same time curved up the rounded fan tail so that its white thumb-mark band showed strikingly—all this as he gave his loud emotional call—*Kroo-kroo'-kroo-kru'*. A rather distant answering call suggested that he was displaying for a prospective mate. Display actually before a female was witnessed a week later by Mrs. Nicholson when I was down in the valley. When the call was given without the emotional display it lapsed almost to monotony, being heard at camp all through the day. Some of the notes were heavily mouthed, while others were muffled. The noise of the flight was volitional. One that I saw, puffed out his chest and started with whacking wings, soaring around, wings and tail spread; but shortly afterward it or another bird was seen flying by silently.

Nesting.—Doctor Wetmore's (1920) account of the nesting habits follows:

On my arrival in June I found them breeding in pairs scattered through the cultivated lands or the open desert, or congregated in large colonies in suitable mesquite *montes* near the Gila River. One or two pairs were found at intervals in cottonwoods beside roads or near ranch houses, but the greatest interest centered in the large congregations to be found in suitable tracts of mesquites. These rookeries were often of considerable extent. One located

three and one-half miles south of Arlington extended over an area a quarter of a mile square, while another three miles beyond occupied a grove nearly half a mile wide and an equal distance in length. The birds maintained regular flights across country and gathered in flocks to feed, so that they were conspicuous figures in the bird life of the region. It was difficult to estimate the number present, as they were scattered about in dense groves of mesquites, but it is believed that there were at least two thousand pairs in the largest colony examined. The total number present in the area was large. It appeared that the period for breeding among these birds was somewhat irregular. A part of them evidently began to nest soon after their arrival, as a number that were feeding young were observed on June 6. Others were nest-building on June 17, so that the entire period of reproduction was somewhat prolonged. In the colonies nests were scattered about irregularly through the mesquites. Sometimes two or three nests were placed in the same tree, or again one pair occupied a tree alone. There was no crowding and apparently the birds, while gregarious, were too truculent to permit close proximity of nests. Often two or three trees, suitable in every way for the primitive needs of these doves, intervened between occupied sites.

In most cases the nest, slight in structure, though usually somewhat larger and bulkier than that of the Mourning Dove, was placed in a mesquite, though a few were observed on the desert in palo verdes. Nests were built on inclined living limbs where forking of small branches gave a firm, broad support. The site varied from six to twenty feet from the ground, with about eight feet as an average height. In most of those that were examined the structure was composed of dead twigs of the mesquite, small in diameter, and from six to ten inches long. For the inner layers small twigs were chosen that had been dead for some time, so that the spines, abundant on mesquite limbs, crumbled at a touch and caused no discomfort to the brooding bird or to the young. The nest was flat and had merely enough depression to receive the eggs that often were visible through the loosely interlaced twigs at the sides.

Gilman (1911), who has had extensive experience with this dove, says:

Nests are always, as far as my observation goes, placed in trees or shrubs at varying distances from the ground. The average height was ten feet and extremes ranged from four to twenty-five feet. The only nest as low as four feet was built in a mesquite tree and placed on top of an old Thrasher's nest. This may have been a shiftless bird; but I found several others using old Cactus Wren's nests as foundation, and one had made use of a deserted Verdin's home.

In choice of nesting sites the bird shows a decided preference for mesquite, as about 70 per cent of nests noted were in that plant. About 20 per cent were in willows, and 3 per cent in cottonwood, *Opuntia fulgida* or tree cholla, and *Prosopis odorata* or screw-bean. *Baccharis gluten* brought up the rear with 1 per cent. The dove is usually very wild on the nest, flying off whenever approached as close as twenty-five feet. Rarely is the broken-wing play made, though I have seen a few mild attempts at it, and occasionally one will allow an approach as close as fifteen feet to the nest before taking flight.

In Arizona we found the white-winged doves nesting in mesquite and hackberry trees in the mesquite forest; the nests were on horizontal branches, 10 or 12 feet above the ground, and were made entirely of grass, weed stems, and straws; a nest found in the San Pedro Valley was 12 feet from the ground in a large willow.

Eggs.—The white-winged dove lays two eggs, very seldom three or only one. Frequently one egg proves to be infertile, resulting in the rearing of only one young. The eggs are elliptical oval or oval in shape, and the shell is smooth but not glossy. Fresh eggs, even after being blown, are often a rich, creamy buff, but this color varies greatly, and many eggs are pale creamy white or nearly pure white; the whitest eggs are probably those that have been incubated longest. The color fades very soon, sometimes within a few days, after the eggs are blown. The measurements of 42 eggs average 31.1 by 23.3 millimeters; the eggs showing the four extremes measure **34** by **24, 27.5** by 21.5, and 33.5 by **21** millimeters.

Young.—The period of incubation, according to Major Bendire (1892), is about 18 days, in which "the male relieves the female somewhat in these duties, but does not assist to any great extent; he, however, assiduously helps to care for the young." Apparently only one brood is raised in a season in the northern part of its range, but farther south in Mexico two or more broods may be raised during a longer breeding season.

Doctor Wetmore (1920) writes:

I was of the opinion that males did not aid in incubation, but this I was unable to ascertain with certainty. Occasionally I saw both parents perched on the sides of a nest that contained young, but all birds that were definitely identified while engaged in incubating were females. Each male chose a perch near the nest site, usually from ten to thirty feet away, and remained there on guard while the female was sitting, save for the times required to secure food. Such perches were selected in situations that were well shaded from the direct rays of the sun during the heat of the day, and when not occupied could be readily located by the collection of ordure, often considerable in quantity, on the ground beneath.

The young birds were fed by regurgitation and at the age of four days received solid food in the form of undigested seeds, in addition to the usual diet of "pigeon's milk." Fledglings left the nest when between three and four weeks old, as nearly as I could ascertain. The first young bird able to fly was noted on June 12, and by June 15 birds of this age were fairly common. These young were still dependent upon their parents for food, and though able to fly well were undeveloped and small. On first leaving the nest they perched about in the mesquites, always seeking shade, but in a few days were often found on the ground, preferably where the soil was sandy. There they walked about in the thin shade of the mesquites, examining bits of sticks and other refuse curiously, often testing such fragments with their bills, or rested quietly, squatting on the earth. In many instances it was found that they were heavily infested with small ticks against whose attacks they seemed inexperienced. No ill results from the presence of these parasites were noted and older birds were free from them.

Plumages.—He says further:

Young White-winged Doves when first hatched were well covered with long, straggling down, that in color was dull white slightly tinged with buff. This natal down was replaced by secondary feather growth so rapidly that it had disappeared for the great part at the end of the first week. The feather quills that followed the down did not burst until they were quite long as that for a time the young were as grotesque as young cuckoos.

Young birds in juvenal plumage are much like adults, but they are grayer on the back and breast and generally paler; there are faint traces of narrow pale edgings on the mantle; and the throat and sides of the head are whiter. I have seen adults, from Arizona, undergoing a complete molt from August 2 to October 11, and adults, from Jamaica, completing the molt of wings and tail from December 25 to January 19.

Food—Doctor Wetmore (1920) gives the following account of the feeding habits of the white-winged dove:

A purple drupe, one-fourth of an inch in diameter, borne by a spiny shrub (*Condalia spathulata*) was a favorite food at this season and the birds also ate the fruits of the giant cactus as rapidly as they ripened. Various seeds were taken also. Harvesting of grain began in this valley about the first of June and continued until the end of the month. Fields of wheat or barley that had been cut recently were attractive to the White-wings as here they found an abundant source of food. The wheat grown in this region shattered (or shelled out) badly during the process of cutting, binding, and shocking, so that kernels of grain were scattered thickly over the fields. Further, there was much additional waste grain from heads matured or stalks too short to be bound that fell to the ground when cut. As may be imagined the White-wings sought this food supply eagerly. They were gregarious in feeding as in nesting so that newcomers passing over the grain fields usually decoyed to those already on the ground until many had gathered in one spot. The grain stubble was cut high and afforded the feeding bands shelter, as the doves were short in leg and walked about with the body bent forward. It was often the case that not a bird was seen in looking across a field of wheat stubble, though several hundred might be feeding there under shelter of the wheat stalks and the low levees thrown up to direct the flow of the water used in irrigation while the crop was being grown. White-wings were wary and easily alarmed while feeding. At times I crawled up under shelter of weeds to watch them at close range. If one of the feeding birds happened to observe some slight motion, the heads of all were up in an instant and all remained motionless, while in a minute or so they usually flew hastily in sudden alarm. Where they were shot at they became even more wary. After feeding, little groups of White-wings often flew up to rest for a time in the shelter of cottonwoods or mesquites.

Occasionally, when feeding in fields where wheat had not been shocked a dove hopped up on one of the bundles of bound grain and pecked at the heads of wheat, choosing, preferably, those that were short so that they were firmly held by the twine. Or a flock of half a dozen dropped down on a shock of wheat and fed on the cap sheaves for a few minutes. Usually, however, the birds preferred to feed in the more secure cover of the stubble and confine their attention to the abundant waste grain as long as this was available. When wheat was not threshed within a short time after it was cut these doves were said to cause serious damage to the grain in the shock. This was particularly true in the case of isolated fields that remained after the surrounding crops had been removed. For this reason the White-winged Doves were in bad repute among many of the ranchers.

Various kinds of waste grain, seeds, berries, mesquite beans, and insects are mentioned by other observers. Dr. Joseph Grinnell (1914) found 33 watermelon seeds and a muskmelon seed in one crop.

Behavior.—Doctor Wetmore (1920) writes:

Combats among males were frequent, but these were bloodless battles, as the birds merely flapped at one another uttering guttural notes, or when near at hand struck quickly with one wing. Often one male was at much trouble to drive all others from some trees, and once I observed one hustle away a pair of Mourning Doves that chanced to intrude upon his domain.

White-winged Doves start in flight with a loud clapping of wings that is accompanied by a whistling noise. When the birds are well under way their passage, while swift and direct, is noiseless. The sound at the start resembles that made by domestic pigeons. The White-wing, like certain tropical doves (for example the White-headed and Scaled Pigeons) in perching in cottonwoods or other trees with dense foliage, usually alights among clumps of leaves on the higher outer branches rather than on dead limbs or in open situations such as those chosen by Mourning Doves. So well did the birds conceal themselves that after I had seen half a dozen fly into such a tree, it was not unusual to be unable to pick out a single dove in spite of their large size. In the mesquite they followed the same practice in perching, so that they were often observed merely as silhouettes through the thin foliage. When perched in trees they remained quiet save when they were calling.

Voice.—Wetmore also gives the best impression of the remarkable vocal performances of these doves, as follows:

In early morning White-winged Doves began to call soon after day break, and when the sun appeared above the horizon were heard cooing in every direction. At this period of the day many males came out to rest on dead limbs in openings in the mesquite *montes*, or flew to more distant perches in mesquites or cottonwoods where they basked in the warm rays of the sun. Others chose perches in the tops of living mesquites where the thin foliage did not cast an appreciable shade. In mid-forenoon when the heat became oppressive they retired again to protected stations. Males had two distinct songs, that were given without apparent choice. One of these efforts may be represented by the syllables *who hoo who hoo-oo'*. The first three notes were gruff and abrupt, the last one strongly accented and somewhat prolonged. The other song, longer and more complicated, may be noted as *who, hoo, whoo, hoo, hoo-ah,' hoo-hoo-ah', who-oo*. In this case the song was separated in five parts. The first section was short and low, the second louder and almost merged with the third; the third and fourth were more musical than the others and were strongly accented on the last syllable, while the last part was lower and was more or less slurred. At times the doves gave one or the other of these two songs in repetition for long intervals, or again alternated them rapidly. The longer song was more varied and pleasing to the ear as the other frequently was given in a burring, guttural tone that was often unpleasant. In addition to these songs males uttered a low, querulous, muttered note resembling *queh queh-eh* that served as a call to the female, or was given when squabbling with other males. No females were observed in the act of cooing and I was unable to ascertain their notes. Although males did not coo in unison the effect produced by hundreds of them calling at the same time was remarkable. Save for one or two birds that might chance to be near at hand, their notes seemed to come from a distance, and were so blended that it was difficult to pick out individual songs. In a large colony the volume of sound produced was so great that it carried readily for a distance of a mile and yet the tone produced was so soft that it was not deafening when near at hand. On the contrary the whole formed an undertone, continuous, and to my ear not unpleasing, that did not intrude sharply on the senses, of so vague a nature that faculties perceptive to sound soon became accustomed to it, so that through constant repetition it might pass unnoted. Although it filled the air with the same effect as that produced by the rushing of water, other sounds, the song of a Redwing or a Lucy's Warbler, the cooing of a Mourning Dove or the stamping of a horse, were heard through it clearly even when such noises originated at some distance. The effect as a whole was most remarkable and, once experienced, lingers long in memory.

Major Bendire (1892) says:

Their call notes are varied, much more so than those of any other species of this family found with us; they are sonorous, pleasing, and rather musical. On this account the natives keep many of them as cage birds, calling them *Paloma cantador*, Singing Dove. They soon become very gentle and reconciled to captivity, feeding readily out of one's hand and allowing themselves to be handled without fear.

One of their most characteristic call notes bears a close resemblance to the first efforts of a young Cockerel when attempting to crow, and this call is frequently uttered and in various keys. While thus engaged the performer usually throws his wings upward and forward above the head and also spreads his tail slightly. Some other notes may be translated into "cook for you," or "cook for two," "cook-kara-coo," besides a variety of calls, one of these a querulous harsh one, resembles somewhat the syllables "chȁȁ-hää."

The commonest note, as I recorded it, might be written "who cooks for you," and I notice that several others have so recorded it. Swarth (1920) very aptly remarks that it is "given with rather insulting emphasis." To my ear it sounds rather like a soft rendering of one of the common notes of the barred owl. Its monotonous repetition becomes rather tiresome, but it is an impressive performance, which once heard can never be forgotten. Dawson (1923) says:

In uttering this note the bird throws his head well forward and closes his eyes ecstatically (thereby disclosing a livid blue eyelid), but he does not open his beak. In defiance of all the masters, he sings through his nose. The effect is charming, it must be admitted, but one can not help wondering what the sound would be if only the bird would "sing out." Chanticleer's effort would surely pale beside it.

Fall.—Gilman (1911) writes:

As soon as the young are grown both they and the parents congregate in large flocks and fly from feeding ground to watering place, thus affording a good chance at wing shooting. One evening in twenty minutes I counted over 700 fly past a bridge over a small irrigating canal.

The gunner, in these birds, has a good test of his skill, as they fly very rapidly with seemingly little effort, and the rate of speed is hard to estimate. They will carry off a large load of shot, too, and all things considered are a fine game bird.

Besides the danger from gunner, the Cooper Hawk is a menace, feeding often on the fat pigeon. I have seen a Marsh Hawk after a White-wing with a broken wing, but do not think any but wounded birds are ever attacked by this species.

Along in August the big flocks begin to grow less, the birds probably scattering out and seeking feeding grounds more distant from the breeding grounds. Toward the first of September they begin to thin out in earnest and by the 15th of the month very few are seen. Individuals may linger a little longer, as in 1909 I saw one as late as October 12, and in 1910 the last seen was on September 25. A few lingered on a sorghum field up till September 10 of this year, but were not seen any later.

COLUMBIGALLINA PASSERINA PASSERINA (Linnaeus)

EASTERN GROUND DOVE

HABITS

The gentle little ground dove is one of the most familiar and confiding dooryard birds in Florida, where it may be seen walking briskly about on its short legs, with a graceful nodding motion of its head, about the houses, gardens, and more quiet streets in nearly every village. It is very tame and will allow a close approach, but, if too hard pressed, it will flit away to the nearest cover with a conspicuous flash of reddish brown in its wings. Besides being very domestic in its habits and attached to the vicinity of human dwellings, it is fond of sandy, cultivated lands, old weedy fields, cottonfields, pea patches, orange groves, and the borders of woods.

Its range extends northward into Georgia and South Carolina, mainly in the coastal counties. Referring to Chatham County, Ga., W. J. Erichsen (1920) says:

A characteristic bird of the Lower Austral zone, this species, while formerly abundant, is now quite uncommon. Its decrease during the past five years has been rapid and the few that now breed are restricted to three or four widely separated localities.

During the period when it was abundant and generally dispersed in the county, I had many opportunities to observe its habits, and while it was to be met with in equal abundance in country of greatly diverse character, its preferred haunts were sparsely timbered woodland containing low and dense undergrowth.

This species is non-migratory, passing its entire life in or very near the locality at which it was hatched. So attached to certain localities does it become that even if the undergrowth is cleared and the land cultivated the bird remains, nesting on the ground among the vegetables.

Courtship.—The courtship of the ground dove is a very simple affair, much like that of the domestic pigeon. The male struts before the female, puffing out his feathers, bowing his head, and making a soft cooing sound. After they are mated the pair may often be seen sitting on a branch with their bodies touching in a most affectionate attitude. Donald J. Nicholson writes to me:

Early in the spring the mating is begun and it is a common sight to see two or three males chasing a female on the ground. The female runs along and the male closely follows, taking short flights to catch her, when she arises and alights just ahead. When she has made her choice, they fly to some elevated spot in a tree and the mating is consummated. I have never seen them copulating on the ground. They utter a soft peculiar note when chasing mates. They have a habit of flitting their wings, when running along.

Nesting.—Erichsen (1920) writes:

In its choice of nesting sites, it exhibits a very wide range. It most frequently selects a low bush, either thinly or densely foliaged. Other situations in which I have found nests include the top of a low stump; high up on a horizontal limb of a large pine; and, frequently, upon the ground. An instance of its nesting on the ocean beach came under my observation May 13, 1915, on Ossabaw Island. In this case there was no attempt at nest building, the eggs being deposited in a slight depression in the sand; and when breeding on the ground in woodland or cultivated fields, little or no material is assembled. In fact, nest building occupies little of the time and attention of this species, as when placed in trees or bushes the nest is simply a slight affair of a few twigs loosely interlaid. Further evidence of this bird's disinclination to build a nest for the reception of its eggs is found in the fact that I once found a set in a deserted nest of the cardinal (*Cardinalis cardinalis cardinalis*).

So gentle and confiding are these birds that it is often possible to touch them while on the nest, especially if incubation is advanced. Upon dropping off the nest they always simulate lameness, dragging themselves over the ground with drooping wings in an effort to draw the intruder away. I am of the opinion that they remain mated for life, since they are observed throughout the year most frequently in pairs.

Mr. Nicholson, who has "examined hundreds of nests," says in his notes:

Nests are built on the ground as frequently as in vines, bushes, or trees, or along the tops of fences. One foot to 10 or 12 feet above the ground is the usual height.

The nests are delicate-looking structures, made usually of fine rootlets or grasses, and seldom any sticks are used, saddled on a limb, or among dead vines. The diameter measures from 2½ to 3 inches across, by 1 inch to 2 inches thick, with scarcely any depression for the eggs, the eggs always showing above the rim of the nest.

A nest that I found on Murrays Key, Bay of Florida, on April 3, 1908, consisted of merely a few straws in a slight hollow in the ground, under and between two tussocks of grass, which were arched over it; it was located in an open space in the brush, with small shrubs and weeds about it. Maynard (1896) "always found the nests in orange groves; the neat domiciles are placed on the lower limbs of trees." Audubon (1840) says that the nest "is large for the size of the bird and compact. Its exterior is composed of dry twigs, its interior of grasses disposed in a circular form;" he found a nest "placed on the top of a cactus not more than two feet high." Dr. T. Gilbert Pearson (1920) writes:

There is no bird in the United States that to my knowledge breeds over so long a period of the year as does the ground dove. In my experience with these birds in Florida, I have found their nests occupying varying situations during different seasons of the year. Thus on February 28 and March 3, I have found nests located on the tops of partially decayed stumps of pine trees, only about 2 feet from the ground. Later in the season I have seen numerous nests placed on the ground, usually in fields of weeds or in standing grain. Fields of oats seem to be especially favored with their presence during midsummer. Late in July, August and on to the latter part of September, I have found their nests on horizontal limbs of large orange trees, on the level fronds of palms, and on the cross-bars or rails, as commonly used for supports of the widespreading scuppernong grape-vines.

Most observers have noted that when a ground dove's nest is approached, the brooding bird quickly leaves the nest and flutters along the ground, attempting to lure the intruder away by feigning lameness. But Doctor Pearson (1920) writes:

Occasionally an individual is found that declines to expose her treasures without an argument. As the inquiring hand comes close to the nest, she does not strike with her bill, nor even indulge in loud scolding, but with ruffled feathers raises her wings in a threatening attitude, as if she would crush the offending fingers if they came too close. Surely a puny, hopeless bit of resistance; nevertheless it shows that a stout heart throbs within the feathered breast of the little mother.

Mr. Nicholson has proved to his satisfaction that the same nest is used for a second or even third brood in a single season, by apparently the same pair of birds.

Eggs.—The ground dove lays almost invariably two eggs, very rarely three and apparently never less than two. The eggs are usually elliptical oval in shape, sometimes oval, and rarely ovate. They show little or no gloss and are pure white in color. The measurements of 34 eggs average 21.9 by 16.2 millimeters; the eggs showing the four extremes measure **24.4** by 16.6, 22.8 by **17.2**, **20.3** by 15.2, and 21.7 by **15.1** millimeters.

Young.—Incubation is said to last from 12 to 14 days, both parents assisting. The breeding season is so prolonged, from February to November, that probably three or four broods, certainly two or three, are raised in a season. The young remain in the nest until they are ready to fly. Nicholson says, in his notes, that "when disturbed the young fly from the nests with a strong flight which is marvelous for the first attempt. After the young have hatched, the nest is very untidy; the droppings piled high on the outer edge of every nest, sometimes to half an inch or more in depth."

Nicholson noted that the young are brooded by one or the other parent, or shaded from the hot sun, until they are well grown and feathered. When somewhat advanced in age they frequently come out from under the brooding parent to sun themselves and stretch their legs and wings.

Feeding begins within two hours after hatching; Nicholson saw a young bird break the shell and watched it until it was fed. The young are fed entirely on regurgitated food, which is given by both parents. Nicholson says it is a common act for a ground dove to feed both young simultaneously and that the young are more often fed this way than singly. He has succeeded in photographing the act.

Plumages.—The nestling ground dove is scantily clothed with long, stringy, hairlike down of a dull gray color. In the juvenal plumage the young bird is much like the adult, but browner with duller markings above and fewer or none below; the upper parts vary in color from "snuff brown" to "cinnamon-brown," brightest on the wing and tail coverts, with conspicuous black spots in the wing coverts; the underparts vary from "buffy brown" to "wood brown." I have been unable to trace the subsequent molts.

Food.—Doctor Pearson (1920) says:

The ground dove's food consists largely of small seeds which it gathers in the garden, on the lawn, by the roadside, in the field, and other places where weeds or grasses are found. Naturally many insects are also picked up in their travels, particularly in the spring and summer. Small wild berries are also consumed. So far as known they never adversely affect the interests of mankind, even in the slightest degree, and wherever found they are protected by statute and by the still stronger law of public sentiment.

Behavior.—When disturbed the ground dove rises on whistling wings; its flight is low and direct, but not protracted to any great distance; it generally amounts to only a short dash into the nearest cover. It is very much attached to certain restricted localities, in which it may be regularly found, and to which it soon returns after being disturbed. It is well named, for it is decidedly terrestrial in its habits, spending most of its time on the ground, where it walks quickly, with a pretty nodding motion of its head and with an elevated tail. It is, however, often seen perched on a fence, the branch of a tree, or the roof of a building.

Nicholson tells me that the incubating or brooding male assumes a fighting attitude when the nest is approached, with wings raised high above his back and uttering an angry, nasal, rasping note. One allowed himself to be lifted from the nest, to which he clung, making angry notes and striking with repeated heavy downward strokes of the wings, but never striking with the bill.

Voice.—The soft, cooing notes of the ground dove are the characteristic sounds that one hears in its Florida haunts; their mournful character has given it the local name of "mourning dove" in many places. Nicholson says that "their cooing is done entirely from an elevated position; a house top, fence, telegraph wire, dead or living trees. I do not know if the female coos or not. For hours at a time it is kept up, but with rest periods, of course. Four or five males can be heard at one time within a 400-yard space."

Enemies.—The gentle little ground dove is too small to be of any account as a game bird, and, because of its sociable and confiding nature, there is a strong sentiment in favor of its protection everywhere. It has little, therefore, to fear from man; but it has plenty of natural enemies, such as cats, foxes, skunks, opossums, hawks, and snakes. It seems well able to take care of itself, however, and its numerous broods, though small, are enough to keep up its numbers.

DISTRIBUTION

Range.—Southern United States and Central America.

The range of the American races of the ground dove extends **north** to southern California (Brawley and opposite Ehrenberg, Ariz.); Arizona (Big Sandy Creek, Fort Verde, Beaver Creek, H Bar Ranch, and Pima); southern New Mexico (Mesilla Park); Texas (Pecos, Devils River, and Seguin); Louisiana (New Orleans); Alabama (Autaugaville and Montgomery); and North Carolina (Davidson County). **East** to North Carolina (Davidson County); South Carolina (Waverly Mills, Sullivans Island, and Frogmore); Georgia (Savannah, Blackbeard Island, and MacIntosh); and Florida (Watertown, San Mateo, De Land, Canaveral, Micco, Eden, Lake Worth, Palm Beach, Fort Lauderdale, and Miami). South to Florida (Miami, Royal Palm Hammock, Vaca Key, Big Pine Key, Key West, Boca Grande, and Marquesas Keys); Guatemala (Duenas and Santa Maria); Guerrero (Chilpancingo); Jalisco (Chapala and Zapotlan); Tepic (San Blas and Tres Marias Islands); and Lower California (Cape San Lucas). **West** to Lower California (Cape San Lucas, San José del Cabo, La Paz, Espiritu Santo Island, and San Felipe); southwestern Arizona (Yuma); and southern California (Winterhaven and Brawley).

The range as above outlined is for only the two American races, the eastern ground dove (*Columbigallina passerina passerina*), and the Mexican ground dove (*C. p. pallescens*), the first of which is found in the South Atlantic and Gulf States west to Louisiana, while *pallescens* is found in Texas, New Mexico, Arizona, southern California, and south to Guatemala. This species has been separated into many other geographic races, extending from Bermuda, through the West Indies, to northern South America (Colombia and Venezuela).

The species is generally nonmigratory, although there is some evidence of a movement in the Southwestern United States that apparently involves only a part of the ground-dove population.

Casual records.—There are several records of occurrence north of the normal range. Among these are: California, one found dead at Salinas in June, 1913, and one obtained at San Francisco in May, 1870; Arkansas, a pair seen regularly for three years (last 1927), at Rogers, according to a report from D. E. Merrill to the Biological Survey; Iowa, one reported as "seen" at Des Moines, on June 10, 1922, by Clifford H. Pangburn; northern Alabama, one shot at Leighton, on May 4, 1889; northern Georgia, one at Rising Fawn, May 9, 1885; western North Carolina, one seen in Buncombe County, May 29, 1891; Virginia, one at Lynchburg on November 4, 1900; Maryland, one near the mouth of Broad Creek, October 14, 1888; District of Columbia, one at Washington, September 1, 1844; Pennsylvania, one shot in Lancaster County in 1844; New Jersey, one taken near Camden in the autumn of 1858; and New York, one taken from a flock of seven, near New York City, in October, 1862.

Egg dates.—Florida: 73 records, February 27 to October 22; 37 records, April 16 to June 2. South Carolina and Georgia: 20 records, February 22 to October 19; 10 records, May 17 to June 10. Texas: 58 records, March 30 to October 1; 29 records, May 3 to June 28. Arizona: 25 records, May 17 to October 8; 13 records, June 2 to August 11. Mexico: 34 records, March 7 to October 18; 17 records, May 3 to August 5.

COLUMBIGALLINA PASSERINA PALLESCENS (Baird)

MEXICAN GROUND DOVE

HABITS

A slightly paler form of the ground dove is found along our southwestern borders and in Mexico. We found it common or abundant in suitable places in southern Texas and in southern Arizona, where its haunts and habits seemed to be like those of the eastern bird. In Arizona it was breeding commonly in the valley of the San Pedro River and I recorded it as abundant in the mesquite forest south of Tucson. It is a common dooryard bird in the small, quiet villages, very tame and confiding, as it runs about in the gardens or along the streets and is equally familiar about the ranches, barnyards, and cultivated fields. It is also common in the well-watered woodlands in the river bottoms and in the willows along the irrigation ditches.

Griffing Bancroft (1930) says of its haunts in Lower California:

The presence of water seems to be the determining factor in the distribution of this little dove. It is common wherever there are irrigation ditches or pools or available water in any form. As a consequence its occurrence is locally concentrated.

The birds breed near open water. They carry their demand for its proximity so far that, assuming my notes to represent a fair average, four-fifths of their nests are within fifty feet of a place to drink. I was surprised when this fact began to develop and I found myself looking about for water whenever I flushed one of the doves from eggs or young. Seldom, indeed, was it not close at hand. There is a marked contrast here with the birds of southern Sonora. There they abound on the open mesas and breed freely twenty to fifty miles or more from water.

Courtship.—W. Leon Dawson (1923) writes:

Business-like always, the ground dove is not less diligent in courtship. The call note *oo woo uk, oo woo uk,* sounds a little hard and unromantic in comparison with that of the larger doves. The sound is very penetrating, but it is so low-pitched that some people fail to observe it. The singer is discreet, and the sound usually ceases upon the appearance of the ever-despicable human. Yet at close quarters with his lady love, the workaday swain knows how to be tender. At such times he trails after his enamorata with trembling wings and cries *kool kooul.* The daily visit to the drinking pool is the recognized occasion for amours.

Nesting.—A typical nest of the Mexican ground dove, which we found and photographed in the San Pedro Valley, Ariz., on May 17, 1922, is shown on Plate 93. It was found while exploring a narrow strip of small willows along the banks of an irrigation ditch. It was well made, for a dove's nest, of coarse dry grasses and placed in the main crotch of a small willow, 6 feet above the ground. The two eggs in it were fresh.

Herbert Brown, as quoted by Major Bendire (1892), says:

They lay two eggs, and nest in trees or bushes. A nest found June 11, 1887, was constructed of a few dead twigs and grass placed on a limb of a willow near the ground; the female was on the nest. One found June 19 was also in a willow tree 20 feet from the ground and out on a limb 15 feet from the body of the tree, and made of a few dried stalks of alfalfa. It contained two eggs and the female was on the nest. A third nest, found June 26, containing two eggs, was made of long stems of dry grass and placed about ten feet from the ground. Whether this was a first laying I can not say. The nests are almost flat. I do not think I ever saw a cavity more than half an inch deep.

Major Bendire (1892) says further:

All of the nests seen by me were placed on bushes or on trees, from 3 to 21 feet off the ground, not a single one was found on the ground.

The first one found, on May 30, was placed in a syringa bush, about 3 feet from the ground. The little platform of small twigs and grass stems was very slight, about 4½ inches in diameter, and almost perfectly flat. The eggs were fresh.

Other nests, subsequently noticed, were placed in various trees and bushes, mostly in mesquite thickets, a few in willows, and two in walnut trees. A nest found July 28 was placed in a tree of this kind, about 20 feet from the ground. The tree was leaning, and some young sprouts had grown out from the main trunk, among which the nest was placed. The eggs were fresh, probably a second laying. All the nests examined by me were found in the creek bottoms or else close by, generally in clumps of mesquite bushes.

Griffing Bancroft (1930) says of its nesting sites in Lower California:

The Mexican ground dove is an unobtrusive little fellow, blending his coloration into his background on every occasion, and carrying his reticence into his choice of nesting sites. He certainly does like concealment for his home, far more so than the birds of the opposite mainland. Trestled grapes are plentiful in this part of Lower California. Commonly a beam of palm wood three or four inches wide is supported on uprights at a height of five feet and a vine is trained to grow over this structure. The favorite nesting site of the dove is on the flat surface of the beam. The bird is snuggled in among the leaves, ideally protected and hidden.

Another popular haunt is in the palm jungles. At a height of from four to ten feet and on the stems of the vertical leaves of the date palms numbers of these birds build. They seek the shadows that come from heavy vegetation or crossed leaves. Most of the fan palms have been trimmed, their leaves being cut largely for roofing material. The stubs left are generally about a foot long, smooth and well cupped. Here, hidden from below and concealed from the sides, many of these doves raise their young. The preferred height is twelve to fifteen feet above the ground.

This dove is tame, flushes at close range, and plays cripple most artistically. The laying season begins the middle of April. The nests are the most substantial of any of the local Columbidae and often attain a thickness of an inch or more. They are built of comparatively long and fine materials, palm fibre and grass stalks being the favorites. They are well matted and the strands are twisted spirally to form a flat disc to which is added somewhat finer material in the center. Fifty-four eggs collected from Santa Agueda to San Joaquin average 21.9×16.3 millimeters.

John C. Fortiner (1920) reports three interesting records of winter nesting of this dove in Imperial County, Calif. On December 21, 1919, he found a dove brooding a single squab in its nest in a eucalyptus tree; the nest was well built and was placed on some lodged bark, well hidden from view from the ground. A second nest was found on January 22, 1920, containing one young bird; "this nest was also in a eucalyptus tree, about 18 feet from the ground, and was a rebuilt mourning dove's nest. This second nest was watched, and on February 14 was seen to have a sitting ground dove on it. The two eggs it contained were collected the next day and found to have been incubated already several days."

Again (1921) he writes:

The Mexican ground dove appears to be partial to old nests, using its own or that of a mourning dove generally; but I have seen a pair trying a Sonora red-winged blackbird's nest; and during 1921 a pair has used an old Abert Towhee's nest for three broods, beginning to sit January 30, on the first eggs, and June 21, on the third set.

M. French Gilman (1911) says:

The nests are fairly well made for doves and are composed mostly of rootlets and small twigs. One nest rather more pretentious than usual was made of rootlets, grass stems and blades, leaf stems with veins attached, small twigs, horse hair, and a few feathers. It was compact and fairly well made, with a decided cup in the center measuring nearly an inch deep, and two inches across from rim to rim. One was an old nest re-vamped, and another was merely a superstructure over an old Abert Towhee's nest. The very late date before mentioned was probably the second brood, as the nest was an old one re-lined, possibly a last year's nest, but more likely an earlier nest of the same year.

Eggs.—The Mexican ground dove lays two eggs, rarely only one. They are just like eggs of the eastern bird. The measurements of 56 eggs average 21.5 by 16.5 millimeters; the eggs showing the four extremes measure **23.5** by 16.5, 23 by **17.5**, and **20** by **15** millimeters.

Young.—Probably two, and perhaps three, broods are raised in a season. Both sexes share in the duties of incubation, which is said to last 14 days. Fortiner (1920) gives the following interesting account of the behavior of the young:

The nesting birds were not disturbed, and two weeks later the two old doves and the young were discovered feeding on the ground. They soon flew to a tree, where the young bird was fed by regurgitation, but by one of the parents only. No time was available for observation until the following Sunday, when the three doves were again seen feeding, and later all three flew to an umbrella tree, where the young dove was fed by both parent doves. The young dove, after being fed once, hopped onto the old bird's back, then down to the limb on which the old dove was perched; then, when not being fed, it extended its wing out over the parent dove and gently tapped the back of its parent until it was fed again. It then flew to where the other parent dove was perched, where it went through the same actions. Whether this is typical of the behavior of young ground doves I am unable to say.

Food.—The food of this dove consists of seeds, waste grain, and various berries.

Voice.—The call notes are described under courtship. William Lloyd, according to Major Bendire (1892), gives them as *pas-cual, pas-cual, pas-cual.* George F. Simmons (1925) describes the voice, which is seldom heard, as "intense cooing; mellow, soft, crooning, floating *coos;* a single long drawn-out ventriloquistic, misleading *woo*, uttered at short intervals. Begins its moaning about mid-afternoon."

COLUMBIGALLINA PASSERINA BAHAMENSIS (Maynard)

BAHAMA GROUND DOVE

HABITS

The form of the ground dove found on the Bermuda and Bahama Islands was described by Charles J. Maynard (1896) as "similar in form and general coloration to the ground dove, but somewhat smaller and paler; the color on the lower parts and on the wings above being much less ruddy and the top of the head is more ashy, this color often extending well down onto the neck above. The bill is wholly black, not red at base."

Its habits apparently do not differ from those of the common ground dove, for he writes:

> This is an exceedingly common and familiar bird throughout all of the Bahama Islands which I have visited, being equally abundant in the grounds about the houses, even in the city of Nassau, in the open spaces in the scrub remote from settlements, as well as on the most desolate and unfrequented keys, provided they are sufficiently wooded to afford the birds shelter. In the city of Nassau, and in other towns and settlements, they are very tame, feeding about the houses either in pairs or in small flocks of from half a dozen to a dozen individuals.
>
> The Bahama ground dove breeds everywhere about the more open portions of the scrub. The nest, as far as I have observed, is always placed on trees or bushes, the latter being most often chosen as a nesting site.

Eggs.—These are also similar to those of the mainland birds. The measurements of the only three eggs available are 24 by 17.5, 20.7 by 15.7, and 21.3 by 15.7 millimeters.

SCARDAFELLA INCA INCA (Lesson)

INCA DOVE

HABITS

CONTRIBUTED BY CHARLES WENDELL TOWNSEND

The charming little Inca dove, sometimes called the scaly dove, or the long-tailed dove, after characteristic features, is a bird of Tropical and Lower Sonoran Zones and occurs in the United States only in Arizona, New Mexico, and Texas. Formerly confined in southern Texas to the region between San Antonio and the Rio Grande, these doves, according to A. E. Schutze (1904), on account of long droughts, have "moved north and eastward to a country where they found food and water in abundance." According to G. F. Simmons (1925), the first record of this dove for Austin, Tex., was in 1889, while, by 1909, they had become common nesters in that region. Wherever found, it is resident, although in Texas, according to the same author, "a few birds move southward in colder, winter weather." Appearing to delight in human companionship, the Inca dove is rarely found at a distance from towns or the neighborhood of houses.

Courtship.—Frank Stephens (1885) says of the courtship: "I saw a little group on the ground, the males strutting around the females, carrying their tails nearly vertical and cooing." Although the Inca dove may be heard cooing in every month in the year, the cooing is most in evidence during the courtship season. At this time, also, the dove is very pugnacious, and the rivalry is intense among them.

M. F. Gilman (1911) says:

The Inca dove could never have inspired the term "dove of peace," as they are pugnacious to a fault and fight like little fiends. Two of them will face each other with one wing on guard, held straight above the body, then close in and mix it, buffeting with wings till the sound of the blows is audible at a distance of fifty yards. The bill is also used with bloody results about the head. I have been told that one will sometimes kill the other, but never saw such an extreme case. When arranging for a fight the combatants utter a sort of growl, if it may be so described; a very guttural, anger-expressing sound.

Bryant (1891), quoting A. J. Grayson, says, "They exhibit the most ardent attachment for their mates and may often be seen caressing each other in a loving manner."

Nesting.—The Inca dove delights to nest near houses or barns even in villages and towns. Indeed, nests are rarely found except in the vicinity of mankind, and this familiarity is shown by the fact, according to M. F. Gilman (1911), that "the birds are generally quite tame on the nest, rarely flying off till the intruder comes closer than arm's length," and, he adds, "they are so accustomed to human presence that the broken-wing subterfuge is rarely resorted to." A curious instance of this familiarity with man and his works is given by George F. Simmons (1925), who reported that "a nest was on a trolley wire at a switch in the eastern part of Austin, where about every seven minutes street-cars raised it from six to twelve inches above its normal position."

The usual location of the nest is on a horizontal fork or flattened limb of a tree or in a bush, and it is generally within 10 or 12 feet of the ground, varying in height from 4 to 25 feet. Shade trees planted about dwellings are commonly used. Umbrella trees, cottonwoods, elms, sycamores, fruit trees, mesquites, live oaks, acacias, thorn bushes, prickly-ash, and even Opuntia cactuses are all used for this purpose, and, according to A. J. Grayson, quoted by W. E. Bryant (1891), "not infrequently they construct a nest under the sheds of the houses, if a suitable beam is found."

A. J. Van Rossem, in his notes from Salvador, says:

Inca doves breed the year round, nor does there appear to be any notable increase or decline of this activity correlated with season. The number of broods raised per year is not known to us, but because of the activity of the species as a whole it is not difficult to conjecture four or five. There is no cessation of nesting because of the fall molt. Males and females alike appear to have no dormant period whatsoever. This statement is based upon specimens taken every month in the year besides others inadvertently shot but not preserved, and observations of numerous nests. Eggs were seen in July, August, September, October, November, February, and April. Nests of the usual slight dove construction were seen in orange trees, balanced on palm fronds, in mimosa thickets, and even in hanging fern baskets around the corridor of an occupied ranch house. Some attempt at concealment was usually noticeable, but this was frequently offset by the unsecretive manner in which the parents left or approached the nest. Two eggs were the invariable rule.

According to Bendire (1892), who quotes Herbert Brown, "the nests are as a rule much better constructed than those of the Mexican ground dove. The cavity is about half an inch deep, and the materials used, fine dead twigs, are much more compactly put together than in the nests of the latter." Simmons (1925), describes the nest construction as follows:

Small, rather compact, firmly matted, almost flat platform or shallow saucer of weed stems, tiny twigs, dried grass, rootlets, a few straws, grass seed stems, bits of Indian tobacco weed, and sometimes bits of Bermuda grass, Spanish and bull moss, mesquite leaves, and a few feathers from the birds; occasionally nests contain string, horse-hair, or strips of cedar bark. Commonly unlined; rarely lined with grass stems, Spanish moss or a few small Inca dove feathers.

Simmons has frequently found Inca doves using the nests of their own species, of the western mourning dove, and of the western mockingbird, after being slightly repaired and relined. Gilman (1911) found two nests each of which was built on top of an old nest of a cactus wren.

The dimensions of the nests, according to Simmons (1925), vary from 1.8 to 3.4 by 3.6 inches, with a height of 1.15 inches and an inside depth of 0.5 inch. He speaks of one nest "about the size of a silver dollar." Bendire (1892) describes a nest he found in a thick mesquite bush as "a slight platform of twigs and grasses about 5 inches in diameter."

Eggs.—[Author's note: The Inca dove lays almost invariably two eggs. These are elliptical oval, smooth with very little gloss, and pure white. The measurements of 34 eggs average 22.3 by 16.8 millimeters; the eggs showing the four extremes measure **24.3** by 16.8, 22.9 by **18**, **20** by 16, and 21.8 by **15.5** millimeters.]

Young.—The duration of incubation is not known. It is probably not far from two weeks, the same as the incubation period of the Mexican ground dove. Two broods are generally raised in a season, sometimes three, and occasionally four. R. W. Quillin and R. Holleman (1918) record a case of a pair rearing four broods in one season in the same yard. Gilman (1911) says: "The past season I noted four cases where two broods were raised in the same nest, and two cases where a last year's nest was relined and used." F. C. Willard observed one pair that laid five sets of eggs; he collected three sets and allowed the pair to raise two broods. Others may have been raised later, as he was away after June 1.

Plumages.—The following observations on plumage changes under artificial conditions are of great interest and possible significance, and are therefore recorded here. William Beebe (1907) subjected with other birds some Inca doves to a warm, superhumid atmosphere and found that the plumage, with each succeeding molt, became darker and developed iridescence. He says:

When the concentration of melanin has reached a certain stage, a change in color occurs, from dull dark brown or black to a brilliant iridescent bronze or green. This iridescence reaches its highest development on the wing coverts and inner secondaries, where, in many genera of tropical and subtropical doves, iridescence most often occurs.

In other words, by subjecting the Inca dove, which belongs in a genus of tropical origin, to the humidity of the Tropics, it reverted back in the lifetime of the individual to an ancestral type. This is certainly a most surprising result and the experiment should be repeated. Mr. Beebe's skins of Inca doves showing these remarkable changes are on exhibition in the park of the New York Zoological Society.

[Author's note: I have never seen the nestling of this dove. A small young bird, about two-thirds grown, in juvenal plumage is much like the adult; it is more heavily barred with black or dusky on the breast and flanks, with more buffy color on the belly; the feathers of the back, scapulars, and wing coverts have a heavier terminal black bar and a subterminal band of "cinnamon-buff."

A. J. Van Rossem tells me that the postjuvenal molt occurs soon after the young bird attains its full size; and that the complete postnuptial molt of adults comes at any time between July and October, inclusive, with individual variation.]

Food.—Weeds abound in back yards and near dwelling houses and barns, and in eating the seeds of these plants Inca doves give good service. They also eat wheat and other small waste grains that have fallen, but they are apparently unable to manage whole Indian corn. Their familiarity permits them to mingle with poultry and partake of their food.

Water is, of course, an essential part of their diet, and reference has already been made to the extension of their range in times of drought in order that they may obtain water. M. F. Gilman (1911) says: "They are rather dainty in their drinking, rarely using the chickens' drinking vessel but perching on the hydrant and catching the drops of water as they leak from the pipe. To do this they nearly have to stand on their heads, but that does not bother them at all."

Behavior.—In summer these birds may be seen singly or in pairs, or rarely in family groups, but in fall and winter they gather in small flocks, which sometimes number as many as 50 individuals.

I shall always remember my first sight of this charming little dove. I was sitting alone in a small park in Tucson, Ariz., at Christmastime when 15 of these birds appeared and walked about within a few feet of me, picking at seeds on the ground. When they stooped over for this purpose, their breasts nearly touched the earth, because their legs were so short. As they walked their long tails generally sloped gently downward so as barely to skim the ground, but at times the tails were cocked up nearly vertically. When disturbed by a passer-by, they flew up rapidly into the tree overhead, their wings making a twittering sound. In the tree they sat in pairs or threes, affectionately snuggled together like love birds, their heads sunk in their breasts, their tails pointing straight down. They went to sleep at once. This was at 4 p. m.; at 4.20 they awoke and dropped to the ground to feed again.

G. F. Breninger (1897) writes:

The strange way in which Inca doves go to roost at night has recently come to my notice. Nearly a month ago, when the air at night was still chilly, I saw seven of these little doves perched on a limb side by side. This in itself was not strange, but directly upon the backs of the first row sat three more doves. At another time I saw five in the lower row and two on top. An examination of the ground beneath showed it to be a resort to which these birds gathered to spend the cold nights of the winter months.

F. C. Willard writes of watching an Inca dove in a grape arbor in a walled-in garden in Tombstone, Ariz., when four others alighted on the same lattice bar:

The newcomer at the far end immediately began to assert herself by sidling up to her next door neighbor and striking it with her wing. She soon forced him to fly and in like manner went down the line forcing each of them to seek another perch. Having thus made room for herself, she crouched as if content to call it a day and take a nap.

They are not all love birds!

The following account of the dove's behavior is by M. F. Gilman (1911):

The vivacious little Inca dove is the cream of the dove family, and is in the public eye or ear most of the time. Whether sitting on a barbed wire fence or on a clothes line, with long tail hanging down perfectly plumb, or marching around in a combative manner with tail erect at right angles to the body, or rushing around busily and hurriedly, not to say greedily, feeding with the chickens in the back yard, it shows a decided individuality and arouses interest and affection.

The faint twittering sound of the wings sometimes heard in flight has already been mentioned, but as a rule the flight is noiseless. It is a quick and jerky flight.

Voice.—Simmons (1925) describes the voice thus:

Monotonous, tiresome, extremely mournful, rather short two-syllabled, hard little *coo*, quite different from the soft, soothing manner of the western mourning dove, a slowly uttered *coe-coo* or *co-o-o-h coo-o-o*, the first slightly shorter, high-pitched, coarser, and with *o* as in *go;* the second lower, with a typical *oo* sound, as in *moon.*

Myron H. and Jane Bishop Swenk (1928) give the *coos* in musical notation, and they describe them as follows:

The call of the Inca dove is a monotonous, unvaried, rather plaintive *coo-oo-coo* or *whoo-oo-whoo*, rapidly repeated over and over. There is a blowing quality to it. We heard the call all through the winter, but it became louder and more insistent as the nesting season approached in March and April. It is very different from the soft, drawled *coo-oo-coo, coo-coo, coo* of the mourning dove.

During the very hot months of July and August the monotonous repetition from morning to night of the Inca's *coo* is much disliked by those with overwrought nerves.

I have quoted under *Courtship* M. F. Gilman's (1911) description of the "growl" of the fighting birds, and he adds that "in animated talk, gossip perhaps, they excitedly utter sounds like *cut-cut-ca-doo-ca-doo.* In all quite a vocabulary is at their command."

Field marks.—The chief field marks of the Inca dove are the long tail with its white edges and the scaled appearance of the feathers over much of the body, which is due to the darker outline of their edges. By these two marks the Inca dove may be distinguished at once from the Mexican ground dove that occurs in the same region. Half of the Inca's length is in the tail. The chestnut-brown of the wing coverts, which is concealed or nearly concealed when the wings are closed, is prominent in flight.

DISTRIBUTION

Range.—Southwestern United States and Central America; non-migratory.

The range of the Inca dove extends **north** to Arizona (Wickenburg, Rice, and Safford); New Mexico (Silver City); and southern Texas (Kerrville, Austin, and Columbus). **East** to Texas (Columbus and Santa Maria); Tamaulipas (San Fernando de Presas, Ciudad Victoria, and Tampico); northern Guatemala (Lake Peten); Honduras (San Pedro); and Nicaragua (Chinandega). **South** to Nicaragua (Chinandega); Salvador (La Libertad); western Guatemala (Duenas); Oaxaca (Tehuantepec); Jalisco (Guadalajara); and southern Sinaloa (Escuinapa, Presidio, and Mazatlan). **West** to Sinaloa (Mazatlan); Durango (Rio Sestin); Sonora (Opodepe); and Arizona (Tubac, Tucson, Sacaton, Phoenix, and Wickenburg).

Casual records.—Although repeatedly listed as a bird of Lower California, there are apparently only two records. Dr. Witmer Stone (1905) reported seeing "a very few in the upper Hardy River region" of the Colorado River delta, in February and March, 1905, and there is a specimen in the British Museum taken at La Paz.

Egg dates.—Arizona: 37 records, February 28 to October 21; 19 records, April 15 to May 25. Mexico: 42 records, March 11 to October 14; 21 records, March 23 to April 21. Texas: 12 records, April 10 to August 10; 6 records, April 19 to May 28.

OREOPELEIA CHRYSIA (Bonaparte)

KEY WEST QUAIL-DOVE

HABITS

Audubon (1840) gives a graphic account of his discovery of the beautiful Key West quail-dove in the dense, tangled, thorny thickets of Key West. He named it the Key West pigeon, but used the specific name *montana*, which is now applied to the ruddy quail-dove. It was originally discovered in Jamaica and now is rarely seen on Key West, as one taken there in 1889 by J. W. Atkins was the only one found by that keen observer in three years of careful field work. It was probably only a straggler, or a rare summer visitor, on Key West, retiring to Cuba in winter. Dr. Thomas Barbour (1923) says of its haunts in Cuba:

The Torito is found in dry upland woods as well as in the low country, and I have flushed a good many in the low but thick forest of the limestone hills, or *sierras*, of Pinar del Rio. I also shot one once on the Sierra de Casas of the Isle of Pines, in low, scrubby second-growth (*manigua*), hardly to be called a forest. In the cayos within the Zapata Swamp it was far less common than the ruddy quail dove; nevertheless we often shot a few for food as well as to skin.

Nowhere abundant, indeed a rather rare bird throughout its considerable range, the Key West quail dove is one of the species which sooner or later will completely disappear.

Nesting.—Audubon (1840) says:

The nest of the Key West pigeon is formed of light dry twigs, and much resembles in shape that of the Carolina dove. Sometimes you find it situated on the ground, when less preparation is used. Some nests are placed on the large branches of trees quite low, while others are fixed on slender twigs.

There is a set of two fresh eggs in Col. John E. Thayer's collection, taken by A. H. Verrill on Inagua Island, Bahamas, April 21, 1905; the nest was on the ground, composed of loose leaves. I have a similar set, taken by Mr. Verrill, at this locality on the previous day; the nest was on the ground, made of leaves and trash. Major Bendire (1892) gives Gundlach's description of the nest, as follows:

The nest, consisting of a slight platform of sticks, is usually placed on the top crown of certain parasitic creepers, found in the more open but shady primitive forests. The eggs are two in number, of a pale ochre yellow color, and measure 31.5 by 24 millimeters. I found nests between the months of February and July.

Dr. Juan Vilaro wrote Major Bendire that "the nest is placed in high trees as a rule, usually *Curujeyes.* It commences breeding in February and lays until July. The eggs are two in number, ochraceous white in color, and measure 31 by 24 millimeters."

Eggs.—Audubon (1840) was evidently mistaken in reporting the eggs of this dove as white, to which no one else agrees. Bendire (1892) evidently never saw the eggs, which he quotes as "ochraceous white." Eggs in my collection, and others that I have seen, are cream color, or a pale shade of "cream-buff"; and this color is fairly permanent in cabinet specimens. The measurements of eight eggs average 30 by 22.7 millimeters; the eggs showing the four extremes measure **32.8** by 22.6, 31.3 by **23.6**, **27.7** by 22.4, and 31.4 by **22.3** millimeters.

Plumages.—C. J. Maynard (1896) describes the nestling as "dark ashy brown, becoming considerably lighter below; feet pink; bill yellow, red at base; and iris red in all stages." Audubon (1840) says: "The young, when fully feathered, are of a dark gray color above, lighter below, the bill and legs of a deep leaden hue. I am inclined to believe that they attain their full beauty of plumage the following spring."

Ridgway (1916) describes the young as

very different in coloration from adults. Above rufous-cinnamon or pecan brown, the scapulars, interscapulars, and wing-coverts narrowly tipped or terminally margined with cinnamon-buff, the pileum and hindneck duller (more brownish) with indistinct, very narrow lighter tips to the feathers, the forehead light grayish brown; a dull white malar-subocular stripe, as in adults; foreneck and chest grayish brown or drab, the feathers margined with dull cinnamon; rest of under parts mostly pale grayish buffy.

Food.—Audubon (1840) says: "Their food consists of berries and seeds of different plants, and when the sea-grape is ripe, they feed greedily upon it." Doctor Vilaro reports "fruits, seeds, and small snails" among the food.

Behavior.—Audubon (1840) writes:

The flight of this bird is low, swift, and protracted. I saw several afterwards when they were crossing from Cuba to Key West, the only place in which I found them. It flies in loose flocks of from five or six to a dozen, with flappings having an interval apparently of six feet, so very low over the sea, that one might imagine it on the eve of falling into the water every moment. It is fond of going out from the thickets early in the morning, for the purpose of cleansing itself in the shelly sand that surrounds the island; but the instant it perceives danger it flies off to the woods, throws itself into the thickest part of them, alights on the ground, and runs off with rapidity until it thinks itself secure. The jetting motions of its tail are much like those of the Carolina dove, and it moves its neck to and fro, forward and backward, as pigeons are wont to do.

The cooing of this species is not so soft or prolonged as that of the common dove, or of the Zenaida dove, and yet not so emphatical as that of any true pigeon with which I am acquainted. It may be imitated by pronouncing the following syllables: *Whoe-whoe-oh-oh-oh.* When suddenly approached by man, it emits a guttural gasping-like sound, somewhat in the manner of the common tame pigeon on such an occasion. They alight on the lower branches of shrubby trees, and delight in the neighbourhood of shady ponds, but always inhabit, by preference, the darkest solitudes.

Doctor Barbour (1923) says:

This ground dove has habits much like those of the Perdiz and is often caught for food by the same means. Its flesh is excellent, although less esteemed than the Perdiz. It is known as "Torito," the little bull, from its habit of bobbing, or "Barbequejo," from the moustache-like markings. This, like the following species, is also called "Boyero," or Ox Driver, for its note, an oft repeated and prolonged monosyllablic *coo*, somewhat resembles the noise constantly made by men urging their oxen to strain to a heavy load.

The Geotrygons, as I still like to call them, walk slowly about on the ground with the head usually pulled in and not extended, and not bobbing except when disturbed or frightened. Then they bob vigorously, as does the Perdiz all the time, and this species is, I think, the shyest and most prone to take flight of any of the group.

L. J. K. Brace (1877), referring to the habits of the Key West quail dove in the Bahamas, writes:

This beautiful bird is frequently met with in the coppices underneath the trees of which it delights to feed, preferring for this purpose those parts which are rather open beneath, and less choked up with undergrowths, its habit being to feed almost exclusively on the ground, on berries and seeds, more particularly on the berries of the "poison wood," on the fruit of which, amongst others, the *Patagioenas leucocephala* feeds also. On being flushed, it scarcely, if ever, flies to any distance, generally alighting after a short curved flight. Its note is peculiarly mournful, being an expiring groan, which is rather startling to hear if the cause of it is not known.

DISTRIBUTION

Range.—Key West, Fla., and the northern West Indies. Nonmigratory.

The range of the Key West quail dove is greatly restricted. It extends **north** to southern Florida (Key West only) and the Bahama Islands (New Providence). **East** to the Bahama Islands (New Providence); San Domingo (Aguacate and Samaná Bay); and possibly Porto Rico (Mona Island). **South** to possibly Porto Rico (Mona Island); and Cuba (Isle of Pines). **West** to Cuba (Isle of Pines) and southern Florida (Key West).

Audubon reported that this species left Key West about the middle of October, but other observers have since recorded their presence as late as the middle of November.

Egg dates.—Bahamas and Cuba: 9 records, April 5 to May 2.

OREOPELEIA MONTANA (Linnaeus)

RUDDY QUAIL-DOVE

HABITS

On December 8, 1888, a boy at Key West, Fla., shot a red dove, which was sold with some mourning doves and plucked. That keen ornithologist, J. W. Atkins, secured the remains, the head and some wing and tail feathers, and sent them to W. E. D. Scott (1889) who published the first Florida record of this West Indian species. A second record for the same locality was published by Ned Hollister (1925), when he reported the receipt at the National Zoological Park of a living specimen of this dove, caught by Ross E. Sawyer, in his backyard in Key West, about May, 1923.

Dr. Thomas Barbour (1923) says of the haunts of the ruddy quail-dove in Cuba:

With habits essentially like those of the preceding, this forest beauty is much more abundant and more confiding. By standing watching some little sunlit glade, or lying flat on one's belly on the damp forest floor, patience was generally rewarded by a shot at the ruddy quail doves, provided one chose a suitable haunt in which to lie in wait. None of the quail doves occur in all situations where, from the character of the terrain, one might expect them. They were really abundant, however, in the low woods between Zarabanda and San Francisco de Morales and the Zapata Swamp, and equally so in a very fine stretch of damp woods which I have visited but once, far to the south of Bolondron.

Dr. Alexander Wetmore (1927), referring to Porto Rico, writes:

The ruddy quail-dove is an inhabitant of dense growths of jungle and finds cover to its liking mainly in the hills and mountains above the coastal plain. It is probable that its distribution is governed somewhat by the abundance of the mongoose near the coast, since from its terrestrial habits the dove is subject to depredation by this mammal.

At times the ruddy quail-dove is seen in coffee plantations, where these are not kept too clean of brush, but it is usually found in areas of dense second growth on the slopes of hills. As such cover becomes restricted in area, these doves grow steadily less abundant. To observe them it is necessary to walk noiselessly along footpaths, crouching low to obtain what vision may be had of the ground beneath the dense brush.

Nesting.—Writing of the tropical wild life of British Guiana, William Beebe (1917) gives a good description of the nesting habits of this dove, as follows:

Though one of the tropical jungle residents, the red mountain dove was seldom seen, for it merged so completely with its surroundings that one passed it by, time after time, without ever knowing that such a bird existed. If it were discovered, careful watch had to be kept or it would seemingly disappear where it sat. The nest was equally difficult to find and usually could only be discovered by frightening the bird from the eggs. If it thought there were a chance to escape undetected, the parent would quietly slip from the nest to the ground, run a few steps and noiselessly flutter to a protecting branch without the hunter being aware of its presence.

The nests were built away from the ground, the distance varying from a foot to five feet. The bird usually selected the head of an old rotted stump or the fork of a low outhanging branch, or possibly the horizontal surface of an old gnarled liana that ran close to the ground. The nest itself was a concave platform of twigs lined with leaves on which rested the two dark, cream-colored eggs. The nest in the accompanying illustration was lined in the same way, but some of the leaves were green and freshly picked so that the whole structure had an effect of not existing at all in the green mass of foliage that grew around it. The habit of mingling green leaves with brown was doubly significant from the fact that other nests found on stumps and lianas, where there was no surrounding green, were lined only with dead brown leaves which made them just as hard to see in their individual locality. The coloring of the eggs was no aid, for they nearly matched the leaves on which they lay.

P. H. Gosse (1847) found two nests in Jamaica, of which he writes:

One day in June I went down with a young friend into a wooded valley at Content to look at a partridge nest. As we crept cautiously toward the spot, the male bird flew from it. I was surprised at its rudeness; it was nothing but a half dozen decayed leaves laid on one another, and on two or three dry twigs, but from the sitting of the birds it had acquired a slight hollowness, about as much as a skimmer. It was placed on the top (slightly sunk among the leaves) of a small bush not more than 3 feet high, whose glossy foliage and small white blossoms reminded me of a myrtle. There were two young recently hatched, callow and peculiarly helpless, their eyes closed, their bills large and misshapen; they bore little resemblance to birds. On another occasion I saw a male shot while sitting; the nest was then placed on a slender bush, about 5 feet from the ground. There were but two eggs, of a very pale buff color; sometimes, however, they are considerably darker.

A set of two eggs in my collection was taken by A. H. Verrill on Dominica, May 2, 1906; the eggs were laid on leaves on the ground.

Eggs.—Two eggs form the usual set. They are oval, smooth with very slight gloss, and vary in color from pale "cream-buff" to "salmon-buff," which fade only slightly. The measurements of 25 eggs average 28.5 by 21.4 millimeters; the eggs showing the four extremes measure **32.6** by 22.2, 30.2 by **23.4**, and **24** by **19.5** millimeters.

Plumages.—I have never seen a nestling of this quail dove. A small but fully feathered young bird, in juvenal plumage in July, has the crown, back, scapulars, and wing coverts "brownish olive," with a greenish luster; the wing coverts and scapulars are tipped with "tawny," most broadly on the median and greater coverts, which are also edged with it; the remiges are edged with "rufous-cinnamon," and the rectrices and upper tail coverts are tipped with it; the breast is "hair brown," with "tawny" tips, the belly buffy white and the chin whitish with buffy tips.

Apparently a postjuvenal molt takes place in fall, which is probably complete, for first winter birds of both sexes are much like the adult female. I have not seen enough material to trace the molts.

Food.—Gosse (1847) writes:

It is often seen beneath a pimento picking up the fallen berries; the physic nut, also, and other oily seeds afford it sustenance. I once observed a pair of these doves eating the large seeds of a mango that had been crushed. With seeds, I have occasionally found small slugs, a species of *Vaginulus*, common in damp places, in its gizzard.

In the Short Cut of Paradise, where the sweet wood abounds, the partridge is also numerous; in March and April, when these berries are ripe, their stomachs are filled with them. Here, at the same season, their cooing resounds, which is simply a very sad moan usually uttered on the ground, but on one occasion we heard it from the limb of a cotton tree at Cave, on which the bird was sitting with its head drawn in; it was shot in the very act.

Doctor Wetmore (1927) says:

During the orange season these doves feed mainly on the seeds of the wild sweet oranges, secured from fruit that has fallen to the ground and has partly decayed, enabling the birds to peck open the skins and reach the seeds at the center. They do not touch this fruit except when on the ground, and can not open oranges except when the skin is soft through decay, so that no injury in orange groves may be charged to them. Near Manati they were eating the fruits of the manchineel.

Behavior.—Regarding the habits of the ruddy quail dove, Doctor Wetmore writes:

If the doves feel that they are liable to observation, they rest motionless, and at such times it is almost impossible to detect them. If approached too closely, they rise and dart into the dense growth. At other times they walk rapidly to one side, with quickly nodding heads, and it is then that they may be momentarily visible. Their flight begins with a loud fluttering of feathers, but after a few feet they set their wings and sail away on noiseless pinions. Occasionally they were seen on low limbs in the trees, perhaps six to ten feet from the ground, but this was unusual.

Voice.—The same writer says that during the nesting season the males "give utterance to a low, resonant note of such character that it seems always to come from a distance, though the singer may be near at hand; this resolves itself into a deep *coo-oo-oo*, with a peculiar undertone as of the humming of wind across the end of a gun-barrel—a striking sound and one whose source is difficult to locate."

DISTRIBUTION

Range.—Northern South America, eastern Central America, and the West Indies, accidental in Key West, Fla., and in western Mexico; nonmigratory.

The range of the ruddy quail dove extends **north to** Hidalgo (Potrero); British Honduras (Orange Walk); Cuba (San Cristobal and Trinidad); San Domingo (Puerto Plata); and Porto Rico (Mona Island, Aguadilla, Manati, and Vieques Island). **East** to Porto Rico (Vieques Island); the Lesser Antilles (Guadeloupe Island, Dominica Island, Grenada Island, and probably Trinidad); British Guiana (Bartica); and eastern Brazil (Para, Capim River, Mirador, Cantagallo, Murungaba, Ypanema, and Iguape). **South** to Brazil (Iguape and Mattogrosso); northwestern Bolivia (Apolo); and Peru (Tarma). **West** to Peru (Tarma, Ucayali River, Yurimaguas, Chamicuros, and Nauta); Ecuador (Sarayacu and Paramba); northern Colombia (Bonda); Panama (Chiriqui); Costa Rica (Boruca, Terraba, Orosi, Angostura, and San Carlos); Nicaragua (Chinandega); southern Vera Cruz (Esperanza); and Hidalgo (Chiquihuite Mountain and Potrero).

Casual records.—Miller (1905), reporting on a collection of birds made in southern Sinaloa, Mexico, cites the capture of a specimen at Arroyo de Limones, on April 21, 1904, and indicates that others ("stragglers") were seen by the collector. There are two records for Key West, Fla., a specimen shot by a hunter on December 8, 1888, and another captured alive in May, 1923. This bird was sent to the National Zoological Park, where it lived until March 5, 1926.

Egg dates.—West Indies: 15 records, March 24 to June 13; 8 records, April 26 to May 11.

STARNOENAS CYANOCEPHALA (Linnaeus)

BLUE-HEADED QUAIL-DOVE

HABITS

The beautiful blue-headed quail-dove, one of the handsomest on our list, is a Cuban species, which occurs, as a rare straggler only, on some of the Florida Keys. No one seems to have recorded it on the keys since Audubon's (1840) experience with it, of which he says:

> A few of these birds migrate each spring from the Island of Cuba to the Keys of Florida, but are rarely seen, on account of the deep tangled woods in which they live. Early in May, 1832, while on a shooting excursion with the commander of the United States Revenue Cutter *Marion,* I saw a pair of them on the western side of Key West. They were near the water, picking gravel, but on our approaching them they ran back into the thickets, which were only a few yards distant. Several fishermen and wreckers informed us that they were more abundant on the "Mule Keys"; but although a large party and myself searched these islands for a whole day, not one did we discover there. I saw a pair which I was told had been caught when young on the latter Keys, but I could not obtain any other information respecting them, than that they were fed on cracked corn and rice, which answered the purpose well.

Major Bendire (1892) quotes from Dr. Jean Gundlach as follows:

It is not uncommon in the extensive forest, especially in such in which the ground is rocky, but is scarcely ever found in cultivated fields or open prairie country. It moves slowly, with the neck contracted and tail erected, while searching for food among the dead leaves on the ground. This consists of seeds of various kinds, berries, and occasionally small snails. After feeding, it usually flies into a tree and perches on a leafless horizontal limb, or on one of the numerous parasitic vines, to rest. In the early mornings, should its plumage, perchance, have become wet while traveling through the dew-laden shrubbery, it selects a sunny spot to dry itself. From time to time this dove utters her call note, consisting of two hollow-sounding notes, *hu-up*, the first syllable long drawn out, the second short and uttered very quickly. Besides this note a low muttering is occasionally heard. Their call notes are deceptive, appearing near when distant, and distant when close by. Its flight is noisy when starting, similar to that of the European partridge, from which it receives its misleading name "Perdiz."

Dr. Thomas Barbour (1923) writes:

The blue-headed quail dove, on account of its brilliant blue crown, can not be confused with any other species. Formerly it was a common denizen of all the lowland forests of the island, where the soil was not too dry. To-day it is greatly reduced in numbers, both because it is so extensively trapped for food and because the forests are being constantly cut away. There are two ordinary methods of trapping ground doves in general use among the country people in Cuba. One involves the use of a *casilla*, a cage made of boughs or twigs, tied one upon the other, but at different intervals so that the complete structure is pyramidal and about two feet square and a foot high. This is put out in the open woods and baited with *tripa de quira*, the mushy inner pulp of the wild calabash, which is full of seeds. The *casilla* is tilted, and a "figure-four" drops the contrivance when it is touched by the bird fussing about inside. Sometimes a small dish of water serves for bait. Another method is to erect a net on hoops of creeper, and put bait beneath, where a decoy either alive or stuffed is often put out conspicuously. The hunter, in hiding, imitates the *hup-up* of the bird by means of a small hollow gourd.

The bird is called "Perdiz" because of its firm white flesh and the noise it makes when flushed. In common with the other ground doves, it prefers to run away from an annoyance rather than take flight. I have collected a good many by lying prone on the forest floor and simply watching for the birds to walk about. Much of the lowland forest in Cuba is flooded during the rainy season, often for several feet, and this eliminates the very low undergrowth, so that one may often see long distances with the eyes near the level of the ground. Standing up, it is impossible to see off at all, so thick are the vines and creepers. In 1915 I found Perdizes very common in the low woods, about five miles inland from Jucaro and Palo Alto. I shot a good many, and the *guajiros* had dozens caged to sell to the planters about Ciego de Avila, who eat them. This forest today is largely gone. About the cayos of the Cienaga where I got the other ground pigeons in numbers, the blue-headed doves were very rare, although I shot a few specimens. In Oriente the bird is still common where it has not been trapped too hard, and here it occurs in the highland forest where also suitable open woods are sometimes to be found.

Nesting.—Major Bendire (1892) states that the blue-headed quail-dove "nests in April and May; the nest is a simple affair, consisting of a few twigs. It is usually placed in the tops of parasitic vines, *Tillandsia*." There are two sets of eggs in Col. John E. Thayer's collection, taken by Oscar Tollin in Cuba on April 18, 1906; one nest was 2 feet above the ground and was made of small sticks; the other nest was in a small bunch of grass.

Eggs.—The eggs of this dove are very scarce in collections; I know of only seven that appear to be authentic. Two eggs in my collection, laid by a bird in captivity in Florida, are between oval and elliptical oval, smooth but not glossy, and pure white, quite unlike other quail-doves' eggs; they measure **31.8** by 24.8 and 32.7 by **24.7** millimeters. The three eggs, one set of two and one set of one, in the Thayer collection are also white and are considerably larger, measuring 37.5 by 30.2, 37.9 by **30.2**, and 37.9 by 28.9 millimeters. Two similar eggs in P. B. Philipp's collection measure **38.5** by 27.7 and 37.7 by 28.9 millimeters. These seven eggs average 36.3 by 28, and the extremes are indicated above.

Plumages.—We know very little about the plumage changes of this dove. The nestling is apparently unknown. In the Museum of Vertebrate Zoology, at Berkeley, Calif., I examined two young birds, about one-third grown, which had been hatched in an aviary. They are fully feathered in juvenal plumage, except that the chin and throat are naked. They are strikingly like adults in color pattern and colors. The colors above are duller; the crown is duller blue; the white stripe below the eye is present; there is a black patch on the lower throat, bordered with white spots; the feathers of the back and wing coverts are narrowly edged with rufous.

DISTRIBUTION

Range.—Cuba, including probably the Isle of Pines; accidental at Key West, Fla.

Very little is known concerning the range of the blue-headed quail-dove. It is apparently restricted chiefly to Cuba (Guama, Trinidad, and Habana). Bangs and Zappey (1905) state that while it has not been seen on the Isle of Pines by a naturalist, the natives report its presence. Cory (1892) states that it is claimed to have been introduced in Jamaica but was exterminated by the mongoose.

Its only claim for inclusion in the North American list is the statement of Audubon that he saw two at Key West, Fla., in May, 1832, and that he also saw a pair in captivity alleged to have been captured on the "Mule Keys."

Egg dates.—Cuba; 4 records, April 14 to June 12.

LITERATURE CITED

ABBOTT, CLINTON GILBERT.

1927. Notes on the nesting of the band-tailed pigeon. The Condor, vol. 29, no. 2, pp. 121–123, illus., Mar.–Apr.

ADAMS, A. LEITH.

1873. Field and forest rambles, with notes and observations on the natural history of eastern Canada, 333 pp., illus. London.

ALEXANDER, MARK LEIGH.

1921. Wild life resources of Louisiana, their nature, value, and protection, 164 pp., illus. New Orleans.

ALLEN, GLOVER MORRILL.

1921. The wild turkey in New England. Bull. Essex County Orn. Club Massachusetts, vol. 3, no. 1, pp. 5–18, Dec.

ALLEN, JOEL ASAPH.

1886. The masked bob-white (*Colinus ridgwayi*) of Arizona, and its allies, Bull. Amer. Mus. Nat. Hist., vol. 1, no. 7, pp. 273–290.

1889. Note on the first plumage of *Colinus ridgwayi*. The Auk, vol. 6, no. 2, p. 189, Apr.

AMERICAN ORNITHOLOGISTS' UNION.

1910. Check-list of North American birds, ed. 3 (revised), 430 pp.

1931. Check-list of North American birds, ed. 4, 526 pp.

ANDERSON, RUDOLPH MARTIN.

1907. The birds of Iowa. Proc. Davenport Acad. Sci., vol. 11, pp. 125–417, 1 map, Mar.

ANTHONY, ALFRED WEBSTER.

1889. New birds from Lower California, Mexico. Proc. California Acad. Sci., ser. 2, vol. 2, pp. 73–82, Oct. 11.

1899. Hybrid grouse. The Auk, vol. 16, no. 2, pp. 180, 181, Apr.

1903. Migration of Richardson's grouse. The Auk, vol. 20, no. 1, pp. 24–27, Jan.

ATKINS, JOHN W.

1899. *Columba corensis* at Key West, Florida. The Auk, vol. 16, no. 3, p. 272, July.

AUDUBON, JOHN JAMES.

1840–1844. The birds of America, 7 vols. New York and Philadelphia.

AUGHEY, SAMUEL.

1878. Notes on the nature of the food of birds of Nebraska. 1st Ann. Rep. U. S. Ent. Comm. for 1877, Appendix 2, pp. 13–62.

BAILEY, ALFRED MARSHALL.

1926. A report on the birds of northwestern Alaska and regions adjacent to Bering Strait, part 9. The Condor, vol. 28, no. 3, pp. 121–126, illus., May–June.

1927. Notes on the birds of southeastern Alaska. The Auk, vol. 44, nos. 1–3, pp. 1–23, 184–205, 351–367, illus., Jan.–July.

BAILEY, FLORENCE MERRIAM.

1902. Handbook of birds of the Western United States, 511 pp., 36 pls., 601 figs. Boston.

BAILEY, FLORENCE MERRIAM.

1918. Wild animals of Glacier National Park. The birds, pp. 103–199, pls. 22–37, figs. 19–94. Washington.

1923. Birds recorded from the Santa Rita Mountains in southern Arizona. Pacific Coast Avifauna no. 15, 60 pp., illus.

1928. Birds of New Mexico, 807 pp., 79 pls., 135 figs., 60 maps. Santa Fe.

BAIRD, SPENCER FULLERTON; BREWER, THOMAS MAYO; and RIDGWAY, ROBERT.

1905. The land birds of North America, 64 pls., 593 figs. Boston.

BANCROFT, GRIFFING.

1930. The breeding birds of central Lower California. The Condor, vol. 32, no. 1, pp. 20–49, illus., Jan.-Feb.

BANGS, OUTRAM.

1912. A new subspecies of the ruffed grouse. The Auk, vol. 29, no. 3, pp. 378, 379, July.

BANGS, OUTRAM, and PENARD, THOMAS EDWARD.

1922. The northern form of *Leptotila fulviventris* Lawrence. Proc. New England Zool. Club, vol. 8, pp. 29, 30, May 8.

BANGS, OUTRAM, and ZAPPEY, W. R.

1905. Birds of the Isle of Pines. Amer. Nat., vol. 39, no. 460, pp. 179–215, illus., Apr.

BARBOUR, THOMAS.

1923. The birds of Cuba. Mem. Nuttall Orn. Club, no. 6, pp. 1–141, 4 pls., June.

BARLOW, CHESTER.

1899. Another chapter on the nesting of *Dendroica occidentalis*, and other Sierra notes. Bull. Cooper Orn. Club, vol. 1, no. 4, pp. 59, 60, illus., July–Aug.

BARLOW, CHESTER (with supplementary notes by W. W. Price).

1901. A list of the land birds of the Placerville-Lake Tahoe Stage Road. The Condor, vol. 3, no. 6, pp. 151–184, illus., Nov.–Dec.

BARROWS, WALTER BRADFORD.

1912. Michigan bird life, 822 pp., 70 pls., 152 figs. Lansing.

BEATTY, HARRY A.

1930. Birds of St. Croix. Journ. Dept. Agr. Porto Rico, vol. 14, no. 3, pp. 135–150, 1 map, July.

BEEBE, WILLIAM.

1907. Geographical variation in birds with especial reference to the effects of humidity. Zoologica, vol. 1, no. 1, pp. 1–41, illus., Sept. 25.

1918–1922. A monograph of the pheasants, 4 quarto vols., pls. (part col.), maps. London.

BEEBE, WILLIAM; HARTLEY, G. INNESS; and HOWES, PAUL G.

1917. Tropical wild life in British Guiana, vol. 1, 504 pp., 143 figs. New York.

BEHR, HERMAN.

1911. Recollections of the passenger pigeon. Cassinia, no. 15, pp. 24–27.

BELDING, LYMAN.

1892. Food of the grouse and mountain quail of central California. Zoe, vol. 3, no. 3, pp. 232–234, Oct.

1903. The fall migration of *Oreortyx pictus plumiferus*. The Condor, vol. 5, no. 1, p. 18, Jan.-Feb.

BENDIRE, CHARLES EMIL.

1892. Life histories of North American birds. U. S. Nat. Mus. Spec. Bull. No. 1, vol. 1, 446 pp., 12 pls.

BENDIRE, CHARLES EMIL.

1894. *Tympanuchus americanus attwateri* Bendire. Attwater's or Southern prairie hen. The Auk, vol. 11, no. 2, pp. 130–132, Apr.

BENT, ARTHUR CLEVELAND.

1912. A new subspecies of ptarmigan from the Aleutian Islands. Smithsonian Misc. Coll., vol. 56, no. 30, 2 pp., Jan. 6.

1912a. Notes on birds observed during a brief visit to the Aleutian Islands and Bering Sea in 1911. Smithsonian Misc. Coll., vol. 56, no. 32, 29 pp., Feb. 12.

BISHOP, LOUIS BENNETT.

1900. Descriptions of three new birds from Alaska. The Auk, vol. 17, no. 4, pp. 113–120, Apr.

BISHOP, SHERMAN CHAUNCY.

1924. A note on the food of the passenger pigeon. The Auk, vol. 41, no. 1, p. 154. Jan.

BISHOP, SHERMAN CHAUNCY, and WRIGHT, ALBERT HAZEN.

1917. Note on the passenger pigeon. The Auk, vol. 34, no. 2, pp. 208, 209, Apr.

BISHOP, WATSON L.

1890. Canada grouse in captivity. Forest and Stream, vol. 34, no. 19, p. 367, May 29.

BOND, FRANK.

1900. A nuptial performance of the sage cock. The Auk, vol. 17, no. 4, pp. 325–327, illus., Oct.

BOWLES, JOHN HOOPER.

1901. Mice as enemies of ground-nesting birds. The Condor, vol. 3, no. 2, p. 47, Mar.–Apr.

BRACE, L. J. K.

1877. Notes on a few birds observed at New Providence, Bahamas, not included in Dr. Bryant's list of 1859. Proc. Boston Soc. Nat. Hist., vol. 19, pp. 240, 241, May 16.

BRADBURY, WILLIAM CHASE.

1915. Notes on the nesting of the white-tailed ptarmigan in Colorado. The Condor, vol. 17, no. 6, pp. 214–222, illus., Nov.–Dec.

BRENINGER, GEORGE FRANK.

1897. A roosting method of the Inca dove. The Osprey, vol. 1, no. 8, p. 111, Apr.

BREWSTER, WILLIAM.

1874. The drumming of the ruffed grouse. American Sportsman, vol. 4, no. 1, p. 7, Apr. 4.

1882. A remarkable specimen of the pinnated grouse (*Cupidonia cupido*). Bull. Nuttall Orn. Club, vol. 7, no. 1, p. 59, Jan.

1885. The heath hen of Massachusetts. The Auk, vol. 2, no. 1, pp. 80–84, Jan.

1885a. Additional notes on some birds collected in Arizona and the adjoining province of Sonora, Mexico, by Mr. F. Stephens in 1884; with a description of a new species of *Ortyx*. The Auk, vol. 2, no. 2, pp. 196–200, Apr.

1887. Further notes on the masked bob-white (*Colinus ridgwayi*). The Auk, vol. 4, no. 2, pp. 159, 160, Apr.

1888. Descriptions of supposed new birds from Lower California, Sonora, and Chihuahua, Mexico, and the Bahamas. The Auk, vol. 5, no. 1, pp. 82–95, Jan.

BREWSTER, WILLIAM.

1889. The present status of the wild pigeon (*Ectopistes migratorius*) as a bird of the United States, with some notes on its habits. The Auk, vol. 6, no. 2, pp. 285–291, Apr.

1890. The heath hen. Notes on the heath hen (*Tympanuchus cupido*) of Massachusetts. Forest and Stream, vol. 35, no. 10, p. 188, Sept. 25.

1895. A remarkable plumage of the prairie hen (*Tympanuchus americanus*). The Auk, vol. 12, no. 2, pp. 99, 100, pl. 2, Apr.

1902. Birds of the Cape region of Lower California. Bull. Mus. Comp. Zool., vol. 41, no. 1, pp. 1–241, 1 map, Sept.

1925. The birds of the Lake Umbagog region of Maine. Bull. Mus. Comp. Zool., vol. 66, pt. 2, 402 pp.

BROOKS, ALLAN CYRIL.

1907. A hybrid grouse, Richardson's and sharp-tail. The Auk, vol. 24, no. 2, pp. 167–169, pl. 4, Apr.

1912. Some British Columbia records. The Auk, vol. 29, no. 2, pp. 252, 253, Apr.

1926. The display of Richardson's grouse, with some notes on the species and subspecies of the genus *Dendragapus*. The Auk, vol. 43, no. 3, pp. 281–287, illus., July.

1927. Notes on Swarth's report on a collection of birds and mammals from the Atlin region. The Condor, vol. 29, no. 2, pp. 112–114, Mar.–Apr.

1929. On *Dendragapus obscurus obscurus*. The Auk, vol. 46, no. 1, pp. 111–113, Jan.

1930. The specialized feathers of the sage hen. The Condor, vol. 32, no. 4, pp. 205–207, July–Aug.

BROOKS, ALLAN CYRIL, and SWARTH, HARRY SCHELWALDT.

1925. A distributional list of the birds of British Columbia. Pacific Coast Avifauna No. 17, 158 pp., 2 col. pls., 38 figs.

BROOKS, WINTHROP SPRAGUE.

1915. Notes on birds from east Siberia and Arctic Alaska. Bull. Mus. Comp. Zool., vol. 59, no. 5, pp. 361–413, Sept.

BROWN, HERBERT.

1885. Arizona quail notes. Forest and Stream, vol. 25, no. 23, p. 445, Dec. 31.

1900. The conditions governing bird life in Arizona. The Auk, vol. 17, no. 1, pp. 31–34, Jan.

1904. Masked bob-white (*Colinus ridgwayi*). The Auk, vol. 21, no. 2, pp. 209–213, Apr.

BRYANT, HAROLD CHILD.

1926. Life history and habits of the western mourning dove. California Fish and Game, vol. 12, no. 4, pp. 175–180, illus., Oct.

BRYANT, HENRY.

1861. A list of birds seen at the Bahamas, from Jan. 20 to May 14, 1859, with descriptions of new or little known species. Proc. Boston Soc. Nat. Hist., vol 7, pp. 102–134.

BRYANT, WALTER E.

1891. Andrew Jackson Grayson. Zoe, vol. 2, no. 1, pp. 34–68, Apr.

BRYANT, WALTER PIERCE.

1889. A catalogue of the birds of Lower California, Mexico. Proc. Calif. Acad. Sci., 2d ser., vol. 2, pp. 237–320.

BURNETT, LEONARD ELMER.

1905. The sage grouse, *Centrocercus urophasianus*. The Condor, vol. 7, no. 4, pp. 102–105, illus., July.

BUTLER, AMOS WILLIAM.
1898. The birds of Indiana. 22d Ann. Rep. Indiana Dept. Geol. and Nat. Res. for 1897, pp. 515–1187, illus.

CAMERON, EWEN SOMERLED.
1907. The birds of Custer and Dawson Counties, Montana. The Auk, vol. 24, nos. 3, 4, pp. 241–270, 389–406, illus., July, Oct.

CAPEN, ELWIN A.
1886. Oology of New England, 116 pp., 25 col. pls. Boston.

CASSIN, JOHN.
1856. Illustrations of the birds of California, Texas, Oregon, British and Russian America, 298 pp., 50 col. pls. Philadelphia.

CHAPMAN, FRANK MICHLER.
1902. List of birds collected in Alaska by the Andrew J. Stone Expedition of 1901. Bull. Amer. Mus Nat. Hist., vol. 16, pp. 231–247.
1904. A new grouse from California. Bull. Amer. Mus. Nat. Hist., vol. 20, art. 11, pp. 159–162.
1908. Camps and cruises of an ornithologist, 432 pp., 250 figs., New York.

CHILDS, JOHN LEWIS.
1905. Personal recollections of the passenger pigeon. The Warbler, ser. 2, vol. 1, no. 3, pp. 71–73, July 20.

CHRISTY, BAYARD HENDERSON, and SUTTON, GEORGE MIKSCH.
1929. The turkey in Pennsylvania. The Cardinal, vol. 2, no. 5, pp. 109–110, illus., Jan.

CLARK, AUSTIN HOBART.
1905. Birds of the southern Lesser Antilles. Proc. Boston Soc. Nat. Hist., vol. 32, no. 7, pp. 203–312, Oct.
1907. Eighteen new species and one new genus of birds from eastern Asia and the Aleutian Islands. Proc. U. S. Nat. Mus., vol. 32, pp. 467–475.
1910. The birds collected and observed during the cruise of the United States Fisheries steamer *Albatross* in the North Pacific Ocean, and in the Bering, Okhotsk, Japan, and Eastern Seas, from April to December, 1906. Proc. U. S. Nat Mus., vol. 38, pp. 25–74, 3 figs., Apr. 30.

CLEAVES, HOWARD HENDERSON.
1920. A partridge Don Quixote. Bird-Lore, vol. 22, no. 6, pp. 329–334, illus., Nov.–Dec.

COLVIN, WALTER.
1914. The lesser prairie hen. Outing, vol. 63, pp. 608–614, illus., Feb.
1927. In the realm of the prairie hen. Outdoor Life, Jan., pp. 17–19, illus.

COOKE, WELLS WOODBRIDGE.
1888. Report on bird migration in the Mississippi Valley in the years 1884 and 1885. U. S. Dept. Agr. Div. Econ. Orn. Bull. 2, 313 pp., 1 map.
1897. The birds of Colorado. State Agr. College Bull. 37, 143 pp., Mar.

CORY, CHARLES BARNEY.
1891. Lists of birds collected by C. L. Winch in the Caicos Islands and Inagua, Bahamas, during January and February, and in Abaco, in March, 1891. The Auk, vol. 8, no. 3, pp. 296–298, July.
1892. Catalogue of West Indian birds, 163 pp., 1 map, Boston.

COTTAM, CLARENCE.
1929. The status of the ring-necked pheasant in Utah. The Condor, vol. 31, no. 3, pp. 117–123, May–June.

COUES, ELLIOTT.

1874. Birds of the Northwest. U. S. Geol. Surv. Terr., Misc. Publ. no. 3, 791 pp.

CRAIG, WALLACE.

1911. The expression of emotion in the pigeons. II, The mourning dove (*Zenaidura macroura* Linn.). The Auk, vol. 28, no. 4, pp. 398–407, Oct.

1911a. The expressions of emotion in the pigeons. III, The passenger pigeon (*Ectopistes migratorius* Linn.). The Auk, vol. 28, no. 4, pp. 408–427, Oct.

CRIDDLE, NORMAN.

1930. Some natural factors governing the fluctuations of grouse in Mani toba. Can. Field-Nat., vol. 44, no. 4, pp. 77–80, Apr.

DANFORTH, STUART TAYLOR.

1925. An ecological study of Cartagena Lagoon, Porto Rico, with special reference to the birds. Journ. Dept. Agr. Porto Rico, vol. 10, no. 1, 136 pp., figs. 2–45, 1 map.

DAWSON, WILLIAM LEON.

1896. Notes on the birds of Okanogan Co., Washington. Wilson Bull., no. 10, pp. 1–4, Sept. 30.

1923. The birds of California, 3 vols., part col. pls. San Diego, Los Angeles, San Francisco.

DAWSON, WILLIAM LEON, and BOWLES, JOHN HOOPER.

1909. The birds of Washington, 2 vols., illus. Seattle.

DEANE, RUTHVEN.

1896. Some notes on the passenger pigeon (*Ectopistes migratorius*) in confinement. The Auk, vol. 13, no. 3, pp. 234–237, July.

DEVANY, J. L.

1921. The spruce drummer. Can. Field-Nat., vol. 35, no. 1, pp. 16, 17, Jan.

DICKEY, DONALD RYDER, and VAN ROSSEM, ADRIAAN JOSEPH.

1923. Description of a new grouse from southern California. The Condor, vol. 25, no. 5, pp. 168, 169, Sept.–Oct.

DIXON, JOSEPH SCATTERGOOD.

1927. Contribution to the life history of the Alaska willow ptarmigan. The Condor, vol. 29, no. 5, pp. 213–223, illus., Sept.–Oct.

1930. Jays or California quail? Nature Mag., vol. 15, no. 1, pp. 42, 43, illus., Jan.

DUTCHER, WILLIAM.

1903. The mourning dove. Nat. Assoc. Audubon Soc. Educ. Leaflet 2, 3 pp., 1 pl., Mar. 2. New York.

DWIGHT, JONATHAN, JR.

1900. The moult of the North American Tetraonidae (quails, partridges, and grouse). The Auk, vol. 17, nos. 1, 2, pp. 34–51, 143–166, illus., Jan., Apr.

EATON, ELON HOWARD.

1910. Birds of New York. New York State Mus. Mem. 12, pt. 1, Water birds and game birds, 501 pp., 42 pls.

EDSON, JOHN MILTON.

1925. The Hooters of Skyline Ridge. The Condor, vol. 27, no. 6, pp. 226–229, Nov.-Dec.

EDWARDS, GEORGE.

1755. A letter to Mr. Peter Collinson, F. R. S., concerning the pheasant of Pennsylvania, and the Otis Minor. Philos. Trans. Roy. Soc. London, vol. 48, pt. 2, for 1754, pp. 499–503, 2 pls.

ELLIOT, DANIEL GIRAUD.
1896. Descriptions of an apparently new species and subspecies of ptarmigan from the Aleutian Islands. The Auk, vol. 13, no. 1, pp. 24–29, pl. 3, Jan.
1897. The gallinaceous game birds of North America. 220 pp., 46 pls., New York.

ERICHSEN, WALTER JEFFERSON.
1920. Observations on the habits of some breeding birds of Chatham County, Georgia. Wilson Bull. no. 113, vol. 32, no. 4, pp. 133–139, Dec.

FIELD, GEORGE WILTON; GRAHAM, GEORGE H.; and ADAMS, WILLIAM C.
1914. Special report of the Board of Commissioners on Fisheries and Game, under Chapter 70 of the Resolves of 1913, relative to the habits of those birds commonly known as pheasants. Commonwealth of Massachusetts, House Rep. no. 2049, pp. 1–14, Jan.

FINLEY, WILLIAM LOVELL.
1896. The Oregon ruffed grouse. Oregon Nat., vol. 3, no. 7, pp. 102, 103, illus., July.

FISHER, ALBERT KENRICK.
1893. Report on the ornithology of the Death Valley Expedition of 1891, comprising notes on the birds observed in southern California, southern Nevada, and parts of Arizona and Utah. North Amer. Fauna no. 7, pp. 7–158, May.

FLEMING, JAMES HENRY.
1903. Recent records of the wild pigeon. The Auk, vol. 20, no. 1, p. 66, Jan.
1907. The disappearance of the passenger pigeon. Ottawa Field Nat., vol. 20, no. 12, pp. 236, 237, Mar.
1930. Ontario bird notes. The Auk, vol. 47, no. 1, pp. 64–71, Jan.

FORBUSH, EDWARD HOWE.
1912. A history of the game birds, wild-fowl and shore birds of Massachusetts and adjacent States, 622 pp., illus., Boston.
1927. Birds of Massachusetts and other New England States, vol. 2, 461 pp., 27 col. pls. Boston.

FORTINER, JOHN C.
1920. Winter nesting of the ground dove. The Condor, vol. 22, no. 4, pp. 154, 155, July–Aug.
1921. The doves of Imperial County, California. The Condor, vol. 23, no. 5, p. 168, Sept.–Oct.

FOWLER, FREDERICK HALL.
1903. Stray notes from southern Arizona. The Condor, vol. 5, no. 3, pp. 68–71, May–June.

FRENCH, JOHN C.
1919. The passenger pigeon in Pennsylvania, its remarkable history, habits, and extinction, with interesting side lights on the folks and forest lore of the Alleghanian region of the old Keystone State, 257 pp., illus. Altoona.

FUERTES, LOUIS AGASSIZ.
1903. With the Mearns quail in southwestern Texas. The Condor, vol. 5, no. 5, pp. 113–116, illus., Sept.–Oct.

GABRIELSON, IRA NOEL.
1922. Short notes on the life histories of various species of birds. Wilson Bull., vol. 34, no. 4, pp. 193–210, illus., Dec.

GARDNER, LEON LLOYD.
1927. Habits of blue jays and doves in central Kansas. The Auk, vol. 44, p. 104, Jan.

GIDDINGS, H. J.
1897. A quail's egg in a towhee's nest. The Osprey, vol. 2, no. 2, p. 26, Oct.

GILMAN, MARSHALL FRENCH.
1903. More about the band-tailed pigeon (*Columba fasciata*). The Condor, vol. 5, no. 5, pp. 134, 135, Sept.–Oct.
1911. Doves on the Pima Reservation. The Condor, vol. 13, no. 2, pp. 51–56, Mar.–Apr.
1915. A forty-acre bird census at Sacaton, Arizona. The Condor, vol. 17, no. 2, pp. 86–90, Mar.–Apr.

GOSS, NATHANIEL STICKNEY.
1891. History of the birds of Kansas, 693 pp., 35 pls. Topeka.

GOSSE, PHILIP HENRY.
1847. The birds of Jamaica, 447 pp. London.

GRAYSON, ANDREW JACKSON.
1871. On the physical geography and natural history of the Tres Marias and of Socorro, off the western coast of Mexico. Proc. Boston Soc. Nat. Hist., vol. 14, pp. 261–302, June 7.

GREEN, CHARLES DEBLOIS.
1928. The fluttering habit of the Richardson's grouse (*Dendragapus obscurus richardsoni*). The Murrelet, vol. 9, no. 3, p. 67, Sept.

GRINNELL, JOSEPH.
1900. Birds of the Kotzebue Sound region. Pacific Coast Avifauna, no. 1, 80 pp., 1 map.
1904. Midwinter birds at Palm Springs, California. The Condor, vol. 6, no. 2, pp. 40–45, Mar.–Apr.
1905. Summer birds of Mount Pinos, California. The Auk, vol. 22, no. 4, pp. 378–391, Oct.
1906. The Catalina Island quail. The Auk, vol. 23, no. 3, pp. 262–265, July.
1908. Catalina quail. The Condor, vol. 10, no. 2, p. 94, Mar.–Apr.
1910. Birds of the 1908 Alexander Alaska Expedition, with a note on the avifaunal relationships of the Prince William Sound district. Univ. California Publ. Zool., vol. 5, no. 12, pp. 361–428, pls. 33, 34, 9 figs., Mar. 5.
1914. An account of the mammals and birds of the lower Colorado Valley. Univ. California Publ. Zool., vol. 12, pp. 51–294.
1916. A new ruffed grouse from the Yukon Valley. The Condor, vol. 18, no. 4, pp. 166, 167, July–Aug.
1926. Another new race of quail from Lower California. The Condor, vol. 28, no. 3, pp. 128, 129, May–June.
1928. A distributional summation of the ornithology of Lower California. Univ. California Publ. Zool., vol. 32, pp. 1–300.
1928a. September nesting of the band-tailed pigeon. The Condor, vol. 30, no. 1, pp. 126, 127, Jan.–Feb.

GRINNELL, JOSEPH; BRYANT, HAROLD CHILD; and STORER, TRACY IRWIN.
1918. The game birds of California, 642 pp., 16 col. pls., 94 figs. Berkeley.

GRINNELL, JOSEPH; DIXON, JOSEPH; and LINSDALE, JEAN M.
1930. Vertebrate natural history of a section of northern California through the Lassen Peak region. Univ. California Publ. Zool., vol. 35, 594 pp., 181 figs. (1 col. map), Oct.

GRINNELL, JOSEPH; STEPHENS, FRANK; DIXON, JOSEPH; and HELLER, EDMUND.
1909. Birds and mammals of the 1907 Alexander Expedition to southeastern Alaska. Univ. California Publ. Zool., vol. 5, no. 2, pp. 171–264, pls. 25, 26, figs. 1–4, Feb. 18.

GRINNELL, JOSEPH; and STORER, TRACY IRWIN.
1924. Animal life in the Yosemite, 752 pp., 60 pls., 65 figs., 2 col. maps. Berkeley.

GRISCOM, LUDLOW.
1926. Notes on the summer birds of the west coast of Newfoundland. The Ibis, ser. 12, vol. 2, no. 4, pp. 656–684, pls. 11–13, Oct.

GROSS, ALFRED OTTO.
1928. The heath hen. Mem. Boston Soc. Nat. Hist., vol. 6, no. 4, pp. 491–588, 12 pls., May.

GURNEY, JOHN HENRY.
1884. Hybrid between *Pedioecetes phasianellus* and *Cupidonia cupido*. The Auk, vol. 1, no. 4, pp. 391, 392, Oct.

HADLEY, PHILIP.
1930. The passenger pigeon. Science, new ser., vol. 71, no. 1833, p. 187, Feb. 14.

HADZOR, RAE T.
1923. A grouse refuge. Bird-Lore, vol. 25, no. 5, p. 318, Sept.–Oct.

HAGERUP, ANDREAS THOMSEN.
1891. The birds of Greenland, 62 pp. Boston.

HALL, HENRY MARION.
1929. An Indian summer day with ring-necks. Forest and Stream, vol. 99, no. 10, pp. 735, 756, 757, illus. Oct.

HANTZSCH, BERNHARD.
1929. Contribution to the knowledge of the avifauna of north-eastern Labrador. Can. Field-Nat., vol. 43, no. 1, pp. 11–18, Jan. (Translated by M. B. A. and R. M. Anderson.)

HARTLEY, GEORGE INNESS.
1922. The importance of bird life, 316 pp., illus. New York.

HENSHAW, HENRY WETHERBEE.
1874. Report upon the ornithological collections made in portions of Nevada, Utah, California, Colorado, New Mexico, and Arizona, during the years 1871, 1872, 1873, and 1874. Wheeler's Rep. Geogr. and Geol. Expl. and Surv. West 100th Meridian, vol. 5, Zoology, chap. 3, pp. 133–507, 15 pls.

HIGGINSON, FRANCIS.
1630. New England's plantations, 11 unpaged leaves, 7th and 8th leaves. London.

HILL, GRACE A.
1922. With the willow ptarmigan. The Condor, vol. 24, no. 4, pp. 105–108, illus., July–Aug.

HODGE, CLIFTON FREMONT.
1905. The drumming grouse. Country Calendar, vol. 1, no. 7, pp. 640–644, illus.
1911. The passenger pigeon investigation. The Auk, vol. 28, no. 1, pp. 49–53, Jan.
1912. A last word on the passenger pigeon. The Auk, vol. 29, no. 2, pp. 169–175, Apr.

HOLLAND, HAROLD MAY.
1917. The valley quail occupying nests of the road-runner. The Condor, vol. 19, no. 1, pp. 23, 24, illus., Jan.–Feb.

HOLLISTER, NED.

1925. Another record of the ruddy quail-dove at Key West. The Auk, vol. 42, no. 1, p. 130, Jan.

HOLZNER, FRANK XAVIER.

1896. Habits of the valley partridge. The Auk, vol. 13, no. 1, p. 81, Jan.

HOWARD, OZRA WILLIAM.

1900. Nesting of the Mexican wild turkey in the Huachuca Mts., Arizona. The Condor, vol. 2, no. 3, pp. 55–57, illus., May–June.

HOWE, CARLTON DURANT.

1904. A tame ruffed grouse. Bird-Lore, vol. 6, no. 3, pp. 81–85, illus, May–June.

HOWELL, ALFRED BRAZIER.

1917. Condition of game birds in east-central California. The Condor, vol. 19, no. 6, pp. 186, 187, Nov.–Dec.

HUDSON, WILLIAM HENRY.

1902. British birds, with their structure and classification, 363 pp., 136 pls. (8 col.).

HUEY, LAURENCE MARKHAM.

1913. With the band-tailed pigeon in San Diego County. The Condor, vol. 15, no. 4, pp. 151–153, July–Aug.

1927. Where do birds spend the night? Wilson Bull., vol. 39, no. 4, pp. 215–217, Dec.

HUNTINGTON, DWIGHT W.

1897. The sage grouse. The Osprey, vol. 2, no. 2, pp. 17, 18, illus., Oct.

1903. Our feathered game, 396 pp., illus. in col. New York.

JEWETT, HIBBARD, J.

1918. Memories of the passenger pigeon. Bird-Lore, vol. 20, no. 5, p. 351, Sept.–Oct.

JOHNSON, R. A.

1929. Summer notes on the sooty grouse of Mount Rainier. The Auk, vol. 46, no. 3, pp. 291–293, pls. 12, 13. July.

JONES, LYNDS.

1903. A bob-white covey. Wilson Bull., no. 45, vol. 10, no. 4, pp. 105, 106, Dec.

JOSSELYN, JOHN.

1672. New-England's rarities discovered, 114 pp., 1 pl., 12 figs. London.

JUDD, SYLVESTER DWIGHT.

1905. The bobwhite and other quails of the United States in their economic relations. U. S. Dept. Agr. Biol. Surv. Bull. 21, 66 pp., 2 cols. pls., 10 figs.

1905a. The grouse and wild turkeys of the United States, and their economic value. U. S. Dept. Agr. Biol. Surv. Bull. 24, 55 pp., 2 pls. (1 col.).

KALM, PEHR.

1911. A description of the wild pigeons which visit the southern English colonies in North America, during certain years, in incredible multitudes. The Auk, vol. 28, no. 1, pp. 53–66 (Translated by S. M. Gronberger from Kongl. Vetenskaps-Akademiens Handlingar för år 1759, vol. 20. Stockholm.)

KELLOGG, LOUISE.

1916. Report upon mammals and birds found in portions of Trinity, Siskiyou and Shasta Counties, California, with description of a new *Dipodomys*. Univ. California Publ. Zool., vol. 12, no. 13, pp. 335–398, pls. 15–18, Jan. 27.

KENTWOOD.
1896. Martha's Vineyard heath hen. Forest and Stream, vol. 47, no. 18, pp. 343, 344, Oct. 31.

KERMODE, FRANCIS.
1914. Report of the Provincial Museum of Natural History for the year 1913, pp. 1–32, illus.

KEYES, CHARLES ROLLIN.
1905. Some bird notes from the central Sierras. The Condor, vol. 7, no. 1, pp. 13–17, illus., Jan.

KNIGHT, ORA WILLIS.
1908. The birds of Maine, 693 pp., illus., 1 map. Bangor.

KOBBÉ, WILLIAM HOFFMAN.
1900. The birds of Cape Disappointment, Washington. The Auk, vol. 17, no. 4, pp. 349–358, Oct.

KUMLIEN, LUDWIG, and HOLLISTER, NED.
1903. The birds of Wisconsin. Bull. Wisconsin Nat. Hist. Soc., new ser., vol. 3, nos. 1–3, Jan.–July.

LACEY, HOWARD.
1911. The birds of Kerrville, Texas, and vicinity. The Auk, vol. 28, no. 2, pp. 200–219, 1 map, Apr.

LAHONTAN, BARON DE.
1703. New voyages to North America, 2 vols. London.

LAING, HAMILTON MACK.
1913. Out with the birds, 249 pp., illus. New York.
1925. Birds collected and observed during the cruise of the *Thiepval* in the North Pacific, 1924. Victoria Mem. Mus. Bull. 40.

LAMB, CHESTER CONVERSE.
1926. The Viosca pigeon. The Condor, vol. 28, no. 6, pp. 262, 263, Nov.–Dec.

LAWRENCE, GEORGE NEWBOLD.
1853. Descriptions of new species of birds of the genera *Ortyx* Stephens, *Sterna* Linn., and *Icteria* Vieillot. Ann. Lyceum Nat. Hist., vol. 6, pp. 1–4.
1874. The birds of western and northwestern Mexico, based upon collections made by Col. A. J. Grayson, Capt. J. Xantus and Ferd. Bischoff, now in the museum of the Smithsonian Institution, at Washington, D. C. Mem. Boston Soc. Nat. Hist., vol. 2, pp. 265–319, Apr. 25.
1889. Remarks upon abnormal coloring of plumage observed in several species of birds. The Auk, vol. 6, no. 1, pp. 46–50, Jan.

LEFFINGWELL, DANA JACKSON.
1928. The ringed-neck pheasant, its history and habits. Occ. Pap. Charles R. Conner Mus. no. 1, pp. 1–35, illus., Apr.

LEWIS, ELISHA JARRETT.
1885. The American sportsman, ed. 3, 553 pp., illus. Philadelphia.

LEWIS, EVAN.
1904. The nesting habits of the white-tailed ptarmigan in Colorado. Bird-Lore, vol. 6, no. 4, pp. 117–121, illus., July–Aug.

LEWIS, HARRISON FLINT.
1928. Notes on the birds of the Labrador Peninsula in 1928. Can. Field-Nat., vol. 42, pp. 191–194.

LINCOLN, FREDERICK CHARLES.
1918. A strange case of hybridism. Wilson Bull. no. 102, vol. 30, no. 1, pp. 1, 2, illus. Mar.
1930. Migratory status of mourning doves is proved by banding. U. S. Dept. Agr. Yearbook for 1930, pp. 386–388, map.

LLOYD, WILLIAM.
1887. Birds of Tom Green and Concho Counties, Texas. The Auk, vol. 4, no. 3, pp. 181–193, July.

MACFARLANE, RODERICK ROSS.
1908. List of birds and eggs observed and collected in the North-West Territories of Canada, between 1880 and 1894, *in* Through the Mackenzie Basin, by Charles Mair, pp. 287–490, illus. London.

MACGILLIVRAY, WILLIAM.
1837–1852. A history of British birds, 5 vols., illus. London.

MACMILLAN, DONALD BAXTER.
1918. Four years in the white north, 426 pp., illus. New York.

MACOUN, JOHN, and MACOUN, JAMES MELVILLE.
1909. Catalogue of Canadian birds, ed. 2, 761 pp. Can. Dept. Mines. Geol. Surv. Branch.

MANNICHE, ARNER LUDVIG VALDEMAR.
1910. The terrestrial mammals and birds of northeast Greenland. Biological observations. Meddelelser om Grønland, vol. 45, no. 1, pp. 1–199, pls. 1–7 (part col.), figs. 1–20.

MAYNARD, CHARLES JOHNSON.
1896. The birds of eastern North America, rev. ed., 721 pp., 40 col. pls., 122 figs. Newtonville, Mass.

MCATEE, WALDO LEE.
1929. Game birds suitable for naturalizing in the United States. U. S. Dept. Agr. Circ. 96, pp. 1–23, figs. 1–14, Nov.

MCATEE, WALDO LEE, and BEAL, FOSTER ELLENBOROUGH LASCELLES.
1912. Some common game, aquatic, and rapacious birds in relation to man. U. S. Dept. Agr. Farmers' Bull. 497, pp. 1–30, 14 figs., May 6.

MCGREGOR, RICHARD CRITTENDEN.
1906. Birds observed in the Krenitzin Islands, Alaska. The Condor, vol. 8, no. 5, pp. 114–122, 1 map, Sept.

MCILHENNY, EDWARD AVERY.
1914. The wild turkey and its hunting, 245 pp., illus. Garden City.

MEARNS, EDGAR ALEXANDER.
1879. Notes on some of the less hardy winter residents in the Hudson River Valley. Bull. Nuttall Orn. Club, vol. 4, no. 1, pp. 33–37, Jan.
1914. Diagnosis of a new subspecies of Gambel's quail from Colorado. Proc. Biol. Soc. Washington, vol. 27, p. 113, July 10.

MERRIAM, CLINTON HART.
1885. Change of color in the wing-feathers of the willow grouse. The Auk, vol. 2, no. 2, pp. 201–203, Apr.
1888. What birds indicate the proximity to water, and at what distance? Auk, vol. 5, no. 1, p. 119, Jan.
1889. Introduced pheasants. Rep. Comm. Agr. for 1888, pp. 484–488.
1890. Results of a biological survey of the San Francisco Mountain region and desert of the Little Colorado, Arizona. Annotated list of birds of the San Francisco Mountain plateau and the desert of the Little Colorado River, Arizona. North Amer. Fauna no. 3, pt. 4, pp. 87–101, Sept. 11.

MERRILL, JAMES CUSHING.
1878. Notes on the ornithology of southern Texas, being a list of birds observed in the vicinity of Fort Brown, Texas, from February, 1876, to June, 1878. Proc. U. S. Nat. Mus., vol. 1, pp. 118–173.

MERSHON, WILLIAM BUTTS.
1907. The passenger pigeon, 225 pp., col. pls. New York.

MILLAIS, JOHN GUILLE.
1909. The natural history of British game birds. 142 pp., 8 pls. London.

MILLER, WALDRON DEWITT.
1905. List of birds collected in southern Sinaloa, Mexico, by J. H. Batty, during 1903–4. Bull. Amer. Mus. Nat. Hist., vol. 21, art. 22, pp. 339–369.

MINOT, HENRY DAVIS.
1877. The land birds and game birds of New England, 456 pp., 1 pl. Salem.

MORSS, CHARLES B.
1923. The signal: An episode of grouseland. Field and Stream, vol. 28, no. 6, pp. 689–691, 757–762, illus., Oct.

MORTON, THOMAS.
1637. New English Canaan, ed. 2, 188 pp. London.

MOSELEY, EDWIN LINCOLN.
1928. Bob-white and scarcity of potato beetles. Wilson Bull., vol. 40, no. 3, pp. 149–151, Sept.

MURPHY, ROBERT CUSHMAN.
1917. Natural history observations from the Mexican portion of the Colorado Desert. Abstract of Proc. Linnaean Soc. New York, nos. 28, 29, pp. 1–114, pls. 1–6, Dec. 11.

NEFF, JOHNSON A.
1923. Some birds of the Ozark region. Wilson Bull., vol. 35, no. 4, pp. 202–215, map, Dec.

NELSON, EDWARD WILLIAM.
1887. Report upon the natural history collections made in Alaska, between the years 1877 and 1881. Arctic ser., Signal Service, U. S. Army, pt. 1, Birds, pp. 19–226, 12 col. pls.
1899. Natural history of the Tres Marias Islands. Birds of the Tres Marias Islands. North Amer. Fauna no. 14, pp. 21–62, Apr. 29.
1900. Description of a new subspecies of *Meleagris gallopavo* and proposed changes in the nomenclature of certain North American birds. The Auk, vol. 17, no. 2, pp. 120–126, Apr.

NEWTON, ALFRED.
1893–1896. A dictionary of birds, 1088 pp., illus. London.

NICE, MARGARET MORSE.
1910. Food of the bobwhite. Journ. Econ. Ent., vol. 3, no. 3, pp. 295–313, June.
1922–1923. A study of the nesting of mourning doves. The Auk, vol. 39, no. 4, pp. 457–474, pl. 18, Oct.; vol. 40, no. 1, pp. 37–58, Jan.

NICHOLSON, DONALD JOHN.
1928. Actions of baby Florida wild turkeys. Florida Nat., vol. 2, no. 1, p. 32, Oct.

NOBLE, GEORGE KINGSLEY.
1919. Notes on the avifauna of Newfoundland. Bull. Mus. Comp. Zool., vol. 62, no. 14, pp. 543–568, Mar.

NUTTALL, THOMAS.
1832. A manual of the ornithology of the United States and of Canada. Land Birds, 683 pp., illus. Cambridge, Mass.

OBERHOLSER, HARRY CHURCH.

1923. Notes on the forms of the genus *Oreortyx* Baird. The Auk, vol. 40, no. 1, pp. 80–84, Jan.

OSGOOD, WILFRED HUDSON.

1901. New subspecies of North American birds. The Auk, vol. 18, no. 2, pp. 179–185, Apr.

1904. A biological reconnaissance of the base of the Alaska Peninsula. North Amer. Fauna no. 24, 86 pp., 7 pls., Nov. 23.

PANGBURN, CLIFFORD HAYES.

1919. A three months' list of the birds of Pinellas County, Florida. The Auk, vol. 36, pp. 393–405.

PATCH, CLYDE A.

1922. A biological reconnaissance on Graham Island of the Queen Charlotte Group. Can. Field-Nat., vol. 36, pp. 101–105, 133–136.

PAYNE, E. B.

1897. Quail's eggs in a meadowlark's nest. The Osprey, vol. 2, no. 4, p. 55, Dec.

PEARSON, THOMAS GILBERT.

1920. The ground dove. Bird-Lore, vol. 22, no. 2, pp. 126–129, illus., Mar.–Apr.

PEMBERTON, JOHN ROY.

1928. The nesting of Howard's grouse. The Condor, vol. 30, no. 6, pp. 347, 348, illus., Nov.–Dec.

PETERS, JAMES LEE.

1923. A new quail from Lower California. Proc. New England Zool. Club, vol. 8, pp. 79, 80, May 16.

PHELPS, FRANK MILLS.

1914. The resident bird life of the Big Cypress Swamp region. Wilson Bull., no. 87, vol. 26, no. 2, pp. 86–101, illus., June.

PHILLIPS, JOHN CHARLES.

1928. Wild birds introduced or transplanted in North America. U. S. Dept. Agr. Techn. Bull. 61, 64 pp.

POKAGON, CHIEF SIMON.

1895. The wild pigeon of North America. Chautauquan, vol. 22, no. 2, pp. 202–206, Nov.

PREBLE, EDWARD ALEXANDER.

1908. A biological investigation of the Athabaska-Mackenzie region. North Amer. Fauna no. 27, 574 pp., 25 pls., 16 figs., map, Oct. 26.

QUILLIN, ROY W., and HOLLEMAN, RIDLEY.

1918. The breeding birds of Bexar County, Texas. The Condor, vol. 20, no. 1, pp. 37–44, Jan.–Feb.

REVOIL, BENEDICT HENRY.

1869. Chasses dans l'Amerique du Nord. (See Bird-Lore, vol. 30, no. 5, pp. 317–320, Sept.–Oct., 1928.)

RICH, WALTER HERBERT.

1907. Feathered game of the Northeast, 432 pp., illus. New York.

1909. The "Hungarian Partridge"—the gray partridge. Journ. Maine Orn. Soc., vol. 11, no. 2, pp. 33–38, illus. June.

RIDGWAY, ROBERT.

1912. Color standards and color nomenclature, 43 pp., 53 col. pls. Washington, D. C.

1915. Descriptions of some new forms of American cuckoos, parrots, and pigeons. Proc. Biol. Soc. Washington, vol. 28, pp. 105–108, May 27.

RIDGWAY, ROBERT.
1916. The birds of North and Middle America, U. S. Nat. Mus. Bull. 50, pt. 7, 543 pp., 24 pls.

RILEY, JOSEPH HARVEY.
1911. Descriptions of three new birds from Canada. Proc. Biol. Soc. Washington, vol. 24, pp. 233–235, Nov. 28.
1912. Birds collected or observed on the expedition of the Alpine Club of Canada to Jasper Park, Yellowhead Pass, and Mount Robson region. Can. Alpine Journ., special number, pp. 47–75, pl. 1, 1912.

ROBERTS, THOMAS SADLER.
1919. A review of the ornithology of Minnesota. Research publications of the Univ. of Minnesota, vol. 8, no. 2, 100 pp., illus., May.

RONEY, H. B.
1879. A description of the Michigan pigeon nesting of 1878 and the work of protection undertaken by the East Saginaw and Bay City game protection clubs. Chicago Field, vol. 10, no. 22, pp. 345–347, Jan. 11.

ROWAN, WILLIAM.
1926. Comments on two hybrid grouse and on the occurrence of *Tympanuchus americanus americanus* in the Province of Alberta. The Auk, vol. 43, no. 3, pp. 333–336, pls. 17, 18, July.

SAGE, JOHN HALL; BISHOP, LOUIS BENNETT; and BLISS, WALTER PARKS.
1913. The birds of Connecticut. Connecticut State Geol. and Nat. Hist. Surv., Bull. 20, Pub. Doc. no. 47, pp. 1–370.

SAMUELS, EDWARD AUGUSTUS.
1883. Our northern and eastern birds, 600 pp., illus., part col. pls. New York.

SANDYS, EDWYN, and VAN DYKE, T. S.
1904. Upland game birds, 429 pp., illus. London.

SAWYER, EDMUND JOSEPH.
1923. The ruffed grouse, with special reference to its drumming. Roosevelt Wild Life Bull., vol. 1, no. 3, pp. 355–384, illus.

SCHUTZE, ADOLF EMIL.
1904. The Inca dove in central Texas. The Condor, vol. 6, no. 6, p. 172, Nov.–Dec.

SCLATER, WILLIAM LUTLEY.
1912. A history of the birds of Colorado, 576 pp., 17 pls., map. London.

SCOTT, WILLIAM EARL DODGE.
1889. Records of rare birds at Key West, Florida, and vicinity, with a note on the capture of a dove (*Geotrygon montana*) new to North America. The Auk, vol. 6, no. 2, pp. 160, 161, Apr.
1890. Description of a new subspecies of wild turkey. The Auk, vol. 7, no. 4, pp. 376, 377, Oct.
1892. Notes on the birds of the Caloosahatchie region of Florida. The Auk, vol. 9, no. 3, pp. 209–218, July.

SENNETT, GEORGE BURRITT.
1878. Notes on the ornithology of the lower Rio Grande of Texas from observations made during the season of 1877. Bull. U. S. Geol. and Geogr. Surv. Terr., vol. 4, art. 1, pp. 1–66.
1879. Further notes on the ornithology of the lower Rio Grande of Texas. Bull. U. S. Geol. and Geogr. Surv. Terr., vol. 5, no. 3, pp. 371–440.
1892. Description of a new turkey. The Auk, vol. 9, no. 2, pp. 167–169, Apr.

SETON, ERNEST E. THOMPSON.
1885. Notes on Manitoban birds. The Auk, vol. 2, no. 3, pp. 267–271, July.

Shaw, William Thomas.

1908. The China or Denny pheasant in Oregon, with notes on the native grouse of the Pacific Northwest, 24 pp., 15 pls. Philadelphia.

Simmons, George Finlay.

1915. On the nesting of certain birds in Texas. The Auk, vol. 32, no. 3, pp. 317–331, pls. 21, 22, July.

1925. Birds of the Austin region, 387 pp., illus. Austin.

Simpson, Gene M.

1914. Pheasant farming. Oregon Fish and Game Comm. Bull. 2, 47 pp., illus., 1 col. pl.

Skinner, Milton Philo.

1927. Richardson's grouse in the Yellowstone Park. Wilson Bull., vol. 39, no. 4, pp. 208–214, Dec.

Slade, Daniel Dennison.

1888. The wild turkey in Massachusetts. The Auk, vol. 5, no. 2, pp. 204, 205, Apr.

Smith, Austin Paul.

1910. Miscellaneous bird notes from the lower Rio Grande. The Condor, vol. 12, no. 3, pp. 93–103, May–June.

1916. Additions to the avifauna of Kerr Co., Texas. The Auk, vol. 33, no. 2, pp. 187–193, Apr.

Smith, Everett.

1883. Birds of Maine. Forest and Stream, vol. 20, pp. 25–61.

Snyder, John Otterbein.

1900. Notes on a few species of Idaho and Washington birds. The Auk, vol. 17, no. 3, pp. 242–245, July.

Spiker, Charles J.

1929. The Hungarian partridge in northwest Iowa. Wilson Bull., vol. 41, no. 1, pp. 24–29, illus., 1 map, Mar.

Spurrell, John A.

1917. Annotated list of the water birds, game birds, and birds of prey of Sac County, Iowa. Wilson Bull., no. 100, vol. 29, no. 3, pp. 141–160, Sept.

Stejneger, Leonhard.

1884. A new subspecies of willow grouse from Newfoundland. The Auk, vol. 1, no. 4, p. 369, Oct.

Stephens, Frank.

1885. Notes on an ornithological trip in Arizona and Sonora. The Auk, vol. 2, no. 3, pp. 225–231, July.

1903. Bird notes from eastern California and western Arizona. The Condor, vol. 5, no. 3, pp. 75–78, May–June.

Stockard, Charles Rupert.

1905. Nesting habits of birds in Mississippi. Auk, vol. 22, no. 2, pp. 146–158, Apr.

Stoddard, Herbert Lee.

1922. Notes on birds from southern Wisconsin. Wilson Bull., vol. 34, no. 2, pp. 67–79, June.

1925. Progress on cooperative quail investigation: 1924.

1926. Report on cooperative quail investigation: 1925–26, with preliminary recommendations for the development of quail preserves, 62 pp., 5 pls.

1931. The bobwhite quail, its habits, preservation, and increase, 559 pp., 69 pls. (part col.), 32 figs. New York.

STONE, WITMER.
1905. On a collection of birds and mammals from the Colorado Delta, Lower California. Proc. Acad. Nat. Sci. Philadelphia, vol. 57, pp. 676–690, Sept.

SUTTON, GEORGE MIKSCH.
1928. The birds of Pymatuning Swamp and Conneaut Lake, Crawford County, Pennsylvania. Ann. Carnegie Mus., vol. 18, no. 1, pp. 19–239, pls. 2–11 (in col.), Mar.
1929. Photographing wild turkey nests in Pennsylvania. The Auk, vol. 46, no. 3, pp. 326–328, pl. 18, July.

SWAINSON, WILLIAM, and RICHARDSON, JOHN.
1831. Fauna Boreali-Americana, vol. 2, Birds, 523 pp., col. pls. and figs. London.

SWARTH, HARRY SCHELWALDT.
1904. Birds of the Huachuca Mountains, Arizona. Pacific Coast Avifauna no. 4, 70 pp.
1909. Distribution and molt of the Mearns quail. The Condor, vol. 11, no. 2. pp. 39–48, illus., Mar.–Apr.
1912. Report on a collection of birds and mammals from Vancouver Island. Univ. California Publ. Zool., vol. 10, no. 1, pp. 1–124, pls. 1–4, Feb. 13.
1920. Birds of the Papago Saguaro National Monument and the neighboring region, Arizona, 63 pp., 8 pls. Washington, D. C.
1921. The Sitkan race of the dusky grouse. The Condor, vol. 23, no. 2, pp. 59, 60, Mar.–Apr.
1922. Birds and mammals of the Stikine River region of northern British Columbia and southeastern Alaska. Univ. California Publ. Zool., vol. 24, no. 2, pp. 125–314.
1924. Birds and mammals of the Skeena River region of northern British Columbia. Univ. California Publ. Zool., vol. 24, no. 3, pp. 315–394, illus.
1926. Report on a collection of birds and mammals from the Atlin region, northern British Columbia. Univ. California Publ. Zool., vol. 30, no. 4, pp. 51–162, pls. 4–8 (1 in col.), 11 figs., Sept. 24.

SWENK, MYRON HARMON, and SWENK, JANE BISHOP.
1928. Some impressions of the commoner winter birds of southern Arizona. Wilson Bull., vol. 40, no. 1, pp. 17–29, Mar.

TABER, WILLIAM BREWSTER, JR.
1930. The fall migration of mourning doves. Wilson Bull., vol. 42, pp. 17–28.

TAVERNER, PERCY ALGERNON.
1914. A new subspecies of *Dendragapus* (*Dendragapus obscurus flemingi*) from southern Yukon Territory. The Auk, vol. 31, no. 3, pp. 385–388, July.
1926. Birds of western Canada. Can. Dept. Mines, Victoria Mem. Mus. Bull. 41, 360 pp., 84 pls.
1929. A study of the Canadian races of rock ptarmigan (*Lagopus rupestris*). Ann. Rep. Nat. Mus. Canada for 1928, pp. 28–38.

TAYLOR, WALTER PENN.
1920. A new ptarmigan from Mount Rainier. The Condor, vol. 22, no. 4, pp. 146–152, illus., July–Aug.

TAYLOR, WALTER PENN, and SHAW, WILLIAM THOMAS.
1927. Mammals and birds of Mount Rainier National Park. U. S. Dept. Int., 249 pp., 109 figs.

TEGETMEIER, WILLIAM BERNHARD.

1911. Pheasants, their natural history and practical management, ed. 5, 276 pp., illus. (part col. plates). London.

THOMPSON, CHARLES S.

1901. Further tape-worm observations. The Condor, vol. 3, no. 1, p. 15, Jan.–Feb.

THOMPSON, ERNEST EVAN (SETON, ERNEST THOMPSON-).

1890. The birds of Manitoba. Proc. U. S. Nat. Mus., vol. 13, pp. 457–643, pl. 38.

THOMPSON, FRANK J.

1881. Breeding of the wild pigeon in confinement. Bull. Nuttall Orn. Club, vol. 6, no. 2, p. 122, Apr.

TODD, WALTER EDMOND CLYDE.

1916. The birds of the Isle of Pines. Ann. Carnegie Mus., vol. 10, no. 12, art. 11, pp. 146–296, pls. 22–27, Jan. 31.

TOWNSEND, CHARLES WENDELL.

1906. Note on the crop contents of a nestling mourning dove. The Auk, vol. 23, no. 3, pp. 336, 337, July.

1912. The case of a crow and a ruffed grouse. The Auk, vol. 29, no. 4, p. 542, Oct.

1920. Supplement to the birds of Essex County, Massachusetts. Mem. Nuttall Orn. Club, no. 5, 196 pp., 1 pl., 1 map.

TURNER, LUCIEN MCSHAN.

1886. Contributions to the natural history of Alaska. Signal Service, U. S. Army, No. 2, pt. 5, Birds, pp. 115–196, 10 col. pls.

TYLER, JOHN GRIPPER.

1913. Some birds of the Fresno district, California. Pacific Coast Avifauna no. 9, 114 pp.

VAN DEWEYER, B.

1919. Pheasant breeding in sparrow hawk's nest. British Birds, vol. 13, no. 3, p. 87, Aug. 1.

VAN DYKE, THEODORE STRONG.

1892. The valley quail of California. Outing, vol. 19, pp. 485–488, Mar.

VAN ROSSEM, ADRIAAN JOSEPH.

1925. Flight feathers as age indicators in *Dendragapus*. The Ibis, ser. 12, vol. 1, no. 2, pp. 417–422, figs. 10–12, Apr.

VREELAND, FREDERICK K.

1918. The drumming of the ruffed grouse. Forest and Stream, vol. 88, no. 4, pp. 199–201, 244; illus., Apr.

WALES, JOSEPH HOWE.

1926. The coo of the band-tailed pigeon. The Condor, vol. 28, no. 1, p. 42, illus., Jan.–Feb.

WARNER, EDWARD PEARSON.

1911. The history of a ruffed grouse's nest. Bird-Lore, vol. 13, no. 2, p. 99, Mar.–Apr.

WAYNE, ARTHUR TREZEVANT.

1910. Birds of South Carolina. Contr. Charleston Mus. no. 1, 254 pp., 1 map.

WETMORE, ALEXANDER.

1916. Birds of Porto Rico. U. S. Dept. Agr. Bull. 326, 140 pp., 10 pls., Mar. 24.

1920. Observations on the habits of the white-winged dove. The Condor, vol. 22, no. 4, pp. 140–146, July–Aug.

WETMORE, ALEXANDER.
1927. The birds of Porto Rico and the Virgin Islands. Scientific Surv. of Porto Rico and the Virgin Islands, vol. 9, pt. 3, pp. 245–406. New York Acad. Sci.

WHEATON, JOHN MAYNARD.
1882. Report on the birds of Ohio. Rep. Geol. Surv. Ohio, vol. 4, pt. 1, Zoology, pp. 187–628.

WHEELOCK, IRENE GROSVENOR.
1904. Birds of California, 578 pp., 10 pls., 78 figs. Chicago.

WICKS, MORIE LANGLEY, JR.
1897. Partnership nesting of valley partridge and long-tailed chat. The Auk, vol. 14, no. 4, p. 404, Oct.

WIDMANN, OTTO.
1907. A preliminary catalog of the birds of Missouri, 288 pp. St. Louis.

WILLARD, FRANCIS COTTLE.
1916. Nesting of the band-tailed pigeon in southern Arizona. The Condor, vol. 18, no. 3, pp. 110–112 illus., May–June.

WILLETT, GEORGE.
1914. Birds of Sitka and vicinity, southeastern Alaska. The Condor, vol. 16, no. 2, pp. 71–91, illus., Mar.–Apr.

WILLIAMS, JOHN J.
1903. On the use of sentinels by valley quail. The Condor, vol. 5, no. 6, pp. 146–148, illus., Nov.–Dec.

WILSON ALEXANDER.
1832. American ornithology, vol. 1, 408 pp., illus. London.

WITHERBY, HARRY FORBES, and others.
1919–1920. A practical handbook of British birds, vol. 1, 532 pp., 17 pls. (part col.). London.

WOOD, WILLIAM.
1635. New England prospect, Chap. 8, pp. 22–27. London.

WRIGHT, ALBERT HAZEN.
1914–1915. Early records of the wild turkey. The Auk, vol. 31, nos. 3, 4, pp. 334–358, and 463–473, July, Oct.; vol. 32, nos. 1, 2, pp. 61–81, and 207–224, Jan., Apr.

WYMAN, LUTHER EVERET.
1921. A very late record of the passenger pigeon (*Ectopistes migratorius*). The Auk, vol. 38, no. 2, p. 274, Apr.

YARRELL, WILLIAM.
1871–1885. A history of British birds, ed. 4, 4 vols., illus., revised and enlarged by Alfred Newton and Howard Saunders. London.

INDEX

THE LAST HEATH HEN

Marthas Vineyard, March 31, 1930. Presented by Dr. Alfred O. Gross.

PLATES

NESTING SITE AND NEST OF BOBWHITE

Berkley, Mass., June 28, 1915; both photographs by the author. Referred to on page 13.

MALE BOBWHITE ON NEST

On the grounds of the Portland Golf Club, Oregon, 1915. Presented by Mr. and Mrs. William L. Finley.

NEST OF BOBWHITE

St. Louis, Mo., June, 1920. Presented by Dr. Frank N. Wilson.

FEMALE BOBWHITE LEAVING NEST
Presented by Dr. Frank N. Wilson.

NEST OF FLORIDA BOBWHITE

Orlando, Fla., April 25, 1925. Presented by Jack H. Connery.

NESTING SITE AND NEST OF PLUMED QUAIL

Under a clump of lupine, at the base of a small pine, Pinehurst, Oreg., May, 1923. Presented by J. E. Patterson.

NEST OF MOUNTAIN QUAIL

Jennings Lodge, Oreg., May 7, 1908. Presented by Mr. and Mrs. William L. Finley.

DOWNY YOUNG MOUNTAIN QUAIL HIDING
Sodaville, Oreg. Presented by Leslie L. Haskin.

IMMATURE MOUNTAIN QUAIL
Jennings Lodge, Oreg. Presented by Mr. and Mrs. William L. Finley.

DOWNY YOUNG SCALED QUAIL
Presented by Miss Angeline M. Keen

NEST OF ARIZONA SCALED QUAIL
Cochise County, Ariz., May 25, 1922. Photograph by the author. Referred to on page 53.

VALLEY QUAIL FEEDING

In yard of J. Eugene Law, Altadena, Calif., March 24, 1929. Photograph by the author. Referred to on page 62.

FEMALE VALLEY QUAIL

On nest in unusually open situation, near Claremont, Calif. Presented by Wright M. Pierce.

NEST OF VALLEY QUAIL

Near Claremont, Calif., May 26, 1916. Presented by Wright M. Pierce.

TWO VIEWS OF VALLEY QUAIL AND HER NEST
Ventura County, Calif. Presented by Donald R. Dickey.

MALE ADULT CALIFORNIA QUAIL

DOWNY YOUNG CALIFORNIA QUAIL

Both photographs taken at Jennings Lodge, Oreg., and presented by Mr. and Mrs. William L. Finley.

NESTING SITE AND NEST OF GAMBEL'S QUAIL

Pima County, Ariz. Presented by Frank C. Willard.

HAUNTS AND NESTING SITE OF MEARNS'S QUAIL
Huachuca Mountains, Ariz. Presented by Frank C. Willard.

HUNGARIAN PARTRIDGE
Corvallis, Oreg. Presented by Olaus J. Murie.

NEST OF HUNGARIAN PARTRIDGE
Waukesha County, Wis., July, 1927. Presented by Dr. Alvin R. Cahn.

FEMALE DUSKY GROUSE
Little Gros Ventre River, Wyo., July 7, 1904. Presented by E. R. Warren.

NEST OF DUSKY GROUSE UNDER SAGEBRUSH
Gunnison County, Colo., June 21, 1900. Presented by E. R. Warren.

TWO VIEWS OF COURTSHIP DISPLAY OF RICHARDSON'S GROUSE

From drawings loaned by Maj. Allan Brooks.

NESTING SITE OF SOOTY GROUSE

Near Dayton, Oreg., May 15, 1905. Presented by William L. Finley and H. T. Bohlman.

NEST OF SOOTY GROUSE

Mulino, Oreg., April 15, 1914. Presented by Alexander Walker.

FEMALE SOOTY GROUSE
Presented by Mr. and Mrs. William L. Finley.

NEST OF SOOTY GROUSE
Sodaville, Oreg., April 22, 1919. Presented by Leslie L. Haskin.

FEMALE CANADA SPRUCE GROUSE ON NEST
New Brunswick, Canada. Presented by P. B. Philipp.

FEMALE CANADA SPRUCE GROUSE ON NEST
Presented by L. W. Brownell.

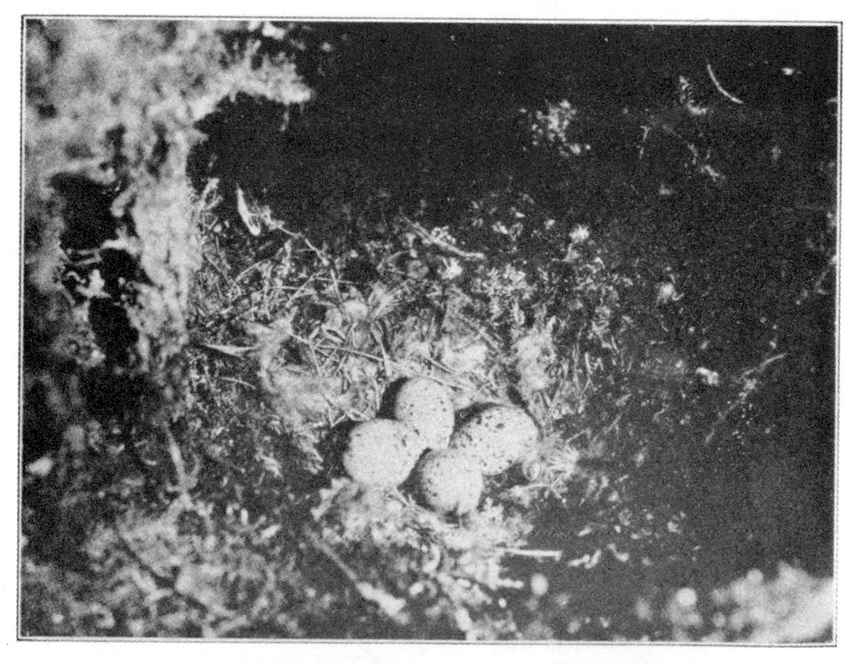

HUDSONIAN SPRUCE GROUSE AND ITS NEST

Belvedere, Alberta, June 13, 1923. Both photographs presented by J. Fletcher Street.

MALE FRANKLIN'S GROUSE
Presented by L. W. Brownell.

FEMALE FRANKLIN'S GROUSE
Glacier National Park, August, 1928. Presented by William L. Finley.

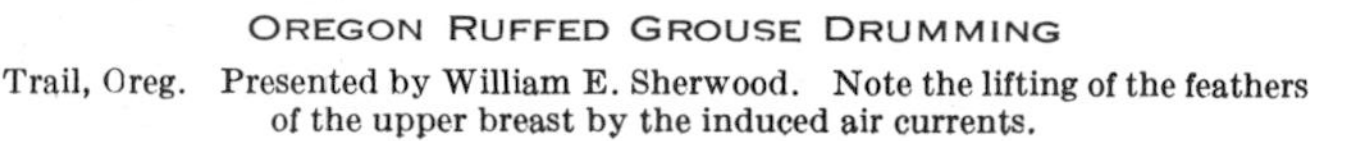
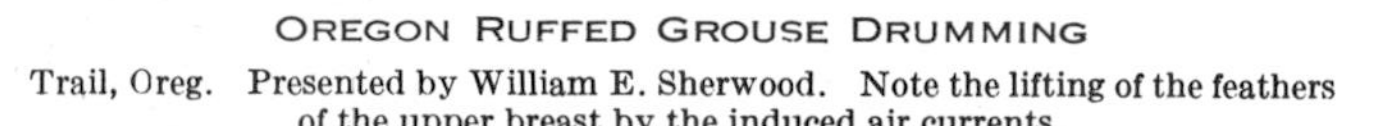

OREGON RUFFED GROUSE DRUMMING

Trail, Oreg. Presented by William E. Sherwood. Note the lifting of the feathers of the upper breast by the induced air currents.

RUFFED GROUSE DRUMMING

A pause between the first slow beats. Presented by W. J. Breckenridge and reproduced by courtesy of the Museum of Natural History, University of Minnesota.

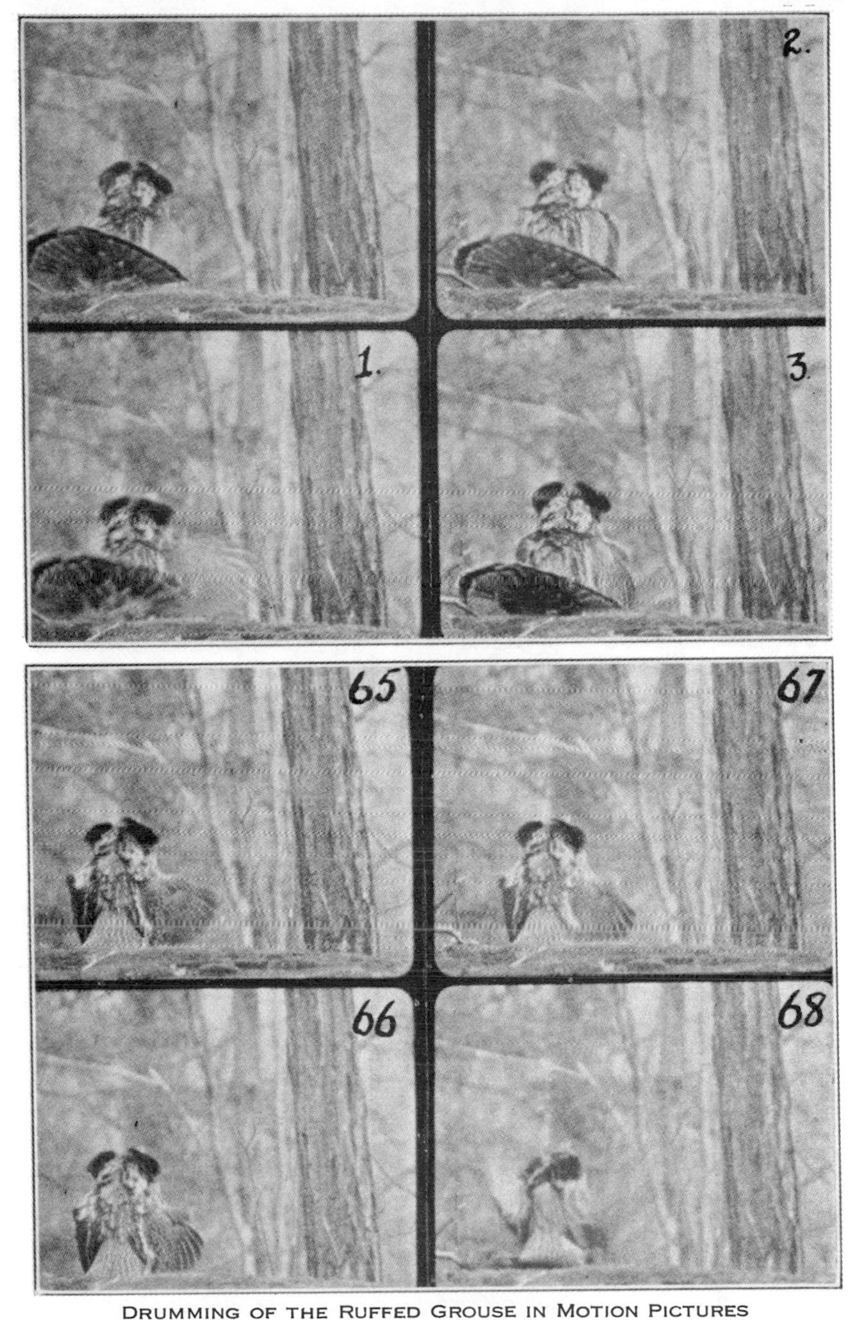

DRUMMING OF THE RUFFED GROUSE IN MOTION PICTURES

Presented by Dr. Arthur A. Allen. Frame 1 shows the first thump; the stroke is forward and downward. Frame 68 shows the eleventh thump, about the middle of the performance; note that the stroke is forward and upward. Referred to on page 145.

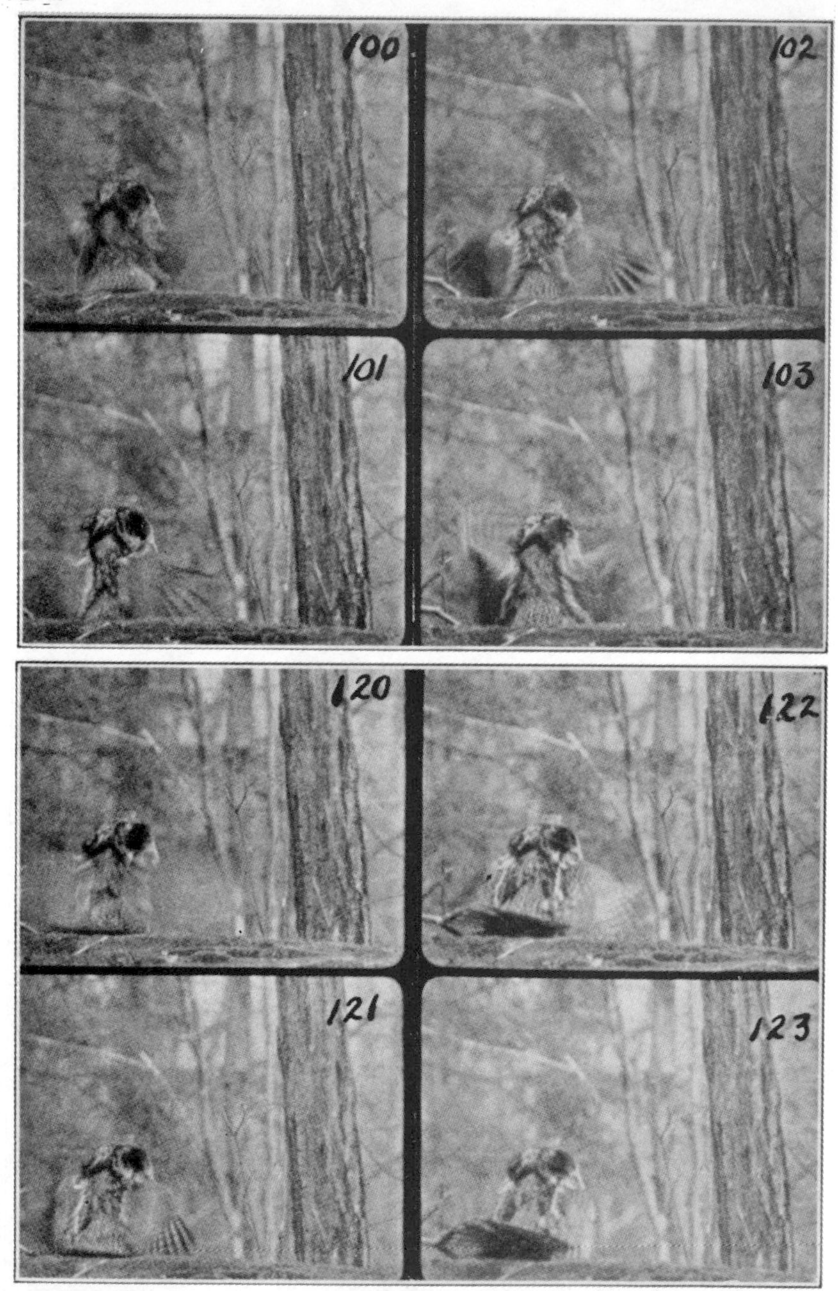

DRUMMING OF THE RUFFED GROUSE—CONTINUED

Presented by Dr. Arthur A. Allen. Frames 100 to 103 show the beginning of the terminal roll, where the wings are blurred. Note the end of the roll, in frame 120, followed by a forward and downward stroke, in frame 122, to help the bird regain its balance, as it pitches forward and the tail springs upward in frames 122 and 123. Referred to on page 145.

FEMALE RUFFED GROUSE ON NEST
Michigan, July 2, 1927. Presented by Dr. Frank N. Wilson.

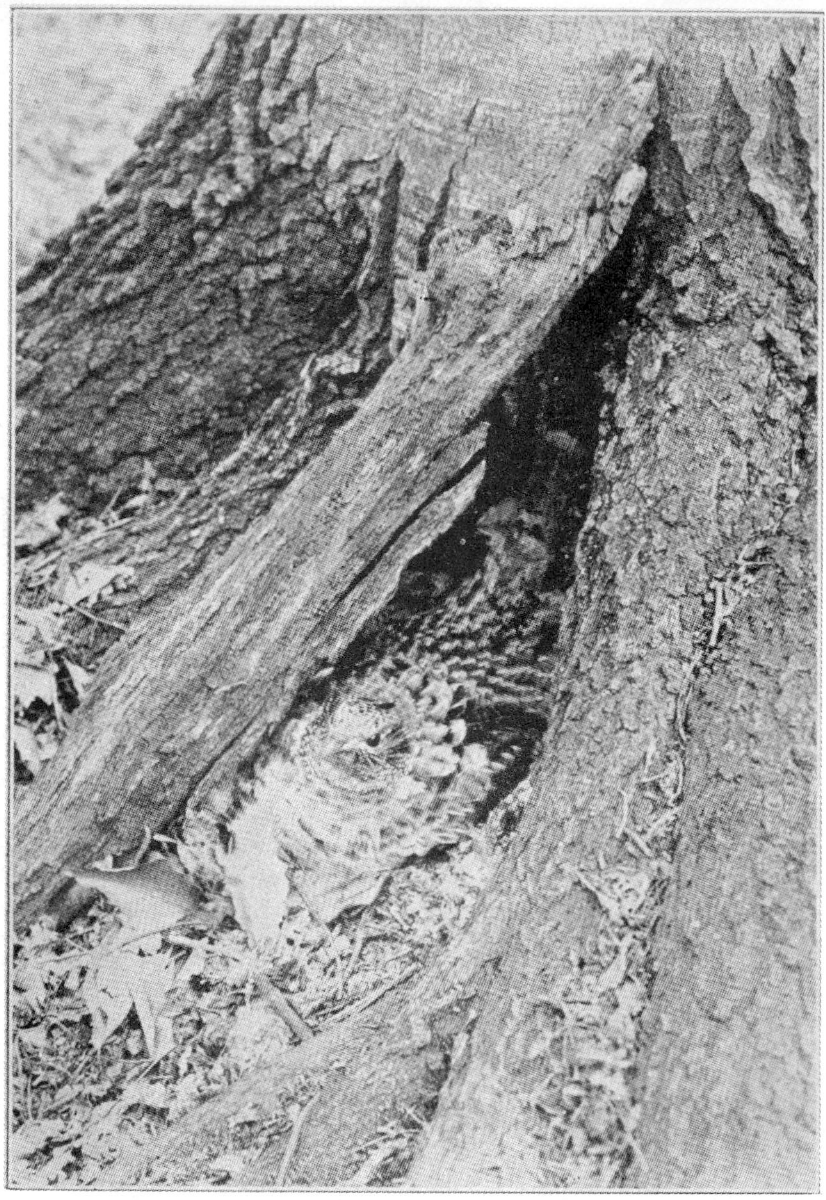

FEMALE RUFFED GROUSE ON NEST
Oakland County, Mich. Presented by Walter E. Hastings.

RUFFED-GROUSE NEST IN PINE SLASH
Berkley, Mass., May 21, 1903.

RUFFED-GROUSE NEST IN OAK WOODS
Rehoboth, Mass., May 18, 1921. Both photographs by the author. Referred to on page 148.

RUFFED-GROUSE CHICK 2 DAYS OLD

RUFFED-GROUSE CHICK 5 DAYS OLD

Both photographs presented by Dr. Arthur A. Allen.

YOUNG RUFFED GROUSE 23 DAYS OLD

YOUNG RUFFED GROUSE 35 DAYS OLD, IN NEARLY COMPLETE JUVENAL PLUMAGE

Both photographs presented by Dr. Arthur A. Allen.

RUFFED-GROUSE CHICK 11 DAYS OLD

RUFFED-GROUSE CHICK 17 DAYS OLD

Both photographs presented by Dr. Arthur A. Allen.

RUFFED-GROUSE CHICKS HIDING

Grand Traverse County, Mich., June 12, 1927. Presented by Walter E. Hastings.

IMMATURE RUFFED GROUSE 130 DAYS OLD, IN FIRST WINTER PLUMAGE

Presented by Dr. Arthur A. Allen.

RUFFED GROUSE BEGINNING TO STRUT

Presented by W. J. Breckenridge and reproduced by courtesy of the Museum of Natural History, University of Minnesota.

RUFFED GROUSE IN FULL DISPLAY

Presented by Dr. Arthur A. Allen.

NEST OF OREGON RUFFED GROUSE

Mulino, Oreg., May 4, 1913. Presented by Alexander Walker.

NESTING SITE OF WILLOW PTARMIGAN

Nome, Alaska, June 21, 1921. Presented by Alfred M. Bailey, by courtesy of the Colorado Museum of Natural History.

PARTIALLY CONCEALED NEST OF WILLOW PTARMIGAN
St. Michael, Alaska, June 8, 1914.

OPEN NEST OF WILLOW PTARMIGAN
Yukon Delta, Alaska, June 14, 1914. Both photographed by F. Seymour Hersey for the author.

WILLOW-PTARMIGAN CHICK
Yukon Delta, Alaska, June 23, 1914.

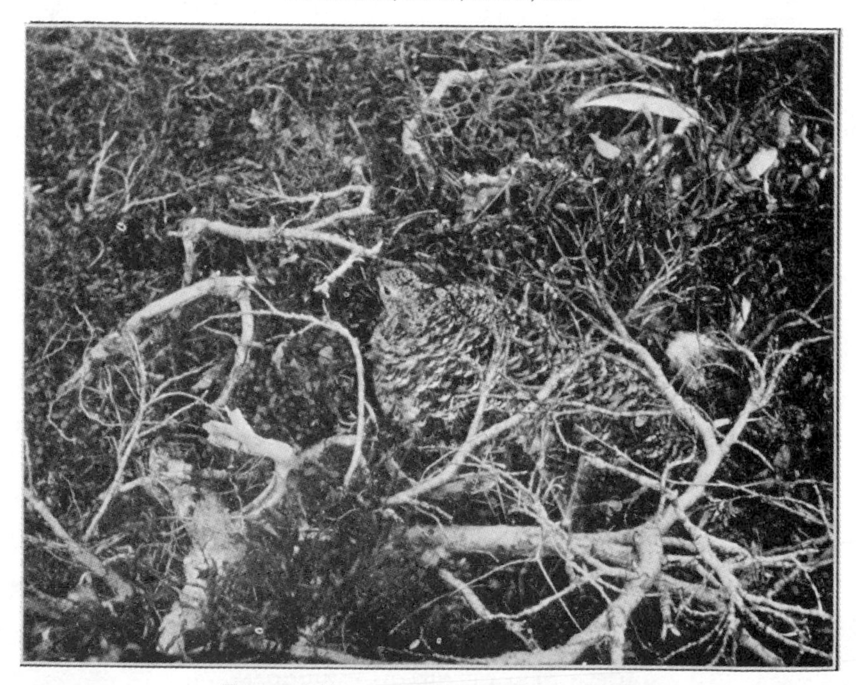

FEMALE WILLOW PTARMIGAN ON NEST
Yukon Delta, Alaska, June 16, 1914. Both photographed by F. Seymour Hersey for the author.

FEMALE WILLOW PTARMIGAN BROODING YOUNG

Presented by Joseph Dixon, by courtesy of the Museum of Vertebrate Zoology, University of California.

MALE WILLOW PTARMIGAN IN LATE-SUMMER PLUMAGE

Mount McKinley National Park, Alaska, September, 1926. Presented by Mr. and Mrs. William L. Finley.

MALE WILLOW PTARMIGAN IN COURTSHIP DISPLAY

Savage River, Alaska, May 24, 1926

MALE WILLOW PTARMIGAN IN NUPTIAL PLUMAGE

Savage River, Alaska, May 25, 1926. Both photographs by Joseph Dixon; presented by courtesy of the Museum of Vertebrate Zoology, University of California.

NESTING SITE OF ROCK PTARMIGAN

Anvil Mountain, Nome, Alaska, June 20, 1922. Presented by Alfred M. Bailey, by courtesy of the Colorado Museum of Natural History.

IMMATURE ROCK PTARMIGAN
Savage River, Alaska, September 6, 1922

YOUNG ROCK PTARMIGAN HIDING
Igiak Bay, Alaska, July 11, 1924. Both photographs by Olaus J. Murie; presented by the United States Biological Survey.

TWO VIEWS OF ROCK PTARMIGAN, CHANGING INTO WINTER PLUMAGE

Mount McKinley National Park, Alaska, September, 1926. Presented by Mr. and Mrs. William L. Finley. Note the concealing coloration in the upper figure.

NESTING SITE AND NEST OF CHAMBERLAIN'S PTARMIGAN

Adak Island, Alaska, June 27, 1911. Photographs by the author. Referred to on page 222.

NESTING SITE AND NEST OF SANFORD'S PTARMIGAN

Tanaga Island, Alaska, June 25, 1911. Photographs by Rollo H. Beck. Referred to on page 225.

FEMALE SOUTHERN WHITE-TAILED PTARMIGAN ON NEST

Idaho Springs, Colo. Photograph taken by Evan Lewis and presented by E. R. Warren. Note concealing coloration.

NEST OF SOUTHERN WHITE-TAILED PTARMIGAN

Presented by the Colorado Museum of Natural History.

SOUTHERN WHITE-TAILED PTARMIGAN CHICK
Gunnison County, Colo., July 8, 1902. Presented by E. R. Warren.

LARGER CHICK OF WASHINGTON WHITE-TAILED PTARMIGAN
Mount Rainier, Wash., August 2, 1929. Presented by Dr. Gayle B. Pickwell.

SOUTHERN WHITE-TAILED PTARMIGAN IN WINTER PLUMAGE

March 13, 1902.

SOUTHERN WHITE-TAILED PTARMIGAN BEGINNING TO ACQUIRE NUPTIAL PLUMAGE

May 30, 1900. Both photographs taken by E. R. Warren in Gunnison County, Colo., and presented by him.

FEMALE SOUTHERN WHITE-TAILED PTARMIGAN BROODING CHICKS
Gunnison County, Colo., July 11, 1927.

WHITE-TAILED PTARMIGAN
Glacier National Park, Mont., June 18, 1927. Both photographs presented by E. R. Warren.

WASHINGTON WHITE-TAILED PTARMIGAN
Mount Rainier, Wash., August, 1919.

NEST OF WASHINGTON WHITE-TAILED PTARMIGAN
Mount Rainier, Wash., July 11, 1919. Both photographs presented by Mr. and Mrs. William L. Finley.

FEMALE WASHINGTON WHITE-TAILED PTARMIGAN AND CHICKS BROWSING ON MOSS

Mount Rainier, Wash., August 2, 1929. Presented by Dr. Gayle B. Pickwell.

Typical Prairie-chicken Habitat in Wisconsin
Presented by Dr. A. O. Gross, by courtesy of the Wisconsin Conservation Commission.

PRAIRIE CHICKEN APPROACHING HER NEST
June 7, 1929.

PRAIRIE CHICKEN ON HER NEST
June 5, 1929. Both photographs presented by Dr. A. O. Gross, by courtesy of the Wisconsin Conservation Commission.

TWO VIEWS OF MALE PRAIRIE CHICKENS FIGHTING

Photographs taken by Walter W. Bennett in northwestern Nebraska and presented by him.

TYPICAL GRASS NEST OF PRAIRIE CHICKEN

Bismarck, N. Dak., May 24, 1925. Presented by Russell Reid.

BRUSH-LAND NEST OF PRAIRIE CHICKEN

Kittson County, Minn., in timber on the outskirts of a village. Presented by the Rev. P. B. Peabody.

Downy Young Prairie Chickens in Nest

Presented by Dr. Alfred O. Gross, by courtesy of the Wisconsin Conservation Commission.

HABITAT OF THE HEATH HEN, MARTHAS VINEYARD, MASS.

June 24, 1923. Presented by Dr. A. O. Gross. Referred to on page 265.

HEATH HEN IN INITIAL STAGE OF BOOMING
April 12, 1923.

HEATH HEN AT CLIMAX OF BOOMING PERFORMANCE
March 29, 1927. Both photographs taken by Dr. A. O. Gross, Marthas Vineyard, Mass., and presented by him. Referred to on page 271.

NESTING SITE AND NEST OF THE HEATH HEN

FEMALE HEATH HEN ON NEST

Both photographs taken by Dr. George W. Field, Marthas Vineyard, Mass., and presented by the Massachusetts Division of Fisheries and Game.

MALE LESSER PRAIRIE CHICKEN
October, 1925.

NEST OF LESSER PRAIRIE CHICKEN
June 2, 1920. Referred to on page 281. Both photographs taken by Walter Colvin in Seward County, Kans., and presented by him.

HABITAT OF PRAIRIE SHARP-TAILED GROUSE

Babcock region, Wisconsin, July 29, 1930. Presented by Dr. A. O. Gross, by courtesy of the Wisconsin Conservation Commission.

GRASS NEST OF PRAIRIE SHARP-TAILED GROUSE

Near Lake Winnipegosis, Manitoba, June 2, 1913. Photographed by the author.

PRAIRIE SHARP-TAILED GROUSE ON HER NEST

BRUSH NEST OF PRAIRIE SHARP-TAILED GROUSE

Both photographs taken by the author near Crane Lake, Saskatchewan, June 6, 1905. Referred to on page 293.

TWO VIEWS OF PRAIRIE SHARP-TAILED GROUSE IN WINTER
Photographs by H. H. Pittman.

THREE VIEWS OF PRAIRIE SHARP-TAILED GROUSE IN FIGHTING ATTITUDES DURING COURTSHIP

Presented by W. J. Breckenridge, by courtesy of the Museum of Natural History, University of Minnesota.

HABITAT OF SAGE HEN, A TYPICAL SAGEBRUSH PLAIN

Weston County, Wyo. Presented by the Rev. P. B. Peabody.

FRONT AND SIDE VIEWS OF COURTSHIP DISPLAY OF SAGE HEN
Near Lower Klamath Lake, Calif., May 12, 1917. Presented by William L. Finley.

NEST OF SAGE HEN

Near Spencer, Idaho. Presented by Henry J. Rust.

SAGE-HEN CHICK

YOUNG SAGE HEN HIDING
Both photographs presented by the Rev. P. B. Peabody.

NEST AND NESTING SITE OF RING-NECKED PHEASANT

Raynham, Mass., May 16, 1920. Photographed by the author.

NEST OF RING-NECKED PHEASANT

Near Dayton, Oreg., May 15, 1905. Presented by William L. Finley and H. T. Bohlman.

RING-NECKED PHEASANT ON HER NEST

Torresdale, Pa., May 12, 1930. Presented by R. L. Coffin.

RING-NECKED PHEASANT CHICK
Dayton, Oreg., May 15, 1905. Presented by William L. Finley and H. T. Bohlman.

MALE RING-NECKED PHEASANT
Corvallis, Oreg. Presented by O. J. Murie.

NEST OF WILD TURKEY

Near Falls Church, Va., May 10, 1903. Presented by Dr. Paul Bartsch.

WILD TURKEY ON HER NEST

Clearfield County, Pa., May 5, 1928. Presented by Dr. George M. Sutton.

PORTRAIT OF A WILD TURKEY
Presented by L. W. Brownell.

WILD TURKEYS IN WINTER
Presented by Norman McClintock.

NEST OF MERRIAM'S TURKEY

At foot of a live-oak on a hillside, Santa Rita Mountains, Ariz., June 15, 1884.
Presented by Frank Stephens.

NEST AND EGGS OF MERRIAM'S TURKEY

Huachuca Mountains, Ariz., July 1, 1900. Photograph taken by O. W. Howard.

FLOCK OF BAND-TAILED PIGEONS IN CHARACTERISTIC RESTING POSE
Presented by Mrs. W. Leon Dawson.

BAND-TAILED PIGEONS
Corvallis, Oreg. Presented by O. J. Murie.

BAND-TAILED PIGEON PORTRAIT
Presented by Joseph H. Wales.

NEST AND EGG OF BAND-TAILED PIGEON

In A. M. Ingersoll's collection. Photograph presented by him.

NEST OF BAND-TAILED PIGEON

Huachuca Mountains, Ariz. Presented by Frank C. Willard.

BAND-TAILED PIGEON ABOUT 10 DAYS OLD

San Diego County, Calif., September 17, 1922. Presented by Clinton G. Abbott.

NESTING SITE AND NEST OF WHITE-CROWNED PIGEON

San Salvador, August 20, 1923. Presented by Dr. Paul Bartsch.

THE LAST LIVING PASSENGER PIGEON IN THE CINCINNATI ZOO

Photograph taken and presented by Dr. William C. Herman.

HABITAT GROUP OF PASSENGER PIGEONS

In Museum of Natural History, University of Minnesota. Genuine nest and egg collected by Dr. Thomas S. Roberts in 1874. Photograph presented by Doctor Roberts and the museum.

YOUNG MOURNING DOVES IN THEIR NEST

NEST AND EGGS OF MOURNING DOVE

Both photographs taken near Buffalo, N. Y., May 8, 1927, and presented by S. A. Grimes.

NEST OF MOURNING DOVE IN OLD HAWK'S NEST
Middleboro, Mass., April 28, 1907. Referred to on page 406.

NEST OF MOURNING DOVE ON THE GROUND
Near Stump Lake, N. Dak., June 5, 1901. Both photographs by the author.

MOURNING DOVE ON HER NEST ON LOW STUMP
Rehoboth, Mass., May 29, 1920. Photograph by the author.

NEST OF WESTERN MOURNING DOVE UNDER FALLEN TREE
Near Brownsville, Oreg. Presented by Mrs. Lillian G. Haskin.

NEST OF WESTERN MOURNING DOVE

In small juniper on an open plateau, Clear Lake, Calif. Presented by J. E. Patterson.

NEST OF WESTERN MOURNING DOVE

In an old great-tailed grackle's nest in Texas. Presented by George F. Simmons.

WESTERN MOURNING DOVE BROODING HER YOUNG ON NEST IN PEPPERTREE

Ventura County, Calif., June 6, 1912. Presented by Donald R. Dickey.

IMMATURE WESTERN WHITE-WINGED DOVES

YOUNG WESTERN WHITE-WINGED DOVES IN NEST

Both photographs taken near Tucson, Ariz., and presented by Mr. and Mrs. William L. Finley.

NEST OF WESTERN WHITE-WINGED DOVE

Near Tucson, Ariz. Presented by Mr. and Mrs. William L. Finley.

MALE GROUND DOVE AT FEEDING STATION

FEMALE GROUND DOVE ENTERING NEST ON THE GROUND UNDER PALMETTO

Both photographs taken at St. Cloud, Fla., and presented by Dr. Arthur A. Allen.

Female Mexican Ground Dove on Nest
Near Tucson, Ariz. Presented by Mr. and Mrs. William L. Finley.

NEST OF MEXICAN GROUND DOVE

Fairbank, Ariz., May 17, 1922. Photograph by the author. Referred to on page 441.

NEST OF MEXICAN GROUND DOVE

Near Tucson, Ariz. Presented by Mr. and Mrs. William L. Finley.

A CATALOG OF SELECTED DOVER BOOKS IN ALL FIELDS OF INTEREST

THE MURDER BOOK OF J.G. REEDER, Edgar Wallace. Eight suspenseful stories by bestselling mystery writer of 20s and 30s. Features the donnish Mr. J.G. Reeder of Public Prosecutor's Office. 128pp. 5⅜ × 8½. (Available in U.S. only) 24374-5 Pa. $3.50

ANNE ORR'S CHARTED DESIGNS, Anne Orr. Best designs by premier needlework designer, all on charts: flowers, borders, birds, children, alphabets, etc. Over 100 charts, 10 in color. Total of 40pp. 8¼ × 11. 23704-4 Pa. $2.25

BASIC CONSTRUCTION TECHNIQUES FOR HOUSES AND SMALL BUILDINGS SIMPLY EXPLAINED, U.S. Bureau of Naval Personnel. Grading, masonry, woodworking, floor and wall framing, roof framing, plastering, tile setting, much more. Over 675 illustrations. 568pp. 6½ × 9¼. 20242-9 Pa. $8.95

MATISSE LINE DRAWINGS AND PRINTS, Henri Matisse. Representative collection of female nudes, faces, still lifes, experimental works, etc., from 1898 to 1948. 50 illustrations. 48pp. 8⅜ × 11¼. 23877-6 Pa. $2.50

HOW TO PLAY THE CHESS OPENINGS, Eugene Znosko-Borovsky. Clear, profound examinations of just what each opening is intended to do and how opponent can counter. Many sample games. 147pp. 5⅜ × 8½. 22795-2 Pa. $2.95

DUPLICATE BRIDGE, Alfred Sheinwold. Clear, thorough, easily followed account: rules, etiquette, scoring, strategy, bidding; Goren's point-count system, Blackwood and Gerber conventions, etc. 158pp. 5⅜ × 8½. 22741-3 Pa. $3.00

SARGENT PORTRAIT DRAWINGS, J.S. Sargent. Collection of 42 portraits reveals technical skill and intuitive eye of noted American portrait painter, John Singer Sargent. 48pp. 8¼ × 11⅛. 24524-1 Pa. $2.95

ENTERTAINING SCIENCE EXPERIMENTS WITH EVERYDAY OBJECTS, Martin Gardner. Over 100 experiments for youngsters. Will amuse, astonish, teach, and entertain. Over 100 illustrations. 127pp. 5⅜ × 8½. 24201-3 Pa. $2.50

TEDDY BEAR PAPER DOLLS IN FULL COLOR: A Family of Four Bears and Their Costumes, Crystal Collins. A family of four Teddy Bear paper dolls and nearly 60 cut-out costumes. Full color, printed one side only. 32pp. 9¼ × 12¼. 24550-0 Pa. $3.50

NEW CALLIGRAPHIC ORNAMENTS AND FLOURISHES, Arthur Baker. Unusual, multi-useable material: arrows, pointing hands, brackets and frames, ovals, swirls, birds, etc. Nearly 700 illustrations. 80pp. 8⅜ × 11¼. 24095-9 Pa. $3.50

DINOSAUR DIORAMAS TO CUT & ASSEMBLE, M. Kalmenoff. Two complete three-dimensional scenes in full color, with 31 cut-out animals and plants. Excellent educational toy for youngsters. Instructions; 2 assembly diagrams. 32pp. 9¼ × 12¼. 24541-1 Pa. $3.95

SILHOUETTES: A PICTORIAL ARCHIVE OF VARIED ILLUSTRATIONS, edited by Carol Belanger Grafton. Over 600 silhouettes from the 18th to 20th centuries. Profiles and full figures of men, women, children, birds, animals, groups and scenes, nature, ships, an alphabet. 144pp. 8⅜ × 11¼. 23781-8 Pa. $4.50

25 KITES THAT FLY, Leslie Hunt. Full, easy-to-follow instructions for kites made from inexpensive materials. Many novelties. 70 illustrations. 110pp. 5⅜ × 8½. 22550-X Pa. $1.95

PIANO TUNING, J. Cree Fischer. Clearest, best book for beginner, amateur. Simple repairs, raising dropped notes, tuning by easy method of flattened fifths. No previous skills needed. 4 illustrations. 201pp. 5⅜ × 8½. 23267-0 Pa. $3.50

EARLY AMERICAN IRON-ON TRANSFER PATTERNS, edited by Rita Weiss. 75 designs, borders, alphabets, from traditional American sources. 48pp. 8¼ × 11. 23162-3 Pa. $1.95

CROCHETING EDGINGS, edited by Rita Weiss. Over 100 of the best designs for these lovely trims for a host of household items. Complete instructions, illustrations. 48pp. 8¼ × 11. 24031-2 Pa. $2.00

FINGER PLAYS FOR NURSERY AND KINDERGARTEN, Emilie Poulsson. 18 finger plays with music (voice and piano); entertaining, instructive. Counting, nature lore, etc. Victorian classic. 53 illustrations. 80pp. 6½ × 9¼. 22588-7 Pa. $1.95

BOSTON THEN AND NOW, Peter Vanderwarker. Here in 59 side-by-side views are photographic documentations of the city's past and present. 119 photographs. Full captions. 122pp. 8¼ × 11. 24312-5 Pa. $6.95

CROCHETING BEDSPREADS, edited by Rita Weiss. 22 patterns, originally published in three instruction books 1939-41. 39 photos, 8 charts. Instructions. 48pp. 8¼ × 11. 23610-2 Pa. $2.00

HAWTHORNE ON PAINTING, Charles W. Hawthorne. Collected from notes taken by students at famous Cape Cod School; hundreds of direct, personal *apercus*, ideas, suggestions. 91pp. 5⅜ × 8½. 20653-X Pa. $2.50

THERMODYNAMICS, Enrico Fermi. A classic of modern science. Clear, organized treatment of systems, first and second laws, entropy, thermodynamic potentials, etc. Calculus required. 160pp. 5⅜ × 8½. 60361-X Pa. $4.00

TEN BOOKS ON ARCHITECTURE, Vitruvius. The most important book ever written on architecture. Early Roman aesthetics, technology, classical orders, site selection, all other aspects. Morgan translation. 331pp. 5⅜ × 8½. 20645-9 Pa. $5.50

THE CORNELL BREAD BOOK, Clive M. McCay and Jeanette B. McCay. Famed high-protein recipe incorporated into breads, rolls, buns, coffee cakes, pizza, pie crusts, more. Nearly 50 illustrations. 48pp. 8¼ × 11. 23995-0 Pa. $2.00

THE CRAFTSMAN'S HANDBOOK, Cennino Cennini. 15th-century handbook, school of Giotto, explains applying gold, silver leaf; gesso; fresco painting, grinding pigments, etc. 142pp. 6⅛ × 9¼. 20054-X Pa. $3.50

FRANK LLOYD WRIGHT'S FALLINGWATER, Donald Hoffmann. Full story of Wright's masterwork at Bear Run, Pa. 100 photographs of site, construction, and details of completed structure. 112pp. 9¼ × 10. 23671-4 Pa. $6.50

OVAL STAINED GLASS PATTERN BOOK, C. Eaton. 60 new designs framed in shape of an oval. Greater complexity, challenge with sinuous cats, birds, mandalas framed in antique shape. 64pp. 8¼ × 11. 24519-5 Pa. $3.50

THE BOOK OF WOOD CARVING, Charles Marshall Sayers. Still finest book for beginning student. Fundamentals, technique; gives 34 designs, over 34 projects for panels, bookends, mirrors, etc. 33 photos. 118pp. 7¾ × 10⅝. 23654-4 Pa. $3.95

CARVING COUNTRY CHARACTERS, Bill Higginbotham. Expert advice for beginning, advanced carvers on materials, techniques for creating 18 projects—mirthful panorama of American characters. 105 illustrations. 80pp. 8⅜ × 11. 24135-1 Pa. $2.50

300 ART NOUVEAU DESIGNS AND MOTIFS IN FULL COLOR, C.B. Grafton. 44 full-page plates display swirling lines and muted colors typical of Art Nouveau. Borders, frames, panels, cartouches, dingbats, etc. 48pp. 9⅜ × 12¼. 24354-0 Pa. $6.00

SELF-WORKING CARD TRICKS, Karl Fulves. Editor of *Pallbearer* offers 72 tricks that work automatically through nature of card deck. No sleight of hand needed. Often spectacular. 42 illustrations. 113pp. 5⅜ × 8½. 23334-0 Pa. $2.25

CUT AND ASSEMBLE A WESTERN FRONTIER TOWN, Edmund V. Gillon, Jr. Ten authentic full-color buildings on heavy cardboard stock in H-O scale. Sheriff's Office and Jail, Saloon, Wells Fargo, Opera House, others. 48pp. 9¼ × 12¼. 23736-2 Pa. $3.95

CUT AND ASSEMBLE AN EARLY NEW ENGLAND VILLAGE, Edmund V. Gillon, Jr. Printed in full color on heavy cardboard stock. 12 authentic buildings in H-O scale: Adams home in Quincy, Mass., Oliver Wight house in Sturbridge, smithy, store, church, others. 48pp. 9¼ × 12¼. 23536-X Pa. $3.95

THE TALE OF TWO BAD MICE, Beatrix Potter. Tom Thumb and Hunca Munca squeeze out of their hole and go exploring. 27 full-color Potter illustrations. 59pp. 4¼ × 5½. (Available in U.S. only) 23065-1 Pa. $1.50

CARVING FIGURE CARICATURES IN THE OZARK STYLE, Harold L. Enlow. Instructions and illustrations for ten delightful projects, plus general carving instructions. 22 drawings and 47 photographs altogether. 39pp. 8⅜ × 11. 23151-8 Pa. $2.50

A TREASURY OF FLOWER DESIGNS FOR ARTISTS, EMBROIDERERS AND CRAFTSMEN, Susan Gaber. 100 garden favorites lushly rendered by artist for artists, craftsmen, needleworkers. Many form frames, borders. 80pp. 8¼ × 11. 24096-7 Pa. $3.50

CUT & ASSEMBLE A TOY THEATER/THE NUTCRACKER BALLET, Tom Tierney. Model of a complete, full-color production of Tchaikovsky's classic. 6 backdrops, dozens of characters, familiar dance sequences. 32pp. 9⅜ × 12¼. 24194-7 Pa. $4.50

ANIMALS: 1,419 COPYRIGHT-FREE ILLUSTRATIONS OF MAMMALS, BIRDS, FISH, INSECTS, ETC., edited by Jim Harter. Clear wood engravings present, in extremely lifelike poses, over 1,000 species of animals. 284pp. 9 × 12. 23766-4 Pa. $8.95

MORE HAND SHADOWS, Henry Bursill. For those at their 'finger ends," 16 more effects—Shakespeare, a hare, a squirrel, Mr. Punch, and twelve more—each explained by a full-page illustration. Considerable period charm. 30pp. 6½ × 9¼. 21384-6 Pa. $1.95

SURREAL STICKERS AND UNREAL STAMPS, William Rowe. 224 haunting, hilarious stamps on gummed, perforated stock, with images of elephants, geisha girls, George Washington, etc. 16pp. one side. 8¼ × 11. 24371-0 Pa. $3.50

GOURMET KITCHEN LABELS, Ed Sibbett, Jr. 112 full-color labels (4 copies each of 28 designs). Fruit, bread, other culinary motifs. Gummed and perforated. 16pp. 8¼ × 11. 24087-8 Pa. $2.95

PATTERNS AND INSTRUCTIONS FOR CARVING AUTHENTIC BIRDS, H.D. Green. Detailed instructions, 27 diagrams, 85 photographs for carving 15 species of birds so life-like, they'll seem ready to fly! 8¼ × 11. 24222-6 Pa. $2.75

FLATLAND, E.A. Abbott. Science-fiction classic explores life of 2-D being in 3-D world. 16 illustrations. 103pp. 5⅜ × 8. 20001-9 Pa. $2.00

DRIED FLOWERS, Sarah Whitlock and Martha Rankin. Concise, clear, practical guide to dehydration, glycerinizing, pressing plant material, and more. Covers use of silica gel. 12 drawings. 32pp. 5⅜ × 8½. 21802-3 Pa. $1.00

EASY-TO-MAKE CANDLES, Gary V. Guy. Learn how easy it is to make all kinds of decorative candles. Step-by-step instructions. 82 illustrations. 48pp. 8¼ × 11. 23881-4 Pa. $2.50

SUPER STICKERS FOR KIDS, Carolyn Bracken. 128 gummed and perforated full-color stickers: GIRL WANTED, KEEP OUT, BORED OF EDUCATION, X-RATED, COMBAT ZONE, many others. 16pp. 8¼ × 11. 24092-4 Pa. $2.50

CUT AND COLOR PAPER MASKS, Michael Grater. Clowns, animals, funny faces...simply color them in, cut them out, and put them together, and you have 9 paper masks to play with and enjoy. 32pp. 8¼ × 11. 23171-2 Pa. $2.25

A CHRISTMAS CAROL: THE ORIGINAL MANUSCRIPT, Charles Dickens. Clear facsimile of Dickens manuscript, on facing pages with final printed text. 8 illustrations by John Leech, 4 in color on covers. 144pp. 8⅜ × 11¼. 20980-6 Pa. $5.95

CARVING SHOREBIRDS, Harry V. Shourds & Anthony Hillman. 16 full-size patterns (all double-page spreads) for 19 North American shorebirds with step-by-step instructions. 72pp. 9¼ × 12¼. 24287-0 Pa. $4.95

THE GENTLE ART OF MATHEMATICS, Dan Pedoe. Mathematical games, probability, the question of infinity, topology, how the laws of algebra work, problems of irrational numbers, and more. 42 figures. 143pp. 5⅜ × 8½. (EBE) 22949-1 Pa. $3.00

READY-TO-USE DOLLHOUSE WALLPAPER, Katzenbach & Warren, Inc. Stripe, 2 floral stripes, 2 allover florals, polka dot; all in full color. 4 sheets (350 sq. in.) of each, enough for average room. 48pp. 8¼ × 11. 23495-9 Pa. $2.95

MINIATURE IRON-ON TRANSFER PATTERNS FOR DOLLHOUSES, DOLLS, AND SMALL PROJECTS, Rita Weiss and Frank Fontana. Over 100 miniature patterns: rugs, bedspreads, quilts, chair seats, etc. In standard dollhouse size. 48pp. 8¼ × 11. 23741-9 Pa. $1.95

THE DINOSAUR COLORING BOOK, Anthony Rao. 45 renderings of dinosaurs, fossil birds, turtles, other creatures of Mesozoic Era. Scientifically accurate. Captions. 48pp. 8¼ × 11. 24022-3 Pa. $2.25

JAPANESE DESIGN MOTIFS, Matsuya Co. Mon, or heraldic designs. Over 4000 typical, beautiful designs: birds, animals, flowers, swords, fans, geometrics; all beautifully stylized. 213pp. 11⅜ × 8¼. 22874-6 Pa. $6.95

THE TALE OF BENJAMIN BUNNY, Beatrix Potter. Peter Rabbit's cousin coaxes him back into Mr. McGregor's garden for a whole new set of adventures. All 27 full-color illustrations. 59pp. 4¼ × 5½. (Available in U.S. only) 21102-9 Pa. $1.50

THE TALE OF PETER RABBIT AND OTHER FAVORITE STORIES BOXED SET, Beatrix Potter. Seven of Beatrix Potter's best-loved tales including Peter Rabbit in a specially designed, durable boxed set. 4¼ × 5½. Total of 447pp. 158 color illustrations. (Available in U.S. only) 23903-9 Pa. $10.50

PRACTICAL MENTAL MAGIC, Theodore Annemann. Nearly 200 astonishing feats of mental magic revealed in step-by-step detail. Complete advice on staging, patter, etc. Illustrated. 320pp. 5⅜ × 8½. 24426-1 Pa. $5.95

CELEBRATED CASES OF JUDGE DEE (DEE GOONG AN), translated by Robert Van Gulik. Authentic 18th-century Chinese detective novel; Dee and associates solve three interlocked cases. Led to van Gulik's own stories with same characters. Extensive introduction. 9 illustrations. 237pp. 5⅜ × 8½. 23337-5 Pa. $4.50

CUT & FOLD EXTRATERRESTRIAL INVADERS THAT FLY, M. Grater. Stage your own lilliputian space battles. By following the step-by-step instructions and explanatory diagrams you can launch 22 full-color fliers into space. 36pp. 8¼ × 11. 24478-4 Pa. $2.95

CUT & ASSEMBLE VICTORIAN HOUSES, Edmund V. Gillon, Jr. Printed in full color on heavy cardboard stock, 4 authentic Victorian houses in H-O scale: Italian-style Villa, Octagon, Second Empire, Stick Style. 48pp. 9¼ × 12¼. 23849-0 Pa. $3.95

BEST SCIENCE FICTION STORIES OF H.G. WELLS, H.G. Wells. Full novel *The Invisible Man*, plus 17 short stories: "The Crystal Egg," "Aepyornis Island," "The Strange Orchid," etc. 303pp. 5⅜ × 8½. (Available in U.S. only) 21531-8 Pa. $3.95

TRADEMARK DESIGNS OF THE WORLD, Yusaku Kamekura. A lavish collection of nearly 700 trademarks, the work of Wright, Loewy, Klee, Binder, hundreds of others. 160pp. 8¾ × 8. (Available in U.S. only) 24191-2 Pa. $5.00

THE ARTIST'S AND CRAFTSMAN'S GUIDE TO REDUCING, ENLARGING AND TRANSFERRING DESIGNS, Rita Weiss. Discover, reduce, enlarge, transfer designs from any objects to any craft project. 12pp. plus 16 sheets special graph paper. 8¼ × 11. 24142-4 Pa. $3.25

TREASURY OF JAPANESE DESIGNS AND MOTIFS FOR ARTISTS AND CRAFTSMEN, edited by Carol Belanger Grafton. Indispensable collection of 360 traditional Japanese designs and motifs redrawn in clean, crisp black-and-white, copyright-free illustrations. 96pp. 8¼ × 11. 24435-0 Pa. $3.95

CHANCERY CURSIVE STROKE BY STROKE, Arthur Baker. Instructions and illustrations for each stroke of each letter (upper and lower case) and numerals. 54 full-page plates. 64pp. 8¼ × 11. 24278-1 Pa. $2.50

THE ENJOYMENT AND USE OF COLOR, Walter Sargent. Color relationships, values, intensities; complementary colors, illumination, similar topics. Color in nature and art. 7 color plates, 29 illustrations. 274pp. 5⅜ × 8½. 20944-X Pa. $4.50

SCULPTURE PRINCIPLES AND PRACTICE, Louis Slobodkin. Step-by-step approach to clay, plaster, metals, stone; classical and modern. 253 drawings, photos. 255pp. 8⅛ × 11. 22960-2 Pa. $7.00

VICTORIAN FASHION PAPER DOLLS FROM HARPER'S BAZAR, 1867-1898, Theodore Menten. Four female dolls with 28 elegant high fashion costumes, printed in full color. 32pp. 9¼ × 12¼. 23453-3 Pa. $3.50

FLOPSY, MOPSY AND COTTONTAIL: A Little Book of Paper Dolls in Full Color, Susan LaBelle. Three dolls and 21 costumes (7 for each doll) show Peter Rabbit's siblings dressed for holidays, gardening, hiking, etc. Charming borders, captions. 48pp. 4¼ × 5½. 24376-1 Pa. $2.00

NATIONAL LEAGUE BASEBALL CARD CLASSICS, Bert Randolph Sugar. 83 big-leaguers from 1909-69 on facsimile cards. Hubbell, Dean, Spahn, Brock plus advertising, info, no duplications. Perforated, detachable. 16pp. 8¼ × 11. 24308-7 Pa. $2.95

THE LOGICAL APPROACH TO CHESS, Dr. Max Euwe, et al. First-rate text of comprehensive strategy, tactics, theory for the amateur. No gambits to memorize, just a clear, logical approach. 224pp. 5⅜ × 8½. 24353-2 Pa. $4.50

MAGICK IN THEORY AND PRACTICE, Aleister Crowley. The summation of the thought and practice of the century's most famous necromancer, long hard to find. Crowley's best book. 436pp. 5⅜ × 8½. (Available in U.S. only) 23295-6 Pa. $6.50

THE HAUNTED HOTEL, Wilkie Collins. Collins' last great tale; doom and destiny in a Venetian palace. Praised by T.S. Eliot. 127pp. 5⅜ × 8½. 24333-8 Pa. $3.00

ART DECO DISPLAY ALPHABETS, Dan X. Solo. Wide variety of bold yet elegant lettering in handsome Art Deco styles. 100 complete fonts, with numerals, punctuation, more. 104pp. 8⅛ × 11. 24372-9 Pa. $4.00

CALLIGRAPHIC ALPHABETS, Arthur Baker. Nearly 150 complete alphabets by outstanding contemporary. Stimulating ideas; useful source for unique effects. 154 plates. 157pp. 8⅜ × 11¼. 21045-6 Pa. $4.95

ARTHUR BAKER'S HISTORIC CALLIGRAPHIC ALPHABETS, Arthur Baker. From monumental capitals of first-century Rome to humanistic cursive of 16th century, 33 alphabets in fresh interpretations. 88 plates. 96pp. 9 × 12. 24054-1 Pa. $3.95

LETTIE LANE PAPER DOLLS, Sheila Young. Genteel turn-of-the-century family very popular then and now. 24 paper dolls. 16 plates in full color. 32pp. 9¼ × 12¼. 24089-4 Pa. $3.50

HOW THE OTHER HALF LIVES, Jacob A. Riis. Journalistic record of filth, degradation, upward drive in New York immigrant slums, shops, around 1900. New edition includes 100 original Riis photos, monuments of early photography. 233pp. 10 × 7⅞. 22012-5 Pa. $7.95

CHINA AND ITS PEOPLE IN EARLY PHOTOGRAPHS, John Thomson. In 200 black-and-white photographs of exceptional quality photographic pioneer Thomson captures the mountains, dwellings, monuments and people of 19th-century China. 272pp. 9⅜ × 12¼. 24393-1 Pa. $12.95

GODEY COSTUME PLATES IN COLOR FOR DECOUPAGE AND FRAMING, edited by Eleanor Hasbrouk Rawlings. 24 full-color engravings depicting 19th-century Parisian haute couture. Printed on one side only. 56pp. 8¼ × 11. 23879-2 Pa. $3.95

ART NOUVEAU STAINED GLASS PATTERN BOOK, Ed Sibbett, Jr. 104 projects using well-known themes of Art Nouveau: swirling forms, florals, peacocks, and sensuous women. 60pp. 8¼ × 11. 23577-7 Pa. $3.00

QUICK AND EASY PATCHWORK ON THE SEWING MACHINE: Susan Aylsworth Murwin and Suzzy Payne. Instructions, diagrams show exactly how to machine sew 12 quilts. 48pp. of templates. 50 figures. 80pp. 8¼ × 11. 23770-2 Pa. $3.50

THE STANDARD BOOK OF QUILT MAKING AND COLLECTING, Marguerite Ickis. Full information, full-sized patterns for making 46 traditional quilts, also 150 other patterns. 483 illustrations. 273pp. 6⅞ × 9⅝. 20582-7 Pa. $5.95

LETTERING AND ALPHABETS, J. Albert Cavanagh. 85 complete alphabets lettered in various styles; instructions for spacing, roughs, brushwork. 121pp. 8¾ × 8. 20053-1 Pa. $3.75

LETTER FORMS: 110 COMPLETE ALPHABETS, Frederick Lambert. 110 sets of capital letters; 16 lower case alphabets; 70 sets of numbers and other symbols. 110pp. 8⅛ × 11. 22872-X Pa. $4.50

ORCHIDS AS HOUSE PLANTS, Rebecca Tyson Northen. Grow cattleyas and many other kinds of orchids—in a window, in a case, or under artificial light. 63 illustrations. 148pp. 5⅜ × 8½. 23261-1 Pa. $2.95

THE MUSHROOM HANDBOOK, Louis C.C. Krieger. Still the best popular handbook. Full descriptions of 259 species, extremely thorough text, poisons, folklore, etc. 32 color plates; 126 other illustrations. 560pp. 5⅜ × 8½. 21861-9 Pa. $8.50

THE DORÉ BIBLE ILLUSTRATIONS, Gustave Doré. All wonderful, detailed plates: Adam and Eve, Flood, Babylon, life of Jesus, etc. Brief King James text with each plate. 241 plates. 241pp. 9 × 12. 23004-X Pa. $6.95

THE BOOK OF KELLS: Selected Plates in Full Color, edited by Blanche Cirker. 32 full-page plates from greatest manuscript-icon of early Middle Ages. Fantastic, mysterious. Publisher's Note. Captions. 32pp. 9¾ × 12¼. 24345-1 Pa. $4.50

THE PERFECT WAGNERITE, George Bernard Shaw. Brilliant criticism of the Ring Cycle, with provocative interpretation of politics, economic theories behind the Ring. 136pp. 5⅜ × 8½. (Available in U.S. only) 21707-8 Pa. $3.00

THE RIME OF THE ANCIENT MARINER, Gustave Doré, S.T. Coleridge. Doré's finest work, 34 plates capture moods, subtleties of poem. Full text. 77pp. 9¼ × 12. 22305-1 Pa. $4.95

SONGS OF INNOCENCE, William Blake. The first and most popular of Blake's famous "Illuminated Books," in a facsimile edition reproducing all 31 brightly colored plates. Additional printed text of each poem. 64pp. 5¼ × 7. 22764-2 Pa. $3.00

AN INTRODUCTION TO INFORMATION THEORY, J.R. Pierce. Second (1980) edition of most impressive non-technical account available. Encoding, entropy, noisy channel, related areas, etc. 320pp. 5⅜ × 8½. 24061-4 Pa. $4.95

THE DIVINE PROPORTION: A STUDY IN MATHEMATICAL BEAUTY, H.E. Huntley. "Divine proportion" or "golden ratio" in poetry, Pascal's triangle, philosophy, psychology, music, mathematical figures, etc. Excellent bridge between science and art. 58 figures. 185pp. 5⅜ × 8½. 22254-3 Pa. $3.95

THE DOVER NEW YORK WALKING GUIDE: From the Battery to Wall Street, Mary J. Shapiro. Superb inexpensive guide to historic buildings and locales in lower Manhattan: Trinity Church, Bowling Green, more. Complete Text; maps. 36 illustrations. 48pp. 3⅞ × 9¼. 24225-0 Pa. $1.75

NEW YORK THEN AND NOW, Edward B. Watson, Edmund V. Gillon, Jr. 83 important Manhattan sites: on facing pages early photographs (1875-1925) and 1976 photos by Gillon. 172 illustrations. 171pp. 9¼ × 10. 23361-8 Pa. $7.95

HISTORIC COSTUME IN PICTURES, Braun & Schneider. Over 1450 costumed figures from dawn of civilization to end of 19th century. English captions. 125 plates. 256pp. 8⅜ × 11¼. 23150-X Pa. $7.50

VICTORIAN AND EDWARDIAN FASHION: A Photographic Survey, Alison Gernsheim. First fashion history completely illustrated by contemporary photographs. Full text plus 235 photos, 1840-1914, in which many celebrities appear. 240pp. 6½ × 9¼. 24205-6 Pa. $6.00

CHARTED CHRISTMAS DESIGNS FOR COUNTED CROSS-STITCH AND OTHER NEEDLECRAFTS, Lindberg Press. Charted designs for 45 beautiful needlecraft projects with many yuletide and wintertime motifs. 48pp. 8¼ × 11. 24356-7 Pa. $1.95

101 FOLK DESIGNS FOR COUNTED CROSS-STITCH AND OTHER NEEDLECRAFTS, Carter Houck. 101 authentic charted folk designs in a wide array of lovely representations with many suggestions for effective use. 48pp. 8¼ × 11. 24369-9 Pa. $1.95

FIVE ACRES AND INDEPENDENCE, Maurice G. Kains. Great back-to-the-land classic explains basics of self-sufficient farming. The one book to get. 95 illustrations. 397pp. 5⅜ × 8½. 20974-1 Pa. $4.95

A MODERN HERBAL, Margaret Grieve. Much the fullest, most exact, most useful compilation of herbal material. Gigantic alphabetical encyclopedia, from aconite to zedoary, gives botanical information, medical properties, folklore, economic uses, and much else. Indispensable to serious reader. 161 illustrations. 888pp. 6½ × 9¼. (Available in U.S. only) 22798-7, 22799-5 Pa., Two-vol. set $16.45

DECORATIVE NAPKIN FOLDING FOR BEGINNERS, Lillian Oppenheimer and Natalie Epstein. 22 different napkin folds in the shape of a heart, clown's hat, love knot, etc. 63 drawings. 48pp. 8¼ × 11. 23797-4 Pa. $1.95

DECORATIVE LABELS FOR HOME CANNING, PRESERVING, AND OTHER HOUSEHOLD AND GIFT USES, Theodore Menten. 128 gummed, perforated labels, beautifully printed in 2 colors. 12 versions. Adhere to metal, glass, wood, ceramics. 24pp. 8¼ × 11. 23219-0 Pa. $2.95

EARLY AMERICAN STENCILS ON WALLS AND FURNITURE, Janet Waring. Thorough coverage of 19th-century folk art: techniques, artifacts, surviving specimens. 166 illustrations, 7 in color. 147pp. of text. 7⅞ × 10¾. 21906-2 Pa. $8.95

AMERICAN ANTIQUE WEATHERVANES, A.B. & W.T. Westervelt. Extensively illustrated 1883 catalog exhibiting over 550 copper weathervanes and finials. Excellent primary source by one of the principal manufacturers. 104pp. 6⅛ × 9¼. 24396-6 Pa. $3.95

ART STUDENTS' ANATOMY, Edmond J. Farris. Long favorite in art schools. Basic elements, common positions, actions. Full text, 158 illustrations. 159pp. 5⅜ × 8½. 20744-7 Pa. $3.50

BRIDGMAN'S LIFE DRAWING, George B. Bridgman. More than 500 drawings and text teach you to abstract the body into its major masses. Also specific areas of anatomy. 192pp. 6½ × 9¼. (EA) 22710-3 Pa. $4.50

COMPLETE PRELUDES AND ETUDES FOR SOLO PIANO, Frederic Chopin. All 26 Preludes, all 27 Etudes by greatest composer of piano music. Authoritative Paderewski edition. 224pp. 9 × 12. (Available in U.S. only) 24052-5 Pa. $6.95

PIANO MUSIC 1888-1905, Claude Debussy. Deux Arabesques, Suite Bergamesque, Masques, 1st series of Images, etc. 9 others, in corrected editions. 175pp. 9⅜ × 12¼. (ECE) 22771-5 Pa. $5.95

TEDDY BEAR IRON-ON TRANSFER PATTERNS, Ted Menten. 80 iron-on transfer patterns of male and female Teddys in a wide variety of activities, poses, sizes. 48pp. 8¼ × 11. 24596-9 Pa. $2.00

A PICTURE HISTORY OF THE BROOKLYN BRIDGE, M.J. Shapiro. Profusely illustrated account of greatest engineering achievement of 19th century. 167 rare photos & engravings recall construction, human drama. Extensive, detailed text. 122pp. 8¼ × 11. 24403-2 Pa. $7.95

NEW YORK IN THE THIRTIES, Berenice Abbott. Noted photographer's fascinating study shows new buildings that have become famous and old sights that have disappeared forever. 97 photographs. 97pp. 11⅜ × 10. 22967-X Pa. $6.50

MATHEMATICAL TABLES AND FORMULAS, Robert D. Carmichael and Edwin R. Smith. Logarithms, sines, tangents, trig functions, powers, roots, reciprocals, exponential and hyperbolic functions, formulas and theorems. 269pp. 5⅜ × 8½. 60111-0 Pa. $3.75

HANDBOOK OF MATHEMATICAL FUNCTIONS WITH FORMULAS, GRAPHS, AND MATHEMATICAL TABLES, edited by Milton Abramowitz and Irene A. Stegun. Vast compendium: 29 sets of tables, some to as high as 20 places. 1,046pp. 8 × 10½. 61272-4 Pa. $19.95

REASON IN ART, George Santayana. Renowned philosopher's provocative, seminal treatment of basis of art in instinct and experience. Volume Four of *The Life of Reason.* 230pp. 5⅜ × 8. 24358-3 Pa. $4.50

LANGUAGE, TRUTH AND LOGIC, Alfred J. Ayer. Famous, clear introduction to Vienna, Cambridge schools of Logical Positivism. Role of philosophy, elimination of metaphysics, nature of analysis, etc. 160pp. 5⅜ × 8½. (USCO) 20010-8 Pa. $2.75

BASIC ELECTRONICS, U.S. Bureau of Naval Personnel. Electron tubes, circuits, antennas, AM, FM, and CW transmission and receiving, etc. 560 illustrations. 567pp. 6½ × 9¼. 21076-6 Pa. $8.95

THE ART DECO STYLE, edited by Theodore Menten. Furniture, jewelry, metalwork, ceramics, fabrics, lighting fixtures, interior decors, exteriors, graphics from pure French sources. Over 400 photographs. 183pp. 8⅜ × 11¼. 22824-X Pa. $6.95

THE FOUR BOOKS OF ARCHITECTURE, Andrea Palladio. 16th-century classic covers classical architectural remains, Renaissance revivals, classical orders, etc. 1738 Ware English edition. 216 plates. 110pp. of text. 9½ × 12¾. 21308-0 Pa. $10.00

THE WIT AND HUMOR OF OSCAR WILDE, edited by Alvin Redman. More than 1000 ripostes, paradoxes, wisecracks: Work is the curse of the drinking classes, I can resist everything except temptations, etc. 258pp. 5⅜ × 8½. (USCO) 20602-5 Pa. $3.50

THE DEVIL'S DICTIONARY, Ambrose Bierce. Barbed, bitter, brilliant witticisms in the form of a dictionary. Best, most ferocious satire America has produced. 145pp. 5⅜ × 8½. 20487-1 Pa. $2.50

ERTÉ'S FASHION DESIGNS, Erté. 210 black-and-white inventions from *Harper's Bazar,* 1918-32, plus 8pp. full-color covers. Captions. 88pp. 9 × 12. 24203-X Pa. $6.50

ERTÉ GRAPHICS, Erté. Collection of striking color graphics: *Seasons, Alphabet, Numerals, Aces* and *Precious Stones.* 50 plates, including 4 on covers. 48pp. 9⅜ × 12¼. 23580-7 Pa. $6.95

PAPER FOLDING FOR BEGINNERS, William D. Murray and Francis J. Rigney. Clearest book for making origami sail boats, roosters, frogs that move legs, etc. 40 projects. More than 275 illustrations. 94pp. 5⅜ × 8½. 20713-7 Pa. $1.95

ORIGAMI FOR THE ENTHUSIAST, John Montroll. Fish, ostrich, peacock, squirrel, rhinoceros, Pegasus, 19 other intricate subjects. Instructions. Diagrams. 128pp. 9 × 12. 23799-0 Pa. $4.95

CROCHETING NOVELTY POT HOLDERS, edited by Linda Macho. 64 useful, whimsical pot holders feature kitchen themes, animals, flowers, other novelties. Surprisingly easy to crochet. Complete instructions. 48pp. 8¼ × 11. 24296-X Pa. $1.95

CROCHETING DOILIES, edited by Rita Weiss. Irish Crochet, Jewel, Star Wheel, Vanity Fair and more. Also luncheon and console sets, runners and centerpieces. 51 illustrations. 48pp. 8¼ × 11. 23424-X Pa. $2.00

YUCATAN BEFORE AND AFTER THE CONQUEST, Diego de Landa. Only significant account of Yucatan written in the early post-Conquest era. Translated by William Gates. Over 120 illustrations. 162pp. 5⅜ × 8½. 23622-6 Pa. $3.50

ORNATE PICTORIAL CALLIGRAPHY, E.A. Lupfer. Complete instructions, over 150 examples help you create magnificent "flourishes" from which beautiful animals and objects gracefully emerge. 8⅛ × 11. 21957-7 Pa. $2.95

DOLLY DINGLE PAPER DOLLS, Grace Drayton. Cute chubby children by same artist who did Campbell Kids. Rare plates from 1910s. 30 paper dolls and over 100 outfits reproduced in full color. 32pp. 9¼ × 12¼. 23711-7 Pa. $2.95

CURIOUS GEORGE PAPER DOLLS IN FULL COLOR, H. A. Rey, Kathy Allert. Naughty little monkey-hero of children's books in two doll figures, plus 48 full-color costumes: pirate, Indian chief, fireman, more. 32pp. 9¼ × 12¼. 24386-9 Pa. $3.50

GERMAN: HOW TO SPEAK AND WRITE IT, Joseph Rosenberg. Like *French, How to Speak and Write It.* Very rich modern course, with a wealth of pictorial material. 330 illustrations. 384pp. 5⅜ × 8½. (USUKO) 20271-2 Pa. $4.75

CATS AND KITTENS: 24 Ready-to-Mail Color Photo Postcards, D. Holby. Handsome collection; feline in a variety of adorable poses. Identifications. 12pp. on postcard stock. 8¼ × 11. 24469-5 Pa. $2.95

MARILYN MONROE PAPER DOLLS, Tom Tierney. 31 full-color designs on heavy stock, from *The Asphalt Jungle, Gentlemen Prefer Blondes,* 22 others. 1 doll. 16 plates. 32pp. 9⅜ × 12¼. 23769-9 Pa. $3.50

FUNDAMENTALS OF LAYOUT, F.H. Wills. All phases of layout design discussed and illustrated in 121 illustrations. Indispensable as student's text or handbook for professional. 124pp. 8⅛ × 11. 21279-3 Pa. $4.50

FANTASTIC SUPER STICKERS, Ed Sibbett, Jr. 75 colorful pressure-sensitive stickers. Peel off and place for a touch of pizzazz: clowns, penguins, teddy bears, etc. Full color. 16pp. 8¼ × 11. 24471-7 Pa. $2.95

LABELS FOR ALL OCCASIONS, Ed Sibbett, Jr. 6 labels each of 16 different designs—baroque, art nouveau, art deco, Pennsylvania Dutch, etc.—in full color. 24pp. 8¼ × 11. 23688-9 Pa. $2.95

HOW TO CALCULATE QUICKLY: RAPID METHODS IN BASIC MATHEMATICS, Henry Sticker. Addition, subtraction, multiplication, division, checks, etc. More than 8000 problems, solutions. 185pp. 5 × 7¼. 20295-X Pa. $2.95

THE CAT COLORING BOOK, Karen Baldauski. Handsome, realistic renderings of 40 splendid felines, from American shorthair to exotic types. 44 plates. Captions. 48pp. 8¼ × 11. 24011-8 Pa. $2.25

THE TALE OF PETER RABBIT, Beatrix Potter. The inimitable Peter's terrifying adventure in Mr. McGregor's garden, with all 27 wonderful, full-color Potter illustrations. 55pp. 4¼ × 5½. (Available in U.S. only) 22827-4 Pa. $1.50

BASIC ELECTRICITY, U.S. Bureau of Naval Personnel. Batteries, circuits, conductors, AC and DC, inductance and capacitance, generators, motors, transformers, amplifiers, etc. 349 illustrations. 448pp. 6½ × 9¼. 20973-3 Pa. $7.95